M000314092

Adaptive Speciation

Unraveling how biological diversity originates through speciation is fundamental to understanding the past, present, and future of life on earth. Promoting an ongoing paradigm shift, *Adaptive Speciation* elucidates how selection driven by biological interactions can trigger the adaptive splitting of lineages. Recent advances in speciation theory are carefully explained and confronted with celebrated empirical examples of speciation under natural selection. With an emphasis on the potentially intricate interplay between geographic patterns and ecological processes of speciation, this book seeks to overcome the default perception of speciation as a mere side effect of geographic isolation. The consequent richer perspective enables adaptive speciation to be appreciated as a major force in the generation of biological diversity. Written for students and researchers alike, this book provides a thorough treatment of the newest developments in speciation science.

ULF DIECKMANN is Project Leader of the Adaptive Dynamics Network at the International Institute for Applied Systems Analysis (IIASA) in Laxenburg, Austria. He is coeditor of *The Geometry of Ecological Interactions: Simplifying Spatial Complexity*, of *Adaptive Dynamics of Infectious Diseases: In Pursuit of Virulence Management*, and of *Evolutionary Conservation Biology*.

MICHAEL DOEBELI is Associate Professor in the Departments of Zoology and Mathematics at the University of British Columbia in Vancouver, Canada.

JOHAN A.J. METZ is Professor of Mathematical Biology at the Institute of Biology, Leiden University, the Netherlands, and Senior Adviser of the Adaptive Dynamics Network at IIASA. He is coeditor of *The Geometry of Ecological Interactions: Simplifying Spatial Complexity*, of *Adaptive Dynamics of Infectious Diseases: In Pursuit of Virulence Management*, and of *The Dynamics of Physiologically Structured Populations*.

DIETHARD TAUTZ is Professor of Evolutionary Genetics at the Institute of Genetics, University of Cologne, Germany. He is editor-in-chief of the journal *Development Genes and Evolution*.

Cambridge Studies in Adaptive Dynamics

Series Editors

ULF DIECKMANN
Adaptive Dynamics Network
International Institute for
Applied Systems Analysis
A-2361 Laxenburg, Austria

JOHAN A.J. METZ
Institute of Biology
Leiden University
NL-2311 GP Leiden
The Netherlands

The modern synthesis of the first half of the twentieth century reconciled Darwinian selection with Mendelian genetics. However, it largely failed to incorporate ecology and hence did not develop into a predictive theory of long-term evolution. It was only in the 1970s that evolutionary game theory put the consequences of frequency-dependent ecological interactions into proper perspective. Adaptive Dynamics extends evolutionary game theory by describing the dynamics of adaptive trait substitutions and by analyzing the evolutionary implications of complex ecological settings.

The *Cambridge Studies in Adaptive Dynamics* highlight these novel concepts and techniques for ecological and evolutionary research. The series is designed to help graduate students and researchers to use the new methods for their own studies. Volumes in the series provide coverage of both empirical observations and theoretical insights, offering natural points of departure for various groups of readers. If you would like to contribute a book to the series, please contact Cambridge University Press or the series editors.

Adaptive Speciation

Edited by

Ulf Dieckmann, Michael Doebeli, Johan A.J. Metz,
and Diethard Tautz

IIASA

CAMBRIDGE
UNIVERSITY PRESS

PUBLISHED BY THE PRESS SYNDICATE OF THE UNIVERSITY OF CAMBRIDGE
The Pitt Building, Trumpington Street, Cambridge, United Kingdom

CAMBRIDGE UNIVERSITY PRESS
The Edinburgh Building, Cambridge CB2 2RU, UK
40 West 20th Street, New York, NY 10011-4211, USA
477 Williamstown Road, Port Melbourne, VIC 3207, Australia
Ruiz de Alarcón 13, 28014 Madrid, Spain
Dock House, The Waterfront, Cape Town 8001, South Africa

http: //www.cambridge.org

© International Institute for Applied Systems Analysis 2004

This book is in copyright. Subject to statutory exception
and to the provisions of relevant collective licensing agreements,
no reproduction of any part may take place without
the written permission of the
International Institute for Applied Systems Analysis.

http://www.iiasa.ac.at

First published 2004

Printed in the United Kingdom at the University Press, Cambridge

Typefaces Times; Zapf Humanist 601 (Bitstream Inc.)   *System* LATEX

A catalog record for this book is available from the British Library

ISBN 0 521 82842 2 hardback

Contents

Contributing Authors

Guy L. Bush (bushfly@pilot.msu.edu) Department of Zoology, Michigan State University, 336 Natural Sciences Building, East Lansing, MI 48824, USA

Roger K. Butlin (r.k.butlin@leeds.ac.uk) School of Biology, University of Leeds, Leeds LS2 9JT, United Kingdom

Diane R. Campbell (drcampbe@uci.edu) Department of Ecology and Evolutionary Biology, University of California, Irvine, CA 92697, USA

David Claessen (david.claessen@bbsrc.ac.uk) Biomathematics Unit, Institute of Arable Crops Research, Rothamsted, Harpenden, Hertfordshire AL5 2JQ, United Kingdom

Ulf Dieckmann (dieckman@iiasa.ac.at) Adaptive Dynamics Network, International Institute for Applied Systems Analysis, A-2361 Laxenburg, Austria

Michael Doebeli (doebeli@zoology.ubc.ca) Departments of Mathematics and Zoology, University of British Columbia, 6270 University Boulevard, Vancouver, BC V6T 1Z4, Canada

Martijn Egas (egas@science.uva.nl) Population Biology Section, Institute for Biodiversity and Ecosystem Dynamics, University of Amsterdam, Kruislaan 320, NL-1090 GB Amsterdam, The Netherlands

Frietson Galis (galis@rulsfb.leidenuniv.nl) Institute of Evolutionary and Ecological Sciences, University of Leiden, P.O.Box 9516, NL-2300 RA Leiden, The Netherlands

Sergey Gavrilets (gavrila@tiem.utk.edu) Department of Ecology and Evolutionary Biology, University of Tennessee, Knoxville, TN 37996, USA

Stefan A.H. Geritz (stefan.geritz@utu.fi) Department of Mathematical Sciences, University of Turku, FIN-20014 Turku, Finland

Tadeusz J. Kawecki (tadeusz.kawecki@unifr.ch) Unit for Ecology and Evolution, Department of Biology, University of Fribourg, Chemin du Musée 10, CH-1700 Fribourg, Switzerland

Éva Kisdi (eva.kisdi@utu.fi) Department of Mathematical Sciences, University of Turku, FIN-20014 Turku, Finland & Department of Genetics, Eötvös University, Múzeum krt 4/A, H-1088 Budapest, Hungary

Eric B. Knox (eknox@bio.indiana.edu) Department of Biology, Indiana University, 1001 E. Third Street, Bloomington, IN 47405, USA

Iza Lesna (lesna@science.uva.nl) Population Biology Section, Institute for Biodiversity and Ecosystem Dynamics, University of Amsterdam, Kruislaan 320, NL-1090 GB Amsterdam, The Netherlands

Jonathan B. Losos (losos@biology2.wustl.edu) Department of Biology, Washington University, Campus Box 1137, St Louis, MO 63130, USA

Anita Malhotra (a.malhotra@bangor.ac.uk) School of Biological Sciences, University of Wales, Bangor, Gwynedd LL57 2UW, United Kingdom

Amy R. McCune (arm2@cornell.edu) Department of Ecology and Evolutionary Biology, Cornell University, E249 Corson Hall, Ithaca, NY 14853, USA

Géza Meszéna (geza.meszena@elte.hu) Department of Biological Physics, Eötvös University, Pázmány Péter sétány 1A, H-1117 Budapest, Hungary

Johan A.J. Metz (metz@rulsfb.leidenuniv.nl) Institute of Biology, Leiden University, Van der Klaauw Laboratory, P.O.Box 9516, NL-2300 RA Leiden, The Netherlands & Adaptive Dynamics Network, International Institute for Applied Systems Analysis, A-2361 Laxenburg, Austria

Ferenc Mizera (mizera@colbud.hu) Department of Biological Physics, Eötvös University, Pázmány Péter sétány 1A, H-1117 Budapest, Hungary

Will Provine (wbp2@cornell.edu) Department of Ecology and Evolutionary Biology, Corson Hall, Cornell University, Ithaca, NY 14853, USA

James T. Reardon (ReardonJ@LandcareResearch.co.nz) Landcare Research, P.O.Box 282, Alexandra, New Zealand

Howard D. Rundle (hrundle@sfa.ca) Department of Biological Sciences, Simon Fraser University, Burnaby, BC V5A 15C, Canada

Maurice W. Sabelis (sabelis@science.uva.nl) Population Biology Section, Institute for Biodiversity and Ecosystem Dynamics, University of Amsterdam, NL-1090 GB Amsterdam, The Netherlands

Dolph Schluter (schluter@zoology.ubc.ca) Department of Zoology, University of British Columbia, Vancouver, BC V6T 1Z4, Canada

Ole Seehausen (o.seehausen@hull.ac.uk) Department of Biological Sciences, University of Hull, Hull HU6 7RX, United Kingdom

Skúli Skúlason (skuli@holar.is) Hólar College, Hólarí Hjaltadal, 551 Sauðrkrókur, Iceland

Sigurður S. Snorrason (sigsnor@hi.is) Institute of Biology, University of Iceland, Grensásvegur 12, Reykjavik 108, Iceland

Andrew Stenson (a.g.stenson@bangor.ac.uk) School of Biological Sciences, University of Wales, Bangor, Gwynedd LL57 2UW, United Kingdom

Diethard Tautz (tautz@uni-koeln.de) Institut für Genetik, Universität zu Köln, Weyertal 121, D-50931 Köln, Germany

Roger S. Thorpe (r.s.thorpe@bangor.ac.uk) School of Biological Sciences, University of Wales, Bangor, Gwynedd LL57 2UW, United Kingdom

Michael Travisano (mtrav@uh.edu) Department of Biology and Biochemistry, University of Houston, Houston, TX 77204, USA

Filipa Vala (f.vala@ucl.ac.uk) Department of Biology, University College London, Wolfson House, London NW1 2HE, United Kingdom

Jacques J.M. van Alphen (alphen@rulsfb.leidenuniv.nl) Section of Animal Ecology, Institute of Evolutionary and Ecological Sciences, University of Leiden, P.O. Box 9516, NL-2300 RA Leiden, The Netherlands

Nickolas M. Waser (waser@citrus.ucr.edu) Department of Biology, University of California, Riverside, CA 92521, USA

Acknowledgments

Development of this book took place at the International Institute of Applied Systems Analysis (IIASA), Laxenburg, Austria, where IIASA's former directors Gordon J. MacDonald and Arne B. Jernelöv, and current director Leen Hordijk provided critical support. A workshop at IIASA brought together all authors to discuss their contributions and thus served as an important element in the strategy of achieving as much continuity across the subject areas as possible.

Financial support toward this workshop by the European Science Foundation's Theoretical Biology of Adaptation Programme is gratefully acknowledged. Michael Doebeli and Ulf Dieckmann received support from the Natural Sciences and Engineering Research Council of Canada. Ulf Dieckmann and Hans Metz received support from the European Research Training Network *ModLife* (Modern Life-History Theory and its Application to the Management of Natural Resources), funded through the Human Potential Programme of the European Commission.

The success of any edited volume aspiring to textbook standards very much depends on the cooperation of the contributors in dealing with the many points the editors are bound to raise. We are indebted to all our authors for their cooperativeness and patience throughout the resultant rounds of revision. The book has benefited greatly from the support of the Publications Department at IIASA; we are especially grateful to Ewa Delpos, Anka James, Martina Jöstl, Eryl Maedel, John Ormiston, and Lieselotte Roggenland for the excellent work they have put into preparing the camera-ready copy of this volume. Any mistakes that remain are our responsibility.

Ulf Dieckmann
Michael Doebeli
Johan A.J. Metz
Diethard Tautz

Notational Standards

To allow for a better focus on the content of chapters and to highlight their interconnections, we have encouraged all the authors of this volume to adhere to the following notational standards:

A, B, C	Locus, with alleles A, a; B, b; C, c
M	Modifier or mating locus, with alleles M, m
p, q	Gene frequency
n	Population density (potentially a vector)
N	Population size (in number of individuals) or
	Total population density (sum of components of n)
E	Condition of the environment
a	Ecological interaction coefficient
r	Per capita growth rate
K	Carrying capacity
m	Migration/movement rate
D	Diffusion coefficient, or dilution rate
x	Phenotypic or allelic trait value
u	Per locus mutation probability or
	Probability of a mutational step in a quantitative trait
U	Gametic mutation probability
f	Fitness in continuous time ($f = 0$ is neutral)
W	Fitness in discrete time ($W = 1$ is neutral)
w	Relative fitness
s	Selection coefficient
h	Heterozygote advantage
D_M	Mahalanobis' (morphological) distance
D_N	Nei's genetic distance (between populations)
d	Genetic distance (between individuals) = number of allele changes
z, z_1, z_2	Spatial coordinates
t	Time
T	Duration
τ	Waiting time
p, q	Probability or relative frequency (subscript indicates type)
σ^2	Variance (subscript indicates type)
i, j	Index
$\ldots'$	Invader
$\hat{\ldots}$	Equilibrium value
$\ldots^*$	Evolutionarily singular value (of a trait)
$\bar{\ldots}$	Average

1

Introduction

Ulf Dieckmann, Johan A.J. Metz, Michael Doebeli, and Diethard Tautz

1.1 A Shift in Focus

Millions of species currently exist on earth, and to secure an understanding of how all this magnificent variety arose is no small task. Biologists have long accepted Darwinian selection as the central explanation of adaptation and evolutionary change; yet, to date, no similar agreement has emerged about evolutionary processes that can create two species out of one. Almost 150 years after Darwin's seminal work *On the Origin of Species* (1859), conditions for and mechanisms of biological speciation are still debated vigorously.

The traditional "standard model" of speciation rests on the assumption of geographic isolation. After a population has become subdivided by external causes – like fragmentation through environmental change or colonization of a new, disconnected habitat – and after the resultant subpopulations have remained separated for sufficiently long, genetic drift and pleiotropic effects of local adaptation are supposed to lead to partial reproductive incompatibility. When the two incipient species come into secondary contact, individuals from one species cannot mate with those of the other – even if they try – or, if mating is still possible, their hybrid offspring are inferior. Further evolution of premating isolation (like assortative mate choice or seasonal isolation) and/or postmating isolation (like gametic incompatibility) eventually ensures that the two species continue to steer separate evolutionary courses.

The trigger for speciation in this standard model is geographic isolation. It is for this reason that the distinction between allopatric speciation (occurring under geographic isolation) and sympatric speciation (without geographic isolation) has taken center stage in the speciation debate. Strictly speaking, this dichotomy characterizes no more than the spatial structure of populations that undergo speciation, as has been pointed out by the originator of the classification, Ernst Mayr:

> [E]ven today some authors confound the mechanisms of speciation – genes, chromosomes, and so forth – with the location of the populations involved in speciation (that is, whether the populations are sympatric or allopatric), not realizing that the two aspects are independent of each other and both are by necessity involved simultaneously. (Mayr 1982, p. 565)

Yet, the common understanding of this classification, widespread in the scientific literature, does not properly distinguish between its biogeographic (or pattern-oriented) and mechanistic (or process-oriented) aspects. Indeed, the term allopatric speciation has come to imply that the primary cause for a speciation event is geographic isolation and its primary mechanism is the emergence of reproductive incompatibility as a by-product of the interrupted gene flow – both implications being in accordance with the standard model. By contrast, the notion of sympatric speciation has become associated with speciation via other causes and different mechanisms. In short, pattern and process have become mixed up.

This confusion has not arisen by chance. Pattern and process are correlated so clearly in the standard model of speciation that no harm seemed to arise from a little conceptual sloppiness. In turn, mechanisms other than genetic drift or pleiotropic effects of local adaptation must be invoked to explain why species can be expected to arise without geographic isolation. Such mechanisms would most likely involve natural or sexual selection and for this reason the notion of sympatric speciation has become almost synonymous with speciation driven by ecological interactions or mate choice.

In this book our focus is on processes of speciation and, in particular, on their causes and mechanisms. To avoid misunderstandings and futile semantic debate, we suggest the terms allopatric and sympatric speciation be used, as far possible, in their original and precise meaning when classifying the biogeography of speciation events. To characterize causes and mechanisms beyond this classic dichotomy, a different terminology is required.

1.2 Adaptive Speciation

Speciation is a splitting process – an ancestral lineage splits into descendant lineages that are differentiated genetically and isolated reproductively. The split may be a consequence of geographic isolation, in which case the chain of cause and effect cannot, in general, be traced further: geographic factors that interrupt the gene flow between populations generally are the result of some coincidental environmental change, for example, in temperature, topography, or in the ranges of other species; or else are linked to chance events, like the incident of a rare colonization.

By contrast, splitting may be an evolutionary consequence of interactions within the speciating population. That is, the splitting itself may be an adaptation. As so often, this idea was foreshadowed in Darwin's work, as the following two quotes illustrate:

> Consequently, I cannot doubt that in the course of many thousands of generations, the most distinct varieties of any one species [...] would always have the best chance of succeeding and of increasing in numbers, and thus of supplanting the less distinct varieties; and varieties, when rendered very distinct from each other, take the rank of species. (Darwin 1859, p. 155)

> Natural selection, also, leads to divergence of character; for more living beings can be supported on the same area the more they diverge in structure, habits, and constitution [...]. Therefore during the modification of the descendants of any one species, and during the incessant struggle of all species to increase in numbers, the more

diversified these descendants become, the better will be their chance of succeeding in the battle of life. Thus the small differences distinguishing varieties of the same species, will steadily tend to increase till they come to equal the greater differences between species of the same genus, or even of distinct genera. (Darwin 1859, p. 169)

Given this precedence, discussions in this book may be seen as contributing to a much-belated renaissance of Darwinian ideas about speciation (Kondrashov 2001; Mallet 2001; Section 2.5). Such a development could have occurred earlier, had it not been for the commitment of major proponents of the Modern Synthesis to reproductive isolation for defining species and to geographic isolation to explain speciation. In a similar vein, the main part of the past century has seen the ubiquity of frequency-dependent selection – which played a key role in Darwin's ideas about speciation – unduly downplayed.

For splitting to be adaptive, a population must be under disruptive selection. Disruptive selection imposed purely by external causes is extremely unlikely, because this implies, as in allopatric speciation, a sudden, and very precisely aimed, change in the environment: otherwise the population would never come to occupy an externally imposed fitness minimum. Therefore, the only realistic scenario for splitting to be adaptive occurs when intraspecific interactions generate disruptive selection. This, in turn, can only happen if such interactions are frequency dependent. That is, these interactions must have the consequence that the fitness of a phenotype (i.e., its expected contribution to future generations) depends on the phenotypic composition of the population in which it occurs.

Obviously, for selection to be frequency dependent ecological contact must occur between the individuals involved. Conversely, it is also true that ecological contact almost invariably leads to frequency-dependent selection: under conditions of ecological contact, other individuals are part of the environment that determines the fitness of a given individual. For the particular phenotypes of these other individuals to be irrelevant in this determination, special, highly nongeneric circumstances would be required (notwithstanding that such circumstances are regularly assumed in simplified evolutionary models). In summary, for all practical purposes ecological contact and frequency-dependent selection are two sides of the same coin.

Strong frequency dependence can generate disruptive selection. If it does, the stage is set for adaptive diversification: a lineage split becomes selectively advantageous, as do adaptations that result in diminished gene flow between the emerging lineages. Under these conditions, the cause for the development of reproductive segregation rests within the species – therefore, such speciation scenarios are more amenable to further investigation. It is this perspective that makes it attractive to view some speciation processes as particular forms of adaptation, driven by selection pressures similar in origin to those that underlie directional evolution. We therefore propose to concentrate on distinguishing speciation processes that are adaptive from those that are nonadaptive and introduce the following definition:

*"Adaptive speciation" refers to speciation processes in which the splitting is
an adaptive response to disruptive selection caused by frequency-dependent
biological interactions.*

Naturally, the question of how often and under which circumstances frequency-dependent interactions are likely to induce disruptive selection is of central importance in the study of adaptive speciation. Traditionally, it is thought that such internally generated disruptive selection can only arise under rather special circumstances. In particular, in classic models of adaptive speciation (Chapter 3), disruptive selection through frequency-dependent interactions typically occurs only for a very restricted range of parameters. However, recent theoretical advances, based on a more dynamic view of the interplay between a population's evolution and its environment, have led to a different picture (Chapter 4).

The basic (and, by itself, well known) observation underlying these new insights is that when selection is frequency dependent, fitness landscapes change dynamically during the evolutionary process, because the phenotypic composition of the population changes. Thus, a population that starts out in a regime of directional selection may, nevertheless, evolve to a state in which it experiences disruptive selection. Indeed, this is not as unlikely as it appears at first sight, as the following metaphor of a gold rush may help to illustrate. Before a gold rush, very few people lived where the gold was found. As news of the gold reached a major city many people moved to the location of the gold find; this corresponds to a regime of directional selection. However, once everybody had ventured to the gold find, things quickly deteriorated, because soon too many people were looking for gold. What initially was an advantageous strategy became severely deleterious, simply because the same strategy was adopted by a plethora of competitors. After the initial regime of directional selection, being caught in the gold rush became the worst option, and resulted in the population of prospectors occupying a "fitness minimum".

If the gold-rush metaphor suggests that the basic cause of diversification is competitive interaction, it should be borne in mind that in any ecology that keeps populations bounded the individuals are necessarily subject to apparent or direct competition. If, moreover, the ecological roles of individuals vary continuously with their traits, similar individuals necessarily compete more strongly than less similar ones. Therefore, all that matters for diversification to be profitable is whether there exists something akin to the location of the gold, and whether at that location competition acts sufficiently narrowly that by behaving differently individuals can temporarily escape from it.

The gold-rush scenario corresponds to an adaptive process during which a trait value gradually converges to a point at which selection turns disruptive. This is illustrated schematically in Figure 1.1, which shows the evolutionary dynamics of the population mean of an arbitrary quantitative trait (thick curve). The figure also shows snapshots of the fitness profiles that generate this dynamics. While selection initially is merely directional, the fitness profiles, because of the adaptation of the mean trait value, soon feature a minimum (thin curve in Figure 1.1).

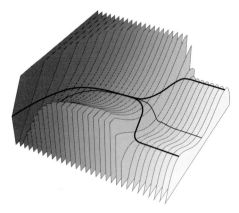

Figure 1.1 Adaptive speciation unfolding. A fitness landscape's shape changes jointly with a population's mean trait value (thick curve; the initial snapshot of the landscape is colored dark gray and the final one white). While the population undergoes directional selection by ascending the fitness landscape, the landscape itself changes because of frequency-dependent selection in such a way that a fitness minimum (thin curve) catches up with the population. Once trapped at the minimum, the population experiences disruptive selection and (under certain conditions) splits into two branches. In the figure, this divergence continues until the two branches arrive at local fitness maxima, at which selection becomes stabilizing.

As long as the mean trait value lies to one side of this minimum, the population still experiences directional selection and accordingly evolves away from the fitness minimum. However, as the evolutionary process unfolds, the fitness landscape continues to change in such a way that the distance between the mean trait value and the fitness minimum decreases. In other words, the fitness minimum catches up with the evolving population. Once the distance has shrunk to zero, the monomorphic population finds itself caught at a fitness minimum: through directional selection it has converged to a state in which it continuously experiences disruptive selection.

In this situation, a splitting of the population becomes adaptive. Adaptive speciation occurs provided the population possesses (or can evolve) a capacity for splitting into two reproductively isolated descendant species, as illustrated in Figure 1.1. Note that splitting induces further changes in the fitness landscape, so that eventually the two descendant species may come to occupy local fitness maxima. Such an outcome underscores that the splitting process itself is adaptive and that the eventually observed two niches do not pre-exist, but instead are generated by the very process of adaptive speciation. In asexual populations, splitting is the immediate consequence of disruptive selection operating at the fitness minimum (Chapter 4). In sexual populations, however, the splitting process is more complicated and requires some mechanism for assortative mating (Chapter 5).

The type of evolutionary dynamics illustrated in Figure 1.1, which comprises gradual convergence to a fitness minimum and subsequent adaptive splitting, has been termed evolutionary branching (Metz *et al.* 1996; Geritz *et al.* 1998). In principle, any continuous trait can undergo evolutionary branching, but despite the intuitive appeal of the gold-rush metaphor, it is not clear *a priori* how ubiquitous evolutionary branching is expected to be. In fact, later chapters in this book show that many different evolutionary models that incorporate frequency-dependent interactions contain the seed for evolutionary branching (Chapters 4, 5, and 7; see also Boxes 9.5, 10.3, 13.3, and 14.3). Moreover, in these models evolutionary branching does not require fine-tuning of the parameters, but instead typically occurs for wide ranges of the parameters. Thus, evolutionary branching appears to correspond to a general process that can occur under a great variety of circumstances.

1.3   Adaptive Speciation in Context

In this book, evolutionary branching is probed as the main theoretical paradigm for adaptive speciation. In sexual populations, evolutionary branching, and hence adaptive speciation, can only occur if assortative mating can latch on to the trait under disruptive selection. In principle, this can happen in a number of different ways, either through direct selection for assortative mating or because assortativeness is linked to the diverging trait as a result of behavioral or physiological constraints. Such linkages can also occur if disruptive selection acts on mating traits themselves, for example through sexual selection or sexual conflict (Chapter 5). Once a population has converged to a fitness minimum, it often experiences selection for nonrandom mating.

In the definition of adaptive speciation given above, the notion of selection encompasses both natural and sexual selection. In the literature, sexual selection is often pitted against natural selection. This convention goes back to Darwin and is meant to highlight a distinction between those causes of selection that exist without mate choice (natural selection) and those that only arise from its presence (sexual selection). We think that, in a general context, this division can mislead: mating traits under sexual selection are special life-history characters and are therefore subject to selection, like any other adaptive trait. In particular, the process of adaptive splitting is not restricted to ecological traits. Instead, adaptive speciation can involve different mixtures of ecological and mating differentiation: on the one extreme are asexual organisms in which speciation results only in ecological differentiation, and on the other extreme are sexual species with very pronounced assortative mating and only minimal ecological differentiation.

It is also worth noting that the scenario of adaptive speciation envisaged in this book contrasts sharply with traditional models for allopatric speciation. Even though selection may lead to divergence between allopatric subpopulations, selection is not disruptive in allopatric scenarios. Thus, in allopatric speciation the splitting may be a by-product of adaptations, but it is not an adaptation itself. This means that reproductive isolation does not evolve through selection for isolating mating mechanisms. Even though it is intuitively appealing to assume that genetic incompatibilities leading to reproductive isolation are an inevitable consequence of prolonged evolution in allopatry, the mechanisms that underlie such incompatibilities are actually poorly understood (as are the ecological and genetic factors that determine the rates at which incompatibilities are expected to accumulate). The same conclusions, in essence, also hold for classic parapatric scenarios with limited gene flow. For example, in speciation models in which sexual selection generates evolutionary runaway processes with directions that differ between populations inhabiting different geographic locations, thus leading to speciation, at no point in time do the speciating populations experience disruptive selection. Thus, even though adaptation obviously plays an important role in such speciation processes, this scenario does not fall in the category of adaptive speciation as defined above, because it does not involve disruptive selection, and thus the splitting itself is not adaptive. Likewise, ecological speciation (Chapter 9) is defined as the consequence of adaptation to different resources or environments, without making explicit the role of frequency dependence in creating disruptive selection. Box 1.1 provides a systematic overview of the relations between adaptive speciation and other speciation concepts prevalent in the literature.

A final question with regard to the definition of adaptive speciation concerns the amount of ecological contact required for a speciation process to be considered adaptive. Since the definition is meant to distinguish speciation by natural and sexual selection from coincidental speciation as a by-product of, for example, spatial segregation, the minimal ecological contact needed for adaptive speciation should prevent, at the considered time scale, speciation by genetic drift and by pleiotropic effects of local adaptation. This also clarifies the relation between adaptive and parapatric speciation. Parapatric speciation occurs under conditions of spatial adjacency between two incipient species. Such a pattern, while it allows for some gene flow and mixing between individuals, may restrict these homogenizing forces to an extent that genetic drift or local adaptation may engender speciation. Alternatively, the spatial proximity in a parapatric setting may preserve the genetic cohesion within a species, and thus only allow for speciation by adaptive mechanisms. In consequence, parapatric speciation can be either adaptive or occur as a by-product of other processes.

The concept of adaptive speciation, of course, does not challenge the need to explain how speciating sexual populations overcome their genetic cohesion. It stresses, however, that there can be internally driven adaptive mechanisms that induce splitting and lead to the cessation of genetic exchange and interbreeding. This is in contrast to the external factors that are assumed to initiate allopatric speciation (although even in this it is believed to be relatively rare that the speciation process achieves completion without some internally driven adaptive mechanisms, such as reinforcement on secondary contact). We may therefore expect to gain a deeper understanding of the biological diversity that surrounds us by careful examination of the relevant forces of frequency- and density-dependent selection as they result from the biological interactions between individuals and their environment. In this

Box 1.1 Notions of speciation

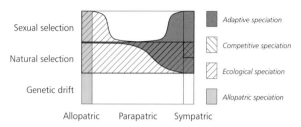

Speciation processes can be broadly categorized by the patterns and mechanisms that underlie the diversification. While the schematic figure above is too coarse to accommodate all the subtleties and multiple stages that may be involved (Box 19.1), it conveniently highlights several basic distinctions. The horizontal axis discriminates between the pattern at the onset of the speciation process being allopatric, parapatric, or sympatric. It can be argued that speciation under fully allopatric or sympatric conditions (left and right columns, respectively) are limiting cases, which, in particular in the case of sympatric speciation, are probably encountered rarely in nature. Although most speciation processes may thus be parapatric (at least initially), they can differ greatly in the level of possible gene flow and ecological contact between the incipient species (from nearly allopatric cases on the left, to nearly sympatric cases on the right). The figure's vertically stacked rows discriminate between the three main mechanisms potentially involved in speciation: genetic drift, natural selection on ecological characters, and sexual selection on mating traits [mixed or layered cases (see Box 19.1) are not represented in the figure].

How can the various notions of speciation suggested in the literature be accommodated on this grid? Within the figure's horizontal rows, the curves describe the propensity for the alternative speciation processes to happen when the assumption about the underlying pattern passes from allopatric, through parapatric, to sympatric.

Adaptive speciation (dark gray region) occurs when frequency dependence causes disruptive selection and subsequent diversification, either in ecological characters (middle row) or in mating traits (top row). Adaptive speciation requires sympatry or parapatry and becomes increasingly unlikely when gene flow and ecological contact diminish toward the allopatric case. Yet, for adaptive speciation in ecological characters to proceed, sufficient ecological contact can, in principle, arise in allopatry, given that such contact is established by other more mobile species that interact with the two incipient species.

Allopatric speciation (light gray region; see Chapter 6) occurs in geographically isolated populations, through genetic drift (bottom row), pleiotropic consequences of local adaptation in ecological characters (middle row), or divergent Fisherian runaway processes in mating traits (top row). When isolation by distance is sufficiently strong (nearly allopatric cases), parapatric speciation can be driven by the same mechanisms as allopatric speciation.   *continued*

Box 1.1 *continued*

Ecological speciation (large hatched region; see Chapter 9) occurs when adaptation to different resources or environments induces divergent or disruptive selection. Ecological speciation can (a) proceed in allopatry, parapatry, or sympatry, (b) result from adaptations to different environments as well as from intraspecific competition for resources, (c) involve by-product reproductive isolation as well as reinforcement, and (d) include speciation through sexual selection. While this definition is meant to encompass all speciation processes driven by natural selection (middle row), ecological speciation by sexual selection (top row) requires the divergence of mating traits to be driven by adaptation to different environments [e.g., by sensory drive (Boughman 2002)], which becomes increasingly unlikely toward the sympatric case. The broad definition of ecological speciation means that such processes can occur through a wide variety of qualitatively different mechanisms.

Competitive speciation (small hatched region; Rosenzweig 1978) results from intraspecific competition in sympatry and leads to the establishment of a stable dimorphism of ecological characters involved in resource utilization. While competitive speciation is a special case of evolutionary branching and thus of adaptive speciation, the latter can also arise from noncompetitive interactions, in parapatry, and through disruptive selection on mating traits.

---

sense the time-honored debate as to the relative importance of allopatric and sympatric speciation may relax in its fervor as discussions shift to elucidate the roles of nonadaptive and adaptive speciation.

1.4 Species Criteria

So far, we have used the notion of species without the usual elaborate qualifications and definitions that tend to be attached to it. There have been so many controversies and misunderstandings about what species "are" that some biologists have become reluctant to engage in or even follow these debates. Also, the purpose of this book – to illuminate the role of selection, driven by intraspecific interactions, in speciation processes – does not seem to benefit too much from refined arguments about the underlying concepts of species. Yet, given the substantial literature that exists on this topic, a few clarifying remarks are in order.

The naive species concept of old refers to a group of individuals, the members of which are relatively similar to each other in terms of their morphology (interpreted in the broadest sense) and clearly dissimilar from the members of any other species that exist at the same time. Species defined in this way are nowadays called morphospecies. A different, though related, perspective is stressed in the concept of ecospecies, defined as groups of ecologically similar individuals that differ in their ecological features from other such groups (Van Valen 1976). As any change in the ecological role of an individual has to be caused by its morphological make-up (in the aforementioned broad sense), we may expect an almost one-to-one correspondence between morpho- and ecospecies.

Sexual populations that differ morphologically or ecologically, but in which the individuals do not differ in their abilities to mate with one another, will hybridize when they share the same habitat. This consideration led Mayr (1963) to replace the naive species definition with the concept of "biological" species: the gene flow of a "biological" species is isolated from that of other species by the existence of intrinsic reproductive barriers. However, consideration of the reverse case reveals a drawback of this species definition: it elevates to the species rank sexual populations that differ in their abilities to mate with one another, but otherwise do not differ morphologically and ecologically. Such ecological sibling species usually are unable to coexist stably when they share the same habitat. So, to adhere to the biological species concept may lead to numerous distinctions that are relevant when addressing very specific questions only. Other, more important, difficulties with the biological species concept arise from the practical problems of testing for interbreeding capacity under "natural conditions" and because the fossil record does not offer direct evidence of reproductive isolation. In addition, the definition of "biological" species does not readily apply to asexual organisms, such as bacteria or imperfect fungi, or to organisms that reproduce clonally, like some plants.

As the concept of biological species attracted increasing criticism, other ideas emerged concerning the specific features of species that could be singled out to define them. The genotypic-cluster species concept, introduced by Mallet (1995) as a direct genetic counterpart to the morphospecies concept, requires that gene flow between species be low enough and disruptive selection strong enough to keep the genotypic clusters separate from one another. The recognition species concept of Paterson (1985) defines species as groups of individuals that share a common fertilization system. The cohesion species concept of Templeton (1989) stresses the gene flow between individuals of a species and their ecological equivalence as characteristic features. Species concepts qualified by attributes like genealogical, phylogenetic, or evolutionary emphasize that individuals of a species share a common evolutionary fate through time, and thus form an evolutionary lineage.

This broad and, as it seems, rather persistent variety of perspectives suggests that some pluralism in species concepts is inevitable and must be regarded as being scientifically justified. The salient criteria championed – variously – by phylogenetic taxonomists, experimental plant systematists, population geneticists, ecologists, molecular biologists, and others legitimately coexist: there are many features in which species can differ and the choice of particular definitions has to be appropriate to the actual research questions and priorities of each circumstance.

For the discussions in this book, perhaps the genotypic-cluster species concept may be most illuminating. It clearly highlights the need for adaptation to counteract gene flow if speciation is to occur outside rigorously allopatric settings. Also, the emphasis of the cohesion species concept on ecological interactions in addition to conditions of reproductive isolation is a welcome contribution to a debate about the prevalence of processes of adaptive speciation. Yet, we believe that biologists can discuss fruitfully causes and mechanisms of speciation processes without reaching, beforehand, a full consensus about their pet species criteria. As pointed out by de Queiroz (1998), such criteria often tend to differ in practice only in where precisely they draw the line between the one-species and the two-species phases of a particular speciation process. In this book we are interested in investigating how processes of speciation advance through time; drawing such lines is therefore not our primary concern.

1.5   Routes of Adaptive Speciation

We now outline some main adaptive speciation routes. As is well known, the ubiquity of frequency-dependent selection prevents the portrayal of evolution as a process of simple optimization. A trait combination that is best in an empty environment may become worst in an environment in which all individuals share that same trait combination. Similarly, directional selection can lead to trait combinations that, once adopted by a whole population, become the worst possible choice, so that selection turns disruptive. As explained above, this self-organized convergence to disruptive selection is the hallmark of evolutionary branching. It allows a phenotypically unimodal asexual population to become bimodal. According to the generally adopted criteria for asexual species, evolutionary branching can thus explain speciation in asexual populations.

In sexual populations, frequency-dependent selection can send evolving populations toward fitness minima. But in this scenario the genetic cohesion of sexual populations prevents their departure from such fitness minima – the continual creation of intermediate types by recombination usually makes it impossible for a randomly mating sexual population to respond to disruptive selection by becoming phenotypically bimodal. However, once individuals start to mate assortatively, the population can escape the trap. If individuals on each side of the fitness minimum happen to choose their partners from the same side, evolutionary branching also becomes possible in sexual populations.

Such assortative mating can come about in a number of ways; here we mention three different possibilities only. In the first scenario, assortative mating comes for free. Such a situation occurs when the ecological setting directly causes increased relative mating rates between partners on the same side of the fitness minimum. An example is the famous apple maggot fly. As a result of the strong spatial and temporal correlations between feeding preferences and mating opportunities, flies that have a slight preference for feeding on apples tend to mate more with partners of the same preference. The situation is analogous for flies with a slight preference for feeding on the traditional host plant, the hawthorn. In the second scenario, assortative mating may already be present, but may be based on traits other than those that vary across the fitness minimum. In such circumstances, the system for mate recognition and preference is already in place; it only has to be latched on to the right trait by the evolution of a genetic correlation. A third scenario is that the population is still mating perfectly randomly when it arrives at the fitness minimum. It can then be shown that such situations tend to give rise to positive selection pressures for the emergence of mate-choice mechanisms. Until assortative mating develops, frequency-dependent selection prevents departure of the population from the fitness minimum, and thus keeps it under a regime of disruptive selection: there is thus ample time for any one out of the plethora of possible mechanisms of assortative mating to develop.

It seems possible that the actual prevalence of nonrandom mating is underrated currently, perhaps because of the widespread dominance of assumptions of panmixia in genetics teaching and modeling, and because of the practical difficulties in empirically testing for assortativeness driven by yet unknown cues. However, independent of any consideration of speciation, choosing a good healthy partner is never a bad idea. Moreover, animals in general have well-developed cognitive abilities, not the least because they often have to cope with interference competition from conspecifics. The need to recognize conspecifics and, even more so, the requirements of social and territorial behavior may easily jump-start the development of mate-recognition systems. Also, if in a group of sexual taxa the processes of adaptive speciation are not uncommon, some mate-recognition mechanisms will have evolved already during preceding speciation events.

The evolution of assortative mating in a population situated at a fitness minimum has some aspects in common with the reinforcement of postmating barriers by the evolution of premating barriers. Yet, concerns about the likelihood of reinforcement do not carry over to the evolution of assortativeness under evolutionary branching. When two only partially isolated species come into secondary contact after allopatric divergence, the time scale at which the underlying bimodal phenotypic distribution again becomes unimodal through the formation of hybrids may be far too short for the relatively slow evolution of premating barriers to take hold. Worse, in the absence of frequency-dependent selection, hybrids may not even experience a selection pressure toward reinforcement. By contrast, in an adaptive-speciation scenario, ecological differentiation between incipient species is regulated dynamically to arise on the same time scale as mate choice emerges. This means that the ecological traits and mating traits evolve in-step: at any moment of the diverging evolutionary process, the current degree of ecological differentiation is sustainable given the current degree of mating differentiation, while – and this is critical – increasing degrees of mating differentiation continue to be selected for.

Although the persistent coexistence of ecological sibling species in sympatry is not expected, under certain conditions processes of adaptive speciation may be driven mainly by sexual selection. In particular, in sexual populations that already have in place a refined system for mate recognition and for which the costs of assortative mating are low, the generation of ecological sibling species by evolutionary branching in mating traits is likely. Here assortativeness comes for free as the differentiating characters are the mate-choice traits themselves. After the initial convergence of a population toward those preferences that would guarantee maximal reproductive success in the absence of mate competition, disruptive selection may favor individuals that avoid this competition by expressing slightly different preferences (Chapter 5). If this occurs in both sexes, the diversity of sympatric sibling species that results from multiple evolutionary branching is only limited by the maximal resolution of mate recognition and the maximal variability of mating signals. This diversity, however, is ephemeral if not accompanied by ecological differentiation or anchored on pronounced spatial heterogeneity in the habitat. And yet, for populations of sufficiently large size, a balance between rates of sibling speciation and extinction through ecological equivalence may lead to the persistence of sizable sympatric flocks of ecological sibling species. In such a situation the appearance of even relatively weak opportunities for ecological differentiation can lead, through evolutionary branching by natural selection, to a fast and bushy adaptive radiation.

1.6   Pattern and Process in Adaptive Speciation

At first sight it seems clear that adaptive speciation always occurs in sympatry and nonadaptive speciation in allopatry. This correlation between pattern and process can probably be expected to hold for a wide range of speciation events. Yet, there are exceptions. Clearly, chromosomal doubling and the emergence of polyploidy are processes of nonadaptive speciation that can take place in sympatry.

There may also be instances of adaptive allopatric speciation, as illustrated by the following hypothetical example. Imagine two disconnected populations of a clonal plant species that can defend itself against herbivory by the metabolism of secondary compounds, like alkaloids or tannins. In the absence of herbivores, both plant populations do not invest in defense. When, however, a mobile herbivore exploits the two plant populations, it pays for the plants to step up their defense. If plant populations in both patches do this by producing the same cocktail of secondary compounds, the herbivore may continue to exploit the two populations, albeit at a reduced level. If, however, one population presents the herbivore with a mixture of defense substances that differs from that adopted by the other population, that deviation will be favored by selection. This leads to the evolution of two different plant ecospecies by a process of adaptive allopatric speciation. The example shows that, in principle, ecological contact, although indirect, can occur in allopatry.

Keeping pattern and process clearly separated is also critical when considering speciation processes that progress via different phases, some of which occur in sympatry, and some in allopatry (Chapter 9; Box 19.1). Indeed, the traditional standard model of speciation, when combined with reinforcement, is already of such a type: postmating barriers emerge in allopatry and could be reinforced by the evolution of premating barriers in sympatry. Simply referring to such a two-stage process as allopatric speciation can be misleading. It is also possible that evolutionary branching in sympatry, followed by further phases of the same speciation process, leads to a biogeographic pattern of parapatry, or even allopatry. For example, we can think of a process in which ecologically differentiated sympatric populations start to latch on to those regions of a habitat with spatial variation to which they are adapted marginally better by a reduction in migration, which thus increases the assortativeness of mate choice. The segregated pattern that results from such a process may be misconstrued easily as evidence for nonadaptive speciation (Chapter 7).

As a last point it should be mentioned that present-day patterns may differ widely from those that occurred during the speciation process, which further complicates the task of inferring back from pattern to process.

1.7 Structure of this Book

The above discussion indicates that the interplay between pattern and process of speciation is potentially much more intricate (and interesting) than the common wisdom seems to suggest. This book is devoted to exploring adaptive speciation in theory and practice; we mean to investigate how far we can push the alternative paradigm. This means that, throughout the empirical parts of the volume, we as editors have strived to highlight the extent to which reported observations are compatible with scenarios of adaptive speciation. This effort must not be misconstrued as implying that in each of the analyzed systems adaptive speciation has been identified as the most likely scenario: such quantitative assessments are mostly still out of reach. Under these circumstances, we have encouraged the authors of this volume to bring out, as sharply as possible, the actual and potential links between their work and the notion of adaptive speciation. This is meant to enable our readers and colleagues to challenge the hypotheses championed in this book, and thus ideally encourage all of us to move forward toward a situation in which the espousal of alternative speciation mechanisms gradually ceases to be largely a matter of tradition and belief.

The book is divided into three parts. Part A outlines the existing theory of adaptive speciation. Part B confronts this theory with reality by exploring the extent to which the mechanisms implicated in models of adaptive speciation have been observed in natural systems. Finally, Part C moves to larger scales in space and time and examines how patterns of speciation inferred from phylogeographic or paleontological data can give insight into the underlying mechanisms of speciation. As we try to show in this book, adaptive speciation is not only an entirely plausible theoretical scenario, but the underlying theory also offers intriguing new perspectives on speciation processes. To make this explicit we start the book with an outline of the theory of adaptive speciation, and thus set the stage for the remainder of the book.

In Part A, recent theoretical developments on adaptive speciation, based on the framework of adaptive dynamics, are discussed in detail. To put matters into perspective, Part A also contains overviews of the classic approaches to modeling sympatric, parapatric, and allopatric speciation. The part ends with Chapter 7, which attempts to synthesize pattern-oriented and process-oriented approaches to understanding speciation through the study of adaptive speciation in geographically structured populations. Chapter 7 shows that parapatric patterns of species distributions may result from intrinsically sympatric ecological processes and provides new perspectives on the role of geographic structure in shaping speciation processes.

Empirical investigations of speciation are often hampered by the problem of long generation times in the organisms under study. Indeed, speciation theory has too often succumbed to speculation, partly because of the paucity of direct empirical tests of hypotheses about mechanisms of speciation. It is therefore imperative to strive for empirical, and in particular experimental, tests of the hypothetical driving forces behind speciation processes. Part B provides an array of examples of natural systems in which mechanisms of frequency-dependent disruptive selection and/or mechanisms of assortative mating are likely to operate. Such systems include fish flocks in young lake systems, insects in the process of host switching or increased specialization, and plants interacting with their pollinators. Perhaps microbes are the class of organisms most amenable to direct observation of the whole process of adaptive diversification originating from a single ancestor. Part B thus ends with an outlook on the great promise that experimental evolution in microorganisms holds for direct empirical tests of hypotheses on adaptive diversification.

Since direct empirical tests are laborious and time consuming, processes of speciation are often inferred from data gleaned from natural speciation experiments, as reflected in phylogeographic patterns and in time series pried from the fossil record. In particular, many closely related species show little overlap in their ranges, which suggests, at first sight, their allopatric origin. However, models of adaptive speciation in geographically structured populations indicate that things may not be that simple, because processes of adaptive speciation under conditions of ecological contact may result in parapatric (and, in the longer run, even allopatric) patterns of species abundance. Thus, extant patterns are not necessarily good indicators of the past processes that brought them about. Moreover, since processes of adaptive speciation are expected to unfold relatively fast on a paleontological time scale (Chapter 18), the conditions under which a phylogenetic split actually occurred may have changed drastically after long periods of subsequent divergence. It is therefore important to interpret phylogeographic patterns in light of the dynamic, and potentially multilayered, nature of speciation processes, and to pay attention to the appropriate time scales. The chapters in Part C examine what phylogeographic or paleontological patterns can tell us about processes of speciation. These chapters show that many of the patterns that arise in a diverse array of taxa are consistent with adaptive speciation processes, and that in many cases adaptive speciation may provide a more parsimonious interpretation of the phylogeographic patterns than does allopatric speciation.

This book has an agenda. We hope to convince the reader that adaptive speciation through frequency-dependent interactions under conditions of ecological contact is a plausible, and perhaps even ubiquitous, evolutionary process. This view is supported both by detailed theories of adaptive diversification and by a growing body of empirical data on patterns and processes of speciation. In our view, the time has come to do away with the notion that allopatric speciation is true until proved wrong, an idea that may prevail mainly because of the deceptive simplicity of allopatric scenarios and the towering scientific stature of its initial proponents. However, how well a mechanistic theory describes reality has little to do with its mathematical complexity; if anything, more detailed theories would appear to be more reliable. On this basis, we think that adaptive speciation should be viewed as an equally valid null hypothesis. Once the bias toward detecting allopatric speciation in empirical data is removed, the data may actually suggest adaptive speciation as the more likely explanation of many speciation events. We hope that the perspectives put forward in this book will spark new empirical work specifically designed to test hypotheses of adaptive speciation. Overall, we hope to contribute to an intellectual process, vaguely akin to adaptive diversification itself, by freeing research on species formation from the constraint of always having to view speciation processes through the allopatric lens. The formation of new species appears to be more complex, and also more fascinating, than the traditional view suggests. Thus, a plea for pluralism: an open mind and a diverse array of perspectives will ultimately be required to understand speciation, the source of our planet's biodiversity.

Acknowledgments   We are indebted to Agusti Galiana for drawing our attention to the quote by Mayr (1982, p. 565) and to Menno Schilthuizen for highlighting the relation between adaptive speciation and the quote by Darwin (1859, p. 155). Franjo Weissing and Sander van Doorn provided valuable assistance in improving the clarity of this chapter.

2

Speciation in Historical Perspective

Will Provine

2.1 Introduction

By the time Darwin died in 1882, evolution by descent was widely hailed, never again to be challenged by the vast majority of biologists. His theory of natural selection as the key to understanding adaptation, however, was less successful. Alfred Russel Wallace's *Darwinism*, which strongly defended natural selection, appeared soon after Darwin's death (Wallace 1889). Detractors of natural selection increased in strength during the 1890s; indeed, historian Peter Bowler termed the period close around the turn of the century as "the eclipse of Darwinism" (Bowler 1983). A contemporary biologist billed this period, more graphically as, "the deathbed of Darwinism" (Dennert 1903). Even Wallace, in his old age, began to backpedal from natural selection toward belief in god-designed organisms (Wallace 1910). Beginning with the "evolutionary synthesis" of the 1930s and 1940s, Darwin's natural selection enjoyed a rebirth, and became known as "Twentieth Century Darwinism," the name made famous by the 1959 Cold Spring Harbor Symposium *Genetics and Twentieth Century Darwinism* (Mayr 1959a).

Evolution produces two major results: adaptations and biodiversity. Natural selection loomed large in the mid-20th century as the explanation of adaptive design, but Darwin's ideas about species and speciation simultaneously were driven to obscurity, where they have remained. The argument here is that "21st Century Darwinian Speciation" is emerging from the darkness of the past half-century.

Study of speciation today addresses most of the same issues debated so intensely by Charles Darwin, Alfred Russel Wallace, Joseph D. Hooker, Henry Walter Bates, Fritz Müller, Moritz Wagner, or Asa Gray:

- What is a species?
- What causes determine the divergence of species from a single ancestral species?
- What is the role of sterility in speciation?
- Why do some kinds of organisms speciate readily and others speciate rarely?
- Is geographic separation required for speciation, or can species form in the same geographic area?
- Do population structure, sexual selection, and recognition of mates play important roles in speciation?
- Are chance factors important in speciation? Is speciation enhanced when open ecological niches are available?

For most of the world's population, gods or purposive forces design adaptations. For biologists, however, adaptation has only one naturalistic explanation: natural selection. Thus, natural selection is a great unifying concept in the study of evolution. Although very few examples of natural selection in nature were revealed in the first century after publication of *On the Origin of Species* (Darwin 1859), they have grown in number since then and a whole book (Endler 1986) devoted to their analysis. Even critics of neo-Darwinism (alias panselectionism), such as Motoo Kimura (1983) and Masatoshi Nei (1989), both advocates of neutral molecular evolution, or Eldredge and Gould (1972), advocates of punctuated equilibrium, nevertheless agree that adaptations are produced by the causes of natural selection.

Diversity, in the form of speciation, has no book such as Endler's, and the examples so far are hardly robust. In theory and practice, speciation is a much more difficult problem than adaptation and far less unified. In this brief chapter only four topics are addressed:

■ Charles Darwin's ideas about species and speciation.
■ Evolutionary synthesis (1930s–1950s) and the work of Ernst Mayr on species and speciation in his *Animal Species and Evolution* (1963).
■ Changes in evolutionary biology and their effects on theories of speciation since 1963.
■ Greater openness of contemporary biologists, by focusing upon Dolph Schluter's book *The Ecology of Adaptive Radiation* (2000) and the review by John Avise (2000b) of species concepts.

2.2   Darwin on Species and Speciation

Darwin's views on species and speciation fare well in comparison with other approaches adopted over the past 150 years.

Darwin on species

Darwin's contemporaries could agree on no clear definition of species, and in the *On the Origin of Species* (1859) and elsewhere Darwin declined to offer one of his own ["No one definition has yet satisfied all naturalists, but every naturalist knows vaguely what he means when he speaks of a species" (Darwin 1859, p. 44)]. This hardly reassuring definition of a species was buttressed by other declarations of vague and arbitrary attributes of species:

> Many years ago, when comparing, and seeing others compare, the birds from the closely neighbouring islands of the Galapagos Archipelago, one with another, and with those from the American mainland, I was much struck how entirely vague and arbitrary is the distinction between species and varieties. (Darwin 1859, p. 48)

> Certainly no clear line of demarcation has as yet been drawn between species and subspecies – that is, the forms which in the opinion of some naturalists come very near to, but do not quite arrive at, the rank of species; or, again, between sub-species and well-marked varieties, or between lesser varieties and individual differences. (Darwin 1859, p. 51)

Darwin, of course, wanted above all to convince his audience that evolution by shared descent had occurred. To have no clear distinctions between species and subspecies, subspecies and well-marked varieties, and lesser varieties and individual differences was crucial to his argument: "These differences blend into each other by an insensible series; and a series impresses the mind with the idea of an actual passage" (Darwin 1859, p. 51).

P. Chalmers Mitchell, who wrote the article on "species" for the 11th edition of the *Encyclopedia Britannica* (1911), argued that Darwin and evolutionists after him had destroyed all possibility of a clear notion of species: "... Systematists no longer regard species as more than an artificial rank in classification to be applied for reasons of convenience ... . A species, in short, is a subjective conception" (Mitchell 1911, pp. 616–617).

Darwin, however, was sure species existed. "I believe that species come to be tolerably well-defined objects, and do not at any one period present an inextricable chaos of varying and intermediate links ..." (Darwin 1859, p. 177). Only when viewed across time, or across geography, did species depart from "tolerably well-defined objects". Thus, Darwin did not torture himself on the species question – he just accepted the assessments of systematists who knew their species, as he trusted his own work on the systematics of barnacles. He used, as would most naturalists of his day, a morphological approach to the differences between species. He would not have accepted the assessment of "species" given by Mr. P. Chalmers Mitchell.

Darwin on speciation

All editions of Darwin's *On the Origin of Species* carried as the first subtitle, "By Means of Natural Selection". Darwin was fascinated by the different ways species were adapted to their ecology:

> The face of Nature may be compared to a yielding surface, with ten thousand sharp wedges placed close together and driven inwards by incessant blows, sometimes one wedge being struck, and then another with greater force. (Darwin 1859, p. 67)

What counted about the species of a genus was how the species ("wedges") differed, and to what extent their fit into to the ecology could be understood by means of natural selection.

Darwin's attempt to understand how closely related species were adapted to their environments presented him with questions about natural selection and the production of species. When Darwin speaks of "sterility," he includes all forms of reproductive isolation:

> Why should the degree of sterility be innately variable in the individuals of the same species? Why should some species cross with facility, and yet produce very sterile hybrids; and other species cross with extreme difficulty, and yet produce fairly fertile hybrids? Why should there often be so great a difference in the result of a reciprocal cross between the same species? Why, it may even be asked, has the production of hybrids been permitted? To grant to species the special power of producing hybrids, and then to stop their further propagation by different degrees of sterility ... seems to be a strange arrangement. (Darwin 1859, p. 260)

These bothersome questions could be answered, however, by attributing the results to something other than natural selection, a process that seemed to have no direct answers to questions about sterility between closely related species:

> On the theory of natural selection the case is especially important, inasmuch as the sterility of hybrids could not possibly be of advantage to them, and therefore could not have been acquired by the continued preservation of successive profitable degrees of sterility. I hope, however, to be able to show that sterility is not a specially acquired or endowed quality, but is incidental on other acquired characters. (Darwin 1859, p. 245)

> There is no more reason to think that species have been specially endowed with various degrees of sterility to prevent their crossing and blending in nature, than to think that trees have been specially endowed with various and somewhat analogous degrees of difficulty in being grafted together in order to prevent them inarching in our forests. (Darwin 1859, p. 276)

Subspecies showed the same range of sterility as species themselves, but with a higher frequency of fertility than in species. So for Darwin it was possible to have two subspecies that were totally sterile when crossed, and two species that were totally fertile when crossed, although neither scenario was a frequent occurrence.

2.3   Mayr on Species and Speciation

In 1964 Mayr published a facsimile edition of Darwin's *On the Origin of Species*, with his own introduction. Population thinking and natural selection, Mayr said, were Darwin's great contributions to evolutionary thought. In other respects, however, Mayr said Darwin was sadly mistaken:

> Though Darwin was wrong in his discussions of inheritance and the origin of variation, confused about varieties and species, and unable to elucidate the problem of the multiplication of species, he was successful in discovering the basic mechanism of evolutionary change. (Mayr 1964, p. viii)

Darwin could hardly be faulted for his lack of appreciation of Mendelian inheritance, but why did he fare so poorly on species and speciation?

Mayr on species

By the mid-1930s, dissatisfaction with the morphological definitions of species was growing. Theodosius Dobzhansky, Ernst Mayr, Sewall Wright, and others pushed for what they called a more "scientific" definition of species, upon which all could agree. The outcome was what Mayr termed the "biological species concept", which defines species by the isolating mechanisms that guarantee the purity of the gene pools of the species. To determine if two populations were two species, morphological differences were useless and ignored in the biological species concept. At last, a satisfyingly universal, *scientific* concept of species had been invented:

The general adoption of the biological species concept has done away with a bewildering variety of "standards" followed by the taxonomists of the past. One taxonomist would call every polymorph variant a species, a second would call every morphologically different population a species, and a third would call every geographically isolated population a species. This lack of a universally accepted standard confused not only the general biologists who wanted to use the work of the taxonomist, but the taxonomists themselves. Agreement on a single yardstick, the biologically defined category species, to be applied by everybody, has been a great advance toward mutual understanding. (Mayr 1963, p. 21)

The quickest way to return to "a bewildering variety of standards" is, according to Mayr, to think about subspecies. Citing the classic paper of Wilson and Brown (1953) on the difficulties of the subspecies category, Mayr declared: "... the subspecies, which conceals so much of the inter- and intrapopulation variation, is an altogether unsuitable category for evolutionary discussions; the subspecies as such is not one of the units of evolution" (Mayr 1963, p. 348). When the subspecies reaches the level of a species (i.e., has an independent gene pool), then it becomes a genuine evolutionary unit, one that anyone can recognize. In a long section on gene pools and homeostasis, he argued that a gene pool was coadapted and resistant to change: "Genetic homeostasis determines to what extent a gene pool can respond to selection" (Mayr 1963, p. 289). With Mayr's great influence, the biological species concept became the most widely accepted definition of species ever, including the present, though now more dissension exists than in the 1960s.

Mayr on speciation

Once the biological species concept is accepted, speciation is easy to envision: one gene pool becomes two separate gene pools. In his 1963 book Mayr addresses the obvious question of how one gene pool with homeostasis becomes two. "The mechanisms that isolate one species reproductively from others are perhaps the most important set of attributes a species has, because they are, by definition, the species criteria" (Mayr 1963, p. 89):

> *Reproductive isolation* refers to the protective devices of a harmoniously coadapted gene pool against destruction by genotypes from other gene pools. These protective devices are known under the term isolating mechanisms. Speciation is characterized by the acquisition of these devices. (Mayr 1963, pp. 546–547)

Calling isolating mechanisms "protective devices" sounds like these devices are primary outcomes of natural selection, but this impression is false as Mayr explains.

In a section of Chapter 17 (The genetics of speciation) entitled "The origin of isolation mechanisms", Mayr (1963) clarifies how these protective devices originate. In the first place, the evolution of isolating mechanisms required geographic separation, perhaps Mayr's most famous and lasting postulate of speciation. After geographic separation,

> The most indispensable step in speciation is the acquisition of isolating mechanisms. Isolating mechanisms have no selective value as such until they are reasonably efficient and can prevent the breaking up of the gene complexes. They are *ad hoc* mechanisms. (Mayr 1963, p. 548)

However, isolating mechanisms are very important for the cohesion of species. How could they have "no selective value" and be merely "*ad hoc*" mechanisms?

Mayr was thinking here about the controversy between Darwin and Wallace (and many later evolutionists) over the evolution of sterility between species. In a major article, Mayr (1959b) had evaluated their positions carefully, and stated his strong support for Darwin's views. As Mayr points out, Darwin used "sterility" to refer mostly to what we now call isolating mechanisms, not mere physiological sterility. So as populations diverge from each other, the isolating mechanisms arise as "an incidental by-product of genetic divergence in isolated populations" or as a "by-product of the total genetic reconstitution of the speciating population" (Mayr 1963, p. 551). For Mayr, the genetic divergence related to morphological differences in adaptive characters had nothing to do with the speciation, which depended upon only the incidental by-products, the isolating mechanisms. Rates of acquisition of isolating mechanisms varied from case to case and "there is no standard rate of speciation. Each case is different and the range between the possible extremes is enormous" (Mayr 1963, p. 581). Even within a genus with only a few several species, one case of speciation might give no good indication of speciation in another case.

So the question is, what kind of "by-products" are these isolating mechanisms? According to Mayr, they comprised by-products of the "genetic divergence" of the populations. There, unfortunately, the story ended and ignorance prevailed. Speciation had become inscrutable, and nothing but vague "correlations" with factors produced by natural selection and observable differences at morphological to genetic levels. However, Darwin at least had not failed Mayr completely on the issue of speciation – they agreed deeply on the issue of sterility factors (or isolating mechanisms) being merely incidental.

2.4   Species Now

A sea change in evolutionary biology has occurred since the 1960s (Provine 2001, Afterword). Thoughts about species and speciation, in particular, have begun to change during the past 40 years. The rise of cladism, DNA sequencing, protein sequencing and analysis of function in proteins, theories of neutral molecular evolution, and other factors have challenged the hope that evolutionary biology is unified across all levels of organization. The evolutionary synthesis has been unraveled since 1980. Now the unity of evolutionary biology in the "synthesis" has given way to a much more intriguing complex of different levels, each with a particular complex of causes. To argue that the DNA-sequence level marches to the same beat as the adaptive-trait level, which seemed natural in the 1960s, seems hopeless at the present.

We have seen how Mayr centralized the ideas of the "gene pool" and "genetic homeostasis" in his biological species concept, but now we can see in hindsight that both concepts are nearly useless. "Gene pool" now appears to be one of the most artificial concepts of population genetics. What exists in the "gene pool" is vague, but perhaps most often either "genes" or "alleles". Other candidates for the gene pool are chromosomes, gametes, and whole organisms. Neither genes nor alleles float free, but are on chromosomes, and do not cleave every generation. To talk about the cohesion, coadaptation, and homeostasis of the gene pool means attributing fancy characteristics to a nonexisting entity.

In small populations, invocation of the gene pool as the intuitive source of the binomial sampling for genetic drift leads to mathematical models that look robust, but are not. The basic problem is that recombination is far too weak to make random binomial sampling of individual genes (or DNA bases) possible over tens or hundreds of generations (Gillespie 1999, 2000a, 2000b).

Genetic homeostasis is an attractive idea, but sadly lacks substance. We hear no more about it these days; instead, we hear about DNA repair mechanisms at the level of the individual genome and nothing about homeostasis at the "gene pool" or population level. For Mayr, the homeostasis of the gene pool held the species together, but now this says nothing biological about species.

In the 1960s, the definition of "species" was mercifully clear. Now, let us try an assignment to a new graduate student in evolutionary biology. Our hypothetical student had a fine background in natural history with undergraduate research in *Drosophila* in the field and in the laboratory, read Doug Futuyma's *Evolutionary Biology* (1979) in her evolution course, and then took a reading course on Mayr's *Animal Species and Evolution* (1963). Untroubled by doubt about either species or speciation, she looks forward to our reading course. In sequence, she reads: Slobodchikoff (1976), Barigozzi (1982), Otte and Endler (1989), Howard and Berlocher (1998), Wheeler and Meier (2000), Schluter (2000), and this volume. At the end of the reading course, species and speciation are discussed. The first question is, "Using your extensive background reading, would you please give your most precise general definition of a species?" She gives, probably, one of two answers. "How do you expect me to give you an answer to this question when the world experts can't agree at all?" or "The only concept that seems general and scientific is the biological species concept." On the one hand, the drive is back to poor P. Chalmers Mitchell and his "artificial" species, but on the other to Mayr's biological species concept with the baggage that it is has to focus on the "by-product" isolating mechanisms and ignore the adaptive radiations that grab our attention as natural historians.

As a historian, I am struck by the continuing struggle of biologists to derive a robust, single, best concept of species. Nearly every book on species read by the hypothetical student expresses over and over the hope of finding *the* definition of species.

John Avise, whose field work and analysis of species and speciation in a wide variety of organisms is well known and erudition wide, has recently written a review of Wheeler and Meier (2000), with the great title, "Cladists in wonderland" (Avise 2000b). Educated in the biological species concept, Avise finds that the four distinctive cladist concepts of species lack biological realism, but at the same time believes that the biological species concept needs revision. Perhaps he should have taken the line that to worry so much about the correct and scientific concept of species was a mistake, but no, he wants that robust concept, a synthesis of the concepts of biological species and phylogenetic species through unification with phylogeographic and coalescent principles:

> Perhaps the ongoing phylogeographic synthesis that tries to wed (rather than divorce) phylogenetic and reproductive concepts in species recognition will yet prove to be only another fantasy. But I doubt it. Instead, I have great hope that the peculiar tea-party banter between the Aliceians and the Mad Hatters over species concepts will eventually clarify, and that a more intelligent dialogue and eventual synthesis will emerge. If so, the 20-year quarrel between proponents of the BSC [biological species concept] and the PSC [phylogenetic species concept], so cogently encapsulated in the Wheeler and Meier volume, will someday be remembered as little more than a "tempest in a teapot". (Avise 2000b)

Perhaps the most insightful single article to deal with species concepts and speciation is Richard G. Harrison's "Linking evolutionary pattern and process: The relevance of species concepts for the study of speciation" (Harrison 1998). Harrison concludes that the BSC remains the only major species concept to escape the withering objections of all phylogenetic and species-recognition concepts. In his review of Howard and Berlocher (1998), Schemske (2000) agrees with Harrison, as do many others; Coyne (1992, 1994), especially, has similar views.

If I were asked the same question posed to the hypothetical student above, and forced to answer, I would probably agree with Harrison. I am not, however, faced with such an unattractive pair of choices. Instead, I can see excellent uses of the wide variety of species concepts, and the crucial mistake is biologists' obsessive search for the best species concept, however limited. As soon as we feel the onus that one species concept has to be chosen for all purposes, from constructing phylogenies to speciation, then we are in a needlessly poor situation. Even armed with a species concept well-suited to its particular use, a biologist might want to carry it lightly.

2.5   Speciation Now

Is, as Harrison suggests, the biological species concept the best species concept for understanding speciation? One place to look is Dolph Schluter's *The Ecology of Adaptive Radiation* (2000), because Schluter focuses upon the same problem as Darwin: how does adaptive radiation take place? Speciation, according to Schluter, is central to adaptive radiation, so he has to address the problem of species concepts and does so early in the book:

> Speciation refers to the evolution of reproductive isolation, defined as the complete absence of interbreeding between individuals from different populations (should they encounter one another), or the strong restriction of gene flow sufficient to prevent collapse of genetically distinct populations that continue to interbreed at a low rate. This is basically the biological species concept (Mayr 1942), but accommodates the fact that a great many sexual species hybridize [many pertinent citations] yet existing levels of assortative mating do not decay. (Schluter 2000, p. 13)

Mention of species' concepts then ceases, and the rest of the book flows along with little attention paid to "isolating mechanisms" or, in Harrison's more neutral language, "barriers to gene exchange". Here is a whole book about speciation and the emphasis is upon adaptive radiations and natural selection. Schluter (2000) does admit that "not everyone's definition of adaptive radiation includes speciation", but that he is happier to focus upon the preponderance of cases in which speciation has occurred.

Compare this book with the many papers published within the past ten years that analyze isolating mechanisms, the key to species and speciation. Although this work is great, I do worry about those doing the labor, because no general theory of speciation will ever come from it, as both Mayr and Harrison emphasize. If J.F.W. Herschel was happy to describe natural selection as the "law of higgeldy, piggeldy", then speciation is best described as "the law of super higgeldy, piggeldy". I now discuss two examples, one famous and one not.

Speciation in cichlids

Tijs Goldschmidt went to Lake Victoria in 1985 to study speciation in the cichlid fish. He wrote about his adventures in his book, *Darwin's Dream Pond: Drama in Lake Victoria* (Goldschmidt 1996), the perfect book for a course on evolution for nonmajors in biology. Indeed, Lake Victoria looked like the place to study speciation. In less than 13 000 years (Goldschmidt thought it was older in 1985), an amazing adaptive radiation of cichlid fish happened. Probably a single introduction of a small number of fish (perhaps only one pregnant female) resulted in from 400 to 800 species by the beginning of the 20th century. The differentiations between species include body size, color patterns, skull shapes, and mouth and jaw parts, along with many differences in feeding habits and adaptations to different ecological niches. Many rift lakes in northern Africa and many other places in the world now and in the fossil record show similar radiations of cichlid fish.

However, "Darwin's dream pond" is hardly that. Goldschmidt himself, after careful collecting and attempts to study speciation of the cichlids, says, "I decided that nothing beat poetic truth and abandoned science" (Goldschmidt 1996, pp. 244b–244c). He became a wonderful writer about the science of studying speciation, but stopped doing it. The most recent survey, "New markers for new species: microsatellite loci and the East African cichlids" (Markert *et al.* 2001), suggests that use of allozymes, nuclear DNA, and mitochondrial DNA (mtDNA) has not been very helpful in the study of speciation in cichlids:

The brief periods between speciation events cause what little variation exists at these markers to be shared among taxa. These types of marker are not definitive in studies of assortative mating, only mildly informative for population studies, and are phylogenetically informative in only the oldest lineages. (Markert *et al.* 2001)

Although using microsatellite DNA gives hints about estimates of divergence, to deduce robust phylogenies for the cichlids in Lake Victoria is still impossible. Nor is it yet possible to examine two closely related species and deduce whether their speciation was "microallopatric" or sympatric. Darwin's dream pond is a tough place to study speciation in cichlids.

Still, what is so interesting about these fishes? Is it the details of the isolating mechanisms? Or is it the adaptive radiation? The recent ecological disturbances of Lake Victoria, especially the introduction of the Nile perch, have so muddied and changed the lake that many of the species of cichlids have disappeared and crossbreeding between remaining ones has occurred. Does this mean that the fish are quickly reverted to the nonbiological category of unscientific "subspecies" and their worth for understanding speciation nothing? To me, the adaptive radiation was no less interesting whether the final "species" had doubtful or certain status as biological species.

North American rat snakes

The other, less famous, example comes from my favorite childhood (and perhaps adult) animal, rat snakes of the USA, and concerns the mtDNA phylogeography of the polytypic North American rat snake *Elaphe obsoleta* (Burbrink *et al.* 2000).

One general pattern in rat snakes was obvious: gray versions occurred in the south, and black versions in the north. Striped versions, north from Florida and up the eastern coast, go from pink to yellow to olive green to black. I have caught black rat snakes from New York State west to Illinois (Pennsylvania, Ohio, Indiana, Illinois), gray in Tennessee, Kentucky, Alabama, and Mississippi, and the striped rat snakes from coastal North Carolina south to Kissimee Lake in central Florida. The map of Burbrink *et al.* (Figure 2.1) fits perfectly my own observations. For a very long time, black rat snakes have been considered to be a single subspecies. The different gray patterns were sometimes given species status.

Into this picture, Burbrink *et al.* (2000) introduced a real surprise, based directly upon their analysis of the phylogeography of mtDNA over the distribution of *E. obsoleta*. Their results show that black rat snakes in North America have descended from three different adaptive radiations, as shown in Figure 2.2.

What a delightful surprise this was. The experts had been completely misled by the apparent similarity of black rat snakes and not even suspected that they could have come from three different radiations. What is the conclusion of the authors regarding species and speciation? The subtitle of the paper ("A critique of the subspecies concept") and its first paragraph both criticize the validity of the subspecies rank. The authors favorably quote Mayr's assessment that "the subspecies was not a concept of evolutionary biology ..." and add: "subspecies

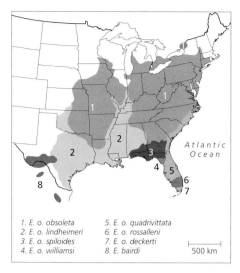

1. *E. o. obsoleta*   5. *E. o. quadrivittata*
2. *E. o. lindheimeri* 6. *E. o. rossalleni*
3. *E. o. spiloides*   7. *E. o. deckerti*
4. *E. o. williamsi*   8. *E. bairdi*   500 km

Figure 2.1 Map of the eastern United States showing the geographic range of the subspecies of *E. obsoleta* and *E. bairdi*. *Source*: Burbrink *et al.* (2000).

have no real taxonomic meaning if they are used to represent arbitrary pattern classes or incipient species" (Burbrink *et al.* 2000).

I find these conclusions perplexing. These authors have resoundingly rejected the long-standing and well-accepted subspecies of *E. obsoleta* with the warning, so clearly expressed in the conclusion:

> This study has demonstrated that the subspecies of *E. obsoleta* do not represent distinct evolutionary lineages and underscores the danger of recognizing subspecies based on few characters, especially coloration. These poorly defined subspecies actually mask the evolutionary history of the group. Therefore, describing or recognizing subspecies from a few characters may not simply be a harmless handle of convenience for museum curators, but may be detrimental to understanding evolutionary history. (Burbrink *et al.* 2000)

The authors suggest that the three clades may possibly be lineages that have evolved independently, or "evolutionary" species. I agree that the previous subspecies were defined poorly, and that the three clades revealed by their studies give a much more accurate taxonomy and evolutionary picture. Giving up upon the entire concept of subspecies from the data given seems unnecessary to me. Whether one calls the clades well-characterized "subspecies" or pre-"evolutionary species" seems a minor issue to me.

Is this article a throwaway if, indeed, these clades are not true "evolutionary species" or true "biological species", but only incipient species? Of course not, the article is equally important whether the clades are full species (evolutionary or biological), or three incipent species of either kind. The phylogenetic order that

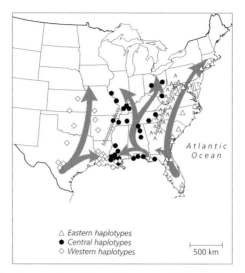

Figure 2.2 Map showing northern dispersal patterns of *E. obsoleta* mitochondrial clades from southern refugia following glacial retreat. *Source*: Burbrink *et al.* (2000).

emerges in this study provides a basis for real investigation of the three adaptive radiations and the differentiation in each clade. The authors mention this possibility.

The rat snake speciation (or subspeciation) problem thus has progressed beyond the case of the Lake Victoria cichlids, for which the entire panoply of molecular evidence has been unable to produce robust phylogenies. However, we know next to nothing about the speciation of the three clades, except that they were probably separate during the last full glaciation. Whether they were separate after the previous glaciation, or the one before that, we have no clue as yet.

Returning to Harrison's belief that the biological species concept is best for understanding speciation, I plead skepticism. If the usual implications of the biological species concept hold true, then to focus upon isolating characters, demotion of subspecies, and dismissal of "incipient species" tends to turn attention away from understanding processes of adaptive speciation. Many chapters in this volume address precisely crucial steps during the incipient divergence of new lineages through adaptive mechanisms.

Our hypothetical student returns to the discussion. She now wants to study speciation as her major interest. She has some requirements for her study. First, she wants complete sequences for nuclear DNA, mtDNA, microsatellite DNA, and any other genetic material for her study organism and its close relatives. She will then see if she can obtain robust phylogenies and, if not, choose another organism for study (but even a perfect, robust phylogeny does not by itself explain speciation). Next, she wants to know about the biology of the study organism, from genes and development to adult morphology and behavior, including sexual selection. She wants to understand the ecological setting not just now, but in recent history. She must, in short, be a natural historian who understands the organism. Finally, she wants a study organism that has recent speciation and incipient species. (A ten-year generous grant from a suitable source would help her project along.)

This volume

The present volume started with a three-day workshop at which the authors outlined their contributions. All the results of previous conferences on speciation that I have read were much involved with species concepts. This conference, refreshingly, was in no way dominated by discussions of "what is a species?" Instead, the processes of adaptive speciation are the main focus of this book. Adaptive speciation, as defined in the Introduction (Chapter 1), occurs when divergence is an adaptive response to interactions within ancestral populations. Isolating mechanisms may then be either a by-product of divergence, or they may be favored by natural selection. In either case, adaptive speciation typically requires some degree of sympatry between the emerging lineages.

Indeed, many participants did talk about adaptive speciation as a kind of code word for sympatric speciation. Even allopatric speciation is tied deeply to adaptive evolution, since isolation factors often move to fixation because of close linkage relationships with parts of the genome that evolve by natural selection. But the emphasis in this volume is on the adaptive significance of divergence *per se*, rather than just upon isolating mechanisms and the biological species concept. As a corollary, the book does not dwell so much on the pattern-oriented debate about allopatric versus sympatric speciation: it is the process of diversification that is at center stage, and biogeographic patterns are of secondary interest.

The book discusses evidence for adaptive speciation from theory (Part A), experiments (Part B), and phylogenetic patterns (Part C). I tend to view the growing trend toward studying speciation and diversification as adaptive processes, a trend that is represented in this volume, as providing a period of "Darwinian speciation" in the 21st century.

Part A

Theories of Speciation

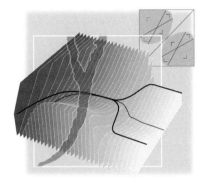

Introduction to Part A

Theories of speciation, in the past often couched in verbal terms, should explain how ecological divergence and genetically determined reproductive isolation evolve between lineages that originate from single, genetically homogeneous ancestral populations. As Will Provine highlights in Chapter 2, the predominant perspective for a long time was that reproductive isolation emerges as a by-product of other evolutionary processes, through the incidental accumulation of genotypic incompatibility between related species. It is easiest to imagine that such incompatibilities arise when subpopulations become geographically isolated and henceforth evolve independently: genetic distance between them is then expected to increase with time. Thus, "given enough time, speciation is an inevitable consequence of populations evolving in allopatry" (Turelli *et al.* 2001). On a verbal level this theory of allopatric speciation appears both simple and convincing. This apparent theoretical simplicity has contributed to the view that the allopatric mode of speciation is the prevalent one – a perspective that has found its most prominent advocate in Ernst Mayr (Chapter 2).

Unfortunately, not only is the simplicity of the usual accounts of allopatric speciation based on the poorly understood concept of genetic incompatibility, but simplicity in itself is no guarantee for ubiquitous validity. Other plausible, but theoretically more intricate, mechanisms for the evolution of reproductive isolation in the absence of geographic isolation have been proposed. Recent approaches have focused attention on adaptive processes that lead to ecological and reproductive divergence as an underlying mechanism for speciation processes – a change in emphasis that occurred concomitantly with a shift in biogeographic focus from allopatric scenarios to parapatric speciation between adjacent populations or fully sympatric speciation. This was foreshadowed by the idea of reinforcement (the evolution of prezygotic isolation through selection against hybrids) and has culminated in theories of sympatric speciation, in which the emergence and divergence of new lineages result from frequency-dependent ecological interactions. Such interactions can induce disruptive selection, which in turn generates indirect selection for a proper choice of mates and thus leads to prezygotic isolation. While these theories of adaptive speciation can also be described verbally, the involved mechanisms are more intricate than those of the basic allopatric scenario. This does not imply that adaptive speciation is an unlikely evolutionary process: it can even be argued that the explicit and detailed inclusion of ecological interactions as driving forces of evolutionary change renders these speciation models more convincing than the purely verbal models.

Part A of this book outlines the existing theory of adaptive speciation. Overviews of the classic approaches to modeling sympatric, parapatric, and allopatric speciation are added for perspective. The material in this part shows that adaptive speciation is a theoretically plausible scenario, and thus sets the stage for the remainder of the book. The most recent theoretical developments on adaptive speciation, based on the framework of adaptive dynamics, are discussed in detail. In these models, evolutionary dynamics are derived explicitly from ecological interactions between individuals. These interactions often result in frequency dependence, and the populations evolve such that they end up at fitness minima, even though they are continually climbing the fitness landscape. During such processes disruptive selection on metric characters emerges gradually and dynamically. This extends traditional ecological theories of sympatric speciation, which descend from the early models of Maynard Smith (1966) and deal with small numbers of alleles and discrete resources or habitat types. To put matters into context, Part A begins with an overview of this classic work on sympatric speciation.

In Chapter 3, Kawecki first reviews models in which protected polymorphisms can arise from density-dependent competition for two discrete resource niches. In some special scenarios, divergent sexual selection alone can lead to reproductive isolation under sympatric conditions, even in the absence of disruptive ecological selection. Disruptive selection creates the conditions for the evolution of assortative mating, which can be based directly on the ecological character under disruptive selection or on selectively neutral traits that enter into linkage disequilibrium with the ecological trait. In the former case the evolution of assortative mating is more likely than in the latter, a line of thought that is taken up again in Chapter 5. Chapter 3 concludes by discussing divergent mate choice that evolves as a by-product of habitat preference, which may be an important mechanism for sympatric speciation through the evolution of divergent host races in insects. The theoretical work discussed in this chapter "will hopefully change the still prevailing attitude that speciation is allopatric until proved sympatric" (Chapter 3, Concluding Comments). This hope is fostered further by recent developments discussed in the subsequent chapters, which take a dynamic perspective on the emergence of disruptive selection and show that a wide range of circumstances makes adaptive speciation a likely event.

In Chapter 4, Geritz, Kisdi, Meszéna, and Metz discuss implications of the recently developed adaptive dynamics framework for the study of speciation mechanisms. In contrast to traditional models of sympatric speciation, adaptive dynamics typically describes the evolution of continuously varying traits or metric characters. For speciation to occur, a population must retain genetic variation under the sustained action of disruptive selection. Without frequency-dependent selection such variation is rapidly lost. By starting at the ecological end, adaptive dynamics models highlight that frequency dependence is not only ubiquitous, but often drives evolving populations to trait combinations at which selection turns disruptive. Frequency dependence can continually reshape the fitness landscape in ways that, after a persistent uphill climb toward so-called evolutionary branching points, the population finds itself at a fitness minimum. Under such conditions clonal populations become dimorphic, with the trait values in the resultant phenotypic branches diverging. This provides a basic paradigm for evolutionary diversification.

In randomly mating Mendelian populations, divergence through evolutionary branching is prevented by recombination: abundance gaps between incipient phenotypic branches are jammed by hybrids. This prevents populations from becoming phenotypically bimodal and instead stabilizes them at the branching point. Chapter 5 by Dieckmann and Doebeli explains how Mendelian populations can undergo evolutionary branching despite this apparent obstacle: at the branching point assortative mating is selected for and, once this has become sufficiently strong, evolutionary branching can proceed. The chapter also presents an overview of the ecological interactions between species that have been shown to cause evolutionary branching. By exploiting an analogy between evolutionary branching driven by natural versus sexual selection, how frequency-dependent disruptive selection can originate either from competition for resources or from competition for mates is explained. The chapter closes with an assessment of the strengths and weaknesses of current models of adaptive speciation through sexual selection and sexual conflict.

Chapter 6 reviews theories of allopatric and parapatric speciation and thus helps to contrast models of these speciation modes, predominantly based on the effects of presumed genetic incompatibilities, with models of adaptive speciation occurring under fully sympatric conditions. By describing how processes of gradual evolution and speciation unfold on adaptive landscapes, Gavrilets provides a unifying account of existing approaches to explain the emergence of reproductive isolation between geographically segregated subpopulations. While evolution on rugged adaptive landscapes would appear to result readily in subpopulations that occupy different adaptive peaks and thus become genetically isolated, selection against descent into adaptive valleys renders such outcomes unlikely. This problem is overcome on adaptive landscapes that feature extended networks of selectively neutral ridges along which evolution by genetic drift proceeds unimpeded. After extending the classic model by Bateson, Dobzhansky, and Muller to genetic architectures with multiple alleles or loci, the chapter ends with a discussion of how different spatial population structures affect evolutionary waiting times for reproductive isolation.

Chapter 7 establishes a link between two important aspects of speciation theory: biogeographic patterns of species distributions and frequency-dependent ecological processes that induce disruptive selection. Linking pattern and process in such a manner, Doebeli and Dieckmann investigate conditions for evolutionary branching in spatially structured populations. Under localized competitive interactions, evolutionary branching occurs much more easily than in nonspatial models, because local adaptations along environmental gradients lead to increased frequency dependence. The chapter also highlights how sympatric evolutionary branching often generates patterns of species abutment, reminiscent of parapatric or allopatric speciation scenarios. This suggests that the classic controversy between allopatric and sympatric speciation may often be ill posed: what eventually appears to be an allopatric (or parapatric) pattern of species distribution may well have been generated by an intrinsically sympatric evolutionary process that results from frequency-dependent ecological interactions. The model presented also suggests an explanation as to why larger areas have higher speciation rates: the likelihood for adaptive speciation increases when ecological interactions are more localized relative to the spatial scale of environmental heterogeneities.

Parts B and C of this book show that, already, a respectable body of empirical work indicates that adaptive speciation occurs in natural systems. Understanding the theoretical underpinnings of speciation, to which Part A contributes, is indispensable for carrying out and interpreting these empirical studies.

3

Genetic Theories of Sympatric Speciation

Tadeusz J. Kawecki

3.1 Introduction

Although populations are usually defined as sympatric if they occur in the same geographic area, application of this definition of sympatry to the classification of genetic mechanisms of population speciation is problematic. Populations that live in the same geographic area, but breed in different habitats or at different times, might be as genetically isolated as populations separated by a mountain range. Consider a one-time colonization of a novel host species by a parasite, with no subsequent gene flow between the parasite population on the novel and usual hosts. This is analogous to a one-time colonization of an isolated island, a classic scenario for allopatric speciation. It does not matter whether or not the parasite populations on the two host species occur in the same geographic area. In contrast, the prospects for and potential mechanisms of speciation would be very different if cross-infection occurred frequently, with a resultant gene flow between the two parasite populations. Overlap of geographic range is thus not a sufficient criterion for defining sympatric speciation; individuals of the incipient species must encounter each other and the potential for gene flow must exist (Futuyma and Mayer 1980; Kondrashov and Mina 1986).

For the purpose of this chapter, speciation is therefore defined as sympatric if:

- The restriction and eventual elimination of gene flow between the two species occur gradually as a consequence of evolutionary (i.e., genetically based) change.
- The entire process takes place diffusely over a larger area, isolation by distance is not important, and all the important events that lead to speciation occur in the area where the ranges of the incipient species overlap.

This definition excludes instant speciation by polyploidization, which, although it occurs within the range of the ancestral species, involves very different mechanisms and thus is not discussed here. The first of the above criteria distinguishes sympatric from allopatric speciation, in which the gene pool is first split by an ecological event. The second criterion distinguishes sympatric from parapatric speciation, in which the two species form on the opposite sides of a cline so their ranges are adjacent, but not overlapping (Lande 1982; Slatkin 1982; Chapter 7). All the speciation scenarios discussed in this chapter are instances of adaptive speciation processes as defined in Chapter 1.

In allopatric speciation, reproductive isolation evolves as a by-product of independent evolution, whereby different gene substitutions occur in the two populations. This divergence may result from differential selection because of the different environments experienced by the nascent species, but may also result from different beneficial mutations that arise in the two populations, different outcomes of sexual selection, or genetic drift. After a period of independent evolution, the new species evolve genetic incompatibilities manifested as hybrid inviability or sterility (postzygotic isolation), or incompatibilities in mating behavior or physiology that prevent successful mating (prezygotic isolation). They now occupy different adaptive peaks (Chapter 6).

In contrast, because of the homogenizing effect of gene flow, genetic differentiation of the incipient species in sympatry can only occur under sustained disruptive (diversifying) selection on some traits, and the intermediate genotypes have lower fitness than both extremes. Thus, sympatric speciation must necessarily be adaptive in the sense used in this book (see Chapter 1). The lower fitness of the intermediates results from ecological factors (exogenous selection; Coyne and Orr 1998) or from sexual selection, not from genetic incompatibility. Selection against the intermediates causes partial postzygotic isolation, which in turn favors the assortative mating that prevents crossing between genotypes from the different "adaptive peaks", which leads to prezygotic isolation. Together with the continuing disruptive selection, the increasing prezygotic isolation leads to elimination of the intermediates and cessation of gene flow between the nascent species, and thus speciation is completed.

Two serious theoretical arguments have been put forth against this scenario, and against the plausibility of sympatric speciation in general. The first suggests that sustained disruptive selection, operating over many generations, is implausible (Mayr 1963; Futuyma and Mayer 1980). The second notes that it is unlikely for assortative mating to be controlled by the same loci as those under disruptive selection, and argues that a lasting association between any other assortative mating system and the loci under disruptive selection will be prevented by recombination (Udovic 1980; Felsenstein 1981). Both these difficulties are most likely to be overcome when the process is driven by specialization of the incipient species on different habitats or resources. First, competition among individuals using the same resource or habitat results in negative frequency-dependent selection, which facilitates the origin and maintenance of polymorphism under disruptive selection. Second, in many biologically realistic cases, partially assortative mating is a by-product of the disruptive selection, particularly if habitat choice is involved. Therefore most theories of sympatric speciation consider divergent ecological specialization as the driving force.

In this chapter an attempt is made to synthesize the theories of sympatric speciation. The scenarios envisioned in these theories can be divided into three groups: sympatric speciation through resource specialization without habitat choice, through sexual selection, and through habitat or host-race formation.

Rather than provide a comprehensive review of the literature, the focus herein is on three crucial processes that form elements of those scenarios:

1. Sustained disruptive selection on a trait (or traits) with ecological relevance;
2. Evolution of divergent mate choice;
3. Evolution of divergent habitat (host) preference.

Process (1) is an essential driving force in sympatric speciation through resource specialization with and without habitat choice (through habitat or host races). Process (2) is included or implied in all scenarios for sympatric speciation. Process (3) is specific to scenarios of sympatric speciation via habitat (host) races, in which prezygotic isolation arises as a by-product of divergent habitat choice. These three processes are described in Sections 3.2 to 3.4; Section 3.5 discusses the interaction between these processes and explains how they facilitate one another. Their role and essential features are explained and illustrated with simple examples; the discussion is largely limited to two ecological niches or habitats and traits affected by a small number of loci. Subsequent chapters describe in detail how stable disruptive selection can originate dynamically through competition in a continuous niche (Chapter 4), and how it can lead to speciation mediated by traits affected by many loci (Chapter 5).

3.2 Sustained Disruptive Selection

Sympatric speciation requires a sustained action of disruptive selection. Using the adaptive landscape metaphor, the new species come to occupy two different adaptive peaks, and the intermediates that fall in the valley between are eliminated. However, if the ancestor species occupies one adaptive peak, corresponding to adaptation to one niche, how is it possible for a part of the original population to cross the valley in sympatry, that is, in the face of a strong gene flow? A rare allele will initially be present mainly in intermediate genotypes (in the heterozygote at the genetic background of the original genotype at other loci), and will thus spread only if the intermediates have a higher fitness than the original genotype (i.e., if there is no valley). The simplest model to illustrate this problem is that of one locus with the heterozygote inferior to both homozygotes (underdominance), in which neither allele can increase in frequency when rare (Hartl and Clark 1997, p. 243); an analogous argument applies to an adaptive valley through the epistatic interaction between loci. Thus, sympatric speciation requires that the intermediate genotypes are first superior and then inferior to the original genotype. This apparent contradiction was seen as a major argument against sympatric speciation (Mayr 1963; Futuyma and Mayer 1980).

A solution to this conundrum lies in the frequency dependence of fitness, which causes the adaptive landscape to change dynamically as the population evolves. Frequency dependence is likely to originate when disruptive selection is imposed by the existence of different ecological niches. The niche difference could involve using a different resource within the same habitat, or different habitats. The distinction between a "resource" and a "habitat" in this context is that the latter implies some spatial segregation such that individuals using the same habitat patch are more likely to interact and potentially mate with each other (see Section 3.4). For many organisms with a parasitic lifestyle (including herbivorous insects), individual host species constitute habitat types. The theories discussed here assume two discrete niches; the theory of sympatric speciation under a continuous niche variation is discussed in the subsequent two chapters.

The frequency dependence that enables stable polymorphism under disruptive selection results from density dependence among individuals using the same niche. Such density dependence in turn results from competition among individuals that use the same resource or habitat, but may also occur via other mechanisms, such as apparent competition through predation or parasitism (Holt 1984), provided that they operate more strongly among individuals that use the same niche. When an allele that improves performance in the second niche first arises, the intermediate genotypes containing it only compete with the original genotypes, over which they are superior, in the new, underused niche (note that this assumes that the new niche is also relatively free of interspecific competition). However, once the allele has become more common, segregation (and recombination if multiple loci are involved) creates genotypes that are better adapted to the new niche than the intermediates. The adaptive valley is created only as the second peak becomes occupied – the adaptive landscape dynamically changes as a result of evolutionary change in the population. Some form of density dependence that operates at least partially independently among individuals using different resources or habitats is thus crucial to initiate sympatric speciation via ecological specialization (for a comprehensive discussion see Rosenzweig 1978; Pimm 1979; Wilson and Turelli 1986).

A form of density dependence often implemented in models of sympatric speciation because of its mathematical simplicity assumes that the total reproductive output of individuals using a particular niche is constant, independent of their number and adaptedness. This assumption, known as "soft selection", implies a very strong form of population regulation (essentially, perfect self-thinning) that operates independently in the two niches. As illustrated in Box 3.1, soft selection permits protected polymorphism with the average fitness of the heterozygote less than that of both homozygotes, particularly for loci of large effects. Under the contrasting assumption of "hard selection", the reproductive output from a given niche is proportional to the number of individuals using it multiplied by their absolute fitness, which implies that population regulation operates only at the level of the entire population. Under hard selection, polymorphism is only maintained if the average fitness of the heterozygote exceeds that of each homozygote (Box 3.1), that is, no stable disruptive selection is possible. Conditions for protected polymorphism under soft selection have been studied extensively (Levene 1953; Prout 1968; Christiansen 1975; Maynard Smith and Hoekstra 1980; Karlin and Campbell 1981; and reviews in Felsenstein 1976; Hedrick *et al.* 1976; Hedrick 1986).

Box 3.1 Multiple-niche protected polymorphism

Protected polymorphism results when each allele has an advantage when rare. This box shows how the frequency dependence generated by the density dependence operating within habitats facilitates the maintenance of protected polymorphism, despite the average fitness of the heterozygote being less than that of either homozygote. Consider an organism that lives in an environment with two habitat types. Each individual completes its development within a single habitat patch. The surviving adults disperse among the habitat patches at random, so that the probability of oviposition in a given habitat type is independent of the habitat of origin (complete mixing). Consider a haploid locus with two alleles that affects the probability of survival in a habitat-dependent way, so that the relative fitnesses are

Genotype	AA	Aa	aa
Fitness in habitat 1	W_{AA}	W_{Aa}	W_{aa}
Fitness in habitat 2	V_{AA}	V_{Aa}	V_{aa}

Under what conditions will selection act to maintain both alleles in the gene pool? The answer to this question depends on how the population is regulated.

Soft selection. Assume first that the subpopulations in the two habitats are independently regulated, so that irrespective of mean fitness each contributes a fixed proportion (k_1, k_2) to the pool of reproducing adults. The frequency p_{t+1} of allele A in the gametes that contribute to the next generation $t + 1$ is now the average of the allele frequencies after selection in generation t in each habitat, weighted by their contribution to the adult pool,

$$
\begin{aligned}
p_{t+1} = k_1 &\frac{p_t^2 W_{AA} + p_t(1 - p_t)W_{Aa}}{p_t^2 W_{AA} + 2p_t(1 - p_t)W_{Aa} + (1 - p_t)^2 W_{aa}} \\
+ k_2 &\frac{p_t^2 V_{AA} + p_t(1 - p_t)V_{Aa}}{p_t^2 V_{AA} + 2p_t(1 - p_t)V_{Aa} + (1 - p_t)^2 V_{aa}} .
\end{aligned}
\tag{a}
$$

This equation implies that allele A will increase when rare if $k_1 W_{Aa}/W_{aa} + k_2 V_{Aa}/V_{aa} > 1$; allele a will increase when rare if $k_1 W_{Aa}/W_{AA} + k_2 V_{Aa}/V_{AA} > 1$. Protected polymorphism requires that both these conditions are simultaneously satisfied [this is a sufficient condition; a necessary condition is slightly broader (Prout 1968; Maynard Smith and Hoekstra 1980)]. Protected polymorphism under soft selection is thus possible even when the heterozygote is less fit in each habitat than the average of the two homozygotes.

The figure below illustrates the outcome of soft selection for a special case with the relative fitnesses given by

Genotype	AA	Aa	aa
Fitness in habitat 1	1	$1 - hs_1$	$1 - s_1$
Fitness in habitat 2	$1 - s_2$	$1 - hs_2$	1

where s_1 and s_2 are the selection coefficients in two habitats and h is the degree of dominance.

continued

Box 3.1 *continued*

For the part of the parameter space labeled "A fixed" or "a fixed" neither allele can invade when rare.

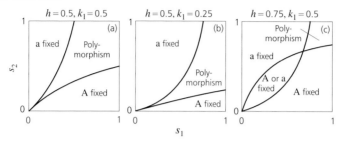

Hard selection. At the other extreme, assume that density dependence operates at the level of the entire population (i.e., there is no habitat-dependent population regulation). In this case the habitat-specific fitness is directly proportional to the contribution to the next generation, and the frequency p of allele A changes from generation to generation according to the equation (Christiansen 1975)

$$p_{t+1} = \frac{p_t^2 \overline{W}_{AA} + p_t(1 - p_t)\overline{W}_{Aa}}{p_t^2 \overline{W}_{AA} + 2p_t(1 - p_t)\overline{W}_{Aa} + (1 - p_t)^2 \overline{W}_{aa}} , \tag{b}$$

using the abbreviation

$$\overline{W}_j = \kappa_1 W_j + \kappa_2 V_j , \qquad j = AA, Aa, aa ,$$

for the average genotype fitness values across the two habitats, with κ_1 and κ_2 being the proportions of individuals that undergo selection in habitats 1 and 2, respectively ($\kappa_1 + \kappa_2 = 1$). Protected polymorphism implies that each allele increases in frequency when rare. Using standard population genetics it is easy to show that each allele is favored when rare if and only if the two inequalities

$$\overline{W}_{AA} < \overline{W}_{Aa} , \tag{c}$$

$$\overline{W}_{aa} < \overline{W}_{Aa} \tag{d}$$

are satisfied. Under hard selection, protected polymorphism is thus possible in this model only if the fitness of the heterozygote averaged over the two habitats is greater than the analogous average fitness of each homozygote (Dempster 1955). This is not satisfied for any of the parameter values assumed in the figure above.

In reality, the regulation of natural populations is somewhere between hard and soft selection – competition usually takes place mostly or entirely between individuals using the same niche, but if only a few individuals use a given niche or they are poorly adapted, the reproductive output from this niche will be smaller. The conditions for protected polymorphism under explicit forms of density dependence in a system of discrete habitat types are usually more restrictive than under soft selection, but can still lead to stable disruptive selection (e.g., Pimm

1979; Christiansen 1985; Wilson and Turelli 1986). Whether one regards the conditions for stable polymorphism under disruptive selection as stringent is a matter of opinion. Generally, the two niches cannot be too different, or the intermediate genotypes would not be able to invade the new niche when they first arise, but also they cannot be too similar, or the selection against intermediates at the equilibrium will not be strong enough. This issue is discussed quantitatively by Kondrashov and Mina (1986) and Udovic (1980). The conditions for maintenance of polymorphism under disruptive selection are less stringent when the reproductive output from each habitat is similar, and the loci involved have large effects. They are also more favorable when offspring are more likely than expected by chance to live in the same habitat as their parent(s), whether because of limited dispersal or because of genetically based habitat preference (e.g., Maynard Smith 1966; Felsenstein 1981).

For simplicity, the above discussion concentrates on cases in which the niche space is discrete. While this is the case for many organisms (e.g., host species are discrete types of resource for parasites or herbivorous insects), for others the main aspects of the ecological niche (e.g., food particle size or foraging depth) are continuous, with no pre-existing gaps. However, even when the niche space is continuous and the resources are most abundant for intermediate phenotypes in the absence of exploitation, competition can dynamically generate an adaptive valley and stable disruptive selection (Rosenzweig 1978; Abrams *et al.* 1993a). This mechanism, which is essential for several major models of sympatric speciation (e.g., Rosenzweig 1978; Pimm 1979; Kondrashov 1986; Dieckmann and Doebeli 1999), is described in detail in Chapters 4 and 5.

3.3   Evolution of Divergent Mate Choice

In sexual species even strongly disruptive selection is unable to split the gene pool, which is held together by recombination and segregation as long as mating occurs between individuals from the two sides of the adaptive valley. Of course, if the offspring of such matings are not viable, postzygotic isolation becomes complete and thus sufficient to declare the two diverging gene pools as species (according to the biological species concept; Mayr 1963). However, as explained above, such polymorphism could not arise in the first place. A theory of sympatric speciation therefore cannot be complete without addressing the evolution of assortative mating, which would prevent mating between individuals from the two incipient species. Such assortative mating should be favored because disruptive selection implies an inferiority of offspring that originate from matings between individuals from the two incipient species. Several mechanisms that produce such inferiority have been implemented in models of sympatric speciation (Box 3.2). However, speciation could, in principle, be achieved by evolution of strong assortative mating by a divergent Fisherian runaway process, in the absence of disruptive selection through ecological factors. Models of this process are summarized in the next subsection; the two subsections after that describe two genetic patterns of assortative mating considered in models of sympatric speciation via ecological specialization.

Box 3.2 Mechanisms of inferiority of intermediates

Several mechanisms responsible for the inferiority of offspring from mating between individuals from the two nascent species have been invoked in theories of sympatric speciation via ecological specialization:

- *Heterozygote inferiority.* As illustrated in Box 3.1, selection with population regulation that operates independently in each niche can maintain polymorphism when the average fitness of the heterozygote is lower than that of either homozygote. For models of sympatric speciation driven by heterozygote inferiority see, for example, Maynard Smith (1966) and Udovic (1980).
- *Recombination breaking up coadapted gene complexes.* If genotypes AABB and aabb are best adapted to niches 1 and 2, respectively, recombinants created through mating between these two genotypes will be inferior in both niches. This mechanism operates in the model by Felsenstein (1981), as well as in most multilocus models (Kondrashov 1986; Dieckmann and Doebeli 1999). An analogous mechanism operates in a model of speciation driven by disruptive selection on habitat choice (Rice 1987), in which intermediate genotypes show preference for a low-quality or nonexisting habitat.
- *Recombination between loci for performance and habitat preference.* If there is genetic variation for both habitat-specific fitness and habitat preference, selection tends to create a linkage disequilibrium between them (see Box 3.3). As a result preference and performance become genetically correlated. Recombination that results from mating between opposite genotypes breaks down this association – descendants of such crosses are more likely to choose a "wrong" habitat (e.g., Johnson *et al.* 1996b).

Evolution of mate preference in the absence of niche divergence

Suppose there is a locus M such that MM females only mate with MM males and mm females only with mm males. In the absence of natural selection on locus M, heterozygotes rapidly become eliminated (whatever their mating preference) and two genetically isolated populations (i.e., species according to the biological species concept) arise in sympatry. Under some (rather stringent) conditions such a split could also occur when the assortative mating trait is polygenic (Kondrashov and Shpak 1998). This is, however, only possible if all genotypes are assumed to have the same mating success. Yet, in reality assortative mating is likely to select against rare genotypes, because their carriers will have difficulties finding a mating partner. In particular, given the limited numbers of males a female can examine, females of a rare genotype are more likely not to find a male of their genotype and thus accept any male, which reduces the mating success of males of the rare genotype. Even if this effect is excluded from the model (i.e., all males are assumed to have the same mating success), polymorphism at the assortative mating loci is at best neutrally stable (Kondrashov and Shpak 1998); otherwise, it is inherently unstable.

This type of assortative mating also requires that an individual "knows" its own phenotype and can compare it with that of a potential mate. This may be satisfied in some cases; for example, when assortative mating results from reproductive phenology (e.g., flowering time), choice of places for mating, or pollinator attraction (e.g., Husband and Schemske 2000). More generally, however, preference is affected by loci other than the preferred trait ("ornament"), and a theory of speciation has to account for recombination, which tends to break up the association between them.

The assumption of separate loci for preference and the preferred character has been made in models of sympatric speciation through the Fisherian runaway process. Natural selection on the preferred character ("ornament") is stabilizing in these models, and thus counteracts divergence. These are often referred to as models of sympatric speciation through sexual selection. Importantly, however, any form of assortative mating that involves mate choice, including those considered in the next two subsections, by definition involves sexual selection (Andersson 1994).

In the Fisherian runaway process of sexual selection, variation in the degree of female preference results in a linkage disequilibrium between loci for male ornament and for female preference: individuals that carry alleles to enlarge the ornament eventually carry alleles for increased preference for the large ornament more often than expected through chance alone. The linkage disequilibrium results in genetic correlation between the ornament and preference. Stronger female preference thus evolves as a correlated response to selection on the ornament, as the selection imposed by the female preference already exists in the population (Kirkpatrick 1982b; Andersson 1994). To facilitate the discussion, assume that the ornament is the color, ranging from white through gray to black. Classic sexual-selection theory predicts that the runaway process may lead to the evolution of either dark or light males, with the corresponding evolution of mean female preference, in the direction determined by the initial conditions and asymmetries of parameters (e.g., Kirkpatrick 1982b). But how can a gene pool be split in sympatry by two runaway processes that operate in different directions?

This problem has been addressed by Higashi *et al.* (1999), who studied an individual-based (stochastic) polygenic model of ornament and preference. The axes of variation for both the ornament and preference were assumed to have a neutral (zero) point (no preference and, say, gray color). The direction of the deviation of the preference trait from this neutral point determined the direction of female preference, and the strength of the preference increased with the magnitude of deviation from the neutral point of both female preference and male ornament. The sexual selection on the ornament that results from this form of female preference is both disruptive and frequency dependent as long as the distribution of the preference trait in the population is not strongly asymmetric about the "no preference" point. If the mean female preference is initially close to zero and the cost of the ornament to male survival is low, the simulation model by Higashi *et al.* (1999) is likely to result in two runaway processes in opposite directions, which leads to speciation. However, if the mean of the preference trait is initially considerably different from zero, or if it becomes so because of drift, the whole population undergoes runaway evolution in one direction. Once that happens, speciation by reversing the direction of the runaway selection in a part of the population is not possible [see also Takimoto *et al.* (2000) for a deterministic version of the model]. In contrast, Turner and Burrows (1995) considered a population that has already undergone the runaway process in one direction and proposed a mechanism that would favor a rare mutant female with a reversed preference. If most females show directional preference for white males, and the population mean of male color is close to pure white, rare mutant females that show a preference for dark males will mate with the darkest available males, which are only a light gray. If there is a high cost of extreme color, the less extreme sons of the mutant females have a higher survival, which also benefits the mutant preference allele they are likely to carry. This may trigger the runaway process in the reverse direction in a part of the population, as indeed happened in a few percent of the simulation runs of the individual-based model by Turner and Burrows (1995). This mechanism, however, apparently hinges on the ornament being a polygenic trait, the direction of female preference being determined by one locus, and on a large cost of extreme ornaments, which is in contrast to the model by Higashi *et al.* (1999). It also requires a small or viscous population, and females that sample many males before mating. The mating disadvantage of rare males would also be reduced if they tended to congregate in places frequented by females that show preference for their phenotype. Such a spatial pattern could arise if males tended to stay in places where they have a high mating success, but to move on if they are shunned by local females (Payne and Krakauer 1997). This would result in a mosaic of patches with opposite combinations of ornaments and preference. This mosaic would not have to be related to any ecologically relevant variation in the environment.

To summarize, although more research is needed, it seems that sometimes sexual selection is able to split a gene pool in sympatry even in the absence of disruptive selection through ecological factors [see also Gavrilets and Waxman (2002) for a model of sympatric speciation through sexual conflict; Section 5.4]. Coexistence of the resultant incipient species, however, requires that they have somewhat different ecological niches. Niche divergence could evolve after they become completely prezygotically isolated (character displacement; Abrams 1986), but competitive exclusion of one is a more likely outcome, given the limited rate of response to natural selection. Thus, even if initiated by sexual selection, sympatric speciation seems to be more likely to go to completion if the two species begin to differentiate ecologically before they have become fully prezygotically isolated. This is the case for the remaining scenarios considered in this section.

Assortative mating according to traits under disruptive selection

Consider an "ecological" trait under disruptive selection and an "assortative mating" allele that makes its carrier more likely to choose a mate with a similar value of the ecological trait. Since most of the inferior, intermediate individuals are produced by matings between individuals from opposite sides of the "adaptive valley", the assortative mating allele is under-represented among intermediate individuals and over-represented among extreme individuals. Since extreme individuals have higher fitness, the assortative mating allele increases in frequency. If the individuals can recognize the "ecological" trait under disruptive selection in potential mates, preference for those mates with a similar phenotype is readily favored (Maynard Smith 1966; Kondrashov 1986; Doebeli 1996a; Dieckmann and Doebeli 1999). It is important that in this type of model the same allele (or alleles) that affect assortative mating becomes fixed in both nascent species. Therefore, recombination between loci that affect the two traits has little influence on the association, and thus does not significantly impede speciation. Felsenstein (1981) refers to such models as "one-allele speciation models", to distinguish from the "two-allele models" considered in the next subsection.

Models that involve this type of assortative mating according to the trait under disruptive selection predict speciation under a broad range of parameters. It remains unclear how often their general assumptions are satisfied. On the one hand, assortative mating based on ecologically important traits is often observed in nature, including a number of putative cases of sympatric speciation (Nagel and Schluter 1998; Rolán-Alvarez et al. 1999; Jiggins and Mallet 2000; Jonsson and Jonsson 2001; see also Chapters 9, 10, and 13). On the other hand, an observation of assortative mating does not necessarily imply preference for similar phenotypes; it is also likely to arise if both sexes show preference in the same direction. For example, assortative mating will be observed if large, bright, or healthy individuals of both sexes prefer to mate with each other, while small, dull or parasitized individuals mate with what remains (i.e., other small, dull or parasitized individuals), even though they would also prefer a large, bright or healthy partner. This, indeed, seems a common mechanism underlying assortative mating (e.g., Elwood et al. 1987; Harari et al. 1999; Thomas et al. 1999). The assortative mating model assumed in the articles cited in the previous paragraph also implies that an individual can inspect its own phenotype and compare it with that of a potential mate. Preference for mates with a similar phenotype of the ecological trait will, however, also be observed if the preference is a genetically determined trait not directly affected by own phenotype, but the two traits are genetically correlated through linkage disequilibrium. This latter case no longer corresponds to Felsenstein's (1981) "one allele model": recombination tends to break down the genetic association between preference and the ecological character. Data on genetic bases of assortative mating observed in nature are needed to resolve this controversy. In the meantime, preference affected by loci other than the preferred ecological character should be integrated into the models of sympatric speciation driven by resource specialization. The generality of the above scenario also depends on how often traits that mediate ecological specialization, which may be physiological or biochemical, can be recognized by potential mates. The above caveats do not apply if assortative mating results automatically as a by-product of divergence in the ecological trait. An important special case – assortative mating according to habitat preference – is extensively discussed in Section 3.4 below.

Assortative mating according to traits not under disruptive selection

Discrimination against heterospecific mates typically involves characters other than the ecological characters that affect performance in specific habitats or the efficiency with which a particular resource is used. A scenario considered in several models of sympatric speciation proposes that:

- A system of assortative mating according to a trait affected by loci other than those under disruptive selection evolves in the population (e.g., through the divergent runaway selection discussed above).
- The assortative mating trait becomes genetically correlated with the trait under disruptive selection.

This correlation can only arise via linkage disequilibrium (nonrandom association of genotypes) between the loci that affect assortative mating and the loci under disruptive selection. The disruptive selection cannot in itself create the linkage disequilibrium if the population is initially at linkage equilibrium. However, if a small amount of disequilibrium originates by drift, the disruptive selection can, under certain circumstances, magnify it (Box 3.3). With such linkage disequilibrium, a reduced frequency of matings between individuals with opposite genotypes at the loci under disruptive selection results as a by-product of assortative mating according to the assortative mating loci. The sign of the linkage disequilibrium is arbitrary, and depends on the initial linkage disequilibrium created by drift. The main problem with this scenario is that the linkage disequilibrium is opposed by recombination, which can thus be seen as the main force to oppose sympatric speciation. This scenario corresponds to Felsenstein's (1981) "two-allele models" of speciation – different assortative mating alleles are fixed within the two nascent species. Several authors (Felsenstein 1976; Udovic 1980; Kondrashov 1986; Kondrashov and Mina 1986) studied the process in detail and concluded that linkage disequilibrium can only be maintained if the disruptive selection is strong and the fidelity of assortative mating is quite high. In other words, under this scenario assortative mating of high fidelity has to evolve before it becomes genetically correlated with the ecologically relevant trait under disruptive selection. On the other hand, recent developments (Dieckmann and Doebeli 1999) suggest that linkage disequilibria between traits under disruptive selection and assortative mating arise more easily in multilocus models. These models are described in Chapter 5. Linkage disequilibrium is also more easily maintained if some prezygotic isolation already exists between the two parts of the gene pool on the two sides of the adaptive valley [selected in opposite directions (Felsenstein 1976; Kondrashov and Mina 1986)]. Such partial prezygotic isolation can originate as a by-product of the evolution of habitat choice, which is the subject of the next section.

Box 3.3 Linkage disequilibrium and assortative mating

This box illustrates with a simple model the build-up of linkage disequilibrium between a locus under disruptive selection and a locus involved in assortative mating. Consider the single-locus model illustrated in the figure in Box 3.1, and simplify it further by assuming $k_1 = k_2 = 0.5$ and $s_1 = s_2 = s$. With these assumptions, a stable polymorphism is maintained under soft selection if $h < 1/(2 - s)$, and the equilibrium frequency of allele A is 0.5 (this follows from the symmetry of parameters). Assume that another locus has two equally common alleles, M and m, such that individuals mate exclusively with partners that carry the same genotype at this locus. If the two loci are at linkage equilibrium, each of the four possible haplotypes (AM, aM, Am, am) occur in the gametes in the same proportion of $\frac{1}{4}$. Assume, however, that for some reason allele A is associated with allele M more often than expected by chance, so that the frequencies of the four gamete types AM, aM, Am, am are $\frac{1}{4} + D$, $\frac{1}{4} - D$, $\frac{1}{4} - D$, $\frac{1}{4} + D$, respectively. D is a measure of linkage disequilibrium ranging from $-\frac{1}{4}$ to $\frac{1}{4}$. To derive the expression for the change in frequencies of the four gamete types in the next generation note first that 50% of the offspring of Mm heterozygotes are homozygous at locus M, but since MM and mm homozygotes only mate with themselves, they do not produce any Mm heterozygotes. The perfect assortative mating assumed above thus leads to elimination of the heterozygotes at locus M (see also Crow and Kimura 1970). Therefore, the frequency of AAMM zygotes will equal the frequency of AM gametes ($\frac{1}{4} + D$) multiplied by the probability that an AM gamete mates with another AM gamete, $2(\frac{1}{4} + D)$. The frequencies of other zygote genotypes can be derived in an analogous way. The following table shows the frequencies of genotypes AAMM, AaMM, and aaMM before selection (in the zygotes) and after selection.

Genotype	AAMM	AaMM	aaMM
Frequency before selection	$2(\frac{1}{4}+D)^2$	$4(\frac{1}{4}+D)(\frac{1}{4}-D)$	$2(\frac{1}{4}-D)^2$
Frequency after selection			
Niche 1	$2(\frac{1}{4}+D)^2/\overline{W}$	$(\frac{1}{4}-4D^2)(1-hs)/\overline{W}$	$2(\frac{1}{4}-D)^2(1-s)/\overline{W}$
Niche 2	$2(\frac{1}{4}+D)^2(1-s)/\overline{V}$	$(\frac{1}{4}-4D^2)(1-hs)/\overline{V}$	$2(\frac{1}{4}-D)^2/\overline{V}$

where $\overline{W} = \overline{V} = 1 - \frac{1}{4}s[1+2h+16D^2(1-2h)]$. The symmetry of all the parameters means that the frequencies of the remaining three genotypes (Mm heterozygotes are neglected) will be a mirror image of those shown in the table. For example, the frequency of genotype AAmm in niche 1 after selection will be the same as the frequency of genotype aaMM in niche 2. Since each niche contributes 50% of the reproducing adults, the allele frequency at both loci will remain at 0.5, and the frequency of haplotype AM (as well as am) among the gametes that contribute to the next generation will be

$$p'_{\text{AM}} = p'_{\text{am}} = (\tfrac{1}{4} + D)\frac{4 + [4D(2h - 1) - 2h - 1]s}{4 + [16D^2(2h - 1) - 2h - 1]s} \ . \tag{a}$$

continued

Box 3.3 *continued*

Inspection of Equation (a) reveals that if $h < \frac{1}{2}$ (average heterozygote fitness is higher than the average fitness of both homozygotes), the deviation of the haplotype frequency from $\frac{1}{4}$ is reduced, and eventually linkage equilibrium is reached. However, if selection is disruptive ($h > \frac{1}{2}$) the deviation from $\frac{1}{4}$, and thus linkage disequilibrium, increases. If this process continues, eventually only genotypes AAMM and aamm (if D was initially positive) or only AAmm and aaMM (if D was initially negative) remain, which in this simple model constitutes sympatric speciation. The sign of linkage disequilibrium (i.e., whether allele A becomes coupled with allele M or m) depends on the initial, fortuitous linkage disequilibrium. Note that the disruptive selection only magnifies an already existing linkage disequilibrium; it does not cause a change if $D = 0$. However, in a finite population some linkage disequilibrium always arises by drift.

The above arguments assume a perfect assortativeness, but assortative mating is likely to be imperfect, especially when it first arises in the population. In which case some Mm heterozygotes are maintained and the build-up of linkage disequilibrium is opposed by recombination, which occurs in the double heterozygotes AaMm. As the probability of mating with a partner that has a different genotype at locus M increases, selection must be increasingly disruptive ($h \gg \frac{1}{2}$) to maintain the linkage disequilibrium, and the amount of the linkage disequilibrium that can be maintained diminishes. However, $h < 1/(2 - s)$ is required to maintain polymorphism, so below a certain precision of assortative mating the linkage disequilibrium is not maintained for any value of h. This has been used as a major argument against the plausibility of sympatric speciation (Felsenstein 1981). Note that this model also assumes a pre-existing system of assortative mating and does not address the question as to why such a system should arise in the first place. Genetic linkage between the loci makes maintenance of the linkage disequilibrium easier [for details see Udovic (1980) and Felsenstein (1981)].

3.4 Evolution of Divergent Habitat or Host Preference

A number of authors (e.g., Rice 1984a, 1987; Diehl and Bush 1989; Johnson *et al.* 1996b; Kawecki 1996, 1997, 1998) propose a sympatric speciation scenario, initiated by partial prezygotic isolation that arises as a by-product of divergent genetically based habitat (host) preference, coupled with mating within habitats (on or near the host). Mating, indeed, typically takes place on or near the host in herbivorous insects, and results in assortative mating (Bush 1994; Caillaud and Via 2000; see Chapter 11), and sympatric host races (specialized subpopulations using different host species) have been found in several insects (Bush 1969; Tauber and Tauber 1989; Craig *et al.* 1997; Via 1999), crustaceans (Duffy 1996) and trematodes (Théron and Combes 1995). Such host races (or, more generally, habitat races) constitute divergent, partially genetically isolated populations and are often regarded as incipient species. How divergent habitat choice interacts with other processes that could potentially lead to speciation is discussed in Section 3.5. This section concentrates on models developed to study why and when divergent habitat specialization that leads to habitat races should evolve; Chapter 11 summarizes the empirical evidence.

Rice (1984a, 1987) considers a model in which habitat preference is the only evolving trait (i.e., there is no variation for fitness within the habitats). He assumes a range of habitats such that extreme habitats are of higher quality than intermediate ones. This results in disruptive selection on habitat choice. Mating within habitats causes assortative mating according to habitat choice (i.e., the character under disruptive selection). Under these assumptions the population readily splits into two specialized, largely genetically isolated races. This prediction has been confirmed by an ingenious experiment (Rice and Salt 1990). The divergence in this model hinges, however, on the assumption that the habitat preference of a single individual is constrained to be unimodal. Otherwise, individuals that prefer both high-quality habitats, but avoid the inferior intermediate ones, would be favored and divergence would not be expected.

In most other models the evolution of divergent habitat preference is driven by genetic variation in habitat-specific fitness. The intuitive logic as to why habitat preference should be favored is as follows. Adult individuals that have grown up in habitat 1, and thus have been exposed to selection in habitat 1, carry a higher frequency of alleles advantageous in habitat 1 than do individuals that have developed in habitat 2, and vice versa. The offspring are thus expected to have, on average, a higher fitness in the same habitat type as their parents. This favors parents that show habitat fidelity. In genetic terms, the evolution of divergent habitat preference involves linkage disequilibrium between loci that affect habitat-specific fitness and loci that determine habitat preference. As illustrated with a simple example in Box 3.4, this disequilibrium readily builds up, even if the population is initially at linkage equilibrium. Furthermore, most of the linkage disequilibrium results from correlated differences of allele frequencies between habitats. Hence, if mating takes place within each habitat, the linkage disequilibrium remains largely unaffected by recombination (Diehl and Bush 1989).

The evolution of divergent habitat choice, which leads to the formation of habitat races, has been shown to be favored under various assumptions about the genetic variation for habitat choice and habitat-specific fitness (Maynard Smith 1966; Diehl and Bush 1989; De Meeûs *et al.* 1993; Bush 1994; Kawecki 1996, 1997, 1998). In contrast to the direct evolution of assortative mating considered in Section 3.3, no minimal strength of selection is needed to create the linkage disequilibrium between the loci for habitat choice and those for habitat-specific fitness. Furthermore, habitat choice is favored even if selection on the fitness locus is not disruptive (i.e., when the mean fitness of the heterozygote is not less than that of both homozygotes).

Although the example in Box 3.4 assumes that different alleles are superior in different habitats, this is not necessary for the habitat choice to be favored. As shown by De Meeûs *et al.* (1993), even if the same genotype is superior in both habitats, there is selection for habitat fidelity if the magnitude of fitness effects of the polymorphism differs between habitats. Kawecki (1994, 1996, 1997, 1998) proposes several scenarios in which divergent habitat (host) specialization is driven

Box 3.4 Evolution of divergent habitat specialization

This box illustrates how variation in habitat-specific fitness promotes the evolution of divergent habitat choice. Habitat choice is likely to be affected by loci other than habitat-specific fitness. Divergence of habitat choice requires linkage disequilibrium – alleles of choice for a given habitat must occur together with alleles that confer high fitness in that habitat more often than expected by chance. To see how such linkage disequilibrium arises, consider a locus A with alleles A and a that are advantageous in habitats 1 and 2, respectively, and a second locus C that determines habitat choice. Consider a rare allele C, which causes its carriers to prefer habitat 1, in a population mostly composed of nonchoosy cc individuals. Suppose that the two loci are initially at linkage equilibrium (i.e., the probability of carrying a given genotype at locus A is independent of the genotype at locus C). Carriers of the nonchoosy genotype cc disperse randomly between the habitats, whereas carriers of allele C are more likely to reach habitat 1, whether habitat choice is made by themselves or by their parents. A greater proportion of allele C carriers are thus exposed to selection in habitat 1, and consequently after selection they are more likely to carry allele A than are cc individuals. Thus, one round of selection is enough to create linkage disequilibrium between loci for habitat choice and habitat-specific fitness. Subsequent recombination reduces the linkage disequilibrium, but by no more than half. The effect of recombination is reduced if mating takes place within the habitats (before or after dispersal). Thus, in the next generation carriers of allele C are, on average, better adapted to habitat 1 and less adapted to habitat 2 than are cc individuals. Since they are more likely to be in habitat 1, they are better off than the nonchoosers, leading to invasion of the rare allele C. As habitat 1 receives most of the carriers of allele C plus its share of the nonchoosy individuals, it becomes relatively overcrowded as allele C increases in frequency. Allele C thus tends toward a certain frequency at which the advantage because of the correlation between habitat choice and preference is balanced by the disadvantage of choosing a more crowded habitat. However, if another allele (say C′) appears such that its carriers choose habitat 2, both alleles tend to fixation within the respective habitats, while allele c is eliminated. In the example in the figure below, alleles C and C′ are assumed dominant to allele c and their carriers show perfect habitat choice; CC′ heterozygotes show no habitat preference. Fitness effects of locus A are as in Box 3.3 with $s = 0.5$ and $h = 0.5$. Allele C is introduced at a frequency of 0.01 at generation zero and in the absence of allele C′ reaches, in habitat 1, a frequency of about 0.05. After allele C′ enters in generation 200, both alleles spread in the respective habitats, and the allele frequencies at locus A diverge between the habitats, which leads to two genetically isolated specialized habitat races.

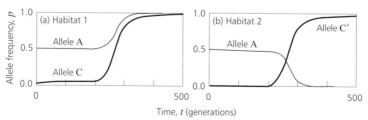

by polymorphism at loci that only affect fitness in one habitat (host). Polymorphism at such loci cannot be maintained by selection alone – if allele a is inferior in one habitat and neutral in the other, it should be eliminated. Such deleterious alleles with habitat-specific effects may be maintained at mutation–selection balance; this results in indirect selection for habitat fidelity, which is favored because a habitat specialist can purge deleterious mutations more effectively (Kawecki 1994, 1997). Loci with habitat-specific effects have also been assumed in the evolution of host races driven by coevolutionary arms races. In such scenarios, host specialization is favored in a herbivorous insect or parasitoid because a genetic lineage that sticks to a single host species can evolve faster in response to the evolving host defenses than can a generalist lineage that uses, and thus has to coevolve with, several host species (Kawecki 1996, 1998).

To summarize, the theoretical results outlined above suggest that habitat preference and host-race formation are likely to play an important role in sympatric speciation, in particular in animals with a parasitic lifestyle. This idea is supported by empirical data reviewed by Bush and Butlin (Chapter 11).

3.5   Concluding Comments: Synergism Between Processes

While evolution of divergent habitat preferences may lead to a considerable reduction of gene flow between the nascent species, the resultant prezygotic isolation is unlikely to be perfect. Mistakes in habitat or host choice are likely to be relatively common, and individuals that fail to find their favorite host are likely to become less choosy. Thus, speciation initiated by the formation of habitat (host) race only goes to completion if other forms of pre- and postzygotic isolation arise. However, the evolution of partial prezygotic isolation as a by-product of divergent habitat preference may form an important stepping-stone in sympatric speciation, because of synergism between processes discussed in the previous three sections. Divergent habitat preference interacts with the other processes in several ways:

- As mentioned above, divergent habitat preference is favored under a broader range of conditions than assortative mating (it does not require that selection on habitat-specific fitness be disruptive).
- Once the population shows some habitat fidelity, conditions become more favorable for protected polymorphism; polymorphisms with a greater asymmetry of the parameters and with greater heterozygote disadvantage become stable (Maynard Smith 1966; Jaenike and Holt 1991). Thus, more loci, and loci with more strongly disruptive selection, may become "recruited" into disruptive selection, which increases both the postzygotic isolation and the strength of selection against intermediates. This in turn feeds back to increase the selection pressure on habitat fidelity and for assortative mating.
- Recombination between habitat preference and fitness loci contributes to the advantage of assortative mating (see Box 3.2).
- Linkage disequilibrium between the loci involved in mate choice and those under disruptive selection builds up more easily if gene flow is limited (Udovic 1980; Felsenstein 1981; Kondrashov and Mina 1986).

These synergistic effects produce positive feedbacks: the effective selection coefficients for alleles responsible for speciation increases as speciation progresses, which means that sympatric speciation occurs with an accelerating pace (e.g., Diehl and Bush 1989; Johnson *et al.* 1996b; Kawecki 1996). This implies that, relatively, many populations will at any one time be at the initial stages of speciation (e.g., will form host races characterized by various degrees of genetic divergence), but it would be difficult to find a system in which a sympatric speciation event is just arriving at completion.

Integration of the various mechanisms should be the aim of future theoretical developments. Johnson *et al.* (1996b) seems the only model that simultaneously incorporates habitat-specific selection, habitat choice, and the evolution of similarity-based assortative mating, and also leads to completion of sympatric speciation. They showed that speciation in this model occurs much more easily than when the evolution of habitat choice is not permitted (see also Bush 1994). Mate choice based on separate loci for preference and the preferred ornament still needs to be integrated into models of sympatric speciation via ecological specialization. Similarly, more realistic genetics of habitat or host preference should be integrated into the models. Another promising direction involves the development of more realistic models of sympatric speciation in continuous niche space, especially given the increasing empirical evidence for this process (Chapters 9, 10, and 13). As first proposed by Rosenzweig (1978), the sustained disruptive selection necessary for speciation may emerge dynamically during the course of evolution. Chapter 4 details a theoretical framework to model this process and applies it to asexual evolution. Chapter 5 describes recent developments that incorporate Rosenzweig's idea into multilocus models of sympatric speciation. Recent advances in computing make such sophisticated multilocus models practical.

Finally, the theory of sympatric speciation should aim to formulate predictions about the patterns of genetic variation and linkage disequilibria that should be observed during sympatric speciation. Predictions that would distinguish ongoing sympatric speciation from secondary contact after incomplete allopatric speciation would be particularly valuable, as they would permit the direct evaluation of putative cases of sympatric speciation in progress in nature.

The importance of sympatric speciation can no longer be dismissed on theoretical grounds. Certainly, there are forces that oppose sympatric speciation, in particular recombination and the difficulty for alternative, initially rare, mate-choice systems to become established. Nonetheless, theoretical developments during the past 20 years, while acknowledging these forces, have shown that they can be overcome under a broad range of biologically reasonable conditions. At the same time, accumulating empirical data (reviewed in the empirical part of this book) provide increasing support that sympatric speciation plays an important role in the generation of biodiversity, at least in some taxa. These developments will hopefully change the still common attitude that any speciation event is allopatric until proved sympatric.

4

Adaptive Dynamics of Speciation: Ecological Underpinnings

Stefan A.H. Geritz, Éva Kisdi, Géza Meszéna, and Johan A.J. Metz

4.1 Introduction

Speciation occurs when a population splits into ecologically differentiated and reproductively isolated lineages. In this chapter, we focus on the ecological side of nonallopatric speciation: Under what ecological conditions is speciation promoted by natural selection? What are the appropriate tools to identify speciation-prone ecological systems?

For speciation to occur, a population must have the potential to become polymorphic (i.e., it must harbor heritable variation). Moreover, this variation must be under disruptive selection that favors extreme phenotypes at the cost of intermediate ones. With disruptive selection, a genetic polymorphism can be stable only if selection is frequency dependent (Pimm 1979; see Chapter 3). Some appropriate form of frequency dependence is thus an ecological prerequisite for nonallopatric speciation.

Frequency-dependent selection is ubiquitous in nature. It occurs, among many other examples, in the context of resource competition (Christiansen and Loeschcke 1980; see Box 4.1), predator–prey systems (Marrow *et al.* 1992), multiple habitats (Levene 1953), stochastic environments (Kisdi and Meszéna 1993; Chesson 1994), asymmetric competition (Maynard Smith and Brown 1986), mutualistic interactions (Law and Dieckmann 1998), and behavioral conflicts (Maynard Smith and Price 1973; Hofbauer and Sigmund 1990).

The theory of adaptive dynamics is a framework devised to model the evolution of continuous traits driven by frequency-dependent selection. It can be applied to various ecological settings and is particularly suitable for incorporating ecological complexity. The adaptive dynamic analysis reveals the course of long-term evolution expected in a given ecological scenario and, in particular, shows whether, and under which conditions, a population is expected to evolve toward a state in which disruptive selection arises and promotes speciation. To achieve analytical tractability in ecologically complex models, many adaptive dynamic models (and much of this chapter) suppress genetic complexity with the assumption of clonally reproducing phenotypes (also referred to as strategies or traits). This enables the efficient identification of interesting features of the engendered selective pressures that deserve further analysis from a genetic perspective.

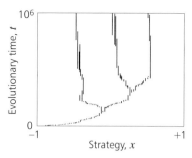

Figure 4.1 Simulated evolutionary tree for the model described in Box 4.1 with $r = 1$, $K(x) = (1 - x^2)_+, a(x, x') = \exp(-\frac{1}{2}(x - x')^2/\sigma_a^2)$ with $\sigma_a = 0.35$. The strategy axis (horizontal) is in arbitrary units; the evolutionary time axis is in units of r^{-1}. For details of the simulation, see Geritz *et al.* (1999) or Kisdi and Geritz (1999).

The analysis begins with the definition of admissible values of the evolving traits (including all trade-offs between traits and other constraints upon them), and the construction of a population dynamic model that incorporates the specific eco-logical conditions to be investigated, along with a specification of how the model parameters depend on the trait values. From the population dynamic model, one can derive the fitness of any possible rare mutant in a given resident population. It is thus possible to deduce which mutants can invade the population, and in which direction evolution will proceed via a sequence of successive invasion and fixation events.

Eventually, directional evolution may arrive at a particular trait value for which a successful invading mutant does not oust and replace the former resident; instead, the mutant and the resident coexist. If the two strategies coexist, and if selection in the newly formed dimorphic population is disruptive (i.e., if it favors new mutants that are more extreme and suppresses strategies between those of the two resi-dents), then the clonal population undergoes *evolutionary branching*, whereby the single initial strategy is replaced by two strategies separated by a gradually widen-ing gap. Figure 4.1 shows a simulated evolutionary tree with two such branching events. With small mutations, such a split can occur when directional evolution approaches a particular trait value called a *branching point*.

Evolutionary branching of clonal strategies cannot be equated with speciation, since clonal models of adaptive dynamics are unable to address the question of reproductive isolation. Chapter 5 discusses adaptive dynamics with multilocus ge-netics and the emergence of reproductive isolation during evolutionary branching. Yet, evolutionary branching itself signals that adaptive speciation is promoted by selection in the ecological system considered.

In this chapter we outline one particular framework of adaptive dynamics that has been developed by Metz *et al.* (1996), Geritz *et al.* (1997, 1998), and, for direc-tional evolution, Dieckmann and Law (1996). This framework integrates concepts from the modern theory of evolutionarily stable strategies (Maynard Smith 1982; Eshel 1983; Taylor 1989; Nowak 1990; Christiansen 1991) and accommodates evolutionary branching. We constrain this summary mainly to a simple graphic approach; the corresponding analytical treatment (which is indispensable if the theory is to be applied to multidimensional traits or to polymorphic populations that cannot be depicted in simple one- or two-dimensional plots; see Box 4.5) can be found in Metz *et al.* (1996) and Geritz *et al.* (1998).

4.2 Invasion Fitness

Invasion fitness is the exponential growth rate of a rare mutant strategy in the environment set by a given resident population (Metz *et al.* 1992). The calculation of invasion fitness depends on the particular ecological setting to be investigated. Here we sketch the basics of fitness calculations common to all models.

Consider a large and well-mixed population in which a rare mutant strategy appears. The change in the density of mutants can be described by

$$n(t + 1) = A(E(t))n(t) . \tag{4.1a}$$

Here n is the density of mutants or, in structured populations, the vector that contains the density of mutants in various age or stage classes. The matrix A describes population growth as well as transitions between different age or stage classes (Caswell 1989); in an unstructured population, A is simply the annual growth rate. In continuous time, the population growth of the mutant can be described by

$$\frac{dn(t)}{dt} = B(E(t))n(t) . \tag{4.1b}$$

The dynamics of the mutant population as specified by $A(E)$ (in discrete time) or $B(E)$ (in continuous time) depends on the properties of the mutant and on the environment E. The environment contains all factors that influence population growth, including the abundance of limiting resources, the density of predators or parasites, and abiotic factors. Most importantly, E contains all the effects the resident population has directly or indirectly on the mutant; generally, E depends on the population density of the residents. As long as the mutant is rare, its effect on the environment is negligible.

The exponential growth rate, or invasion fitness, of the mutant strategy is defined by comparing the total density $N(t)$ of mutants, after a sufficiently long time, with the initial density $N(0)$, while keeping the mutant's environment fixed. In structured populations N is the sum of the vector components of n, whereas in unstructured populations there is no difference between the two. Formally, the invasion fitness is given by (Metz *et al.* 1992)

$$f = \lim_{t \to \infty} \frac{1}{t} \ln \frac{N(t)}{N(0)} . \tag{4.2}$$

The long time interval is taken to ensure that the population experiences a representative time series of the possibly fluctuating environment $E(t)$, and that a structured mutant population attains its stationary distribution. For a nonstructured population in a stable environment (which requires a stable resident population), there is no need to consider a long time interval: the invasion fitness of the mutant

---

Box 4.1 Invasion fitness in a model of competition for a continuous resource

Consider the Lotka–Volterra competition model

$$\frac{1}{n_i}\frac{dn_i}{dt} = r\left[1 - \frac{\sum_j a(x_i, x_j)n_j}{K(x_i)}\right], \tag{a}$$

where the trait value x_i determines which part of a resource continuum the ith strategy can utilize efficiently (e.g., beak size determines which seeds of a continuous distribution of seed sizes are consumed). The more similar two strategies are, the more their resources overlap, and the more intense the competition. This can be expressed by the commonly used Gaussian competition function $a(x_i, x_j) = \exp(-\frac{1}{2}(x_i - x_j)^2/\sigma_a^2)$ (see Christiansen and Fenchel 1977). We assume that the intrinsic growth rate r is constant and that the carrying capacity K is unimodal with a maximum at x_0; K is given by $K(x) = (K_0 - \lambda(x - x_0)^2)_+$, where $(...)_+$ indicates that negative values are set to zero. This model (or a very similar model) has been investigated, for example, by Christiansen and Loeschcke (1980), Slatkin (1980), Taper and Case (1985), Vincent *et al.* (1993), Metz *et al.* (1996), Doebeli (1996b), Dieckmann and Doebeli (1999), Drossel and McKane (1999), Day (2000), and Doebeli and Dieckmann (2000).

As long as a mutant strategy is rare, its self-competition and impact on the resident strategies are negligible. The density of a rare mutant strategy x' thus increases exponentially according to

$$\frac{1}{n'}\frac{dn'}{dt} = r\left[1 - \frac{\sum_j a(x', x_j)\hat{n}_j}{K(x')}\right], \tag{b}$$

where $\hat{n}_j$ is the equilibrium density of the jth resident. These equilibrium densities can be obtained by setting Equation (a) equal to zero and solving for n_i. The right-hand side of Equation (b) is the exponential growth rate, or invasion fitness, of the mutant x' in a resident population with strategies $x_1, ..., x_n$. Specifically, in a monomorphic resident population with strategy x, the equilibrium density is $K(x)$ and the mutant's fitness simplifies to

$$f(x', x) = r\left[1 - a(x', x)\frac{K(x)}{K(x')}\right]. \tag{c}$$

Figures 4.1 and 4.2, and the figure in Box 4.5, are based on this model.

---

is then simply $f = \ln A(\hat{E})$ in discrete time and $f = B(\hat{E})$ in continuous time, with $\hat{E}$ being the environment as set by the equilibrium resident population. A positive value of f indicates that the mutant strategy can spread in the population, whereas a mutant with negative f will die out. Box 4.1 contains an example of how to calculate f for a concrete model.

At the very beginning of the invasion process, typically only a few mutant individuals are present. As a consequence, demographic stochasticity plays an important role so that the mutant may die out despite having a positive invasion fitness f. However, the mutant has a positive probability of escaping random extinction whenever its growth rate f is positive (Crow and Kimura 1970; Goel and Richter-Dyn 1974; Dieckmann and Law 1996). Once the mutant has grown sufficiently in number so that demographic stochasticity can be neglected, its further invasion dynamics is given by Equation (4.1) as long as it is still rare in frequency. Equation (4.1) ceases to hold once the mutant becomes sufficiently common that it appreciably influences the environment E.

Henceforth the fitness of a rare mutant strategy with trait value x' in a resident population of strategy x is denoted by $f(x', x)$ to emphasize that the fitness of a rare mutant depends on its own strategy as well as on the resident strategy, since the latter influences the environment E. This notation suppresses the associated ecological variables, such as the equilibrium density of the residents. It is essential to realize, however, that the fitness function $f(x', x)$ is derived from a population dynamic model that appropriately incorporates the ecological features of the system under study.

4.3 Phenotypic Evolution by Trait Substitution

A single evolutionary step is made when a new strategy invades the population and ousts the former resident. The phenotypes that prevail in the population evolve by a sequence of invasions and substitutions. We assume that mutations occur infrequently, so that the previously invading mutant becomes established and the population reaches its population dynamic equilibrium (in a deterministic or statistical sense) by the time the next mutant arrives, and also that mutations are of small phenotypic effect (i.e., that a mutant strategy is near the resident strategy from which it originated).

Consider a monomorphic resident population with a single strategy x. A mutant strategy x' can invade this population if its fitness $f(x', x)$ is positive. If strategy x has a negative fitness when strategy x' is already widespread, then the mutant strategy x' can eliminate the original resident. We assume that there is no unprotected polymorphism and thus infer that strategy x' can replace strategy x if and only if $f(x', x)$ is positive and $f(x, x')$ is negative. On the other hand, if both strategies spread when rare, that is, if both $f(x', x)$ and $f(x, x')$ are positive, then the two strategies form a protected dimorphism.

In the remainder of this section, as well as in Section 4.4, we focus on the evolution of strategies specified by a single quantitative trait in monomorphic resident populations. To visualize the course of phenotypic evolution it is useful to depict graphically those mutant strategies that can invade in various resident populations and those strategy pairs that can form protected dimorphisms. Figure 4.2a shows a so-called *pairwise invasibility plot* (Matsuda 1985; Van Tienderen and de Jong 1986): each point inside the gray area represents a resident–mutant strategy combination such that the mutant can invade the population of the resident. Points inside the white area correspond to mutant–resident strategy pairs such that the mutant cannot invade. A pairwise invasibility plot is constructed by evaluating the mutant's fitness $f(x', x)$ for all values of x and x' and "coloring" the corresponding point of the plot according to whether $f(x', x)$ is positive or negative. In

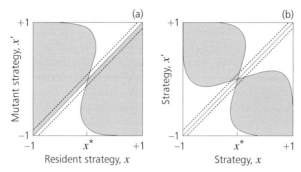

Figure 4.2 Course of phenotypic evolution for the model described in Box 4.1 with $r = 1$, $K(x) = (1 - x^2)_+$, $a(x, x') = \exp(-\frac{1}{2}(x - x')^2/\sigma_a^2)$ with $\sigma_a = 0.5$. (a) Pairwise invasibility plot. Gray areas indicate combinations of mutant strategies x' and resident strategies x for which the mutant's fitness $f(x', x)$ is positive; white areas correspond to strategy combinations such that $f(x', x)$ is negative. (b) The set of potentially coexisting strategies. Gray areas indicate strategy combinations for which both $f(x', x)$ and $f(x, x')$ are positive; protected coexistence outside the gray areas is not possible. In both (a) and (b), the dotted lines schematically illustrate the narrow band of mutants near the resident that can arise by mutations of small phenotypic effect. The singular strategy is denoted by x^*.

Figure 4.2b the gray area indicates that both $f(x', x)$ and $f(x, x')$ are positive, and hence the two strategies are able to coexist. This plot is obtained by first mirroring the pairwise invasibility plot along its main diagonal $x' = x$ [which amounts to reversing the roles of the mutant and the resident and gives the sign plot of $f(x, x')$] and then superimposing the mirror image on the original. The overlapping gray areas correspond to strategy pairs that form protected dimorphisms.

With small mutations, x and x' are never far apart, so that only a narrow band along the main diagonal $x' = x$ is of immediate interest. The main diagonal itself is always a borderline between "invasion" (gray) and "noninvasion" (white) areas, because residents are selectively neutral among themselves, and therefore $f(x) = 0$ for all x. In Figure 4.2a, resident populations with a trait value less than x^* can always be invaded by mutants with slightly larger trait values. Coexistence is not possible, because away from x^* any combination of mutant and resident strategies near the main diagonal lies within the white area of Figure 4.2b. Thus, starting with a trait value left of x^*, the population evolves to the right through a series of successive substitutions. By the same argument, it follows that a population starting on the right of x^* evolves to the left. Eventually, the population approaches x^*, where directional selection ceases. Trait values for which there is no directional selection are called *evolutionarily singular strategies* (Metz *et al.* 1996; Geritz *et al.* 1998).

The graphic analysis of Figure 4.2 is sufficient to establish the direction of evolution in the case of monomorphic populations in which a single trait is evolving, but gives no explicit information on the speed of evolution. In Box 4.2, we outline a quantitative approach that assesses the speed of mutation-limited evolution.

Box 4.2 The speed of directional evolution

The speed of mutation-limited evolution is influenced by three factors: how often a new mutation occurs; how large a phenotypic change this causes; and how likely it is that an initially rare mutant invades. If the individual mutational steps are sufficiently small, and thus long-term evolution proceeds by a large number of subsequent invasions and substitutions, the evolutionary process can be approximated by the *canonical equation* of adaptive dynamics (Dieckmann and Law 1996),

$$\frac{dx}{dt} = \frac{1}{2}\alpha(x)\mu(x)\hat{N}(x)\sigma_M^2(x) \left.\frac{\partial f(x', x)}{\partial x'}\right|_{x'=x} . \tag{a}$$

Here μ is the probability of a mutation per birth event, and $\hat{N}$ is the equilibrium population size: the product $\mu\hat{N}$ is thus proportional to the number of mutations that occur per unit of time. The variance of the phenotypic effect of a mutation is σ_M^2 (with symmetric unbiased mutations, the expected phenotypic effect is zero and the variance measures the size of "typical" mutations). The probability of invasion consists of three factors. First, during directional evolution, either only mutants with a trait value larger than the resident, or only mutants with a trait value smaller than the resident, can invade (see Figure 4.2a); in other words, half of the mutants are at a selective disadvantage and doomed to extinction. This leads to the factor $\frac{1}{2}$. Second, even mutants at selective advantage may be lost through demographic stochasticity (genetic drift) in the initial phase of invasion, when they are present in only small numbers. For mutants of small effect, the probability of not being lost is proportional to the selective advantage of the mutant as measured by the fitness gradient $\partial f(x', x)/\partial x'|_{x'=x}$. Finally, the constant of proportionality α is proportional to the inverse of the variance in offspring number: with the same expected number of offspring, an advantageous mutant is more easily lost through demographic stochasticity if its offspring number is highly variable. The constant α equals 1 for a constant birth–death process in an unstructured population, as considered by Dieckmann and Law (1996).

Other models of adaptive dynamics agree that the change in phenotype is proportional to the fitness gradient, that is

$$\frac{dx}{dt} = \beta \left.\frac{\partial f(x', x)}{\partial x'}\right|_{x'=x} \tag{b}$$

(e.g., Abrams *et al.* 1993a; Vincent *et al.* 1993; Marrow *et al.* 1996). This equation leads to results similar to those from quantitative genetic models (Taper and Case 1992) and, indeed, can be derived as an approximation to the quantitative genetic iteration (Abrams *et al.* 1993b). Equations (a) and (b) have a similar form, though the interpretation of their terms is different: in quantitative genetics, β is the additive genetic variance and thus measures the standing variation upon which selection operates; it is often assumed to be constant. In the canonical equation, β depends on the probability and distribution of new mutations; also, β generally depends on the prevalent phenotype x, if only through the population size $\hat{N}(x)$. In quantitative genetics, evolutionary change is proportional to the fitness gradient, because stronger selection means faster change in the frequencies of alleles that are present from the onset. In mutation-limited evolution, a higher fitness gradient increases the probability that a favorable mutant escapes extinction by demographic stochasticity.

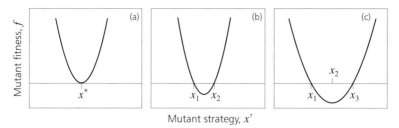

Figure 4.3 Evolutionary branching and a mutant's fitness as a function of its strategy. (a) Mutant fitness in a monomorphic resident population at a branching point x^*. (b) Mutant fitness in a dimorphic resident population with strategies x_1 and x_2, both similar to the branching point x^*. Notice that only those mutants outside the interval spanned by x_1 and x_2 have a positive fitness and hence can invade. (c) Mutant fitness in a dimorphic resident population with strategies x_1 and x_3. The former resident x_2 now has a negative fitness, and hence is expelled from the population.

4.4 The Emergence of Diversity: Evolutionary Branching

Although the evolutionarily singular strategy x^* in Figure 4.2a is an attractor of monomorphic directional evolution, it is not evolutionarily stable in the classic sense (Maynard Smith 1982), that is, it is not stable against invading mutants. In fact, mutants both smaller and larger than x^* can invade the resident population of x^*. Unlike in directional evolution, in the neighborhood of x^* the invasion of a mutant results in coexistence of the resident and mutant strategies (Figure 4.2b). As the singularity is approached by small but finite mutational steps, the population actually becomes dimorphic as soon as the next mutant enters the area of coexistence (i.e., a little before exactly reaching the singular strategy, Figure 4.2b).

To see how evolution proceeds in the now dimorphic population, it is useful to plot the mutant's fitness as a function of the mutant trait value (Figure 4.3). In the resident population of the singular strategy x^*, all nearby mutants are able to invade (i.e., they have positive fitness), except for the singular strategy itself, which has zero fitness. The fitness function thus attains a minimum at x^* (Figure 4.3a). In a dimorphic population with two strategies x_1 and x_2, both similar to x^*, the fitness function is also similar, but with zeros at x_1 and x_2, because residents themselves are selectively neutral (Figure 4.3b).

According to Figure 4.3b, a new mutant that arises in the dimorphic population with strategies x_1 and x_2 similar to x^* has a positive fitness, and therefore can invade, if and only if it is outside the interval spanned by the two resident trait values. By contrast, mutants between these values have a negative fitness and therefore must die out. A mutant cannot coexist with both former residents, because the parabolically shaped fitness function cannot have three zeros to accommodate three established resident strategies. It follows that the successfully invading mutant will oust the resident that has become the middle strategy (Figure 4.3c).

Since the initial dimorphic population is formed of the most recent monomorphic resident and its mutant, with small mutations these two strategies are very

Box 4.3 How to recognize evolutionary branching points

One can easily search for evolutionary branching points in a model once the mutant fitness function $f(x', x)$ has been determined. If $f(x', x)$ is known analytically, then the following criteria must be satisfied by an evolutionary branching point x^* (Geritz *et al.* 1998):

1. x^* must be an evolutionary singularity, i.e., the fitness gradient vanishes at x^*,

$$\left. \frac{\partial f(x', x)}{\partial x'} \right|_{x'=x=x^*} = 0 . \tag{a}$$

2. x^* must be an attractor of directional evolution (Eshel 1983),

$$\left. \frac{\partial^2 f(x', x)}{\partial x \partial x'} + \frac{\partial^2 f(x', x)}{\partial x'^2} \right|_{x'=x=x^*} < 0 . \tag{b}$$

3. In the neighborhood of x^*, similar strategies must be able to form protected dimorphisms (Geritz *et al.* 1998),

$$\left. \frac{\partial^2 f(x', x)}{\partial x^2} + \frac{\partial^2 f(x', x)}{\partial x'^2} \right|_{x'=x=x^*} > 0 . \tag{c}$$

4. x^* must lack evolutionary stability (Maynard Smith 1982), which ensures disruptive selection at x^* (Geritz *et al.* 1998),

$$\left. \frac{\partial^2 f(x', x)}{\partial x'^2} \right|_{x'=x=x^*} > 0 . \tag{d}$$

As can be verified by inspection of all the generic singularities (see Box 4.4), the second-order criteria (2)–(4) are not independent for the case of a single trait and an initially monomorphic resident population; instead, criteria (2) and (4) are sufficient to ensure (3) as well. This is, however, not true for multidimensional strategies or for coevolving populations (Geritz *et al.* 1998). These criteria are thus best remembered separately.

Alternatively, a graphic analysis can be performed using a pairwise invasibility plot (Figure 4.2a). Although drawing the pairwise invasibility plot is practical only for the case of single traits and monomorphic populations, it is often used when the invasion fitness cannot be determined analytically. In a pairwise invasibility plot, the evolutionary branching point is recognized by the following pattern:

■ The branching point is at a point of intersection between the main diagonal and another border line between positive and negative mutant fitness.
■ The fitness of mutants is positive immediately above the main diagonal to the left of the branching point and below the main diagonal to its right.
■ Potentially coexisting strategies lie in the neighborhood of the branching point (this can be checked on a plot similar to Figure 4.2b, but, as highlighted above, in the simple case for which pairwise invasibility plots are useful, this criterion does not have to be checked separately).
■ Looking along a vertical line through the branching point, the mutants immediately above and below are able to invade.

Box 4.4 Types of evolutionary singularities

Eight types of evolutionary singularities occur generically in single-trait evolution of monomorphic populations, as in the figure below (Geritz *et al.* 1998). As in Figure 4.2a, gray areas indicate combinations of mutant strategies and resident strategies for which the mutant's fitness is positive.

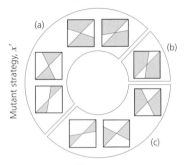

Resident strategy, *x*

These types can be classified into three major groups:

- *Evolutionary repellors, (a) in the figure above.* Directional evolution leads away from this type of singularity, and therefore these types do not play a role as evolutionary outcomes. If the population has several singular strategies, then a repellor separates the basins of attraction of adjacent singularities.
- *Evolutionary branching points, (b) in the figure above.* This type of singularity is an attractor of directional evolution, but it lacks evolutionary stability and therefore evolution cannot stop here. Invading mutants give rise to a protected dimorphism in which the constituent strategies are under disruptive selection and diverge away from each other. Evolution can enter a higher level of polymorphism by small mutational steps only via evolutionary branching.
- *Evolutionarily stable attractors, (c) in the figure above.* Singularities of this type are attractors of directional evolution and, moreover, once established the population cannot be invaded by any nearby strategy. Such strategies are also called continuously stable strategies (Eshel 1983). Coexistence of strategies may be possible, but coexisting strategies undergo convergent rather than divergent coevolution such that eventually the dimorphism disappears. Evolutionarily stable attractors act as final stops of evolution.

---

similar. After the first substitution in the dimorphic population, however, the new resident population consists of two strategies with a wider gap between. Through a series of such invasions and replacements, the two strategies of the dimorphic population undergo divergent coevolution and become phenotypically clearly distinct (see Figure 4.1).

The process of convergence to a particular trait value in the monomorphic population followed by gradual divergence once the population has become dimorphic

Box 4.5 Polymorphic and multidimensional evolution

If the theory of adaptive dynamics were only applicable to the caricature of one-dimensional trait spaces or monomorphic populations, it would be of very limited utility. Below we therefore describe how this framework can be extended.

We start by considering polymorphic populations. By assuming mutation-limited evolution we can ignore the possibility of simultaneous mutations that occur in different resident strategies. Two strategies x_1 and x_2 can coexist as a protected dimorphism if both $f(x_2, x_1)$ and $f(x_1, x_2)$ are positive (i.e., when both can invade into a population of the other). For each pair of resident strategies x_1 and x_2 we can construct a pairwise invasibility plot for x_1 while keeping x_2 fixed, and a pairwise invasibility plot for x_2 while keeping x_1 fixed. From this we can see which mutants of x_1 or of x_2 could invade the present resident population and which could not (i.e., in what direction x_1 and x_2 will evolve by small mutational steps).

In the example shown in the figure below, the arrows indicate the directions of evolutionary change in x_1 and in x_2. On the lines that separate regions with different evolutionary directions, selection in one of the two resident strategies is no longer directional: each point on such a line is a singular strategy for the corresponding resident, if the other resident is kept fixed. The points of these lines, therefore, can be classified similarly to the monomorphic singularities in Box 4.4. Within the regions of coexistence in the figure below continuous lines indicate evolutionary stability and dashed lines the lack thereof. At the intersection point of two such lines, directional evolution ceases for both residents. Such a strategy combination is called an evolutionarily singular dimorphism. This dimorphism is evolutionarily stable if neither mutants of x_1 nor mutants of x_2 can invade (i.e., if both x_1 and x_2 are evolutionarily stable); in the figure this is the case.

continued

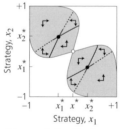

Adaptive dynamics in a dimorphic population for the model described in Box 4.1 with $r = 1$, $K(x) = (1 - x^2)_+$, and $\sigma_a = 0.5$. Gray areas indicate strategy pairs (x_1, x_2) that can coexist as a protected dimorphism. Lines inside the gray areas separate regions with different evolutionary directions for the two resident strategies, as illustrated by arrows. On the steeper line, which separates strategy pairs evolving either toward the left or toward the right, directional evolution in x_1 ceases. Likewise, on the shallower line, which separates strategy pairs evolving upward from those evolving downward, directional evolution in x_2 ceases. Continuous lines indicate that the corresponding strategy is evolutionarily stable if evolution in the other strategy is arrested. After branching at the branching point x^* (open circle), the population evolves into the gray area toward the evolutionarily stable dimorphism (x_1^*, x_2^*) (filled circles), where directional evolution ceases in both strategies and both strategies also possess evolutionary stability.

Box 4.5 *continued*

For a singular dimorphism to be evolutionarily attracting it is neither necessary nor sufficient that both strategies are attracting if the other resident is kept fixed at its present value (Matessi and Di Pasquale 1996; Marrow *et al.* 1996). With small evolutionary steps, we can approximate the evolutionary trajectories by utilizing the canonical equation (see Box 4.2) simultaneously for both coevolving strategies. Stable equilibria of the canonical equation then correspond to evolutionarily attracting singular dimorphisms. If such a dimorphism is evolutionarily stable, it represents a final stop of dimorphic evolution. However, if one of the resident strategies at the singularity is not evolutionarily stable and, moreover, if this resident can coexist with nearby mutants of itself, the population undergoes a secondary branching event, which leads to a trimorphic resident population (Metz *et al.* 1996; Geritz *et al.* 1998). An example of such a process is shown in Figure 4.1.

Next we consider the adaptive dynamics framework in the context of multidimensional strategies. In natural environments, strategies are typically characterized by several traits that jointly influence fitness and that may be genetically correlated.

Though much of the basic framework can be generalized to multidimensional strategies, these also pose special difficulties. For example, unlike in the case of scalar strategies, a mutant that invades a monomorphic resident population may coexist with the former resident also away from any evolutionary singularity. This coexistence, however, is confined to a restricted set of mutants, such that its volume vanishes for small mutational steps proportionally to the square of the average size of mutations. With this caveat, directional evolution of two traits in a monomorphic population can be depicted graphically in a similar way to coevolving strategies. There are two important differences, however. First, the axes of the figure on the previous page no longer represent different residents, but instead describe different phenotypic components of the same resident phenotype. Second, if the traits are genetically correlated such that a single mutation can affect both traits at the same time, then the evolutionary steps are not constrained to being either horizontal or vertical. Instead, evolutionary steps are possible in any direction within an angle of plus or minus 90 degrees from the selection gradient vector $\partial f(x', x)/\partial x'|_{x'=x}$.

For small mutational steps, the evolutionary trajectory can be approximated by a multidimensional equivalent of the canonical equation (Dieckmann and Law 1996; see Box 4.2), where dx/dt and $\partial f(x', x)/\partial x'|_{x'=x}$ are vectors, and the mutational variance is replaced by the mutation variance–covariance matrix $C(x)$ (the diagonal elements of this matrix contain the trait-wise mutational variances and the off-diagonal elements represent the covariances between mutational changes in two different traits that may result from pleiotropy). With large covariances, it is possible that a trait changes "maladaptively", that is, the direction of the net change is opposite to the direct selection on the trait given by the corresponding component of the fitness gradient (see also Lande 1979b).

An evolutionarily singular strategy x^*, in which all components of the fitness gradient are zero, is evolutionarily stable if it is, as a function of the traits of the mutant strategy x', a multidimensional maximum of the invasion fitness $f(x', x^*)$. If such a singularity lacks evolutionary stability, evolutionary branching may occur.

Box 4.6 The geography of speciation

Evolutionary branching in a spatially subdivided population based on a simple model by Meszéna *et al.* (1997) is illustrated here. Two habitats coupled by migration are considered. Within each habitat, the population follows logistic growth, in which the intrinsic growth rate is a Gaussian function of strategy, with different optima in the two habitats. The model is symmetric, so that the "generalist" strategy, which is exactly halfway between the two habitat-specific optima, is always an evolutionarily singular strategy. Depending on the magnitude of the difference Δ between the local optima relative to the width of the Gaussian curve and on the migration rate m, this central singularity may either be an evolutionarily stable strategy, a branching point, or a repellor [(a) in the figure below; see also Box 4.4].

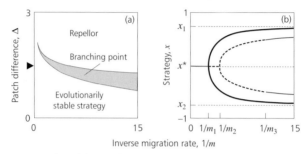

Evolutionary properties of singularities in a two-patch model with local adaptation and migration. (a) A generalist strategy that exploits both patches is an evolutionary repellor, a branching point, or an evolutionarily stable attractor, depending on the difference Δ between the patch-specific optimal strategies and the migration rate m, as indicated by the three different parameter regions. (b) Evolutionary singularities as a function of inverse migration rate. The difference between the patch-specific optimal strategies was fixed at $\Delta = 1.5$. For comparison, the two thin dotted horizontal lines at x_1 and x_2 denote the local within-patch optimal strategies ($x = \pm \Delta/2$). Monomorphic singular strategies are drawn with lines of intermediate thickness, of which the continuous lines correspond to evolutionarily stable attractors, the dashed lines to branching points, and the dotted line to an evolutionary repellor. The monomorphic generalist strategy is indicated by x^*. Along a cross-section at $\Delta = 1.5$ in (a), indicated by an arrow head, the generalist strategy changes with increasing $1/m$ from an evolutionarily stable attractor into a branching point and then into an evolutionary repellor. The two branches of the bold line indicate the strategies of the evolutionarily stable dimorphism. *Source*: Meszéna *et al.* (1997).

There are four possible evolutionary scenarios [(b) in the figure above]: at high levels of migration (inverse migration rate smaller than $1/m_1$), the population effectively experiences a homogeneous environment in which the generalist strategy is evolutionarily stable and branching is not possible. With a somewhat lower migration rate (inverse migration rate between $1/m_1$ and $1/m_2$), the generalist is at an evolutionary branching point, and the population evolves to a dimorphism that consists of two habitat specialists. Decreasing migration further (inverse migration rate between $1/m_2$ and $1/m_3$), the generalist becomes an evolutionary repellor, but

continued

Box 4.6 *continued*

there are two additional monomorphic singularities, one on each side of the generalist, both of which are branching points. Finally, in case of a very low migration rate (inverse migration rate greater than $1/m_3$), these two monomorphic attractors are evolutionarily stable and branching does not occur, even though there also exists an evolutionarily stable dimorphism of habitat specialists. (A similar sequence of transitions can be observed if, instead of decreasing the migration rate, the difference between the habitats is increased.) Evolutionary branching is also possible if the environment forms a gradient instead of discrete habitats, provided there are sufficiently different environments along the gradient and mobility is not too high (Mizera and Meszéna 2003; Chapter 7).

---

is called evolutionary branching. The singularity at which this happens (x^* in Figure 4.2a) is an evolutionary branching point. In Box 4.3 we summarize how to recognize branching points by investigating the fitness function $f(x', x)$.

The evolutionary branching point, though perhaps the most interesting with regard to speciation, is not the only type of singular strategy. In Box 4.4 we briefly summarize the basic properties of all singularities that occur generically. Throughout this section, we constrain our discussion to single-trait evolution in an initially monomorphic population. A brief summary on how to extend these results to polymorphic populations (including further branching events as in Figure 4.1) and to multiple-trait evolution is given in Box 4.5; more details can be found in Metz *et al.* (1996) and Geritz *et al.* (1998, 1999), and, concerning directional evolution, in Dieckmann and Law (1996), Matessi and Di Pasquale (1996), Champagnat *et al.* (2001), and Leimar (2001 and in press).

For the adaptive dynamics framework to be applicable to spatially subdivided populations, sufficient dispersal must occur between subpopulations for the stationary population distribution to be attained on an ecological time scale. Full sympatry is, however, by no means a necessary condition, and the framework has been used to analyze evolution in spatially structured populations as well (e.g., Meszéna *et al.* 1997; Day 2000; see Box 4.6).

So far we have considered clonally inherited phenotypes. The very same model can be applied, however, to the evolution of alleles at a single diploid locus in a Mendelian population [Box 4.7; Kisdi and Geritz 1999; see also Christiansen and Loeschcke (1980) for a related approach] when assuming that a continuum of allele types is possible, and that the mutant allele codes for a phenotype similar to that of the parent allele. Evolutionary branching in alleles then occurs similarly to clonal phenotypes and produces two distinct allele types that may continue to segregate within the species. Since intermediate heterozygotes are at a disadvantage under disruptive selection, selection occurs for dominance and for assortative mating (Udovic 1980; Wilson and Turelli 1986; Van Dooren 1999; Geritz and Kisdi 2000).

Box 4.7 Adaptive dynamics of alleles and stable genetic polymorphisms

As an example for the adaptive dynamics of alleles, consider the classic soft-selection model of Levene (1953; see Box 3.1). We assume that not just two alleles may segregate (A and a, with fixed selection coefficients s_1 and s_2), as in the classic models, but instead that many different alleles may arise by mutations and that they determine a continuous phenotype in an additive way (i.e., if the phenotypes of AA and aa are, respectively, x_A and x_a, then the heterozygote phenotype is $(x_A + x_a)/2$). Within each habitat, local fitness is a function of the phenotype: $\varphi_i(x)$ in habitat i. For example, in the first habitat local fitness values are $W_{AA} = \varphi_1(x_A)$, $W_{Aa} = \varphi_1((x_A + x_a)/2)$, and $W_{aa} = \varphi_1(x_a)$. For any two alleles A and a, drawn from the assumed continuum, the dynamics and equilibrium of allele frequencies can be obtained as described in Box 3.1. In particular, the frequency of a rare mutant allele a increases in a population monomorphic for allele A at a per-generation rate of $k_1 \varphi_1((x_A + x_a)/2)/\varphi_1(x_A) + k_2 \varphi_2((x_A + x_a)/2)/\varphi_2(x_A)$, where k_i is the relative size of habitat i, with $k_1 + k_2 = 1$. If this expression is greater than 1 [or, equivalently, if its logarithm, $f(x_a, x_A)$ in the notation of the main text, is positive], then the mutant allele can invade.

Assuming that mutations only result in small phenotypic change (x_a is near x_A), we can apply the adaptive dynamics framework to the evolution of alleles (Geritz *et al.* 1998). Invasion by a mutant allele usually leads to substitution (i.e., the new allele replaces the former allele, just as in clonal adaptive dynamics). The ensuing directional evolution, however, leads to singular alleles in which protected polymorphisms become possible. Evolutionary branching of alleles means that the homozygote phenotypes diverge from each other, and results in a genetic polymorphism of distinctly different alleles that segregate in a randomly mating population (Kisdi and Geritz 1999).

Assuming a more flexible genetic variation sheds new light on the old question of whether stable genetic polymorphisms are sufficiently robust to serve as a basis for sympatric speciation. Recall from Box 3.1 that if selection coefficients s_i are small, then polymorphism is possible only in a very narrow range of parameters (the parameter region that allows for polymorphism actually has a cusp at $s_1 = s_2 = 0$). Given two arbitrary alleles A and a, and therefore given selection coefficients s_1 and s_2, polymorphism results only if the environmental parameters, in this case the relative habitat sizes k_1 and $k_2 = 1 - k_1$, are fine-tuned. This means that a polymorphism of two particular alleles is not robust under weak selection (Maynard Smith 1966; Hoekstra *et al.* 1985), and this property appeared a significant obstacle to sympatric speciation.

By contrast, the assumption of more flexible genetic variation (a potential continuum of alleles rather than only two alleles) considerably facilitates the evolution of stable genetic polymorphisms. Here we focus on polymorphisms of similar alleles (which may arise by a single mutation at the onset of evolutionary branching); this immediately implies that the selection coefficients are small and that the two alleles cannot form a polymorphism without fine-tuning of the environmental parameters. With many potential alleles, however, the requirement of fine-tuning may be turned around: given a certain environment (k_1 and k_2), polymorphism will result if the *alleles* are chosen from a narrow range. This narrow range turns out to

continued

Box 4.7 *continued*

coincide with the neighborhood of an evolutionarily singular allele. Thus, starting with an arbitrary allele A, population genetics and adaptive dynamics agree in that the invasion of a mutant allele a usually results in substitution rather than polymorphism. Repeated substitutions, however, lead toward a singular allele, in the neighborhood of which stable polymorphisms are possible. In other words, evolution by small mutational steps proceeds exactly toward those exceptional alleles that can form polymorphisms: long-term evolution itself takes care of the necessary fine-tuning (Kisdi and Geritz 1999; see figure below). If the singularity is a branching point, then the population not only becomes genetically polymorphic, but also is subject to disruptive selection, as is necessary for sympatric speciation. Of course, it remains to be seen whether reproductive isolation can evolve [see Chapters 3 and 5 and references therein; see also Geritz and Kisdi (2000) for an analysis of the evolution of reproductive isolation through the adaptive dynamics of alleles]. Fine-tuning is necessary not only in multiple-niche polymorphisms [such as Levene's (1953) model and its variants, see Hoekstra *et al.* (1985)], but also in any generic model in which frequency dependence can maintain protected polymorphisms; long-term evolution then provides the necessary fine-tuning whenever many small mutations incrementally change an evolving trait.

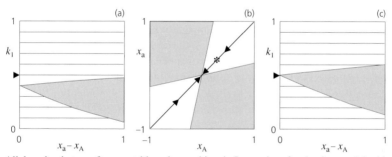

Allele pairs that can form a stable polymorphism in Levene's soft selection model with stabilizing selection within each habitat (φ_i is Gaussian with unit width and the two peaks are located at a distance of $\Delta = 3$). (a) An arbitrary allele $x_A = 0.3$ can form a polymorphism with allele x_a within the shaded area. Notice that if $x_a - x_A$ is small (i.e., if the two alleles produce similar phenotypes and hence selection is weak), then polymorphism is possible only in a narrow range of the environmental parameter k_1 (fine-tuning). In one particular environment $k_1 = 0.5$ (indicated by arrowheads in the left and right panels), the allele $x_A = 0.3$ cannot form a polymorphism under weak selection. (b) The set of allele pairs that can form polymorphisms in the particular environment $k_1 = 0.5$. The thick line corresponds to identical alleles $x_a = x_A$; similar alleles with small difference $x_a - x_A$ thus lie in the neighborhood of the thick line. Again, the allele $x_A = 0.3$ (denoted by an asterisk) cannot form a polymorphism with alleles similar to itself in this particular environment. Allele substitutions in the monomorphic population, however, lead to the evolutionary singular allele $x^* = 0$, where similar alleles are able to form polymorphisms in this particular environment. (c) With $x_A = x^*$, the narrow range of k_1 that permits polymorphism with small $x_a - x_A$ shifts to the actual value of the environmental parameter, $k_1 = 0.5$. Note that x^* depends on the actual value of the environmental parameter; it happens to be the central strategy only for the particular choice $k_1 = 0.5$.

4.5   Evolutionary Branching and Speciation

The phenomenon of evolutionary branching in clonal models may appear very suggestive of speciation. First, there is directional evolution toward a well-defined trait value, the evolutionary branching point. As evolution reaches the branching point, selection turns disruptive. The population necessarily becomes dimorphic in the neighborhood of the branching point, and disruptive selection causes divergent coevolution in the two coexisting lineages. The resultant evolutionary pattern is of a branching evolutionary tree, with phenotypically distinct lineages that develop gradually by small evolutionary steps (Figure 4.1).

Naturally, clonal models of adaptive dynamics are unable to account for the genetic details of speciation, in particular how reproductive isolation might develop between the emerging branches (see Dieckmann and Doebeli 1999; Drossel and McKane 2000; Geritz and Kisdi 2000; Matessi *et al.* 2001; Meszéna and Christiansen, unpublished; see Chapter 5). What evolutionary branching does imply is that there is *evolution toward disruptive selection* and, at the same time, *toward polymorphism* in the ecological model in which branching is found. These are the ecological prerequisites for speciation and set the selective environment for the evolution of reproductive isolation. Evolutionary branching thus indicates that the ecological system under study is prone to speciation.

Speciation by disruptive selection has previously been considered problematic, because disruptive selection does not appear to be likely to occur for a long time and does not appear to be compatible with the coexistence of different types (either different alleles or different clonal types or species). For disruptive selection to occur, the population must be at the bottom of a fitness valley (similar to Figure 4.3). In simple, frequency-independent models of selection, the population "climbs" toward the nearest peak of the adaptive landscape (Wright 1931; Lande 1976). The fitness valleys are thus evolutionary repellors: the population is unlikely to experience disruptive selection, except possibly for a brief exposure before it evolves away from the bottom of the valley.

As pointed out by Christiansen (1991) and Abrams *et al.* (1993a), evolution by frequency-dependent selection often leads to fitness minima. Even though in each generation the population evolves "upward" on the fitness landscape, the landscape itself changes such that the population eventually reaches the bottom of a valley. This is what happens during directional evolution toward an evolutionary branching point.

Disruptive selection has also been thought incompatible with the maintenance of genetic variability (e.g., Ridley 1993). In simple one-locus models disruptive selection amounts to heterozygote inferiority, which, in the absence of frequency dependence, leads to the loss of one allele. This is not so under frequency dependence (Pimm 1979): at the branching point, the heterozygote is inferior only when both alleles are sufficiently common. Should one of the alleles become rare, the frequency-dependent fitness of the heterozygote increases such that it is no longer at a disadvantage, and therefore the frequency of the rare allele increases again.

Except for asexual species, evolutionary branching corresponds to speciation only if reproductive isolation emerges between the nascent branches. There are many ways by which reproductive isolation could, in principle, evolve during evolutionary branching (see Chapter 3). Assortative mating based on the same ecological trait that is under disruptive selection automatically leads to reproductive isolation as the ecological trait diverges (Drossel and McKane 2000), and this possibility appears to be widespread in nature (e.g., Chown and Smith 1993; Wood and Foote 1996; Macnair and Gardner 1998; Nagel and Schluter 1998; Grant *et al.* 2000). For example, in the apple maggot fly *Rhagoletis pomonella*, there is disruptive selection on eclosion time; different timing of reproduction helps to prevent hybridization between the host races (Feder 1998; see Chapter 11). If differences in the ecological trait are associated with different habitats, as is the case for the apple maggot fly, then reduced migration, habitat fidelity, or habitat choice ensures assortative mating (Balkau and Feldman 1973; Diehl and Bush 1989; Kawecki 1996, 1997). Assortative mating based on a neutral "marker" trait (e.g., different flower colors that attract different pollinators) can lead to reproductive isolation between the emerging branches only if an association (linkage disequilibrium) is established between the ecological trait and the marker. This is considered to be difficult because of recombination (Felsenstein 1981; but see Dieckmann and Doebeli 1999; Chapter 5), and possible only if there is strong assortativeness, strong selection on the ecological trait, or low recombination (Udovic 1980; Kondrashov and Kondrashov 1999; Geritz and Kisdi 2000).

The degree of assortativeness in any mate choice system may be sufficiently high at the onset, or else it may increase evolutionarily while the population is at a branching point. Since disruptive selection acts against intermediate phenotypes, assortative mating between phenotypically similar individuals is selectively favored at the branching point. Adaptive increase of assortativeness may be suspected if mating is more discriminative in sympatry (Coyne and Orr 1989, 1997; Noor 1995; Sætre *et al.* 1997) or if some unusual preference appears as a derived character (Rundle and Schluter 1998). In models, increased assortativeness readily evolves if it amounts to the substitution of the same allele in the entire population [a "one-allele mechanism" in the sense of Felsenstein (1981)], such as an allele for increased "choosiness" when selecting mates based on the ecological trait (Dieckmann and Doebeli 1999; Chapter 5; but see Matessi *et al.* 2001; Meszéna and Christiansen, unpublished) or on an allele for reduced migration (Balkau and Feldman 1973). By contrast, two-allele mechanisms depend on the replacement of different alleles in the two branches and thus involve the emergence of linkage disequilibria, a process counteracted by recombination (Felsenstein 1981). Yet such mechanisms have been shown to evolve under certain conditions: when different alleles in the two branches code for different habitat preferences, the process is aided by spatial segregation (Diehl and Bush 1989; Kawecki 1996, 1997), and when the different alleles code for ecologically neutral marker traits, linkage disequilibria can arise from the deterministic amplification of genetic drift in finite populations (Dieckmann and Doebeli 1999; see Chapter 5). Partial reproductive isolation by one mechanism facilitates the evolution of other isolating mechanisms (Johnson *et al.* 1996b), whereby the remaining gene flow is further reduced and the divergent subpopulations attain species rank.

Alternatively, reproductive isolation may arise for reasons independent of disruptive natural selection on the ecological trait. Such mechanisms include sexual selection (Turner and Burrows 1995; Payne and Krakauer 1997; Seehausen *et al.* 1997; Higashi *et al.* 1999) and the evolution of gamete-recognition systems (Palumbi 1992). If the emergent species experience directional or stabilizing natural selection, they remain ecologically undifferentiated and hence they are unlikely to coexist for a long time. Evolutionary branching, however, can latch on such that the two species evolve into two branches, which ensures the ecological differentiation necessary for long-term coexistence (Galis and Metz 1998; Van Doorn and Weissing 2001). Once reproductive isolation has been established between the branches in any way, further coevolution of the species proceeds as in the clonal model of adaptive dynamics.

4.6 Adaptive Dynamics: Alternative Approaches

In this chapter, we concentrate on the adaptive dynamics framework developed by Metz *et al.* (1996) and Geritz *et al.* (1997, 1998). This is by no means the only approach to adaptive dynamics [see Abrams (2001a) for a review]. We focus on this particular approach because the concept of evolutionary branching may help in the study of nonallopatric speciation. Alternative approaches consider the number of species fixed [and hence do not consider speciation at all; see Abrams (2001a) for references to many examples], or assume invasions of new species from outside the system [the invading species in such cases may be considerably different from the members of the present community and its phenotype is more or less arbitrary; e.g., Taper and Case (1992)], or establish that polymorphism will occur at fitness minima, but do not investigate the subsequent coevolution of the constituent strategies (e.g., Brown and Pavlovic 1992). An important exception is the work of Eshel *et al.* (1997), which paralleled some results of Metz *et al.* (1996) and Geritz *et al.* (1997, 1998). A recent paper by Cohen *et al.* (1999) gives similar results to those presented in this chapter. This approach uses differential equations to describe the convergence to the branching point in a monomorphic population and divergence in a dimorphic population; to incorporate the transition from monomorphism to dimorphism, the model has to be modified by adding a new equation at the branching point (see, however, Abrams 2001b). Most models of adaptive dynamics agree on the basic form of the equation that describes within-species phenotypic change over time [see Box 4.2, Equation (b)].

In the present framework, analytical tractability comes at the cost of assuming mutation-limited evolution, that is, assuming mutations that occur infrequently and, if successful, sweep through the population before the next mutant comes along. Simulations of the evolutionary process (similar to that shown in Figure 4.1, but with variable size and frequency of mutations) demonstrate that the qualitative patterns of monomorphic evolution and evolutionary branching are robust with respect to relaxing this assumption. With a higher frequency of mutations, the next mutant arises before the previous successful mutant has become fixed, and therefore there is always some variation in the population. The results are robust with respect to this variation because the environment, E in Equations (4.1a) and (4.1b), generated by a cluster of similar strategies is virtually the same if the cluster is replaced by a single resident. Therefore, there is no qualitative difference in terms of which strategies can invade.

4.7 Concluding Comments

Recent empirical research has highlighted the significance of adaptive speciation (e.g., Schluter and Nagel 1995; Schluter 1996a; Losos *et al.* 1998; Schneider *et al.* 1999; Schilthuizen 2000); in many instances, natural selection plays a decisive role in species diversification. It is a challenge for evolutionary theory to construct an appropriate theoretical framework for adaptive speciation. Adaptive dynamics provides one facet as it identifies speciation-prone ecological conditions, in which selection favors diversification with ecological contact.

Classic speciation models (e.g., Udovic 1980; Felsenstein 1981; Kondrashov and Kondrashov 1999) emphasize the population genetics of reproductive isolation, and merely assume some disruptive selection, either as an arbitrary external force or by incorporating only the simplest ecology [very often a version of Levene's (1953) model with two habitats; see Chapter 3]. By contrast, adaptive dynamics focuses on the ecological side of speciation. It offers a theoretical framework for the investigation in complex ecological scenarios as to whether, and under which conditions, speciation can be expected. Beyond the prediction of a certain speciation event, adaptive dynamics can analyze various patterns in the development of species diversity (see Box 4.8).

On a paleontological time scale, evolution driven by directional selection and, presumably, adaptive speciation is very fast (McCune and Lovejoy 1998; Hendry and Kinnison 1999). It is thus tempting to think of the paleontological record as a series of evolutionarily stable communities, the changes being brought about by some physical change in the environment (Rand and Wilson 1993). The emerging bifurcation theory of adaptive dynamics (Geritz *et al.* 1999; Jacobs, *et al.* unpublished; see Box 4.6 for an example) is capable of studying the properties of evolutionarily stable communities as a function of environmental parameters.

Evolutionary branching has been found in many diverse ecological models including, for example, resource competition (Doebeli 1996a; Metz *et al.* 1996; Day 2000), interference competition (Geritz *et al.* 1999; Jansen and Mulder 1999; Kisdi 1999), predator–prey systems (Van der Laan and Hogeweg 1995; Doebeli and Dieckmann 2000), spatially structured populations and metapopulations (Doebeli and Ruxton 1997; Meszéna *et al.* 1997; Kisdi and Geritz 1999; Parvinen 1999; Mathias *et al.* 2001; Mathias and Kisdi 2002; Mizera and Meszéna 2003), host–parasite systems (Boots and Haraguchi 1999; Koella and Doebeli 1999), mutualistic interactions (Doebeli and Dieckmann 2000; Law *et al.* 2001), mating systems (Metz *et al.* 1992; Cheptou and Mathias 2001; de Jong and Geritz 2001;

Box 4.8 Pattern predictions

In this box, we collect predictions about macroevolutionary patterns derived from adaptive dynamics. No claim is intended that those predictions are all equally hard, or that they cannot be derived through different arguments.

First assume that the external environment exhibits no changes on the evolutionary time scale. [Note that fluctuations on the ecological time scale, like weather changes, are incorporated in the invasion fitness; see Metz *et al.* (1992) and Section 4.2 of this chapter.] Adaptive dynamics theory then predicts that

- Speciation only occurs at specific, and in principle predictable, trait values; here these are called evolutionary branching points.
- The ensuing gradual phenotypic differentiation is initially slow compared to the preceding and ensuing periods of directional evolution. Populations sitting near a branching point experience a locally flat fitness landscape, i.e., far weaker selective pressure than during directional selection. Weak selection slows divergence even when assortative mating is readily established. This prediction rests on the assumption that phenotypic variation is narrow compared to the curvature of the fitness function (see Abrams *et al.* 1993b).
- Speciation typically is splitting into two (i.e., not three or more). As argued in the main text, for one-dimensional phenotypes the geometry of the fitness landscape near branching points precludes the coexistence of more than two types. With multi-dimensional traits, three or more coexisting types can arise in but a few mutational steps (Metz *et al.* 1996). However, we recently showed that the coevolving incipient species generically align in one dominant direction, making the process effectively one-dimensional.
- Starting with low diversity, many models for adaptive speciation show a quick decrease of the rate of speciation over evolutionary time as the community moves toward a joint ESS (see Box 18.2 for a heuristic explanation). Note, however, that other evolutionary attractors, e.g., evolutionary limit cycles (Dieckmann *et al.* 1995; van der Laan and Hogeweg 1995; Khibnik and Kondrashov 1997; Kisdi *et al.* 2001; Dercole *et al.* 2002; Mathias and Kisdi, in press) are also possible.

If the external environment does change on a time scale comparable to the initial divergence of new species, it generally precludes speciation from taking off (Metz *et al.* 1996), which can be understood as follows. Speciation only occurs at special trait values. On these points abut cones within which incipient species can, and outside of which they cannot, coexist (see Figure 4.2 and Box 4.7). Externally caused environmental changes move those cones around, away from the current pairs of incipient species, and by snuffing out one branch abort the speciation process before its completion.

If the environment changes sufficiently slowly, species keep tracking their adaptive equilibria till the equilibrium structure undergoes a qualitative change. This brings us in the domain of the bifurcation theory of adaptive dynamics (Box 4.6). Many phenomena seen in the fossil record may be of this type. Two special bifurcations deserve attention. First, if an ESS disappears in a merger with an evolutionary repellor, a punctuation event occurs: the species goes through a fast evolutionary transient toward another evolutionary attractor (Rand and Wilson 1993). Second, if an ESS changes into a branching point, a punctuation event starting with speciation is seen in the fossil record (Metz *et al.* 1996; Geritz *et al.* 1999).

Maire *et al.* 2001), prebiotic replicators (Meszéna and Szathmáry 2001), and many more. The evolutionary attractors that correspond to fitness minima found, for example, by Christiansen and Loeschcke (1980), Christiansen (1991), Cohen and Levin (1991), Ludwig and Levin (1991), Brown and Pavlovic (1992), Brown and Vincent (1992), Abrams *et al.* (1993a), Vincent *et al.* (1993), Doebeli (1996b), and Law *et al.* (1997) are all evolutionary branching points.

An important insight that emerges from adaptive dynamics is that evolution to a fitness minimum occurs frequently in eco-evolutionary models, suggesting that diversification by evolutionary branching may be common in nature. However, there are a number of caveats. Obviously, the accuracy of the prediction hinges on the assumptions made about the (physiological and other) trade-offs and other constraints on the evolving traits, as well as about the ecological interactions and population dynamics as determined by these traits. Most models predict evolution to a fitness minimum only in some parameter regions but not in others; it is usually difficult to make quantitative estimations of critical model parameters. In view of the often ingenious adaptations in nature, it seems unlikely that many species are persistently trapped at fitness minima, but empirical difficulties hinder measuring the actual shape of the fitness function. Divergence from the fitness minimum by evolutionary branching in diploid multilocus systems requires reproductive isolation, i.e., speciation (Doebeli 1996a; Dieckmann and Doebeli 1999; Chapter 5). If evolution to fitness minima are indeed common, and persistently maladapted species are indeed rare, then adaptive speciation may be prevalent.

Acknowledgments   This work was supported by grants from the Academy of Finland, from the Turku University Foundation, from the Hungarian Science Foundation (OTKA T 019272), from the Hungarian Ministry of Education (FKFP 0187/1999), and from the Dutch Science Foundation (NWO 048-011-039). Additional support was provided by the European Research Training Network *ModLife* (Modern Life-History Theory and its Application to the Management of Natural Resources), funded through the Human Potential Programme of the European Commission (Contract HPRN-CT-2000-00051).

5

Adaptive Dynamics of Speciation: Sexual Populations

Ulf Dieckmann and Michael Doebeli

5.1 Introduction

When John Maynard Smith (1966) wrote on sympatric speciation more than 35 years ago, he acknowledged that the argument "whether speciation can occur in a sexually reproducing species without effective geographical isolation" was an old problem and voiced his opinion that the "present distribution of species is equally consistent either with the sympatric or the allopatric theory." Yet, from the heyday of the Modern Synthesis until relatively recently, the importance of sympatric speciation has been downplayed, and the corresponding hypotheses remained obscure well beyond Maynard Smith's seminal study.

Looking back from today's perspective, it is astounding that, for such a long period, the research community at large essentially turned a blind eye to sympatric speciation. Given the widely acknowledged difficulties involved in inferring past process from present pattern, one can only feel uneasy about a logic that claims to find evidence for the prevalence of allopatric speciation in the present-day distribution of species. To a large extent it seems to have been the scientific community's perception of the *theory* of sympatric speciation that has brought about a profound skepticism toward the broader empirical relevance of this speciation mode. Scientific attempts to overcome this skepticism have come and gone in waves. In the 1960s, luminaries of North American evolutionary biology pulled no punches when assessing the merit of such attempts. Displaying a characteristic hint of restrained intimidation, Ernst Mayr (1963) wrote on sympatric speciation, "One would think that it should no longer be necessary to devote much time to this topic, but past experience permits one to predict that the issue will be raised again at regular intervals. Sympatric speciation is like the Lernaean Hydra which grew two new heads whenever one of its old heads was cut off." And also Theodosius Dobzhansky's verdict was categorical when he remarked, in 1966, that "sympatric speciation is like the measles; everyone gets it and we all get over it" (Bush 1998). Sometimes models of sympatric speciation were interpreted to imply that such speciation was only possible under very special and narrow conditions, while at other times the same models were called into question as they allegedly predicted sympatric speciation to happen too easily. In the words of Felsenstein (1981), "one might come away from some of these papers with the disturbing impression that [sympatric speciation] is all but inevitable." Such concerns were echoed again recently, by Bridle and Jiggins (2000), for example.

Sympatric speciation is contingent on constraints that are both ecological and genetic. An appreciation of the complementary character of these constraints helps to explain how the orientation of researchers toward ecological or genetic detail influenced their views on sympatric speciation. For instance, while Maynard Smith (1966) emphasizes that the "crucial step in sympatric speciation is the establishment of a stable polymorphism", Felsenstein (1981) contends that progress toward sympatric speciation ought to be measured in terms of the evolution of prezygotic isolation. Felsenstein even goes so far as to argue that, without genetic constraints on speciation, we ought to expect "a different species on every bush", and thus implies that the corresponding ecological constraints are fulfilled trivially. It seems that this latent disciplinary divide has not helped the subject, and lingers on today.

Perhaps the majority view of the past decades can be summarized crudely by four brief statements. First, most biologists remained deeply skeptical about sympatric speciation, in recognition that its relevance was highly contentious. Second, they believed that the ecological constraints for speciation to occur in sympatry were still somewhat more restrictive than admitted by proponents of this mode. Third, they insisted that the true and central challenge was to explain the emergence of reproductive isolation in sexual species and, fourth, there was a widespread impression that models of sympatric speciation failed as soon as genetic constraints were accounted for adequately.

Recent theoretical research is contributing to overcoming these concerns. Chapter 4 describes why the phenomenon of evolutionary branching through frequency-dependent selection provides a unifying framework within which to understand the ecological constraints on sympatric speciation – by explaining the dynamic emergence and subsequent perpetuity of disruptive selection in speciation events.

This chapter, in turn, shows how the genetic constraints on sympatric speciation are overcome more readily than earlier work had us believe. Thus, we claim no less than that previous reservations based on theoretical difficulties turn out to be unfounded upon closer, and more elaborate, inspection of the issues. We note that virtually all known theoretical examples, old and new, of sympatric speciation arise in the context of disruptive selection induced by frequency-dependent selection. Therefore, in adherence with our tenet that speciation research would benefit from concentrating on processes and mechanism, rather than on biogeographic patterns alone (Chapter 1), all the examples reviewed in this chapter must be recognized as representing instances of adaptive speciation. However, out of respect for the tradition of the field we retain, for the most part, the classic terminology of "sympatric" speciation. Section 5.2 starts out with a detailed review of the relevant genetic constraints and thus highlights the particular challenges that models of sympatric speciation have to meet if they are to be applied to sexual populations. After a short historical overview, we discuss ecologically and genetically explicit models that extend beyond the simplicity of earlier genetic studies. Two extensions are key:

- Earlier models involved only a few loci (usually two or three), each of which coded for a different phenotypic component through a very small number of alleles (usually just two). These simple approaches occasionally allowed some analytical treatment (e.g., reviewed in Christiansen and Feldman 1975), and were geared usefully to the capacities available for numerical simulation 20 years ago. Today, however, multilocus models that involve quasi-continuous quantitative characters seem more suitable and realistic.

■ The deterministic nature of earlier models also turned out to be problematic. Today, individual-based stochastic approaches are becoming increasingly tractable numerically. These latter models can exhibit dynamics qualitatively different from those predicted by their deterministic counterparts (Dieckmann and Doebeli 1999). This is not surprising since deterministic models, strictly speaking, only correctly describe infinitely large populations that do not contain ecological and genetic drift.

Section 5.2 concludes with a demonstration of how avoidance of these two pitfalls of over-simplification changes our views about the restrictiveness of genetic constraints on sympatric speciation. Underscoring the same general message again from the ecological end, Section 5.3 illustrates how easily a wide variety of ecological settings, which involve all three fundamental types of ecological interaction, can induce evolutionary branching in sexual populations.

Evolution in sexual populations is not always more restricted than in asexual ones. For models of sympatric speciation, the additional options for the emergence of prezygotic isolation presented by adaptation in sexual traits are especially interesting. In particular, it has been suggested that divergent Fisherian runaway processes in sexual traits can cause sympatric speciation, even in the absence of disruptive selection on any ecological character. Section 5.4 features a summary of these recent developments and explains how sexual selection and sexual conflict can facilitate speciation in sympatry. Also in such processes, frequency-dependent disruptive selection plays a central role, which reveals a fundamental similarity of the underlying mechanisms. Sympatric speciation driven by such frequency-dependent selection on ecological and sexual traits is adaptive (Chapters 1 and 19), and allows either an entire population or the separate sexes to escape from fitness minima.

5.2 Adaptive Speciation in Sexual Populations

The evolutionary force that favors sympatric speciation is disruptive selection, while the forces that oppose it, generally speaking, are segregation and recombination (Felsenstein 1981). The ecological constraints on sympatric speciation are therefore essentially given by the conditions required for a population to be exposed to disruptive selection for a sufficiently long period (Chapter 4), whereas the genetic constraints originate from a need to overcome segregation and recombination before a sexual population can split sympatrically (and potentially undergo character displacement). After examining in some detail the specific challenges posed to sympatric speciation by segregation and recombination, and after providing an overview of earlier work on this topic, this section describes the corresponding remedies.

The obstructive role of segregation and recombination

Speciation in sympatry requires reproductive isolation to arise between two incipient species, and so overcome the cohesion of a species' gene pool caused by segregation and recombination. In the initial phases of sympatric speciation, the divergent subpopulations are often separated by postzygotic isolation in the form of hybrid inferiority through disruptive selection. Enhancement of such initial, ecologically inflicted postzygotic isolation would typically occur through prezygotic isolation (see Box 6.1). Hence, the evolution of prezygotic isolation, through assortative mating in the broadest sense, lies at the core of the problem of segregation and recombination obstructing sympatric speciation in sexual populations. Assortativeness can either be tied directly to the species' ecology that causes the disruptive selection, or evolve independently. A well-known example of the former case is the famous *Rhagoletis* system (Bush 1975; Feder *et al.* 1988), in which the preferences of maggot flies for apple or hawthorn trees not only determine their feeding grounds, but also simultaneously restrict the types of mate they are likely to encounter. A typical example for the latter case is an ecologically neutral mating preference based on courtship behavior.

Below we consider general models characterized by two sets of loci that code, respectively, for an ecologically relevant phenotype (E-loci) and for assortative mating behavior (A-loci). Felsenstein (1981) focused on models that involved one E-locus and one A-locus. On this basis he introduced an important distinction between "speciation in which the reproductive isolating mechanisms come into existence by the substitution of different alleles in the two nascent species, and speciation in which the same alleles are substituted in both species." Felsenstein (1981) refers to these cases as "two-allele" and "one-allele" models, respectively. In "one-allele" models of sympatric speciation that involve two patches, the A-locus could code, for example, for the degree of patch philopatry or for the probability with which individuals settle in the patch they are best adapted to. By either means, two incipient species might evolutionarily restrict their mixing between the two patches, and thus enhance prezygotic isolation by exhibiting the same allele at their A-locus. In both cases, the degree of choice is determined by the A-locus, while the choice itself is pleiotropically affected by the E-locus. By contrast, in "two-allele" models the A-locus codes for the target of mate choice itself, such as for the chosen mating patch or for the partner's ecological phenotype, which requires incipient species to carry different alleles at this locus.

Despite its importance, the allele-centric terminology Felsenstein (1981) chose for his simple models is not ideal for a variety of reasons:

▪ When considering sympatric speciation in multilocus models, allele substitutions may be required at many loci, and thus necessarily involve more than just one or two alleles.
▪ The term "two-allele" model is potentially quite confusing in the context of models with diallelic loci.

■ It is actually immaterial whether alleles involved in substitutions are genotypically identical or different – it is only their effects on the mating phenotype that matter.

For these reasons it would be more accurate and less confusing to refer to Felsenstein's "one-allele" models as requiring *concordant* allele substitutions on the A-loci, while his "two-allele" models can lead to sympatric speciation only through *discordant* substitutions on these loci. Yet, since Felsenstein's terminology has become well established in the thinking of speciation scientists, we continue to refer to "one-allele" and "two-allele" models below, but the three provisos and the extended multilocus meaning of these traditional terms must be kept firmly in mind.

Returning to the general case that involves arbitrary numbers of E- and A-loci, we can now unravel the role of segregation and recombination in sympatric speciation processes into three components:

■ *Segregation and recombination among E-loci.* Intermediate ecological phenotypes can arise whenever two or more ecological alleles are involved per individual: they are generated through recombination if there is more than one E-locus, and through segregation if inheritance is diploid. For most genotype-to-phenotype maps a depression in the frequency of intermediates is swamped rapidly by offspring from parents with ecological phenotypes from either mode, which thus causes the modes to coalesce. Too much segregation or recombination, therefore, is bound to prevent the evolution of any bimodality in the ecological character. At the same time, however, some intermediates are required for the population to continue to experience the consequences of hybrid inferiority, which potentially increases prezygotic isolation through evolution on the A-loci. Therefore, a certain degree of segregation and/or recombination is indispensable for the evolution of prezygotic reproductive isolation during processes of sympatric speciation.

■ *Segregation and recombination between E- and A-loci.* In "two-allele" models – in which the E-loci affects the degree of postzygotic isolation and the A-loci that of prezygotic isolation – recombination and segregation tend "to break down the association between the prezygotic and postzygotic isolating mechanisms, so that it is always eroding the degree of progress toward speciation" (Felsenstein 1981). Again, therefore, segregation and recombination obstruct the speciation process. Sometimes, however, segregation and/or recombination between E- and A-loci are even needed for sympatric speciation. This is the case in a model by Dieckmann and Doebeli (1999) in which E-loci code for an ecological character under disruptive selection and A-loci determine an ecologically neutral marker character on which mate choice is based. In a process of selection-enhanced symmetry breaking (as explained below in the context of Figure 5.4a), segregation and recombination cause the correlation between ecological and marker characters to fluctuate enough for selection to maximize that linkage disequilibrium.

▩ *Segregation and recombination among A-loci.* Also for the loci involved in assortative mating, recombination and segregation can lead to the formation of intermediates. In "one-allele" models, recombination or segregation between the A-loci is not an issue since the corresponding alleles are expected to undergo concordant substitutions throughout the speciation process. The situation is very different in "two-allele" models, since here recombination among A-loci can create intermediates that suffer from decreased mating frequencies with partners from *both* incipient species, which thus introduces a cost to assortativeness that may prevent its evolutionary emergence.

This decomposition of effects allows us to draw the following conclusions:

▩ It is only in trivial haploid models with a single E-locus and a "one-allele" mechanism at the A-loci that segregation and recombination do not impede sympatric speciation.
▩ In all other "one-allele" models, segregation and recombination selectively favor the evolution of assortative mating because of hybrid inferiority, but hinder the evolution of bimodality in the ecological character. This means, incidentally, that the assertion by Felsenstein (1981) that only in "two-allele" models "does recombination act as a force to retard or block speciation" does not carry over to multilocus models.
▩ In all "two-allele" models, sympatric speciation is further obstructed by segregation and recombination between E- and A-loci, as well as among A-loci. At the same time, however, in some of these models the emergence of linkage disequilibria through selection-enhanced symmetry breaking relies on the fluctuations caused by segregation and recombination in finite populations.

Notice that the effects of segregation and recombination become more subtle if there is epistasis between the E- and A-loci, or if the E- or A-loci code for more than one quantitative character each. In the latter case, sympatric speciation may require some of these quantitative characters to evolve through concordant allele substitutions, while for other characters substitutions ought to be discordant. In such general cases, segregation and recombination have two effects: first, they potentially obstruct bimodalities in the individual characters that ought to undergo discordant substitutions, and second, they may weaken linkage disequilibria in pairs of such characters.

An overview of earlier studies

Many key ideas about sympatric speciation can be traced back to landmark papers by Maynard Smith (1966), Rosenzweig (1978), and Felsenstein (1981). Especially when the effects of segregation and recombination, described above, are considered these and other analyses offered divergent perspectives on the relative importance and implications of ecological and genetic constraints on the feasibility of sympatric speciation.

The pioneering study by Maynard Smith (1966) emphasized the ecological conditions required for the evolution, through hybrid inferiority, of stable polymorphisms between incipient species in two-niche models of soft selection (Levene 1953; see also Chapter 3 and Kisdi 2001). Based on diploid genetics with one E-locus and at most one A-locus, it was concluded that "the conditions which must be satisfied are [...] severe": density regulation must operate separately in the two niches, and the advantages of local adaptation to either niche must be large. Maynard Smith (1996) also put forward four mechanisms for the evolution of reproductive isolation:

■ Habitat choice – implying assortative mating as a by-product of an individual's fidelity to or preference for the habitat experienced after birth, and envisaged in terms of a "one-allele" model.
■ Pleiotropism – in which an allele that affects the ecological character under disruptive selection itself causes assortative mating, an option that does not require any A-loci and that Maynard Smith considered unlikely to occur in nature.
■ Modifier genes – a "one-allele" mechanism of assortative mating on the ecological character, in which the responsible allele is assumed to be dominant.
■ Assortative mating genes – a "two-allele" mechanism of assortative mating in which one of the two involved alleles is again assumed to be dominant.

Assessing the resultant genetic constraints for sympatric speciation, Maynard Smith admitted that for the "one-allele" and "two-allele" mechanisms to work, they must be supported by habitat choice. Also, without analyzing the "two-allele" model in full, he concluded that in such models habitat choice and direct assortative mating must be accurate enough and disruptive selection must be sufficiently strong for sympatric speciation to be initiated.

Maynard Smith's focus on Levene-type models of soft selection has been retained in most subsequent studies of sympatric speciation, as reviewed in Chapter 3. In particular, the synergistic interactions between habitat choice and "two-allele" mechanisms of assortative mating observed by Maynard Smith are highlighted in Section 3.5; see also Kawecki (1996, 1997) and Johnson *et al.* (1996b). Note that Maynard Smith himself had not yet presented a perspective in which the issue of segregation or recombination between E- and A-loci (especially in "two-allele" models) was placed at center stage.

"Two-allele" mechanisms of assortative mating also did not figure in an analysis by Dickinson and Antonovics (1973), which probed Maynard Smith's extension of Levene's model in greater detail. Instead, these authors explored, again based on simple diploid two-niche models of soft selection, the evolution of polygenic ecological characters coded for by up to three diallelic E-loci, of linkage between the involved loci, and of dominance relations between the involved alleles. They also considered an extra diallelic A-locus that determined selfing or assortative mating, but only according to "one-allele" mechanisms. Dickinson and Antonovics (1973) showed that a polymorphism could be maintained with strong selection and/or low

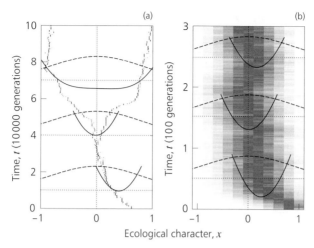

Figure 5.1 Evolutionary convergence to a fitness minimum. Gray scales indicate the frequency of phenotypes (the darkest shade corresponds to the highest frequency). (a) The asexual population first evolves toward $x_0 = 0$, the maximum of the carrying capacity $K(x)$ (dashed curves). Once this maximum is reached, directional selection turns disruptive, and the population finds itself exposed to a fitness minimum. As a result, the previously unimodal population becomes bimodal, and splits into two morphs through evolutionary branching. The overlays show the invasion fitness of mutant phenotypes (continuous curves) as generated by the ecological interactions with the resident phenotypes at three different points in time (indicated by horizontal dotted lines). (b) As in (a), but with multilocus genetics for the ecological character and with random mating. Shades of gray represent the phenotype distributions: with $l_E = 5$ diploid and diallelic loci, 11 ecological phenotypes can arise. Despite the continual action of disruptive selection at $x_0 = 0$, as shown by the invasion fitness of asexual mutants (continuous curves), segregation and recombination among the E-loci prevent the occurrence of evolutionary branching. For this reason, the sexual population remains trapped at the fitness minimum. Other parameters: $r = 1$, $K_0 = 500$, $\sigma_K = 2$, $\sigma_a = 0.8$, $\sigma_{as} = 2 \cdot 10^{-2}$, $u_{as} = u_s = 10^{-3}$. *Source:* Dieckmann and Doebeli (1999).

levels of gene flow between the niches, and that modifiers for linkage, dominance, selfing, and assortativeness could spread.

Speciation models based on multiple discrete niches are exposed occasionally to the criticism that what they really model is not sympatric, but at best microallopatric, or – if the niches considered are not finely entangled spatially – even parapatric speciation. A central contribution to the debate about the wider relevance of sympatric speciation was therefore Rosenzweig's (1978) model of competitive speciation, in which he argued that – even in a continuously structured niche space and when considering the evolution of a quantitative character in such a space – intraspecific resource competition could be expected to induce speciation. Rosenzweig (1978) stressed the fundamental capacity of frequency-dependent selection to re-mold the shape of fitness landscapes (e.g., as illustrated in Figures 1.1 and 5.1a): when an evolving population has attained a location on a fitness landscape that would be a peak in the absence of intraspecific competition, such competition can impose a dimple on the fitness landscape at the crowded location, as a result of which selection there becomes disruptive. This mechanism considerably broadens the range of ecological settings that could be expected to facilitate sympatric speciation. Taking an ecological perspective, Rosenzweig (1978) remained characteristically unconcerned about the evolution of reproductive isolation: mechanisms of assortative mating are not even mentioned, and effects of habitat choice are only discussed in passing.

A change in tack occurred with the studies of Udovic (1980), Felsenstein (1981), and Seger (1985a). All three articles focused on the conditions required for the evolution of prezygotic reproductive isolation driven by disruptive selection, and all explored "two-allele" mechanisms of assortative mating in a potentially panmictic population. The investigations show further similarities as they all relied on simple genetic models that involve one or two diallelic E- and A-loci, and thus on a coarse array of phenotypes. Another parallel between the setup of these models is that no provision was made to let the *degree* of assortative mating evolve gradually: usually, one allele was assumed to code for random mating behavior, while the other caused assortativeness at an externally fixed level. While Seger (1985a) focused on haploid inheritance, Udovic (1980) analyzed diploid genetics. Felsenstein (1981) mostly concerned himself with modeling haploids, but also presented results for the diploid case.

For one E-locus and one A-locus, Udovic (1980) investigated in detail how, in a diploid model, the reproductive isolation between two incipient species depended on the degrees of imposed disruptive selection, of frequency-dependent selection, of assortative mating, and of recombination. He concluded that the evolution of prezygotic isolation is constrained by a lower bound on the intensity of disruptive selection, and that the threshold value increased with the assumed recombination fraction and decreased with the amount of assortativeness assumed to be conferred by one of the alleles.

The seminal analysis by Felsenstein (1981) set the agenda for discussing sympatric speciation models for the subsequent two decades. Felsenstein numerically analyzed a haploid model with two E-loci and one A-locus by determining the degrees of disruptive selection and of assortative mating that would allow prezygotic isolation to become established. His results are in agreement with those of Udovic (1980), especially with regard to a "complex trade-off" between the strength of disruptive selection and the accurateness of assortative mating required for speciation. Felsenstein (1981) also looked at the robustness of his results with regard to recombination fractions, migration rates between the two niches, reversing the sequence of mating and migration in the species' life cycle, epistasis between the ecological loci, diploidy, and evolution at a modifier locus, which resulted in two diallelic A-loci and allowed three different levels of assortative mating. He concluded that "selection is at risk of being overwhelmed by recombination" and that "selection can proceed only when there is sufficiently strong selection at [the ecological loci], or sufficiently weak gene flow between the two nascent species."

Seger (1985a) studied a haploid model that also featured two diallelic E-loci and one A-locus. Perhaps inspired by Rosenzweig (1978), and in contrast to the analyses by Udovic (1980) and by Felsenstein (1981), Seger considered a more explicit and mechanistic ecological underpinning, based on intraspecific resource competition and involving ecological phenotypes derived from the corresponding genotypes by the imposition of environmental variation. Seger (1985a) assumed that this variation broadened the range of three ecological genotypes to five ecological phenotypes, which corresponded to adjacent intervals along a continuous resource axis. He also assumed that resource competition would only operate within these intervals, and considered different resource distributions across the intervals. He showed that the resultant frequency-dependent selection pressures could lead to sympatric speciation if the distribution of resources was overdispersed slightly relative to the phenotypic distribution that would occur under panmixia. Under this condition, "one-allele" and "two-allele" mechanisms of assortative mating could evolve. Seger (1985a) also investigated an extended scheme with two A-loci, in which one diallelic A-locus coded for an ecologically neutral marker trait while the other A-locus determined whether or not mating was assortative on that character. Although sympatric speciation could occur under the latter scheme (albeit very slowly), Seger expressed reservations about the conditions required and concluded that prezygotic isolation under such circumstances could not be expected to arise spontaneously.

It seems likely that Seger's results were affected by the artificial discretization of the resource space and by the low resolution of his set of genotypes. Two remedies are available for this problem: either a multitude of alleles can be considered, or many loci can be assumed to affect the characters under selection. When Kisdi and Geritz (1999) and Geritz and Kisdi (2000) re-analyzed the evolution of specialization in Levene's model of soft selection, they assumed a single E-locus, but (unlike previous work) they considered an infinite range of potential alleles on that locus. Envisaging that the evolution of these alleles would proceed by rare mutations that cause small steps in the ecological character only, Kisdi and Geritz (1999) showed the ecological constraints on sympatric speciation to be much less restrictive than previous research based on diallelic loci had suggested (see Box 4.7).

Kondrashov (1983a, 1983b) was the first to introduce diploid multilocus genetics systematically into models of sympatric speciation (see also Kondrashov 1986; Kondrashov and Mina 1986). Kondrashov's models do not use an explicit ecological embedding to provide a mechanistic and dynamic basis for regimes of disruptive selection: instead, an unspecified cause is assumed to favor marginal phenotypes at all times. Kondrashov's early studies considered only E-loci: the type and degree of assortative mating was not modeled as a quantitative character, and thus the evolution of assortativeness was not actually investigated. Kondrashov and Shpak (1998) showed that the types of assortative mating that can cause sympatric speciation when the ecological character is turned into an ecologically neutral trait are rather limited. Kondrashov and Kondrashov (1999) incorporated A-loci into the earlier models. In one scenario, these extra loci code for two quantitative characters that describe an ecologically neutral male trait (A1-loci) and a female preference for a male trait value (A2-loci). In a simplified scenario, only the male trait was modeled, while the male trait preferred by a female was supposed to be the trait it would display as a male. Sympatric speciation was found to occur in both scenarios, facilitated by strong disruptive selection, a high number of E-loci, and a low number of A-loci. Notice that in none of these models was the choosiness of female preference allowed to evolve – instead, a preexisting mechanism of assortative mating was assumed to operate before, throughout, and after the speciation process.

Work by Dieckmann and Doebeli (1999) integrated previous advances in the theory of sympatric speciation into a single framework. To investigate conditions for sympatric speciation in sexual populations, Dieckmann and Doebeli (1999) used a generic ecological embedding with frequency-dependent selection that arises from intraspecific resource competition (like Rosenzweig and Seger), considered mechanisms of assortative mating that required either concordant or discordant allele substitutions (like Felsenstein and Udovic), employed explicit multilocus genetics (like Kondrashov) for separate ecological and mating characters, and incorporated these elements into individual-based stochastic population dynamics. These models and the resultant findings are summarized in the following three subsections.

Asexual adaptive speciation through resource competition

We start from ecological assumptions that are likely to be satisfied in many natural populations. Individuals vary in an ecological character x that characterizes their resource utilization, such as when beak size in birds determines the size of seeds they can best consume. Populations that consist of individuals of a given trait value x have density-dependent logistic growth with carrying capacity $K(x)$, $dn/dt = rn[1 - n/K(x)]$, where n is the population's density. Individuals give birth at a constant rate r and die at a rate determined by frequency- and density-dependent competition. The resource distribution and thus the carrying capacity $K(x)$ are assumed to be unimodal (a multimodal resource distribution can be partitioned into unimodal segments, which may be analyzed separately). The carrying capacity depends on the ecological character x and varies according to a Gaussian function, $K(x) = K_0 \exp(-\frac{1}{2}(x - x_0)^2/\sigma_K^2)$, which peaks at an intermediate phenotype x_0 and has variance σ_K^2. The stable equilibrium density of a population monomorphic in x is thus $\hat{n} = K(x)$. Without loss of generality, we chose $x_0 = 0$.

In polymorphic populations that consist of subpopulations with different trait values x_i and population densities n_i, dissimilar individuals interact only weakly, as, for example, when birds with different beak sizes eat different types of seeds. This implies that competition is not only density-dependent, but also frequency-dependent, and rare phenotypes experience less competition than common phenotypes. Specifically, we assume that the strength of competition between individuals declines with phenotypic distance Δx according to a Gaussian function

$a(\Delta x) = \exp(-\frac{1}{2}\Delta x^2/\sigma_a^2)$ that peaks at 0 and has variance σ_a^2. Polymorphic population dynamics are then described by

$$\frac{dn_i}{dt} = rn_i[1 - \tilde{n}_i/K(x_i)] , \tag{5.1}$$

where the effective population density that affects individuals with ecological character x_i is given by $\tilde{n}_i = \sum_j a(x_j - x_i)n_j$, that is, by summing over all other competitors while weighting their impact in accordance with the competition function a (see Box 4.1 for an analysis of this model, as well as for references to the extensive earlier literature).

As a first step in our investigation, these assumptions are integrated into an asexual individual-based model, in which each individual is characterized by its trait value x (for implementation details, see Box 7.1). Evolutionary dynamics occur because offspring phenotypes may deviate slightly from parent phenotypes (offspring have the same ecological character as their parent, except when a mutation occurs at rate u_{as}, in which case their character is chosen from a normal distribution that peaks at the parent's character and has variance σ_{as}^2). The quantitative character first evolves to the value $x_0 = 0$, which confers maximal carrying capacity. After that, two things can happen: either x_0 is evolutionarily stable and evolution halts at this point, or x_0 is actually a fitness minimum and can be invaded by all nearby phenotypes. In the latter case, evolutionary branching occurs as shown in Figure 5.1a. This happens for $\sigma_a < \sigma_K$, that is, if the curvature of the carrying capacity at its maximum is less than that of the competition function. Under this condition the advantage of deviating from the crowded optimal phenotype x_0 more than compensates for the disadvantage of a lower carrying capacity.

The incidence of evolutionary branching observed in the individual-based asexual model can be predicted as follows. When a rare mutant x' appears in a population that is monomorphic for the ecological character x at carrying capacity $K(x)$, it competes with the discounted density $a(x - x')K(x)$, and the per capita growth rate $f(x', x)$ of the rare mutant x' (i.e., its invasion fitness, see Section 4.2) is given by $r[1 - a(x - x')K(x)/K(x')]$ (see also Box 4.1). The derivative $g(x)$ of $f(x', x)$ with respect to the mutant character x', evaluated at the resident character x, $g(x) = \partial f(x', x)/\partial x'|_{x'=x} = rK'(x)/K(x)$, is positive for $x < x_0$ and negative for $x > x_0$. Therefore, x_0 is an attractor for the monomorphic adaptive dynamics. In addition, if $f(x', x)$ has a minimum at $x' = x_0$, then x_0 is an evolutionary branching point (Box 4.3). The fitness minimum occurs if and only if $\sigma_a < \sigma_K$.

As mentioned above, a verbal account of this scenario of sympatric speciation in an asexual population had already been provided by Rosenzweig (1978), who remarked on the ecological character x_0 focusing utilization on the carrying capacity's peak: "[I]f its degree of specialization is sufficient relative to the breadth of its original Wrightian peak, it has created a dimple in the surface: the fitness of phenotypes to its left and right are higher than its own." The "surface" here has become formalized as the invasion fitness f in adaptive dynamics theory (see Chapter 4). The "breadth of its original Wrightian peak" has to be equated with the width σ_K

of the carrying capacity function K, which remains unaffected by density regulation. And finally, the character's "degree of specialization" is measured by the width σ_a of the competition function a. Although Rosenzweig's pioneering work did not anticipate the abstraction and generality associated with the phenomenon of evolutionary branching, which can originate from many causes other than resource competition, he had already captured the conceptual essence of evolution in the resource utilization model described above.

No adaptive speciation under random mating

As a second step, sexual reproduction is incorporated into the model by assuming that the ecological character $-1 \leq x \leq +1$ is determined by l_E equivalent diploid and diallelic E-loci with additive effects and free recombination. With a small probability u_s, a mutation occurs in the inherited alleles and reverses their value. Only female individuals are modeled, while the genotypic distribution of males across the considered loci is assumed to match that of females. Alternatively, one could look upon this model as describing hermaphrodites.

In both the sexual and the asexual case, the population evolutionarily shifts its mean phenotype to x_0. However, if mating in the sexual population is random, evolutionary branching does not occur for any values of σ_K and σ_a (Figure 5.1b). As explained above, the evolution of two modes in the frequency distribution of the ecological character is prevented by the continual generation of intermediate phenotypes through segregation and recombination – even though, as shown in Figure 5.1b, the population continues to experience the disruptive selection associated with its exposure to a fitness minimum.

Adaptive speciation through evolution of assortative mating

To model the evolution of assortative mating, and thus of prezygotic isolation, we assume, as a third step, that individuals express additional quantitative characters that determine mating probabilities according to two scenarios:

- In the first scenario, mating probabilities are based on similarity in the ecological character x. The degree of choosiness involved in assortative mating is described by a mating character y that is modeled as a second quantitative trait and is free to evolve.
- In the second scenario, mating probabilities are based on similarity in a third independent quantitative character $\tilde{x}$ that describes an ecologically neutral marker trait. In this latter scenario, both the level of choosiness in the mate choice and the marker trait itself can evolve.

Mating character y and marker trait $\tilde{x}$ are underpinned by the same kind of multilocus genetics as the ecological character described before and are determined by l_{A1} A1-loci and l_{A2} A2-loci, respectively. The effect of the mating character is detailed in Figure 5.2; it allows the mating behavior of individuals to change gradually from negative assortativeness over random mating to positive assortativeness. Considering a mating character that describes the degree of choosiness

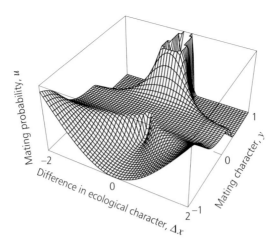

Figure 5.2 Modeling the degree of assortative mating. The mating character $-1 \le y \le +1$ is given by the difference between the number of $+$ and $-$ alleles divided by the total number of alleles at the A1-loci. For $y > 0$, mating probabilities in the first scenario depend on the difference Δx in the ecological trait according to $\exp(-\frac{1}{2}\Delta x^2/\sigma_+^2)$, with $\sigma_+ = 1/(20y^2)$. Hence, individuals that carry mostly $+$ alleles at their A1-loci mate assortatively: the mating probability increases with phenotypic similarity to the partner. For $y = 0$, mating is random (i.e., independent of the partner's ecological phenotype). For $y < 0$, mating probabilities increase with the difference Δx according to $1 - \exp(-\frac{1}{2}\Delta x^2/\sigma_-^2)$, with $\sigma_- = y^2$. Individuals that carry mostly $-$ alleles at their A1-loci thus mate disassortatively, being more likely to mate with partners with ecological phenotypes different from their own. In the second scenario, mating probabilities depend on the difference $\Delta \tilde{x}$ in an ecologically neutral marker trait instead of on Δx. To avoid a bias against marginal phenotypes, mating probabilities are normalized, so that their sum over all potential partners equals 1 for all phenotypes. *Source*: Dieckmann and Doebeli (1999).

enables us to study sympatric speciation without having to assume a preexisting level of assortative mating: initially our population is genetically coded to mate completely randomly.

Sympatric speciation in the first scenario requires concordant substitutions at the A1-loci toward higher degrees of choosiness, similar to Felsenstein's (1981) "one-allele" models. By contrast, the second scenario also necessitates concordant substitutions at the A1-loci, but in addition requires discordant substitutions at the A2-loci, at which a bimodal distribution of marker traits has to evolve and become correlated with the ecological character. The need for the incipient species to undergo such discordant allele substitutions means the second scenario is more akin to Felsenstein's (1981) "two-allele" models.

For the first scenario, Figure 5.3a shows the evolutionary dynamics of an initially randomly mating population that starts away from the evolutionary branching point. While the ecological character evolves toward x_0, the mating character

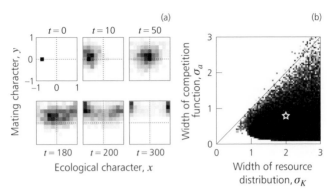

Figure 5.3 Adaptive speciation through resource competition, with assortative mating acting on the ecological character under disruptive selection. (a) The mean ecological character first evolves to the evolutionary branching point (50 generations); then the mean mating
character increases to positive values (180 generations), which allows frequency-dependent
disruptive selection to cause speciation (300 generations). Gray scales indicate the frequency of phenotypic combinations (black corresponds to the highest frequency). *Source*:
Dieckmann and Doebeli (1999). (b) Combinations of standard deviations of carrying capacity and competition function that allow for adaptive speciation. Gray scales indicate the time
to speciation (black corresponds to 0 generations and white to 20 000 or more generations).
In the corresponding asexual model, evolutionary branching happens for combinations that
lie below the diagonal. Parameters are as in Figure 5.1, with $l_{A1} = 5$. The asterisk in (b)
shows the parameter combination used in (a).

initially changes only slowly, but picks up speed and evolves toward positive assortativeness when the mean of the ecological character reaches x_0. Once assortativeness is strong enough, the population splits into two ecologically different
morphs, which eventually become almost completely reproductively isolated. As
explained above, near the dynamically emerging fitness minimum at x_0, selection
favors mechanisms that allow for a split in the phenotype distribution and hence for
an escape from the fitness minimum: assortative mating is such a mechanism, because it prevents the generation of intermediate offspring phenotypes from extreme
parent phenotypes. Figure 5.3a shows that alleles which induce a high degree of
positive assortative mating concordantly evolve in both incipient species.

Parameter requirements for evolutionary branching in the first scenario for the
evolution of sexual populations are not much more demanding than those in the
asexual case, as is shown in Figure 5.3b. There is one qualification, however.
Selection for a polymorphic sexual population is disruptive only when the con-
volution $n * a$ of the phenotypic distribution $n(x)$ and the competition function
$a(\Delta x)$ are narrower than the carrying capacity function $K(x)$ (G. Meszéna, personal communication). With Gaussian functions, this translates into the condition
$\sigma_a^2 + \sigma_n^2 < \sigma_K^2$, where σ_n^2 is the variance of the phenotypic distribution when its
mode has converged to x_0. Therefore, when σ_a is just slightly smaller than σ_K,

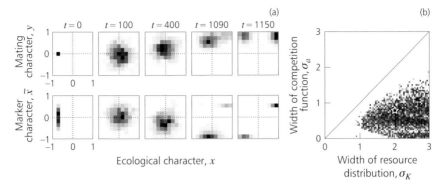

Figure 5.4 Adaptive speciation through resource competition, with assortative mating acting on an ecologically neutral marker trait. (a) The mean ecological character first evolves to the evolutionary branching point (100 generations). As a result of temporary correlations between marker trait and ecological character, assortative mating increases, which in turn magnifies these correlations (generations 400 to 1090). This positive feedback eventually leads to speciation (1150 generations). Gray scales indicate the frequency of phenotypic combinations (black corresponds to the highest frequency). *Source*: Dieckmann and Doebeli (1999). (b) Combinations of standard deviations of carrying capacity and competition function that allow for adaptive speciation. Gray scales indicate the time to speciation (black corresponds to 0 generations and white to 20 000 or more generations). In the corresponding asexual model, evolutionary branching happens for combinations that lie below the diagonal. Parameters are as in Figure 5.1, with $l_{A1} = l_{A2} = 5$. The asterisk in (b) shows the parameter combination used in (a).

(i.e., when disruptive selection in the asexual model is weak), the sexual population may not actually experience disruptive selection at all (Matessi *et al.* 2001). Yet, as we can see in Figure 5.3b, in the setting we study here this effect does not lead to much of a reduction in the conditions that result in adaptive speciation in the first scenario.

Although Felsenstein's (1981) criticism of the biological relevance of "one-allele" models seems exaggerated – in particular because host races seem so widespread among insects (e.g., Berlocher and Feder 2002), and because assortativeness on body size is ubiquitous in animals (e.g., Schliewen *et al.* 2001) – it is nevertheless interesting to check whether the results described above carry over to the "two-allele" model in the second scenario. Felsenstein's (1981) general conclusion that sympatric speciation requiring discordant allele substitutions is very difficult was based crucially on his analysis of simple deterministic models that involved only two loci and two alleles per locus. We thus have to ask whether Felsenstein's (1981) time-honored conclusion stands up in the less restrictive context of quasi-continuous characters and stochastic multilocus genetics. As is shown below, the answer is negative.

For the second scenario, Figure 5.4a illustrates that in this case also the selective amplification of ecological and genetic drift in finite populations readily leads to speciation, despite the opposing forces of segregation and recombination. Such drift temporarily results in small and localized linkage disequilibria between some A2-loci and some E-loci. Both positive and negative correlations select for assortative mating, which in turn magnifies the local disequilibria into a global linkage disequilibrium between ecological character and marker trait. This feedback eventually induces a sympatric split into reproductively isolated phenotypic clusters. Thus, stochastic fluctuations in finite populations can spontaneously break the symmetry of locally stable linkage equilibria observed in deterministic models. Notice that even though the linkage equilibrium in the second scenario is stable, it is surrounded by but a small basin of attraction. Once this small domain is left through fluctuations, the resultant linkage disequilibrium is deterministically and swiftly amplified by selection. This result highlights a trade-off involved in using deterministic population models (e.g., like the one in Drossel and McKane 2000), which cannot easily capture processes of symmetry breaking that crucially rely on finite fluctuations for their initiation. In the present context, deterministic models could, in principle, be used to assess whether an initially small but finite linkage disequilibrium becomes large enough to allow for speciation, but they cannot be used to investigate the initial appearance of linkage disequilibria through stochastic effects, and thus cannot capture the speciation process in full. Note that in the second scenario, parameter requirements for evolutionary branching in sexual populations are somewhat more restrictive than those in the asexual case; information on this is summarized in Figure 5.4b.

Dieckmann and Doebeli (1999) also showed that adaptive speciation in the second scenario occurs more rapidly when fewer loci underlie the three quantitative characters (Figures 5.5b and 5.5c). This is because fewer loci allow larger phenotypic effects of drift and are thus more likely to trigger the process of selection-enhanced symmetry breaking. An exception to this general tendency occurs for the extreme case $l_E = 1$, where only a single diallelic locus codes for the ecological character, which results in no more than three ecological phenotypes. Ironically, it was this special case that the influential analysis by Felsenstein (1981) relied on. However, Felsenstein did not even have the opportunity to realize that sympatric speciation was possible, even for $l_E = 1$, since he restricted his attention to purely deterministic models. Felsenstein's conclusions – putting "two-allele" mechanisms of sympatric speciation into disrepute for the next two decades – thus originated from an unfortunate confluence of restrictive modeling assumptions. Since the relative effects of drift that triggered the selection-enhanced symmetry breaking are most pronounced when populations are small, waiting times until adaptive speciation in the second scenario grow as population size increases. In the first scenario, the effect of loci numbers on speciation speed exhibits a pattern roughly similar to that in the second scenario (without a special role for $l_E = 1$). Here, however, times to speciation are considerably shorter – on the order of hundreds, rather than thousands, of generations (Figure 5.5a). It also turns out that in the first scenario there is no significant effect of population size on speciation speed.

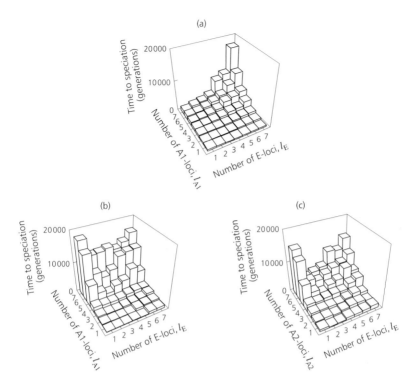

Figure 5.5 Dependence of time to speciation on the number of loci that affect the in-volved quantitative traits. (a) First scenario (assortative mating acts on ecological charac-ter); dependence on the number l_E of ecological loci and on the number l_{A1} of mating loci. (b) Second scenario (assortative mating acts on marker trait); dependence on the number l_E of ecological loci and on the number l_{A1} of mating loci for a fixed number of marker loci, $l_{A2} = 5$. (c) Second scenario; dependence on the number l_E of ecological loci and on the number l_{A2} of marker loci for a fixed number of mating loci, $l_{A1} = 5$. Other parameters are as in Figure 5.1. Notice that the vertical scale in (a) differs from that in (b) and (c), and that – with the important exception of the special case $l_E = 1$ – more loci always imply slower speciation. *Source (b), (c)*: Dieckmann and Doebeli (1999).

Notice that the model described above differs from the majority of previous theoretical studies on sympatric speciation in that it does not require two discrete patches or habitats, since resources in the continuum of types considered in the model are not spatially segregated. This means that there is no opportunity for populations to become allopatric (or parapatric) by very strong habitat choice or through very low migration. This makes it easier to appreciate that this model does not deal with allopatric (or parapatric) speciation in a "microalloptric" guise. Notice also that the results above remain qualitatively unchanged when assortative mating involves a cost of choosiness, as long as this cost is not too large. Such a cost of choosiness comes on top of the already present cost of rarity: when mating is assortative, rare phenotypes suffer from being chosen as mates with reduced probability. Obviously, large costs to assortative mating make its evolution less likely.

In summary, results for the two scenarios demonstrate that the genetic obstacles to sympatric speciation can, indeed, be overcome by sexual populations with stochastic multilocus genetics. What is more, the analysis above suggests sympatric speciation that requires either concordant or discordant allele substitutions can occur under realistic ecological and genetic conditions, and that even the degree of assortative mating gradually evolves as required for sympatric speciation.

5.3   Coevolutionary Adaptive Speciation in Sexual Populations

While classic models used to study the processes of diversification focused on competition (e.g., MacArthur and Levins 1967), other types of ecological interactions have received less attention. We now illustrate how selection regimes that lead to evolutionary branching are expected to arise readily from a wide variety of different ecological interactions within and between species.

Specifically, Doebeli and Dieckmann (2000) demonstrated the potential for evolutionary branching in models for mutualism and for predator–prey interactions. In the models reviewed below, coevolutionary dynamics of quantitative characters in two separate species are driven by interspecific ecological interactions; frequency-dependent selection can then result in convergence to an evolutionary attractor at which either one or both species find themselves at fitness minima. This may lead to evolutionary branching in only one species, or in both species simultaneously, or in both species sequentially.

As shown by Law *et al.* (1997) and Kisdi (1999), combinations of intra- and interspecific competition can also lead to evolutionary branching. When competition is asymmetric, the resultant evolutionary dynamics are particularly prone to cycles of evolutionary branching and selection-driven extinction.

Adaptive speciation through mutualistic interactions

Extending population dynamic studies by Vandermeer and Boucher (1978) and by Bever (1999), Doebeli and Dieckmann (2000) analyzed the potential for evolutionary branching in the following Lotka–Volterra system,

$$\frac{dn_{1i}}{dt} = r_1 n_{1i}\left[1 - \sum_j n_{1j}/K_1(x_{1i}) + \sum_j a_1(x_{1i}, x_{2j})n_{2j}\right], \tag{5.2a}$$

$$\frac{dn_{2i}}{dt} = r_2 n_{2i}\left[1 - \sum_j n_{2j}/K_2(x_{2i}) + \sum_j a_2(x_{2i}, x_{1j})n_{1j}\right], \tag{5.2b}$$

where n_{1i} and n_{2i} are the population densities of mutualists in the two species with ecological characters x_{1i} and x_{2i}, respectively; r_1 and r_2 are birth rates, and K_1 and K_2 are trait-dependent carrying capacities. The indices i and j range over all different ecological characters present in the two species, so that the second and third terms in the square brackets in Equations (5.2a) and (5.2b) determine, respectively, a death rate that results from intraspecific competition and an extra birth rate that results from interspecific mutualism. The mutualistic interaction between the two species is determined by the interaction functions a_1 and a_2, which depend on the ecological characters, as follows

$$a_1(x_{1i}, x_{2j}) = a_{10}\frac{\alpha(x_{1i} - x_{2j})}{\sum_k \alpha(x_{1k} - x_{2j})n_{1k}} , \tag{5.2c}$$

with an analogous expression for $a_2(x_{2i}, x_{1j})$. Here α is a normal distribution with mean 0 and variance σ_a^2 – this shape implies that the mutualistic support between individuals decreases when their ecological characters move apart, with a tolerance given by σ_a. Following Kiester *et al.* (1984), it is assumed that the total mutualistic support an individual from one species provides to the other is fixed and given by a_{10} and a_{20}, respectively. This assumption gives rise to the denominator in the equation above.

The trait-dependent carrying capacities are given by

$$K_1(x_{1i}) = K_{10} + K_{11}\beta(x_{1i}) , \tag{5.2d}$$

with an analogous expression for $K_2(x_{2i})$. Here β is a normal distribution with maximum 1, mean x_{01}, and variance σ_{K1}^2: this shape implies that the carrying capacity in the first species possesses a baseline at the background level K_{10}, and increases, with a weighting factor K_{11} and a tolerance σ_{K1}, when the ecological character x_{1i} moves toward the carrying capacity's peak at x_{01}.

When the maxima of the mutualists' carrying capacities are different ($x_{01} \neq x_{02}$), finite tolerances σ_a, σ_{K1}, and σ_{K2} result in a tension between each species' advantage of adapting its ecological character to the peak of its own carrying capacity and the mutualistic benefit that it can reap from minimizing the distance between its ecological character and that of its mutualistic partner. When this tension is strong enough, gradual adaptations in x_1 and x_2 can trap either one or both species at fitness minima, and thus set the ecological stage for evolutionary branching. Introducing stochastic individual-based dynamics, multilocus genetics, and the evolution of assortative mating (as described in Section 5.2), Doebeli and Dieckmann (2000) confirmed that these situations can readily lead to sympatric speciation in sexual populations of the two mutualists.

As illustrated in Figure 5.6, the evolutionary outcome that results from such a speciation process is interesting: evolutionary branching creates two species pairs, in each of which one of the species is close to the peak of its carrying capacity and can thus provide much mutualistic support to the other species in the pair, which in turn is far away from the peak of its carrying capacity and therefore can give little support. Thus, in each of the two original species, speciation brings about a "supportive branch" and an "exploitative branch", and thereby breaks the community's symmetry prior to branching in each resultant species pair.

Adaptive speciation through predator–prey interactions

Evolutionary branching can also be induced by the ecological interactions between predators and their prey. Extending work by Brown and Vincent (1987, 1992), Marrow *et al.* (1992, 1996), Saloniemi (1993), Dieckmann *et al.* (1995),

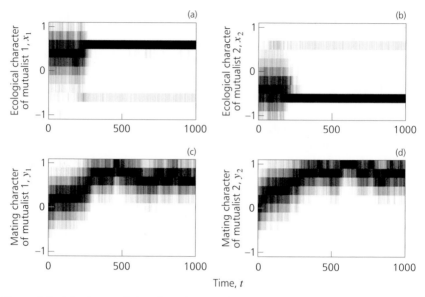

Figure 5.6 Adaptive speciation through mutualistic interactions. (a, c) Evolution of ecological characters jointly affects the compatibility of the two mutualists and their carrying capacities. (b, d) Evolution of the degree of assortative mating on the ecological characters. Both mutualists speciate simultaneously once their degree of assortative mating becomes high enough. Parameters: $r_1 = r_2 = 1$, $K_{10} = K_{20} = 300$, $K_{11} = K_{21} = 400$, $x_{01} = -0.5$, $x_{02} = +0.5$, $\sigma_{K1} = \sigma_{K2} = 0.5$, $a_{10} = a_{20} = 0.00016$, $\sigma_a = 0.2$, $l_E = l_{A1} = 5$, $u_s = 10^{-3}$. *Source*: Doebeli and Dieckmann (2000).

and Doebeli (1997), Doebeli and Dieckmann (2000) assumed that predation efficiency depends on two ecological characters, x_1 and x_2, the first in the prey and the other in the predator. These characters are scaled such that the interaction is the stronger the more similar these characters are. The ecological dynamics of a polymorphic predator–prey community can be described by the Lotka–Volterra equations

$$\frac{dn_{1i}}{dt} = rn_{1i}\left[1 - \sum_j n_{1j}/K(x_{1i}) - \sum_j a(x_{1i} - x_{2j})n_{2j}\right],\tag{5.3a}$$

$$\frac{dn_{2i}}{dt} = n_{2i}\left[-\delta + c\sum_j a(x_{1j} - x_{2i})n_{1j}\right].\tag{5.3b}$$

Here the predation efficiency a depends on the character difference between prey and predator according to a normal distribution with maximum a_0, mean 0, and variance σ_a^2, while the prey's carrying capacity function K is a normal distribution with maximum K_0, mean x_0, and variance σ_K^2. The densities of prey and predator populations with ecological characters x_{1i} and x_{2i} are denoted by n_{1i} and n_{2i}, respectively, while r, δ, and c measure the prey's intrinsic birth rate, the predator's intrinsic death rate, and the predator's conversion efficiency of prey resources. The

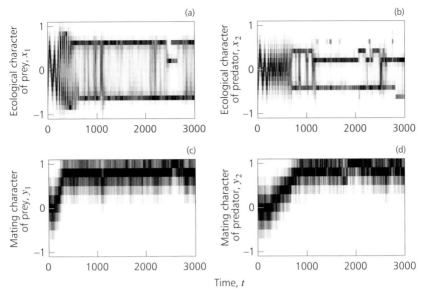

Figure 5.7 Adaptive speciation through predator–prey interactions. (a, c) Evolution of ecological characters jointly affects the prey harvesting efficiency of the predator and the carrying capacity of the prey. (b, d) Evolution of the degree of assortative mating on the ecological characters. After the prey has been evolutionarily chased and caught by the predator, and its degree of assortative mating has become high enough, it speciates, thus temporarily reducing the predator's harvesting efficiency. With the prey having split up, the predator's degree of assortative mating also increases, allowing it to speciate as well, which again increases its (now separate) harvesting efficiencies on each of the prey species. Parameters: $r = 1$, $\delta = 1$, $c = 2$, $K_0 = 2000$, $x_0 = 0$, $\sigma_K = 0.7$, $a_0 = 0.001$, $\sigma_a = 0.4$, $l_E = 10$, $l_{A1} = 5$, $u_s = 10^{-3}$. *Source*: Doebeli and Dieckmann (2000).

indices i and j again range over all different ecological characters present in the two species. The second and third terms in the prey equation determine death rates that result, respectively, from intraspecific competition and from the predator's harvesting. Correspondingly, the sum in the predator equation determines a birth rate that results from harvesting the prey.

Notice that we do not assume any frequency dependence in the competitive interactions among the prey, which could be an independent cause for prey diversification even in the absence of predators. Yet, the existence of the predator imposes frequency-dependent selection on the prey through apparent competition, because common prey phenotypes have the disadvantage that the predator phenotype that preys upon them most efficiently is thriving. Doebeli and Dieckmann (2000) showed that this frequency dependence leads to evolutionary branching in the prey if predation efficiency decreases sufficiently fast with increasing distance between the ecological characters of prey and predator.

Figure 5.7 shows the resultant process of sympatric speciation, based on the same extension to sexual populations as described in Section 5.2. Notice that in this example it is the primary evolutionary branching in the prey that induces evolutionary branching in the predator. This happens if the phenotypic distance between the two emerging prey species becomes large relative to the width σ_a of the predation efficiency function (Doebeli and Dieckmann 2000). In this case the prey's evolution establishes the predator at a fitness minimum (see also Abrams and Matsuda 1996). The slightest polymorphism in the prey, maintained by mutation–selection balance, allows the coexistence of predator traits that straddle this fitness minimum, and thus allows the predator to undergo secondary evolutionary branching (É. Kisdi, personal communication). By contrast, if the two prey species remain sufficiently close to each other, the predator population does not experience disruptive selection and instead becomes a generalist (with predation efficiencies on each of the two emerging prey species significantly lower than the one it had on the single prey species prior to branching).

5.4   Adaptive Speciation through Sexual Selection

Darwin (1859, 1871) introduced the notion of sexual selection to explain the conspicuous sexual dimorphisms found in many species. Sexual selection arises from differences in reproductive success caused by competition for fertilization (Andersson 1994; see also Maynard Smith 1991); such selection can have both intrasexual and intersexual components. The former arise from the interaction between individuals of the same sex in the course of their competition for mates, while the latter originate from the interaction between the sexes when mates are actually chosen. Intersexual selection can be caused by sensory bias or drive (Ryan and Rand 1990; Endler 1992; Boughman 2001, 2002), which results in an organism's intrinsic preference for mates whose phenotypes stimulate its sensors where the sensors' sensitivity is highest.

Sexual selection is sometimes envisaged as being fundamentally different from natural selection. One reason might be Darwin's belief, still shared by some biologists, that only sexual selection can lead to the evolution of traits that appear irrelevant or even disadvantageous to their bearers, whereas "natural selection will never produce in a being anything injurious to itself" such that "any variation in the least degree injurious would be rigidly destroyed" (Darwin 1859). Today, we understand that both natural and sexual selection can harm a population by letting it converge to fitness minima (Sections 4.4 to 4.7 and 5.2 to 5.3) and even by actively driving it to extinction (Matsuda and Abrams 1994a, 1994b; Ferrière 2000; Dercole et al. 2002; Gyllenberg et al. 2002; Dieckmann and Ferrière 2004). It is therefore appropriate to look upon the contrast between sexual and natural selection as being no more than a distinction that concerns the particular life-history component exposed to selection (Barnard 1998).

A key phenomenon that can arise from the interplay of intrasexual and intersexual selection is Fisher's "runaway process" (Fisher 1915, 1930). At the onset of such a process, some females – for instance, because of a sensory bias – happen to mate preferentially with males that possess a certain trait or ornament: the resultant offspring inherits the genes both for the mother's preference and for the father's ornament, which results in a positive feedback that can fuel the further concerted evolution of male trait and female preference. While the male benefit that drives the runaway process is simply access to females, the female benefits can be more diverse – ranging from direct protection through the male to indirectly improved fitness for the female's offspring. Such indirect female benefits can arise either through an increased likelihood of mating with males that can afford to sport the preferred trait because of their above-average fitness ("handicap principle"; Zahavi 1975, 1977; Hamilton and Zuk 1982; Zahavi and Zahavi 1997) or through a greater attractiveness of their male offspring to other choosy females ("sexy son hypothesis"; Weatherhead and Robertson 1979). While the first effect relies on a correlation between male trait and male fitness, the second effect can involve a completely arbitrary, ecologically neutral male trait. A Fisherian runaway process could come to an end when genetic variance is exhausted (Kirkpatrick and Ryan 1991), developmental constraints are encountered, or costs start to exceed benefits – in particular when the male trait cannot be enhanced any further without exposing the male to strongly detrimental natural selection. With Fisher himself having provided a verbal argument only, mathematical models of Fisherian runaway were introduced and analyzed by O'Donald (1962, 1967, 1977, 1980), Lande (1980, 1981), and Kirkpatrick (1982b); see also Seger (1985b), Lande and Kirkpatrick (1988), Endler (1989), Pomiankowski *et al.* (1991), and Iwasa *et al.* (1991).

Mechanisms by which, not just one, but two or more Fisherian runaway processes can be triggered in a population are of particular interest in speciation theory, because of their potential to explain the evolution of prezygotic isolation. In the next two subsections we therefore briefly review speciation models based on the divergent evolution of mating preferences and preferred traits (see also Section 3.3). Speciation models based on sexual selection in spatially extended populations – including those by Lande (1982) and by Payne and Krakauer (1997) – are reviewed in Chapter 7. For a survey of empirical evidence that implicates sexual selection in speciation, see Panhuis *et al.* (2001).

Adaptive speciation through mate competition

Lande's original model of Fisherian runaway with a line of neutral equilibria (Lande 1981), as well as many subsequent investigations (e.g., Kirkpatrick 1982b and Seger 1985b), only led to genetic drift stochastically moving the coadapted combinations of male trait and female preference along this line. Soon it was discovered, however, that any slight cost to female choosiness reduced such neutral lines of equilibria to one or two asymptotically stable equilibrium points (Pomiankowski 1987; Bulmer 1989). With neutrality gone, populations that undergo separate Fisherian runaway can no longer diverge through genetic drift, which obliterates the potential of Lande's model to explain speciation. Also, being crucially based on neutral equilibria, Lande's speciation mechanism could provide no basis for the coexistence of the resultant species in stable polymorphisms. Other evolutionary forces were thus required to explain reproductive isolation between, and coexistence of, populations that undergo separate Fisherian runaway.

Such other forces were introduced in models by Turner and Burrows (1995) and by Higashi *et al.* (1999). Both studied the evolution through sexual selection of a polygenic male ornamental trait in finite populations of diploid individuals. The loci underlying the male trait were assumed to act additively and undergo free recombination, with intermediate male phenotypes living longer, and extreme phenotypes on either side of the male spectrum experiencing highest mortality. Populations were assumed to be polygynous, such that females examined many males before choosing one to mate with, and the resultant offspring had a 1:1 sex ratio.

In the study by Higashi *et al.* (1999), female preference for a particular male trait had the same genetic underpinning as the male trait itself, and females could examine the entire population of males before choosing a mate; mating probabilities decreased for males that did not meet their preference. Higashi *et al.* showed that, when starting from males with intermediate phenotypes and with females that preferred these males, both the male and the female character distribution could become bimodal, which resulted in reproductive isolation and thus speciation. The results by Higashi *et al.* (1999) are robust under the introduction of a cost of female choosiness, since, once the distributions of male trait and female preference have become bimodal, this cost can be compensated by the benefit of not producing unmated hybrid offspring. For a deterministic version of this model, see Takimoto *et al.* (2000).

By contrast, Turner and Burrows (1995) modeled female preference for either of two extreme male phenotypes through a single dominant allele; they allowed females to choose the best-matching male out of a subset of males ("best-of-*n*" rule), and assumed that Fisherian runaway had already brought about an extreme male phenotype together with the matching female preference. Turner and Burrows (1995) showed that, occasionally, a mutant female, which preferred the other end of the male spectrum to the resident females, could invade; this resulted, once again, in two reproductively isolated species with matching male traits and female preferences. The mutant females could invade since they were driven to mate with less extreme males, and therefore their male offspring enjoyed better survival than that of the resident females.

Both of these models are not without problems. The model by Turner and Burrows (1995) crucially relies on female preference being reversible through a single mutation – an assumption that, although perhaps justified in some natural systems, does not seem very general. The model by Higashi *et al.* (1999) suffers from the constraint that the initial distributions of male trait and female preference must be fine-tuned for speciation to occur. Unless female preference is more or less symmetrically distributed about the mode of the male distribution, no disruptive selection on males occurs and no speciation ensues. An interesting difference between the two approaches is that a reduced survival cost for extreme male phenotypes hinders speciation in the model by Turner and Burrows, but promotes speciation in the model by Higashi *et al.* (1999). Both models only work well if the females can examine a great number of males before committing themselves to a mate.

Fundamental problems of the sort just described would be overcome if plausible models of sexual selection could be devised that give rise to a (two-dimensional) evolutionary branching point (Box 4.3) for male trait and female preference. Evolution of the two sexual characters would then first converge to this point, at which both characters would experience frequency-dependent disruptive selection:

■ Through its convergence stability an evolutionary branching point would solve the problem of the initial fine-tuning required in the model by Higashi *et al.* (1999).

■ Through its lack of local evolutionary stability an evolutionary branching point would solve the problem of a large step in female preference being required in the model by Turner and Burrows (1995).

To have the option of encountering evolutionary branching points in models of sexual selection does not seem too far-fetched once it is realized that competition for mating partners is much akin to competition for ecological resources (Box 4.1 and Section 5.2). Specifically, there is a close formal resemblance between the distributions of an ecological resource and of female preference types available to males, and between a resource utilization spectrum and a function that describes the mating tolerance of females around a preferred male trait. Denoting, as in Section 5.2, the width of the former pair of distributions by σ_K and of the latter pair by σ_a, recall that evolutionary branching of the ecological character happens if resource utilization is specific enough, $\sigma_K > \sigma_a$. Van Doorn and Weissing (2001) derived that, likewise, evolutionary branching of the male trait happens if female choice is specific enough, $\sigma_K > \sigma_a\sqrt{1 + \sigma_a^2/\sigma_s^2}$, where $1/\sigma_s$ measures the strength of stabilizing selection on the male trait at the evolutionary branching point. In the absence of such extra selection, this condition reduces to $\sigma_K > \sigma_a$, which confirms the strong analogy between evolutionary branching through mate competition and resource competition. Evolutionary branching through mate competition can also occur when male trait and female preference are high-dimensional genetic traits (Van Doorn *et al.* 2001).

The mechanism considered so far enables evolutionary branching in the one sex (usually the males) that experiences competition for mates. Yet, for speciation to occur, the other sex (usually the females) also has to undergo evolutionary branching. This means that frequency-dependent disruptive selection must act on female preference, which can occur in two ways (Van Doorn *et al.*, unpublished): either through pleiotropic effects of frequency-dependent disruptive selection on an ecological character that evolves jointly (as in Van Doorn and Weissing 2001), or through female competition for males. In the former case speciation is not driven by sexual selection primarily; it also has to be underpinned by a suitable ecology. In the latter case, models ought to depart from the "typical sex-role assumption" that underlies most classic approaches to sexual selection. Under this assumption, only males are limited in their mating opportunities, whereas females have access to mates *ad libitum*, which implies that mate competition cannot cause frequency-dependent selection in females. Compared to reality, this remains an approximation, since, if there are far too many females around, their mating opportunities diminish. Especially if males have to engage in any time- or energy-intensive activities, like courtship or parental care, this approximation will often not be suitable.

Leaving the "typical sex-role assumption" aside, it can be shown that evolutionary branching of female preference becomes possible (Van Doorn et al., unpublished). This can lead, in principle, to the convergence of male trait and female preference, through directional selection, to a combination at which selection turns disruptive for both characters. This produces two Fisherian runaway processes that diverge from each other through frequency-dependent directional selection, which implies speciation. Frequency dependence means that the resultant species coexist in a stable polymorphism – a feature not encountered in many other models of speciation through sexual selection.

It turns out, however, that conditions for such two-dimensional evolutionary branching are quite restrictive (Van Doorn et al., unpublished). This is because parameter requirements for male-trait branching and female-preference branching do not overlap unless some additional source of frequency-dependent selection, other than mere competition for mates, is involved. This mutual exclusiveness arises since male fitness increases when a male mates with more females, whereas in the same situation female fitness decreases, which translates into opposing selective forces: when it pays males to diversify and undergo evolutionary branching, females experience stabilizing selection, and vice versa (Van Doorn et al., unpublished).

These results underscore the utility of investigating speciation by sexual selection through analysis of the potential for evolutionary branching, and highlight how the two sexes can first converge to and then escape from fitness minima induced by sexual selection. Adaptive speciation driven by sexual selection clearly is possible – even though perhaps less likely than previously believed.

Adaptive speciation through sexual conflict

Sexual conflict arises when traits that enhance the reproductive success of one sex reduce reproductive success of the other sex. It is appreciated increasingly that sexual conflict can be a strong driving force of evolutionary dynamics (Parker and Partridge 1998; Gavrilets 2000a; Martin and Hosken 2003). In particular, sexual conflict can lead to an evolutionary chase between the sexes, an idea substantiated both empirically (Arnquist and Rowe 2002) and theoretically (Gavrilets 2000a). Here we review a recent model by Gavrilets and Waxman (2002), which shows that sexual conflict over mating rates can give rise to adaptive speciation. This happens if, rather than staging a unidirectional evolutionary escape from males, females diversify into separate genotypic clusters. This can generate disruptive selection in the males, and so cause adaptive speciation with each female cluster being chased by a separate male cluster. The two pairs of genotypic clusters then represent incipient species.

Gavrilets and Waxman (2002) consider a sexual haploid population in which a multi-allelic locus A1 with alleles $x_1 = 0, \pm 1, \pm 2, \ldots$ determines female mating characteristics, while a multi-allelic locus A2 with alleles $x_2 = 0, \pm 1, \pm 2, \ldots$ determines male mating characteristics. The probability that a female with allele x_1 is compatible with a male that has allele x_2 is given by a normal function $a(x_1 - x_2) = \exp(-\frac{1}{2}(x_1 - x_2)^2/\sigma_a^2)$, which peaks at 0 and has variance σ_a^2. The weighted number of males compatible with a female that has allele x_{1i} is then $\tilde{n}_{2i} = \sum_j a(x_{2j} - x_{1i})n_{2j}$, where n_{2j} is the number of males with allele x_{2j} and the sum extends over all male alleles present in the population. It is assumed that the reproductive success of a female with allele x_{1i} is a function $W_1(\tilde{n}_{2i})$ with a maximum at some intermediate value $\check{n}_2$ – which means that too few as well as too many matings are detrimental to female reproductive success. Too many matings can be harmful if males employ aggressive mating strategies that damage the female, and it is this assumption that leads to sexual conflict in the model. Notice that this model is based on the "typical sex-role assumption" of limiting females and limited males. It also assumes that a reduced compatibility between males and females is sufficient to prevent the females from the harmful consequences of excessive mating (Panhuis *et al.* 2001).

To determine male reproductive success is more involved, because the success of a male depends not only on the frequency with which the various female alleles x_{1i} occur in the population, but also on the number of males $\tilde{n}_{2i}$ compatible with these female alleles, as well as on the females' reproductive success $W_1(\tilde{n}_{2i})$. Details can be found in Gavrilets and Waxman (2002), which also presents the equations that govern the dynamics of allele frequencies in males and females. For the present purpose it suffices to indicate the important conceptual components of this model:

- Reproductive success in both males and females depends on the genotypic composition of the population (i.e., reproductive success is frequency dependent).
- In a population in which females are monomorphic for allele x_1, it is best for males to carry allele $x_2 = x_1$.
- In a population in which males are monomorphic for allele x_2, it is best for females to carry allele $x_1 = x_2 \pm \Delta x$, where the optimal distance Δx is determined by the functions a and W_1.

This situation leads to an evolutionary chase of females by males. Male alleles that increase male compatibility with females are favored, while female alleles that move the number $\tilde{n}_2$ of compatible males closer to the optimal value $\check{n}_2$ are favored. This means that, evolutionarily, females tend to escape from males, which in turn try to catch up with them. Under certain conditions (for details, see Gavrilets 2000a), this can result in a continual evolutionary chase, as illustrated in Figure 5.8a, which schematically depicts the evolutionary dynamics of the average allelic values x_1 and x_2 in females and males. Depending on the initial conditions, these values evolutionarily converge on one of two lines, $x_2 = x_1 \pm \delta x$, and the evolutionary chase moves them along these lines toward $\pm\infty$; these two

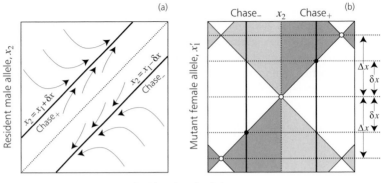

Figure 5.8 Adaptive speciation through sexual conflict. (a) Evolutionary chase of females by males. Any initial combination of female alleles x_1 and male alleles x_2 quickly evolves toward $x_2 = x_1 \pm \delta x$, which results in an intermediate level of male–female compatibility. The evolved distance δx lies between the optimal distance 0 for males and the optimal distance Δx for females, $0 < \delta x < \Delta x$, and leads to the two alternative configurations of evolutionary chase shown by the two thick lines. *Source*: Gavrilets (2000a). (b) Pairwise invasibility plot for female alleles (x_1', x_1) when the resident male allele is x_2. Combinations (x_1', x_1) of mutant and resident female alleles for which the mutant can invade are highlighted in gray (light gray indicates those mutants that can be reached only through evolutionary tunneling). The two alternative relative positions of resident female alleles expected after some period of evolutionary chase are marked by thick lines. The two evolutionary singularities (indicated by the intersection points of the boundaries of the gray areas) in the lower-left and upper-right corners of the diagram represent locally evolutionarily stable female strategies $x_1 = x_2 \pm \Delta x$ for the given male trait x_2, while the evolutionary singularity in the diagram's center represents the combination $x_1 = x_2$, which would be optimal for the males and acts as a repellor for evolution in the female trait.

cases are highlighted as Chase$_+$ and Chase$_-$ in Figure 5.8a. The allelic distance between males and females thus equilibrates at a positive value δx that is larger than the optimal distance 0 for males, but smaller than the optimal distance Δx for females: $0 < \delta x < \Delta x$.

Gavrilets and Waxman (2002) show that when selection on females is reduced by increasing $\check{n}_2$, and when mutation rates are increased, the evolutionary chase can be interrupted by diversification in the females through an evolutionary "tunneling" effect: a population of females that escapes from the males in one mutational direction at an average distance $+\delta x$ can occasionally give rise to successful mutant female alleles that are at a distance larger than $-\delta x$ from the current males, and thus escape from the males in the other direction. As only the absolute value, and not the direction of the distance to the males, is important, these mutant female alleles have similar reproductive success as the resident female allele. Such evolutionary tunneling can lead to the establishment of a polymorphism in the females.

In general, evolutionary tunneling happens whenever a fitness valley is crossed by a single, large mutational step. Figure 5.8b illustrates this through a pairwise invasibility plot for the female alleles, portrayed at any point in time throughout the evolutionary chase. In the situation shown, the population consists of males that are monomorphic for allelic value x_2. For any resident females allele x_1, all mutant female alleles x_1' that increase the distance to the males to a value closer to Δx can invade (depicted in gray). For resident females alleles x_1 close to x_2, this means that two groups of mutant female alleles x_1' can invade:

- A first range of mutant female alleles x_1' that is immediately adjacent to x_1 (dark gray in Figure 5.8b) can invade.
- A second range of mutant female alleles x_1' that is situated on the opposite side of x_2 relative to the resident female allele x_1 (light gray in Figure 5.8b) can invade.

For example, if the distance of the resident female allele to the males is δx, as is expected after the evolutionary chase has settled on the line $x_2 = x_1 + \delta x$, female mutant alleles that are either a bit larger than x_1 or a bit smaller than $x_2 - \delta x$ can invade (the latter case requires evolutionary tunneling). Likewise, for $x_2 = x_1 - \delta x$, female mutant alleles can invade that are either a bit smaller than x_1 or a bit larger than $x_2 + \delta x$ (the latter again through tunneling). The preceding two cases are highlighted, respectively, as Chase$_+$ and Chase$_-$ in Figure 5.8b.

For the evolving population with allelic values x_2 in males and $x_2 + \delta x$ in females it is clear that – as long as selection on females is sufficiently weak (so that δx is small) and mutation rates and step sizes are large enough (so that mutations of sufficient size are available) – it is likely that female mutant alleles will appear in the population that allow for evolutionary tunneling. Once this happens, the female population becomes polymorphic, and consists of two female allelic clusters on either side of the single male cluster (Gavrilets and Waxman 2002). Scope for the stable coexistence of two female alleles (which is not evident from the monomorphic analysis in Figure 5.8b) is created by any slight degree of polymorphism in the male population that leads to frequency dependence in the females: if one of the female branches becomes more abundant, this is counteracted by the growth of the corresponding tail in the distribution of males. After diversification in the females, selection ceases to be directional in the males, and the evolutionary chase comes to a halt. Trapped between two equidistant female clusters the males can experience a situation that Gavrilets and Waxman (2002) liken to that of Buridan's Ass, forever trapped in the middle of two equally appealing bales of hay.

However, if the distance between the female clusters and the trapped males is large enough, the male situation is not evolutionarily stable. More precisely, if d is the allelic distance between each of the two female clusters and the males, then the males experience disruptive selection for mating compatibility if $d^2 a / dd^2 > 0$ (Gavrilets and Waxman 2002). In this case, the males also diversify evolutionarily and form two clusters, each of which starts a new evolutionary chase with the corresponding female cluster. Once these separate chases have progressed far enough, the males that chase one female cluster become incompatible with the females from the other cluster, and vice versa: sympatric speciation has occurred.

The whole process of trapping males through evolutionary tunneling in females and subsequent adaptive male diversification that leads to speciation can repeat itself within the newly emerging species (Gavrilets and Waxman 2002).

In sum, evolution in the above model is characterized by three phases:

■ First is a phase of evolutionary chase between males and females, which takes the females close to the evolutionarily unstable singularity in the center of Figure 5.8b.

■ Second, evolutionary tunneling in the females leads to a state in which males are trapped in the middle of two female clusters, thus ending the evolutionary chase. This tunneling effect results from mutational chance events and does not occur in the limit of very small and very rare mutations.

■ Third, males diversify if the allelic distance between the two female clusters is large enough to generate a fitness minimum for the trapped males.

The model by Gavrilets and Waxman (2002) thus shows how sexual conflict can explain the dynamic emergence of disruptive selection through frequency-dependent interactions between the two sexes, and hence how sexual conflict can give rise to adaptive speciation. Notice the evolutionary dynamics of sequential evolutionary branching, first in females and then in males, that occurs in this model is similar to that exhibited by the coevolutionary predator–prey model discussed in Section 5.3.

Interestingly, in the high-dimensional trait spaces sexual selection is likely to operate in, evolutionary tunneling across fitness valleys might not even be required: the larger the number of dimensions, the higher the likelihood that these valleys do not actually have to be trespassed by large mutational jumps, but can instead be circumvented by sequences of small mutational steps [following an "extra-dimensional bypass" (Conrad 1990) along a "neutral network" (Schuster 1996), as explained in detail in Section 6.6 on "holey adaptive landscapes" (Gavrilets 1997)].

More generally, however, we think that more work is needed to determine whether sequential evolutionary branching is, indeed, a likely mechanism for adaptive speciation through sexual selection. There are two reasons for this:

■ As for any other speciation mechanism, segregation and recombination are bound to destroy the bimodality required in the two sexes (Section 5.2). The corresponding problems are, of course, avoided when there is no segregation or recombination, as in the model of Gavrilets and Waxman (2002), which uses haploid genetics (no segregation) and a single locus for the traits involved (no recombination).

■ When evolutionary branching in the two sexes is sequential, rather than simultaneous, the establishment of a sufficient degree of reproductive isolation between the incipient species is particularly difficult: if only females exhibit a bimodal distribution of their mating trait, while males are still unimodal, too much gene flow between the two female modes is likely to occur via the male population because of segregation and recombination, which thus collapses the female bimodality (G.S. van Doorn, personal communication).

It is thus an interesting open question as to how robust the results by Gavrilets and Waxman (2002) are when more realistic genetic assumptions are allowed. Consequently, mechanisms of sympatric speciation through sexual selection based on mate competition, as described in the previous subsection – involving frequency-dependent selection in *both* sexes and based on *simultaneous* evolutionary branching – may turn out to be of wider relevance.

5.5   Concluding Comments

Key challenges in the theory of sympatric speciation are twofold. First, ecological conditions must induce disruptive selection in such a way that the evolving population does not become monomorphic for one of the favored phenotypes. Second, given such ecological conditions, the mating system must evolve such that reproductive isolation ensues between the phenotypes favored by disruptive selection. In the history of the theory of sympatric speciation, positions have shifted as to the rank of these two difficulties. In this chapter, we present an integrative framework with which to study sympatric speciation that simultaneously addresses both issues.

The ecological dynamics considered in Sections 5.2 and 5.3 – diverse as they are as to featured detail – share one common characteristic: they all give rise to evolutionary branching points as described by the theory of adaptive dynamics. Such ecological settings naturally allow for directional selection to converge to fitness minima (Rosenzweig 1978; Eshel and Motro 1981; Eshel 1983; Taylor 1989; Christiansen 1991; Brown and Pavlovic 1992; Metz *et al.* 1992, 1996; Abrams *et al.* 1993a; Geritz *et al.* 1997, 1998) where disruptive selection can initiate sympatric speciation. As highlighted by Tregenza and Butlin (1999), "This provides an explanation for the nagging problem in other models of how the initial population comes to be in a state in which all phenotypes are intermediate and adaptation to the environment is suboptimal." Or, as put by Kawecki (Section 3.2), "sympatric speciation requires that the intermediate genotypes are first superior and then inferior to the original genotype. This apparent contradiction was seen as a major argument against sympatric speciation (Mayr 1963; Futuyma and Mayer 1980). A solution to this conundrum lies in the frequency dependence of fitness, which causes the adaptive landscape to change dynamically as the population evolves." There are four important points to appreciate here:

- A great variety of generic ecological settings robustly induces evolutionary branching points.
- Evolutionary branching points explain how populations come to be situated at fitness minima, without the need for exceedingly unlikely external events.
- Evolutionary branching points also retain populations at such minima for whatever time it takes them to evolutionarily overcome genetic constraints that prevent escape from such a trap.

▓ Since the underlying disruptive selection is not imposed externally, but rather is generated internally through frequency-dependent selection, no fine-tuning of conditions and parameters is required to stabilize the resulting dimorphism.

The genetic dynamics considered throughout this chapter illustrate that previous misgivings about the potential for sexual populations to escape disruptive selection by becoming bimodal in the selected trait may have been overrated. Both for "one-allele" and "two-allele" mechanisms of assortative mating, genetic constraints imposed by segregation and recombination are readily overcome in stochastic multilocus models, provided disruptive selection is not too weak. Felsenstein (1981) has conceded already that "It is not clear *a priori* whether the results found here are sensitive to the number of loci assumed involved in the traits." As described in Section 5.2, we now know that Felsenstein's hesitation was indeed justified: multilocus models of sympatric speciation do behave quite differently from earlier models that involved only one locus for each character. And when the traditional deterministic models are replaced by more realistic individual-based stochastic models, the scope for sympatric speciation widens even further. What is more, the stochastic multilocus models described in this chapter demonstrate that even when starting out from a population that is determined genetically to mate randomly, the degree of assortativeness can increase gradually because of the selection pressures encountered at an evolutionary branching point. This means that sympatric speciation in sexual populations can evolve completely from scratch, without the artificial assumption either that the population miraculously starts out situated at an externally imposed fitness minimum or that, from the very start, assortative mating occurs whenever ecologically different phenotypes encounter each other.

The reason why assortative mating evolves under branching conditions is easy to understand. Under frequency-dependent disruptive selection, segregation and/or recombination trap the population at a fitness minimum for as long as mating remains random. Assortative mating reduces the generation of intermediate offspring phenotypes from extreme parent phenotypes. As their offspring are thus less likely to exhibit a phenotype at the fitness minimum, individuals that mate assortatively are favored by natural selection. Once assortative mating has evolved, evolutionary branching can occur in sexual populations. In "one-allele" models of assortative mating, this mechanism alone allows for sympatric speciation. In "two-allele" models, which require the emergence of a linkage disequilibrium, the evolution of increasing levels of assortativeness additionally involves a process of selection-enhanced symmetry breaking (as illustrated in Figure 5.4).

A wide range of particular mechanisms for assortative mating exists in nature, and a large set of ecologically neutral marker phenotypes is always available for assortativeness to act on. It is thus important to realize that any single one of these mechanisms can enable a sexual population to escape from its trap at an evolutionary branching point. Viscous populations – in which a local component of a large population may succeed in escaping the fitness minimum through speciation, and subsequently conquer the population's remainder – further diminish the propensity of populations to remain stuck at evolutionary branching points. In some settings (including, most prominently, the evolution of host races), assortative mating even comes for free with the resource utilization of organisms. In addition, earlier rounds of sympatric speciation that involved evolutionary branching are bound to have left behind a legacy of assortative mating options, upon which subsequent speciation processes can freely draw; thus the escape of trapped populations is even more likely.

With the genetic constraints on sympatric speciation thus looking considerably less severe than many studies and discussions over the past few decades suggested, it is the ecological constraints that take center stage in determining the likelihood of sympatric speciation. Evaluation of these ecological constraints reduces to no more than addressing the simple and general question as to whether or not a particular ecology features an evolutionary branching point. As this question has been answered affirmatively for a wide variety of fundamental ecological models, not much doubt can remain that, from the vantage point of contemporary theory, sympatric speciation is entirely feasible. Since the speciation mechanism described here enables a population to escape from a fitness minimum, the speciation process itself must be considered adaptive. In the words of Seger (1985a), "sympatric speciation could provide, figuratively speaking, an easy way out of a difficult ecological and developmental bind." In general, adaptive speciation unfolds as a solution to a particular evolutionary problem posed by a population's frequency-dependent ecology (Chapters 1 and 19). This means that adaptive speciation is just as good a paradigm for asexual species as it is for sexual species – no fundamentally disparate mechanism have to be invoked to explain why both asexual and sexual species come in discrete chunks rather than in an unstructured continuum (Dobzhansky 1951).

It is exciting to realize that models of speciation by sexual selection also exhibit the characteristics of adaptive speciation. As shown in Section 5.4, an evolutionarily unstable singularity (Box 4.4) is always implicated in the reviewed models. We must thus regard processes of adaptive speciation driven by natural and sexual selection as two sides of the same coin: whether the traits that experience frequency-dependent disruptive selection affect a population's ecological interactions or its mating system is a matter of detail, not of principle. The eco-evolutionary feedback that can cause evolutionary branching (Chapter 1) may operate equally well through components of an individual's environment affecting the abundance of preferred resources or mates. Yet, adaptive speciation through sexual selection involves some additional problems, different from those relevant to adaptive speciation through natural selection. The reason is that under pure mate competition it pays males to diversify whenever females experience stabilizing selection, and vice versa (since, when a male mates with more females, male fitness increases, but female fitness decreases; Van Doorn *et al.*, unpublished). As explained in Section 5.4, this means that simultaneous evolutionary branching in both sexes cannot happen unless, as well as mate competition, additional sources of frequency-dependent selection contribute to the process. A notable alternative is adaptive speciation through sexual conflict: here evolutionary branching in the two sexes happens sequentially rather than simultaneously, driven by mate competition under the divergent mating interests of the two sexes.

We close this chapter by highlighting some compelling directions for future research. On the theoretical side, we should try to do better justice to the complexity of evolving mating systems and study models that can simultaneously describe adaptations in at least four different quantitative traits: the ecological character under frequency-dependent disruptive selection, a male marker trait, a female trait that determines which male trait females prefer to mate with, and a fourth trait that specifies the choosiness or specificity of female preference. While the analyses by Kondrashov and Kondrashov (1999) and by Dieckmann and Doebeli (1999) have already covered two different subsets, each containing three out of these four traits, investigation of a full four-trait model is pending. Also, studies that target specific types of assortative mating in more mechanistic detail are needed to enhance our understanding of how these affect the likelihood of adaptive speciation. Eventually, we need to bring together studies of the spatial patterns and temporal processes involved in speciation. Most speciation processes, strictly speaking, are parapatric to a larger or smaller extent – ranging from, as the extremes along a realistic continuum, cases with almost no gene flow between incipient species to cases with almost complete mixing. For this reason (and also since the traditional divide between allopatric and sympatric speciation is based on spatial pattern) it is compelling to incorporate a spatial dimension into the models described in this chapter – a program that is carried out in Chapter 7.

On the empirical side, unbiased assessments of the relative frequency at which (mostly) sympatric or allopatric speciation process occur in nature pose an essential challenge. A recent review by Turelli *et al.* (2001) concluded that "comparative analyses (Barraclough and Nee 2001) show that, in several taxa, the most recently evolved species generally have allopatric ranges, supporting Mayr's view that allopatric speciation might be most common." We believe that, unfortunately, such conclusions are as yet unwarranted: if sympatric mechanisms can induce spatial segregation between emerging species readily (as demonstrated in Section 7.4; Doebeli and Dieckmann 2003), abutting distributions of even recently evolved species can offer no clues as to the underlying speciation mechanism – unless a geographic barrier to gene flow at the time of speciation is demonstrated explicitly. It therefore seems fruitful to refocus the time-honored debate about the sympatric versus allopatric mode on the specific mechanisms that drive natural speciation processes, and, in particular, on evaluating the extent to which these processes are adaptive (Chapters 1 and 19). For such purposes, new approaches and methods will be needed (Berlocher 1998) to prevent problematic implicit assumptions – about how past processes can be inferred from present patterns – causing a bias in such assessments in the future. We suggest that this can be achieved reliably only by combining spatially explicit genetic, ecological, and environmental data and by focusing attention on speciation in gestation. Steps toward this ambitious goal are discussed in Chapter 15.

Acknowledgments   Ulf Dieckmann gratefully acknowledges financial support from the Austrian Science Fund, from the Austrian Federal Ministry of Education, Science, and Cultural Affairs, and from the European Research Training Network *ModLife* (Modern Life-History Theory and its Application to the Management of Natural Resources), funded through the Human Potential Programme of the European Commission.

Michael Doebeli gratefully acknowledges financial support from the National Science and Engineering Council (NSERC) of Canada, and from the James S. McDonnell Foundation, USA.

6

Genetic Theories of Allopatric and Parapatric Speciation

Sergey Gavrilets

6.1 Introduction

In this chapter, first some key ideas and notions relevant to allopatric and para-patric speciation are outlined briefly and then mathematical models that describe these processes are discussed. The logic behind these scenarios of speciation, as well as the data that support them, are thoroughly discussed in numerous excellent reviews (e.g., Mayr 1942, 1963; White 1978; Brooks and McLennan 1991; Ridley 1993; Rosenzweig 1995; Futuyma 1998), which should be consulted for additional information. Three questions are reviewed:

- Which models are available to explain allopatric and parapatric speciation?
- What are their predictions regarding the dynamics of speciation?
- How robust are these conclusions under realistic biological settings?

With regard to models' predictions, a major focus is on the waiting time to speciation, which represents a fundamental characteristic of the speciation process; also, the waiting time to speciation is the quantity that we might be able to estimate from the fossil record. Here, consideration is given to how this time is expected to depend on different evolutionary factors and parameters, such as the rates of migration and mutation, the strength of selection for local adaptation, population size, spatial structure of the population, and the genetic architecture of reproductive isolation. A knowledge of these dependences enables the plausibility of different mechanisms of speciation to be evaluated. The focus here is mostly on *simple* models, mainly because simple mathematical models allow analytical investigations (rather than complex and/or numerically analyzed models) that lead to the most transparent conclusions and thus form the basis of our scientific under-standing.

In addition to adaptive landscapes in genotype space, models of allopatric and parapatric speciation have to be concerned with real landscapes in physical space, since these determine the spatial proximity of individuals and populations. The models reviewed in this chapter assume that populations are subdivided into a number of discrete subpopulations with limited migration between neighboring subpopulations; this is the stepping-stone model of Kimura and Weis (1964). Al-though modeling speciation in continuous space is very important (see Chapter 7 for such an approach), at present discrete-space models of allopatric and parapatric speciation are the most advanced.

Box 6.1 Mechanisms of genetic isolation

Biological differences between populations of different species that prevent gene flow between them are called reproductive barriers (or isolating mechanisms). Reproductive barriers are usually classified into prezygotic and postzygotic:

- *Prezygotic* barriers include seasonal and habitat differences that prevent potential mates from meeting each other, behavioral isolation (prevents mating), mechanical isolation (prevents transfer of male gametes during copulation), and gametic incompatibility (prevents fertilization), etc.
- *Postzygotic* barriers include embryo inviability, hybrid inviability and sterility, hybrid breakdown (reduced viability or fertility in F2 or backcross generations), etc.

Further details on this fundamental distinction can be found, for example, in Dobzhansky (1937), Mayr (1963), and Futuyma (1998).

A prerequisite for speciation in sexual populations is the evolution of reproductive isolation between divergent forms. It is for this reason that a chapter concerned with the allopatric and parapatric modes of speciation, and thus with patterns on real landscapes, has to focus on adaptive landscapes. Reproductive isolation is defined as reduction or prevention of gene flow between populations by the genetically determined differences between them (e.g., Futuyma 1998). Numerous mechanisms can lead to reproductive isolation (Box 6.1). Although very different biologically, these mechanisms can be treated within a unifying model framework provided by the metaphor of adaptive landscapes in genotype space. Within this framework, modeling the dynamics of speciation is equivalent to modeling the dynamics of genetic divergence on an adaptive landscape. In this chapter, the metaphor of adaptive landscapes is explained in detail and applied to different speciation scenarios.

6.2 Modes of Speciation

As already highlighted in Chapters 1 and 3, three major modes of speciation correspond to three geographic settings in which speciation may occur: allopatric, parapatric, and sympatric. Notice that, as discussed in more detail in Chapters 1 and 19, these terms refer to pattern-based rather than process-based distinctions. The former become important when (as in this chapter and the next) the spatial structure of speciation processes is considered. As a basis for the further discussion, the specific characteristics of each of the three modes are highlighted briefly.

Allopatric speciation

Allopatric speciation is the evolution of reproductive isolation between populations that are completely isolated spatially. Allopatric speciation is widely regarded as the most common form of speciation (e.g., Brooks and McLennan 1991;

Ridley 1993; Rosenzweig 1995; Futuyma 1998). The chain of events typically implied in this scenario is:

1. The appearance of spatially isolated populations of the same species;
2. Genetic divergence of these populations;
3. Increasing reproductive isolation (as a by-product of genetic divergence);
4. Eventual emergence of one (or more) new species.

For each of these steps, different evolutionary factors and mechanisms can be of importance. There are three general mechanisms by which populations of the same species can become spatially isolated: vicariance events, extinction of intermediate links in a chain of populations, and dispersal. In the first case, a new extrinsic barrier, such as a river or mountain chain, isolates parts of a formerly continuous population. In the second case, a continuous population becomes fragmented when some "intermediate" subpopulations become extinct and the remaining subpopulations do not expand quickly to fill the resultant "gap". In the third case, new isolated populations are formed when some individuals (or propagules) manage to overcome an existing barrier and succeed in colonizing a new territory. All these mechanisms are important in natural systems (see, e.g., Mayr 1942, 1963; White 1978; Brooks and McLennan 1991; Ridley 1993; Rosenzweig 1995; Futuyma 1998).

Once spatially isolated, the populations are expected to diverge genetically. In general, genetic divergence is driven by a combination of stochastic and deterministic factors. The former include random genetic drift and the random order in which new genes are supplied by mutation. The latter include natural selection (resulting from abiotic and/or biotic factors) and sexual selection. Stochastic factors are more important in small populations, whereas deterministic factors are more important in large populations. If the environment of an evolving population is variable, natural selection can have a stochastic component. A by-product of genetic divergence can be reproductive isolation. In general, reproductive isolation can arise from genetic factors (such as incompatibility of specific genes, or specific combinations of genes, or whole chromosomes, or specific traits) or from ecological factors (such as the lack of an appropriate ecological niche or competition with parental forms). Experimental work has shown that both stochastic and deterministic factors can result in genetic divergence and some degree of reproductive isolation (Rice and Hostert 1993; Templeton 1996). Eventually, reproductive isolation becomes complete and the new species no longer interbreeds with its ancestor species if the two come into contact.

If two genetically divergent populations come into contact before their reproductive isolation is complete, a hybrid zone is formed. Hybrid zones are geographic regions in which genetically distinct populations meet and interbreed to some extent, and produce some individuals of mixed ancestry. The analysis of hybrid zones provides insights into the nature of species, the strength and mode of natural selection, the genetic architecture of species differences, and the dynamics of the speciation process (e.g., Endler 1977; Harrison 1990; Barton and Gale

1993). Many hybrid zones exhibit a gradual change ("cline") in a quantitative character or in some allele frequency along a geographic transect. Theoretical studies of hybrid zones concentrate on the shape of clines and the ability of genes to penetrate hybrid zones (e.g., Barton 1979; Bengtsson 1985; Barton and Bengtsson 1986; Gavrilets and Cruzan 1998).

Parapatric speciation

Parapatric speciation is the process of species formation in the presence of some gene flow between diverging populations. The chain of events, as well as mechanisms and factors typically implied in the parapatric speciation scenario, are similar to those of allopatric speciation, except that the spatial isolation of diverging subpopulations is partial rather than complete. Whether parapatric speciation is more likely at the periphery or at the center of a species' range has been a matter of debate (see Box 6.2).

From a theoretical point of view, parapatric speciation represents the most general scenario of speciation and includes both allopatric and sympatric speciation as special cases (which correspond to zero gene flow and very large gene flow, respectively). The geographic structure of most species, which are often composed of many local populations that experience little genetic contact for long periods of time (Avise 2000a), is thought to be characteristic for parapatric speciation. Despite this, parapatric speciation has received relatively little attention compared to the large body of empirical and theoretical work devoted to allopatric and sympatric modes of speciation (but see Endler 1977; Ripley and Beehler 1990; Burger 1995; Friesen and Anderson 1997; Rolán-Alvarez *et al.* 1997; Frias and Atria 1998; Macnair and Gardner 1998).

Sympatric speciation

Sympatric speciation is the origin of a new species within a population that initially mates randomly. A prerequisite for sympatric speciation is strong disruptive selection, which promotes the evolution of assortative mating within the population. Although this mode was favored by Darwin (1859, Ch. 4) originally, until recently support for sympatric speciation has been very weak. Chapters 3, 4, and 5 summarize the current state of theories of sympatric speciation.

6.3 Adaptive Landscapes

Speciation is an extremely complex process influenced by a large number of genetic, ecological, environmental, developmental, and other factors. To understand such a very complex phenomenon, it is helpful to have available a simple metaphor for its description.

The standard metaphor for discussing adaptive evolution and speciation is that of adaptive landscapes (Wright 1932). Different versions of adaptive landscapes are known (e.g., Gavrilets 1997). In what follows, an adaptive landscape represents fitness of specific combinations of genes (or traits). Each individual is

Box 6.2 Peripatric and centrifugal speciation

An important question concerns the relative importance of central versus peripheral populations in the emergence of new species:

■ Mayr's (1942, 1963) theory of *peripatric* speciation claims that peripheral populations are more likely to give rise to new species. According to Mayr, peripheral populations have a higher probability of splitting off because they consist of smaller numbers and experience different selection regimes. This theory puts special emphasis on dispersal as a mechanisms for the formation of spatial isolates.

■ In contrast, Brown's (1957) theory of *centrifugal* speciation argues that new species often emerge from central populations because central populations have higher genetic variation. Brown's theory places special emphasis on the extinction of subpopulations as a mechanism for generating spatial fragmentation of a species' range.

Each of these two theories is supported by numerous examples (Mayr 1942, 1963; Brown 1957; Lynch 1989; Frey 1993; Briggs 1999).

The scenarios implied by the two alternative approaches are illustrated in the figure below. Each panel shows the initial and final states of a system that initially consists of one "central" population (large square) and four "peripheral" populations (small squares). The new species is shown by a gray square and emerges either peripherally (left panel) or centrally (right panel).

Recent theoretical studies (Gavrilets *et al.* 1998, 2000a) show that both of these theories can be treated within a single unifying framework and that whether the major source of new species is a central or peripheral population depends on specific limiting factors (see Box 6.5).

envisaged as a point in a genotype space (defined as the set of all possible genotypes). Accordingly, a population can be thought of as a cloud of points on the adaptive landscape, and different populations (or species) are represented by different clouds. Selection, mutation, recombination, random drift, and other factors change the size, location, and structure of these clouds.

Different components of fitness and different forms of selection and reproductive isolation can be treated within this conceptual framework. For example, "fitness" can be an organism's viability (in the case of viability selection), the probability of successful mating between a pair of organisms (in the case of sexual selection), their average fertility (in the case of fertility selection), or the invasion fitness that underlies models of adaptive dynamics (Chapter 4). Note that because

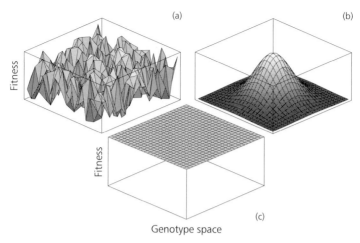

Figure 6.1 Adaptive landscapes. (a) Rugged adaptive landscape. (b) Single-peak adaptive landscape. (c) Flat adaptive landscape.

genotype spaces are discrete, adaptive landscapes as defined here are not continuous surfaces, but rather sets of points. Yet, for the purpose of visualization, it is often convenient to represent both genotype spaces and adaptive landscapes as continuous.

Following Wright (1932), adaptive landscapes are usually imagined as "rugged" surfaces that have many local "adaptive peaks" of different height separated by "adaptive valleys" of different depth (Figure 6.1a). Adaptive peaks are interpreted as corresponding to different (potential) species, adaptive valleys between them are interpreted as characterizing unfit hybrids (Barton 1989a), and adaptive evolution is considered as local "hill climbing" (Kauffman and Levin 1987). Speciation is then thought of as occurring by a "peak shift" (Wright 1932); that is, by a population somehow crossing an adaptive valley and moving onto a local peak that is different from the one it occupied initially.

The utility of Wright's rugged adaptive landscapes has been questioned. In particular, Fisher (see Provine 1986, pp. 274–275; Ridley 1993, pp. 206–207) pointed out that as the number of dimensions in an adaptive landscape increases, local peaks in lower dimensions tend to become saddle points in higher dimensions. In this case, according to Fisher, natural selection is able to move the population to a single global peak rather than to one of the numerous local peaks emphasized by Wright. A typical adaptive landscape implied by Fisher's views has a single peak (see Figure 6.1b for a simple low-dimensional rendering). However, recent work has shown that Fisher's criticism is not always warranted: the peaks that transform to saddle points can easily be outnumbered by new local peaks brought into existence by an increasing dimensionality of genotype space (Kauffman and Levin 1987). Thus, the number of local peaks can increase with the dimensionality of genotype space. From a different angle, Kimura (1983) argued that most evolutionary changes are neutral. A typical adaptive landscape implied by Kimura's neutral theory of molecular evolution is flat (as in Figure 6.1c). However, recent advances in molecular evolution show, convincingly, that in many cases evolutionary dynamics at the molecular level cannot be treated as strictly neutral (e.g., Gillespie 1991; Ohta 1992, 1998; Li 1997).

Below I argue from yet another perspective that the metaphor of rugged adaptive landscapes has been misleading in the context of speciation. Nevertheless, because of its dominant position in the near past, I start by considering the mathematical models of allopatric and parapatric speciation that are based on this metaphor.

6.4 Rugged Adaptive Landscapes

In general, the metaphors and simple models we use not only help to answer the questions we have, but sometimes also define those questions that we believe should be answered. Accepting the metaphor of rugged adaptive landscapes as a suitable basis for describing speciation processes immediately leads to a fundamental problem that has to be solved: how can a population evolve from one adaptive peak to another across an adaptive valley? Several models of peak shift have been proposed. None of these, however, models the whole process of speciation, but instead only describes some steps toward increased reproductive isolation.

Stochastic transitions between isolated adaptive peaks

The mechanism for escaping a local peak that has received most attention is random genetic drift, which is always present if the population size is finite (e.g., Lande 1979a, 1985; Barton and Charlesworth 1984). However, the two examples described in Box 6.3 illustrate the difficulties with such stochastic peak shifts on rugged adaptive landscapes. These examples show that, although stochastic transitions across very shallow adaptive valleys may sometimes occur within a reasonable time, in general the waiting time involved in a stochastic transition is extremely long. This is especially so if the population size is larger than a few hundred individuals and if the stochastic transition results in significant reproductive isolation. Natural populations are usually much larger than a few hundred individuals, and reproductive isolation, even between closely related species, is often strong. Therefore, stochastic transitions across valleys of maladaptation cannot be a major mechanism of genetic divergence of population on a large scale.

Shifting-balance theory

To solve the problem of stochastic transitions between different adaptive peaks, Wright (1931, 1932) proposed the shifting-balance theory. He considered populations to be subdivided into a large number of small subpopulations (demes) connected by migration. As local subpopulations are small and there are many of them, the probability that at least one of them is driven across an adaptive valley by stochastic factors (phase one of the shifting balance) and then pushed by selection to a new peak (phase two of the shifting balance) is non-negligible. Wright reasoned that once established in a subpopulation, the new adaptive combination of genes can take over the whole population by differential migration (phase three of the shifting balance).

Wright's argument was mainly verbal. Recent formal analyses of different versions of the shifting-balance theory led to the conclusion that, although the mechanisms that underlie this theory can, in principle, result in a shift of the population to a new peak, the conditions are rather strict. The main problem lies in the third phase of the shifting-balance process – the spread of a new combination of genes from a local subpopulation to the whole system (Gavrilets 1996; Coyne *et al.* 1997; but see Wade and Goodnight 1998).

Founder-effect speciation

Another possible way to escape a local adaptive peak is provided by founder-effect speciation (Mayr 1942, 1963; Carson 1968; Templeton 1980). In this scenario a few individuals found a new population that is geographically isolated from the ancestral species and that quickly expands to fill a new area. Here, a stochastic transition to a new adaptive peak happens during a short initial time interval when the size of the expanding population is still small. An advantageous feature of the scenario of founder-effect speciation relative to the shifting-balance process is that the new combination of genes does not have to compete with old combinations of genes that outnumber it.

The proponents of this scenario proposed only verbal schemes with no attempt to formalize them. Later, formal analyses of founder-effect speciation on rugged adaptive landscapes using analytical models and numerical simulations showed, however, that stochastic transitions between isolated adaptive peaks after a founder event cannot result in a sufficiently high degree of reproductive isolation with a sufficiently high probability to be a reasonable explanation for speciation (Barton and Charlesworth 1984; Barton 1989a).

Peak shifts by selection

Under certain conditions, transition from one adaptive peak to another can be accomplished by selection. This could happen if the adaptive landscape has multiple peaks, but the mean fitness of the population has a single peak. For example, in Kirkpatrick's (1982a) and Whitlock's (1995) models, individual fitness is a bimodal function of a quantitative trait value x (as in the second figure in Box 6.3), while the mean fitness $\overline{w}$ of the population changes from a bimodal to a unimodal form as genetic variance σ_G increases. Therefore, if the variance σ_G becomes sufficiently large, the population can make a rapid deterministic transition from one peak to a higher peak, despite the presence of an intervening valley in the individual fitness function (details are given in the second part of Box 6.3). Such a transition can be initiated either by a change in the external environment (affecting the individual fitness function), by a change in the internal (e.g., mutational or developmental) properties of the character that deterministically increases the character's variance in the population, or by an increase of the variance σ_G through stochastic fluctuations.

Box 6.3 Stochastic transitions between isolated adaptive peaks

Fixation of an underdominant mutation. Consider a finite diploid population of size N. Let $w_{AA} = 1$, $w_{Aa} = 1 - s$, and $w_{aa} = 1$ be the viabilities of genotypes AA, Aa, and aa, respectively, and $s > 0$ measures the strength of selection against the heterozygote. This model can also be applied to describe negatively heterotic chromosomal rearrangements.

The resultant adaptive landscape has two "peaks" represented by the homozygotes AA and aa; these peaks are separated by a "valley" represented by the heterozygote Aa.

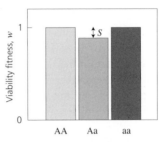

We now assume that initially all N individuals are homozygotes of genotype AA and that the rate u of mutation from A to a is very small. Then the expected waiting time until the peak shift takes place – that is, until the fixation of allele a – is approximately (Lande 1979a)

$$\tau = \frac{1}{u} \frac{e^{Ns}\text{erf}(\sqrt{Ns})}{\sqrt{4Ns/\pi}}. \tag{a}$$

Here $\text{erf}(x)$ is the error function (see, e.g., Gradshteyn and Ryzhik 1994). Assuming a mutation rate of $u = 10^{-5}$, for $Ns = 5$, we obtain $\tau \approx 0.59 \; 10^7$, for $Ns = 10$, $\tau \approx 0.62 \; 10^9$, and for $Ns = 20$, $\tau \approx 0.96 \; 10^{13}$.

Peak shift in a quantitative character. Now consider a diploid population of size N and assume that a single additive quantitative trait x controls fitness $w(x)$. Let the distribution of the trait in the population be normal with a constant (genetic) variance, and further assume that the fitness function $w(x)$ has two "peaks" at $x = a$ and $x = b$ with a "valley" between at $x = v$.

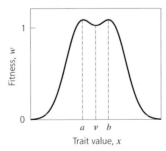

continued

Box 6.3 *continued*

If, initially, the population is at one peak, the expected time until the peak shift takes place through random genetic drift is approximately

$$\tau = \frac{2\pi}{\sigma_G}(-c_a c_v)^{-1/2} \left[\frac{\overline{w}(a)}{\overline{w}(U)} \right]^{2N} , \tag{b}$$

where $\overline{w}(x)$ and c_x are the average fitness of the population and the curvature of the fitness function at x (Barton and Charlesworth 1984; Lande 1985). For example, if the initial adaptive peak is 5% higher than the valley (that is, if $\overline{w}(a)/\overline{w}(v) = 1.05$), then, using realistic values for the other parameters, τ is of the order 10^6 generations for $N = 100$, while for $N = 200$ it is already of the much larger order of 10^{10} to 10^{11} generations (Lande 1985).

Other models exhibit similar features. In the model by Price *et al.* (1993), a first character that has two alternative optima can be dragged from one optimum to the other as a correlated response to selection on a second character. In this model, high genetic correlation between the two characters makes the peak shift easier. In the model by Pál and Miklós (1999), it is an increase in the epigenetic variability that creates an opportunity to evolve around an adaptive valley by selection alone.

Divergence driven by selection for local adaptation

The models discussed above assume the existence of multiple adaptive peaks in a single environment (i.e., they do not account for the fact that the shape of adaptive landscapes may change over physical space). Specifically, if the environment that a population experiences varies between different spatial locations, selection regimes may differ accordingly. Such differences can drive genetic divergence and lead to allopatric or parapatric speciation.

Allopatric speciation. Let us consider a one-locus two-allele population with genotypes aa, Aa, and AA. We assume that the population inhabits two "patches" that have become completely isolated. It is then natural to assume that environmental conditions differ between the patches and that these differences can result in different genotypes being better adapted to different conditions. Let the genotype fitnesses in patch 1 be 1, $1 - s$, and $1 - 2s$ for aa, Aa, and AA, respectively (with a selection coefficient $s > 0$). Allowing for mutation, the equilibrium frequency of A at mutation–selection balance is $\hat{p}_1 = u/s$, where u is the allele's mutation rate. Let the genotype fitnesses in patch 2 be $1 - 2s$, $1 - s$, and 1 for aa, Aa, and AA, respectively. Allowing for mutation, the equilibrium frequency of A at mutation–selection balance is $\hat{p}_2 = 1 - u/s$. In this model, the two populations thus have different genetic compositions, with allele A close to fixation in the first population and allele a close to fixation in the second population. If, somehow, a hybrid between the two most common genotypes AA and aa is produced, it will

Figure 6.2 Patterns of existence and stability of different equilibria in Slatkin's model of parapatric speciation. Stable and unstable equilibria are depicted, respectively, by filled and open circles. Arrows indicate the direction of change in the frequency of allele A. Note that the critical migration rates m_c and the allele frequencies at the stable polymorphic equilibria $\hat{p}$ differ between the cases (a) $s_f < s_v$ and (b) $s_f > s_v$. *Source*: Slatkin (1981).

have reduced fitness $1 - s$. This scheme can be viewed as describing a step toward allopatric speciation.

Parapatric speciation. Slatkin (1981) introduced a simple model of evolution of reproductive isolation driven by selection for local adaptation in spite of gene flow. Consider a very large population of haploid individuals with discrete nonoverlapping generations. [The restriction to haploid organisms is for the sake of mathematical simplicity. The main conclusions are the same in the case of diploid organisms (Slatkin 1981).] There is a single diallelic locus with alleles A and a. We assume that the population is subject to gene flow from another population fixed for the "ancestral" allele a, while the "new" allele A is advantageous in the environment the population experiences. Let the viabilities of A and a be $w_A = 1 + s_v$ and $w_a = 1$, respectively. The coefficient $s_v > 0$ measures the intensity of viability selection against migrants that carry the ancestral alleles. We assume that the surviving adults mate randomly, but that the A × a pairs produce only a fraction $1 - s_f$ of offspring relative to the A × A and a × a pairs. The coefficient $s_f > 0$ measures the intensity of fertility selection against mixed matings that results from pre- or postzygotic factors (see Box 6.1).

Let p and q be the frequencies of A and a at the beginning of a generation. Slatkin (1981) showed that for small migration rate m, s_v, and s_f, the change in p_v over a single generation can be approximated as

$$\Delta p = -mp + s_v pq + s_f pq(p - q) , \qquad (6.1)$$

where the first, second, and third terms describe changes in p through migration, viability selection, and fertility selection, respectively. The easiest way to understand the properties of the model is to assume that selection coefficients s_v and s_f are fixed, but that migration rate m can vary:

■ First, let fertility selection against mixed matings be weaker than viability selection against migrants ($s_f < s_v$). There is then a critical migration rate $m_c = s_v - s_f$ such that for $m > m_c$ the state with the ancestral allele fixed ($p = 0$) is globally stable (Figure 6.2a left). In biological terms this means that a strong gene flow swamps alleles that may help a population to adapt to local conditions. For $m < m_c$, selection partially overcomes the flow of deleterious alleles and maintains the advantageous allele at a potentially high frequency $\hat{p}$ (Figure 6.2a right). As a by-product of local adaptation the population evolves some reproductive isolation with regard to the immigrants. The extent of reproductive isolation can be measured by the average reduction in the number of offspring produced by a cross between a migrant and a randomly chosen member of the population $s_f \hat{p}$.

■ Second, let fertility selection against mixed matings be stronger than viability selection against migrants ($s_f > s_v$). The critical migration rate is then $m_c = (s_v + s_f)^2/(8 s_f)$, such that for $m > m_c$ the state with the ancestral allele fixed ($p = 0$) is again globally stable (Figure 6.2b, left). For $m < m_c$, the state $p = 0$ is locally stable, and there is another locally stable polymorphic equilibrium, which, however, cannot be reached deterministically from low frequencies (Figure 6.2b, right).

To summarize the results of Slatkin's model, an allele that adapts the population to local conditions can deterministically increase from low frequency only if the sacrifice in fertility s_f that results for carriers of this allele is small relative to the benefit s_v of improved adaptation that results from selection viability. Reduction in fertility is a measure of reproductive isolation. Consequently, strong reproductive isolation cannot evolve in Slatkin's parapatric model.

6.5 Bateson–Dobzhansky–Muller Adaptive Landscapes

As the previous section illustrates, models of evolution on rugged adaptive landscapes have not described speciation very successfully. A major problem of these models is that to speciate populations must overcome selection that "holds" them in the neighborhood of an adaptive peak (unless selection is frequency dependent, see Chapter 4). However, evolution of reproductive isolation does not necessarily have to be opposed by selection. As was first shown by Bateson (1909), Dobzhansky (1937), and Muller (1940, 1942), reproductive isolation can result after a series of genetic substitutions, each of which is unopposed by selection. For example, reproductive isolation arises if different populations that were initially identical genetically experience substitutions at different loci and the derived alleles (or genotypes) happen to be "incompatible". To illustrate this, consider a two-locus two-allele diploid model with alleles A and a at the first locus and alleles B and b at the second locus. Assume that alleles a and B are incompatible in the sense that all organisms simultaneously carrying these two alleles are inviable or sterile (Figure 6.3a). Then consider two geographically isolated populations, initially monomorphic, for the genotype AAbb. Now, if in one population's allele A is substituted for allele a (by mutation, random genetic drift, selection, etc.) and if, in the other population, allele b is substituted for allele B, the resultant populations is reproductively isolated. Alternatively, if both populations are initially monomorphic for the genotype aabb, reproductive isolation follows if one of them experiences two substitutions: first, of allele a for allele A and then of allele b for allele B.

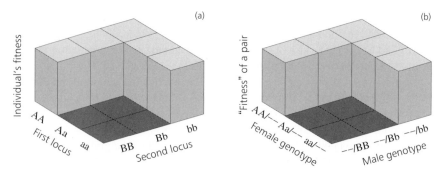

Figure 6.3   Adaptive landscape of the Bateson–Dobzhansky–Muller (BDM) model.
(a) Viability or fertility selection against individuals with incompatible genes a and B.
(b) Prezygotic reproductive isolation or fertility selection against mixed matings.

This example concerns reproductive isolation in the form of viability or fertility selection against hybrid organisms (postzygotic isolation). However, exactly the same scheme is applicable to prezygotic reproductive isolation or fertility selection against specific mating pairs. For example, assume females that bear allele a do not mate at all with males that bear allele B, or that these matings are infertile. Let no other type of mating be prevented and let all matings be fertile (Figure 6.3b). In this case, complete reproductive isolation can evolve via exactly the same pathways as in the case of selection against specific genotypes.

The models depicted in Figure 6.3 assume that the reduction in fitness does not depend on the number of incompatible alleles present. For example, in Figure 6.3a the fitnesses of genotypes AaBb and aaBB are assumed to be the same. This need not be the case: see Turelli and Orr (2000) for a discussion of more general situations in which fitness depends on the number of incompatible alleles present.

A long time ago, Dobzhansky (1937, p. 282) observed that the scheme he proposed "may appear fanciful, but it is worth considering further since it is supported by some well-established facts and contradicted by none." The model has both an attractive simplicity and an intuitive appeal. Indeed, "the loss in fitness to species hybrids is no more surprising than the fact that a carburetor from a car manufactured in the USA does not function in an engine made in Japan" (Charlesworth 1990, p. 103). By now, the BDM model is widely accepted and is supported by a growing amount of data (e.g., Cabot *et al.* 1994; Wu and Palopoli 1994; Orr 1995; Coyne and Orr 1998; Gavrilets 2002). The BDM model provides a basis for describing reproductive isolation through natural and sexual selection. In fact, all existing models of sexual selection explicitly incorporate the genes (or traits) that are perfect in the appropriate genetic (or phenotypic) background but become "incompatible" when brought together. Thus, sexual selection can be interpreted in terms of the BDM model (Figure 6.3b).

The BDM model postulates a specific genetic architecture of reproductive isolation. There are at least three general mechanisms that lead to this kind of architecture:

■ Genes brought together in hybrid genotypes may be "incompatible" because of negative epistatic interaction (e.g., Dobzhansky 1937; Muller 1942; Thompson 1986; Cabot *et al.* 1994; Wu and Palopoli 1994; Whitlock *et al.* 1995; Coyne and Orr 1998). A number of recent mathematical models have shown that negative epistatic interaction is a generic feature of gene regulatory networks (e.g., Johnson and Porter 2000; Kaneko and Yomo 2000; Omholt *et al.* 2000).

■ Hybrid genotypes may lack specific functions present in the parental organisms (e.g., Werth and Windham 1991; Lynch and Force 2000).

■ Maternal and paternal genes (or traits) may lack "matching" (e.g., Tregenza and Wedell 2000). Matching of some genes or traits is required at all levels of the reproduction process, including sperm–egg interactions (e.g., Howard 1999; Palumbi 1998; Vacquier 1998), mate recognition and mating pair formation (e.g., Paterson 1985; Grant and Grant 1997; Grant 1999), copulation process (e.g., Arnqvist 1998), and development after fertilization.

As the BDM model is central, those approaches toward describing allopatric and parapatric speciation processes that rely on this model are now considered in some more detail. The description below is based on Gavrilets (2000a). Other aspects of the BDM model (such as the maintenance of genetic variation, hybrid zones, the distribution of effects among offspring, etc.) are considered in a number of other studies (e.g., Wills 1977; Bengtsson and Christiansen 1983; Bengtsson 1985; Barton and Bengtsson 1986; Gavrilets 1997; Phillips and Johnson 1998; Turelli and Orr 2000).

Allopatric speciation in the BDM model

Consider an isolated diploid population that starts with the haplotype ab fixed and evolves to a state with haplotype AB by fixing allele B first, and then fixing allele A. Assume that mutations a $\rightarrow$ A and b $\rightarrow$ B are irreversible and occur at a rate u per generation.

No selection for local adaptation. Without selection for local adaptation, the process of allele fixation is neutral. The average waiting time to fixation of a neutral allele that is initially absent is $1/u$ (e.g., Nei 1976). Because the population needs to fix two mutations, the average waiting time to speciation τ is approximately

$$\tau = \frac{2}{u} . \tag{6.2a}$$

Selection for local adaptation. It is possible that alleles A and B have selective advantages over alleles a and b (through pleiotropy or other causes). With population size N, the number of mutations per generation is $2Nu$. The probability that a slightly advantageous mutation is fixed in a diploid population is approximately $2s/(1 - e^{-4Ns})$ (Kimura 1983). Thus, the average rate of fixation is $uS/(1 - e^{-S})$, where $S = 4Ns$. This results in an average waiting time to speciation approximately given by

$$\tau = \frac{2}{u} \frac{1 - e^{-S}}{S} .$$ (6.2b)

For example, increasing S from 0 to 10 decreases the waiting time to speciation to approximately $1/10$ of what it would be without selection for local adaptation.

Parapatric speciation in the BDM model

Consider a finite population of diploid organisms with discrete nonoverlapping generations. The population is subject to immigration from another population. For example, imagine a peripheral population (or an island) that receives immigrants from a central population (or the mainland). All immigrants are homozygous and have a fixed "ancestral" haplotype ab. Mutation supplies new alleles A and B, which may be fixed by random genetic drift and/or selection for local adaptation, while migration introduces ancestral genes. As before, assume that alleles b and A are incompatible. Speciation occurs when the genotype AB becomes fixed.

To proceed, use a weak-mutation and weak-migration approximation and neglect within-population variation (e.g., Slatkin 1981). Under this approximation the only role of mutation and migration is to introduce new alleles that quickly become either fixed or lost. The dynamics of speciation in this case can then be modeled as a random walk on a set of states 0, 1, and 2, where state 0 corresponds to a population fixed for the ancestral haplotype ab, state 1 corresponds to a population fixed for the "intermediate" haplotype Ab, and state 2 corresponds to a population in which haplotype AB is fixed. In this model, reaching state 2 represents a speciation event. The average time to speciation τ is thus defined as the average time required to reach state 2 starting from state 0.

No selection for local adaptation. Assume that there is no selection for local adaptation. Let u be the mutation rate per locus per generation. The process of fixation is approximately neutral, with the probability of fixation of an allele being equal to its initial frequency. Thus, the transition rates from state 0 to state 1 and from state 1 to state 2 are equal to the mutation rate u, and the transition rate from state 1 to state 0 is equal to the migration rate m. In this model, the average waiting time to speciation is

$$\tau = \frac{2 + \Theta}{u} \approx \frac{m}{u^2} ,$$ (6.3a)

where $\Theta = m/u$ characterizes the strength of migration relative to that of mutation, and the approximation is valid if $\Theta \gg 1$. For example, if $m = 0.01$ and $u = 10^{-5}$, then $\tau \approx 10^8$ generations.

Selection for local adaptation. Assume that new alleles A and B have selective advantage over ancestral alleles a and b. As before, the number of mutations per locus per generation is $2Nu$, and the probability of fixation of a slightly advantageous mutation is approximately $2s/(1 - e^{-4Ns})$. Thus, the transition probabilities from state 0 to state 1 and from state 1 to state 2 are $uS/(1 - e^{-S})$, where $S = 4Ns$. For a population in state 1, the number of ancestral alleles a introduced by migration is $2Nm$ per generation. These alleles are deleterious in the environment the population inhabits; the corresponding probability of fixation is approximately $2s/(e^{4Ns} - 1)$ (Kimura 1983). Thus, the probability of transition from state 1 to state 0 is $m/(e^S - 1)$. The average waiting time to speciation is then

$$\tau = \frac{2 + \Theta}{u} \frac{1 - e^{-S}}{S} \approx \frac{m}{u^2} e^{-S} \frac{1 - e^{-S}}{S} , \tag{6.3b}$$

where $\Theta = m/(ue^S)$ characterizes the strength of migration relative to that of mutation and selection, and the approximations are good if $\Theta \gg 1$. Equation (6.3b) shows that even relatively weak selection for local adaptation can decrease the waiting time to speciation by orders of magnitude. For example, with $s = 1$ and the same values of m and u as above, $\tau \approx 2.34 \ 10^7$ generations; with $s = 3$, $\tau \approx 1.64 \ 10^6$ generations, and with $s = 5$, $\tau \approx 1.73 \ 10^5$ generations.

In the process of evolution toward speciation, many unsuccessful speciation attempts will occur when the population substitutes allele for A only to lose it again quickly and thus return to the ancestral state. Another important characteristic is the average duration of speciation, defined as the time that it takes the population to evolve from its ancestral state to the state of complete reproductive isolation *without returning to the ancestral state*. (The duration of speciation is similar to the conditional time that a new allele, destined to be fixed, segregates before fixation.) In the model considered in this section, the average duration of speciation is $\tau/(1 + \Theta)$ (Gavrilets 2000a). Since Θ is typically large (e.g., with the same parameter values as above, $\Theta = 368, 50$, and 7 for $s = 1, 3$, and 5, respectively), the duration of speciation is usually much shorter than the waiting time to speciation. This suggests a very small probability of observing – say, in the fossil record – incipient species that are in transition to the state of complete reproductive isolation (see also Chapter 15). This feature of the models considered here is compatible with the patterns observed in the fossil record, which form the empirical basis of the theory of punctuated equilibrium (Eldredge 1971; Eldredge and Gould 1972).

6.6 Holey Adaptive Landscapes

How can one place the BDM model within the context of adaptive landscapes? The BDM model, as originally described, has two important and somewhat independent features (Orr 1995):

- It suggests that, in some cases, reproductive isolation can be reduced to understanding interactions of specific "incompatible" genes. For example, in Figure 6.3, it is genes a and B that are incompatible.
- It postulates the existence of "ridges" of high-fitness genotypes that "connect" reproductively isolated genotypes in genotype space. These "ridges" make it possible for a population to evolve from one state to a reproductively isolated state without passing through any maladaptive states ("adaptive valleys"). For example, in Figure 6.3a such a ridge is formed by genotypes aabb, Aabb, AAbb, AABb, AABB.

Neutral networks and holey adaptive landscapes

At first sight, ridges appear to be an unlikely feature of adaptive landscapes. Indeed, from our own experience with real landscapes in three dimensions, we know that one cannot move too far along ridges in mountain regions without having to lose or gain in altitude. However, a more thorough analysis shows that ridges are a general feature of *high-dimensional* adaptive landscapes (Gavrilets 1997, 2002).

The dimensionality of a genotype space can be defined as the number of new genotypes that can be obtained by changing single elements in the sequence of genetic units that represent a genotype. For example, if there are L haploid loci with H alleles each, the dimensionality of the genotype space is $L(H - 1)$. Even the simplest organisms have on the order of a thousand genes and on the order of a million DNA base pairs. Each of the genes can be at least in several different states (alleles). Thus, the dimensionality of genotype space is at least on the order of thousands. It is on the order of millions if one considers DNA base-pairs instead of genes. This results in an astronomically large number of possible genotypes (or DNA sequences), which is much higher than the number of organisms present at any given time, or even cumulatively since the origin of life. However, the number of different fitness values is limited. For example, if the smallest fitness difference one can measure (or that is important biologically) is, say, 0.001, then only 1000 different fitness classes are possible.

There is an important consequence of this observation. The redundancy in the genotype-to-fitness map requires that many different genotypes are bound to have very similar (identical from any practical point of view) fitness values. Unless there is a strongly "nonrandom" assignment of fitness (because, say, all high-fitness genotypes are put together in a single "corner" of the genotype space), a possibility exists that high-fitness genotypes can form connected networks that might extend, to some degree, throughout the genotype space. If this were so, populations could evolve along such networks by single substitutions and diverge genetically, without passing through any adaptive valleys. These ideas are reflected in the notions of neutral and nearly neutral networks and holey adaptive landscapes (e.g., Gavrilets 1997, 2002). A "neutral network" is a contiguous set of genotypes (sequences) that possess the same fitness. A "nearly neutral network" is a contiguous set of genotypes (sequences) that possess approximately the same fitness. A "holey adaptive landscape" is an adaptive landscape in which relatively infrequent high-fitness genotypes form a nearly neutral network that expands throughout the genotype space. An appropriate three-dimensional image of such an adaptive landscape is a flat surface with many holes, which represent genotypes that do not belong to the nearly neutral network (Figure 6.4).

Indeed, extensive neutral and nearly neutral networks have been discovered in numerical studies of ribonucleic acid (RNA) fitness landscapes (Fontana and Schuster 1987; Schuster *et al.* 1994; Grüner *et al.* 1996a, 1996b; Huynen 1996; Huynen *et al.* 1996), and also for protein fitness-landscapes (Babajide *et al.* 1997). In analytical studies of different general classes of adaptive landscapes, the existence of connected networks of high-fitness genotypes has been shown to be

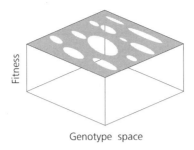

Genotype space

Figure 6.4 A holey adaptive landscape.

inevitable under fairly general conditions (Gavrilets and Gravner 1997; Reidys 1997; Reidys *et al.* 1997). Such networks were also encountered in models of multiplicative selection (Woodcock and Higgs 1996), models of stabilizing selection on additive quantitative characters (Barton 1989b; Mani and Clarke 1990), and Kaufmann's NK model (Newman and Engelhardt 1998). The origin of connected networks of high-fitness genotypes is illustrated in Box 6.4 using a very simple model. The existence of connected networks (or ridges) of high-fitness genotypes that allow for nearly neutral divergence appears to be a very general property of adaptive landscapes that possess a sufficiently large number of dimensions.

In the metaphor of holey adaptive landscapes, populations diverge (by mutation, drift, selection, etc.) along ridges of high-fitness genotypes and become reproductively isolated when they happen to arrive on opposite sides of a "hole" in a holey adaptive landscape. Below, a number of models based on this general idea are considered.

Multiallele versions of the BDM model

The original BDM model was formulated for the case of two alleles at each of the two loci. Nei *et al.* (1983) introduced a series of multiallele versions of the BDM model:

- *One-locus model of postmating reproductive isolation.* With multiple alleles, the idea that underlies the BDM model works even in the case of a single locus. Let there be an ordered sequence of alleles ..., A_{-2}, A_{-1}, A_0, A_1, A_2, ... such that allele A_i can mutate to either A_{i-1} or A_{i+1} with a certain probability. This is the stepwise mutation model of Ohta and Kimura (1973). Assume that alleles separated by more than one step are incompatible in the sense that the corresponding heterozygous genotypes (e.g., genotypes $A_i A_{i-3}$, or $A_i A_{i-2}$, or $A_i A_{i+2}$, or $A_i A_{i+3}$, etc.) are inviable or sterile (Figure 6.5a). Another version of this model uses Kimura and Crow's (1964) infinite-site model and assumes that each allele A_i is compatible only with its ancestral allele A_{i-1}, its descendant alleles A_{i+1}, and with itself.
- *Two-locus model of postmating reproductive isolation.* Here, the step-wise mutation model is used for each of the loci A and B. It is assumed that genotype

Box 6.4 Neutral networks in the Russian roulette model

Consider haploid individuals that can differ with respect to L diallelic loci. Assume that genotype fitnesses are generated randomly and independently and are equal to either 1 (viable genotype) or 0 (inviable genotype) with probabilities p_v and $1 - p_v$, respectively. This is like the set of all possible genotypes playing one round of Russian roulette, with p_v being the probability that an empty barrel will be encountered.

A counterintuitive feature of this model is that viable genotypes form neutral networks in genotype space if $p_v > 1/L$. It is very easy to see why this is so. As there are L diallelic loci, each genotype has L one-step neighbors (that is, other genotypes that differ at only a single locus). A viable genotype, say genotype 0, is acquired. If at least one of its one-step neighbors, say genotype 1, is viable, there is a neutral network with at least two genotypes (0 and 1). Next, consider genotype 1, which also has L one-step neighbors, of which one is genotype 0. If at least one of the remaining $L - 1$ one-step neighbors, say genotype 2, is viable, there is a neutral network with at least three genotypes (0, 1, and 2). Next, consider genotype 2, which also has L one-step neighbors of which one is genotype 1. If at least one of the remaining $L - 1$ one-step neighbors, say genotype 3, is viable, there is a neutral network with at least four genotypes (0, 1, 2, and 3). The point is that if there is one viable genotype out of $L - 1$ one-step neighbors, the size of the neutral network increases by one. Thus, as $L \to \infty$, the probability of an infinite neutral network started at genotype 0 is approximately the same as the probability of survival of the probabilistic branching process with $L - 1$ successors, where each successor is viable with probability p_v. This branching process is illustrated in the figure below. It does not die out if the expected number of viable successors, which is $p_v(L - 1)$, exceeds one. Thus, a neutral network of viable genotypes expands throughout the genotype space if $p_v > 1/(L - 1) \approx 1/L$. The existence of an extensive neutral network is therefore guaranteed even for very small values of p_v, provided that the number of loci L is sufficiently large.

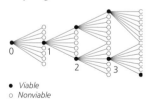

● *Viable*
○ *Nonviable*

$A_i B_j$ is compatible only with the nine genotypes that can be formed by alleles A_{i-1}, A_i, A_{i+1} at the first locus and alleles B_{i-1}, B_i, B_{i+1} at the second locus (Figure 6.5b). Diploid zygotes that consist of compatible genotypes are assumed to be viable and fertile, whereas all others are inviable or sterile. In this model, any two substitutions cause reproductive incompatibility.

■ *Two-locus model of premating reproductive isolation.* In this model, two unlinked loci A and B control male-limited and female-limited characters, respectively. The population is haploid. The stepwise mutation model is used for

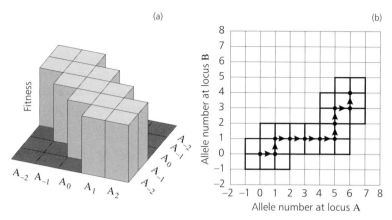

Figure 6.5 Multiallele models. (a) Adaptive landscape in the one-locus multiallele model of postmating reproductive isolation by Nei *et al.* (1983). Only part of the genotype space is shown, equivalent to five alleles. (b) An imaginary example of the evolution of reproductive isolation in the two-locus multiallele model by Nei *et al.* (1983). Two-locus haplotypes are represented by the intersections of the corresponding grid lines. A population starts out with haplotype A_0B_0 fixed and evolves to a state with haplotype A_6B_4 fixed by a sequence of ten mutational steps (six at locus A and four at locus B). Each haplotype along the evolutionary path from A_0B_0 to A_6B_4 is shown by a circle. The 2×2 squares centered on each of the circles include all nine haplotypes that are "compatible" with the haplotype represented by the circle.

both loci A and B. It is assumed that males that carry allele A_i mate only with females that carry alleles B_{i-1}, B_i, and B_{i+1}, and females that carry allele B_i mate only with males that carry alleles A_{i-1}, A_i, and A_{i+1}.

In these three models, complete reproductive isolation between different populations arises if the populations fix incompatible alleles (or chromosomes). To study allopatric speciation, Nei *et al.* (1983) used numerical simulations. Their findings can be summarized as follows:

- Reproductive isolation evolves faster in small populations than in large populations. This happens because mutant alleles that are two or more steps apart are selected against when rare, and random genetic drift is essential to fix mutant alleles.
- In spite of the continuous presence of incompatible alleles in the population, the intrapopulation fertility or viability is always very high (meaning that genetic load is low).
- Evolution of reproductive isolation is a very slow process, and it would take thousands to millions of generations for a mutation rate of the order of 10^{-5} per locus per generation. However, once started, the substitutions of incompatibility genes in a population occur rather quickly.
- Reproductive isolation evolves faster with two loci than with one locus.

■ The dynamics of postmating reproductive isolation and premating reproductive isolation are similar.

Nei *et al.* (1983) also used a one-locus stepwise mutation model of postmating reproductive isolation to study parapatric speciation in a system of two populations. They noticed that even a small amount of gene migration considerably retards the evolution of reproductive isolation. However, using a similar two-locus model of premating reproductive isolation, Wu (1985) demonstrated the possibility of sympatric speciation driven by mutation and random genetic drift. In Wu's simulations, however, the waiting time to speciation was very long.

Multilocus versions of the BDM model

In this subsection, some simple multilocus models that utilize the idea underlying the BDM model are considered and their possible generalizations are discussed.

Accumulation of genetic incompatibilities. Orr (1995; Orr and Orr 1996) introduced a series of models that describe the accumulation of genetic incompatibilities between allopatric populations. Consider two isolated diploid populations that initially are genetically identical. Assume that in both populations the "ancestral" alleles (represented by lower case letters) are substituted by new or "derived" alleles (represented by upper case letters) as a result of mutation, random genetic drift, or selection. As before, genetic variation within populations is neglected. In accordance with the BDM model, each substitution event does not result in within-population incompatibility, yet derived alleles in one population may happen to be incompatible with some alleles in the other population. Assume that initially all organisms have the haplotype abcde ... and that a number of derived alleles have been fixed in the system. Each derived allele in one population could be incompatible with each derived allele in the other population. In addition, each derived allele in one population could be incompatible with each ancestral allele still present in the other population at the loci that have experienced a substitution in the first population at the locus under consideration. For example, let the history of substitutions in the first population be abcde $\rightarrow$ Abcde $\rightarrow$ AbCde $\rightarrow$ AbCdE and that in the second population be abcde $\rightarrow$ aBcde $\rightarrow$ aBcDe. Then allele E fixed in the first population could be incompatible with alleles B and D fixed in the second population. Allele E could also be incompatible with alleles a and c. In general, if there have been $d_1 - 1$ substitutions in population 1 and d_2 substitutions in population 2, then the next substitution in population 1 can be incompatible with d_2 derived and $d_1 - 1$ ancestral alleles in population 2. Thus, the dth overall substitution ($d = d_1 + d_2$) may be incompatible with $d - 1$ alleles.

Assume that each derived allele has a fixed probability q_1 of being incompatible with each of the loci that have experienced a substitution previously. Then the ith substitution has a probability $(1 - q_1)^{i-1}$ of not causing reproductive isolation and thus speciation. Consequently, the probability that two populations at genetic distance d (that is, differing in d loci) are not reproductively isolated is

$$Q(d) = \prod_{i=1}^{d} (1 - q_1)^{i-1} = (1 - q_1)^{d(d-1)/2} \approx e^{-q_1 d^2/2} , \qquad (6.4)$$

assuming that q_1 is small and d is large. Note that this expression does not depend on the proportion of substitutions that occur in each population, nor on the order of substitutions. The cumulative probability of reproductive isolation $P(d)$ is defined as the probability that at least one incompatibility occurs and is thus given by $P(d) = 1 - Q(d)$. The average number of substitutions required for speciation can also be determined

$$k \approx \int_0^\infty d\frac{dP}{dd} \, dd = \sqrt{\frac{\pi}{2q_1}} . \qquad (6.5)$$

For example, with $q_1 = 10^{-5}$, $k \approx 400$ substitutions.

Gavrilets and Gravner (1997) studied how the probability $P(d)$ of complete reproductive isolation increases with genetic distance d in a diploid version of the Russian roulette model described in Box 6.4. In their analysis, $Q(d)$ is assumed to be given by a simple exponential function $e^{-\alpha d}$, rather than by the Gaussian curve derived in Equation (6.4).

A common feature of models based on the BDM model is that the probability of successful mating of two individuals – that is, the probability $Q(d)$ of no reproductive isolation – is a decreasing function of the genetic distance d between the individuals. This feature corresponds to the intuitively appealing idea that mating and the development of viable and fertile offspring is possible only between organisms that are not too different cumulatively over a specific set of loci (or traits).

Allopatric speciation. If genetic divergence of populations is driven by mutation and random genetic drift and if within-population genetic variation is neglected, the process of genetic divergence is approximately neutral. The rate of accumulation of substitutions does not then depend on population size and is equal to the mutation rate (Kimura 1983). Thus, the average waiting time to speciation is

$$\tau = \frac{k}{u} , \qquad (6.6a)$$

where u is the mutation rate per locus per generation and k is the average number of substitutions required for speciation. For example, in Orr's (1995) model of pairwise incompatibilities, k is given by Equation (6.5).

If some of the substitutions are advantageous rather than neutral, the waiting time to speciation decreases. For example, if each substitution is advantageous with selection coefficient s, the average waiting time to speciation is

$$\tau = \frac{k}{u} \frac{1 - e^{-S}}{S} , \qquad (6.6b)$$

where $S = 4Ns$ and N is the size of the diploid population.

Parapatric speciation. The dynamics of parapatric speciation are sensitive to the detailed shape of the function $Q(d)$ that specifies how the degree of reproductive isolation increases with genetic distance, see Equation (6.4). Here, a simple threshold function is used,

$$Q(d) = \begin{cases} 1 & \text{for } d < k , \\ 0 & \text{for } d \geq k , \end{cases} \tag{6.7}$$

(Gavrilets *et al.* 1998, 2000a; Gavrilets 1999, 2000a). This description of reproductive isolation is similar to that used by Nei *et al.* (1983, Figure 6.5a). The difference is that in the model by Nei *et al.* the degree of genetic divergence was defined by the number of mutations within a locus, whereas in Equation (6.7) it is determined by the number of mutations over a set of different loci.

The dynamics of speciation can thus be modeled as a random walk performed by the genetic distance d on a set of integers $0, 1, ..., k$. Speciation occurs when d hits the boundary k. Neglecting within-population variation and using the same procedure as employed for the BDM model, the time to speciation is approximately

$$\tau \approx \frac{1}{U} \left(\frac{m}{U} \right)^{k-1} (k-1)! , \tag{6.8a}$$

where U is the mutation rate per gamete and generation (Gavrilets 2000a). For example, if $m = 0.01$, $U = 0.001$, and $k = 6$, the waiting time to speciation, $\tau = 1.2 \; 10^{12}$ generations, is very long. (By contrast, the duration of speciation is of the order $1/U$, i.e., much shorter.)

If each substitution, rather than being neutral, is advantageous with selection coefficient s, the expression for τ given above is multiplied by the factor $(1 - e^{-S})/(Se^{kS})$,

$$\tau \approx \frac{1}{U} \left(\frac{m}{U} \right)^{k-1} (k-1)! \frac{1 - e^{-S}}{Se^{kS}} , \tag{6.8b}$$

where $S = 4Ns$ and N is the population size. For example, with $s = 2$, $m = 0.01$, $U = 0.001$, and $k = 6$, τ is 0.00002 of that in the neutral case. Thus, as noted before, even very weak selection pressures can reduce the waiting time to parapatric speciation by orders of magnitude.

Extensions of the models. The analytical results described in the preceding sections are based on a number of simplifications. Here some extensions of these models are summarized:

- The threshold function in Equation (6.7) represents an obvious (over)simplification. Gavrilets (1999, 2000a) has considered alternative functions to describe reproductive compatibility. For example, Gavrilets (2000a) analyzed the case of a linearly decreasing function $Q(d)$, a modification that resulted in a shortening of the waiting time to speciation.
- The approximations used above to model the dynamics of speciation neglect within-population genetic variation. Gavrilets (1999) has begun to develop a

more general approach that incorporates genetic variation and has shown that, in general, the process of accumulation of incompatibilities cannot be treated as neutral because the alleles that produce reproductive isolation are weakly selected against when rare (see also Nei *et al.* 1983; Gavrilets *et al.* 1998). As a consequence, in the case of divergence driven by mutation and random genetic drift, small populations accumulate reproductive isolation faster than large populations, contrary to the claim of Orr (1995).

- All models discussed above concentrate on one or two populations. Gavrilets *et al.* (1998, 2000a) used individual-based simulations to study the dynamics of parapatric speciation in multideme models. A summary of the corresponding results is provided in Box 6.5.

- Gavrilets *et al.* (2000b) have begun to incorporate the extinction of local populations into the modeling framework. In this approach, speciation and diversification are modeled as a continuous process of accumulation of genetic (or morphological) differences, accompanied by species and subpopulation extinction and/or range expansion.

- As discussed above, based on isolated adaptive peaks, founder-effect speciation is unlikely (Barton and Charlesworth 1984; Barton 1989a). By contrast, the existence of "ridges" in the adaptive landscape makes founder-effect speciation plausible. This was shown by Gavrilets and Hastings (1996) within the context of viability selection, and by Gavrilets and Boake (1998) within the context of postzygotic reproductive isolation.

Models based on quantitative characters

In the framework of quantitative genetics, individuals are characterized by continuously varying quantitative traits, rather than by sequences of genes. For quantitative characters, an adaptive landscape represents "fitness" of combinations of traits (Simpson 1953; Lande 1976).

Gavrilets and Hastings (1996) introduced a series of viability selection models based on the BDM model for quantitative traits and applied it to the scenario of founder-effect speciation using numerical simulations.

A crucial component of most sexual-selection models that operate in terms of quantitative characters is a "preference function" Ψ that controls the probability of mating between females and males with specific phenotypes. For example, in Lande's (1981) classic model of absolute preferences, this function is chosen as $\Psi = \exp(\frac{1}{2}(x - y)^2/v^2)$, where x and y are a female's and a male's trait values, and v is a parameter that measures mating tolerance. Preference functions can be interpreted as fitness functions of potential mating pairs. Figure 6.6 illustrates the resultant adaptive landscape: most possible mating pairs x and y have low fitness, whereas a small proportion of mating pairs that have high fitness (mating pairs with $x \approx y$) form a continuous "ridge" across the two-dimensional phenotype space. This adaptive landscape fits the definition of holey adaptive landscapes, with holes formed by all low-fitness mating pairs.

Box 6.5 Individual-based model of parapatric speciation

Gavrilets *et al.* (1998, 2000a) used large-scale individual-based simulations to study the dynamics of parapatric speciation in multideme systems. They considered one- and two-dimensional stepping-stone models with up to 24 subpopulations. Each population comprised up to a few hundred individuals, and each individual had a few hundred mutable linked diallelic loci. Simulations were based on the threshold model described by Equation (6.7), and genetic divergence was driven by mutation and random drift only.

The major conclusions from these studies are briefly summarized here:

■ For some parameter values, rapid parapatric speciation on the time scale of a few hundred to a few thousand generations is possible even when neighboring populations exchange several individuals in each generation. Typically, the waiting time to speciation was much shorter than expected from the simple estimates described in Equations (6.8). This suggests that those factors that are neglected in the analytical models play an important role for controlling the time scale of speciation.

■ The simulations substantiated the claims that species with smaller ranges (which are characterized by smaller local densities and reduced dispersal ability) are expected to have higher speciation rates than species with larger ranges.

■ If mutation rates are small, or local abundances are low, or if substantial genetic changes are required for reproductive isolation, then central populations should be the place at which most speciation events take place. This is in accord with Brown's (1957) theory of centrifugal speciation.

■ By contrast, for high mutation rates or high local densities, or if moderate genetic changes suffice for reproductive isolation, speciation events are expected to involve mainly peripheral populations. This is in accordance with Mayr's (1942, 1963) theory of peripatric speciation.

An important open question concerns the extent to which these conclusions are robust to changes in the assumptions and parameter values of the underlying model. This question requires careful additional investigation.

In Lande's (1981) model, random genetic drift can drive the divergence of different populations along a ridge of high-fitness pairs of quantitative traits, which leads to allopatric speciation. The addition of spatially varying selection for local adaptation to the model can result in genetic divergence in spite of gene flow, and thus potentially leads to parapatric speciation (Lande 1982). In the model by Iwasa and Pomiankowski (1995), mating preferences drive cycling in quantitative characters, which can result in the divergence of allopatric populations. In the model by Gavrilets (2000b), allopatric divergence is driven by selection that arises from sexual conflict in which the preferred combinations of mating traits differ between males and females. In that model and for biologically plausible parameter values, strong reproductive isolation can evolve on the time scale of a few hundred to a few thousand generations.

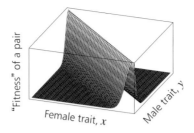

Figure 6.6 Adaptive landscape of mating pairs in Lande's (1981) model of absolute preferences.

To a large degree, conclusions from models of speciation driven by sexual selection parallel those for models of speciation driven by natural selection, which is not surprising given the inherent similarity (from a mathematical point of view) of both types of selection.

6.7 Concluding Comments

Classic population genetics concentrated exclusively on microevolutionary problems below the species level. Speciation processes, which provide a link between micro- and macroevolution, were not incorporated into the mathematical foundations of what has become known as the Modern Synthesis.

Also, more recent attempts to develop a mathematical theory of speciation based on the metaphor of rugged adaptive landscapes were not very successful. The framework of rugged adaptive landscapes is very useful for thinking about adaptation. However, for speciation this framework is not only not helpful, but also is even rather misleading. In particular, the problem arising from the crossing of adaptive valleys, which was considered of fundamental importance within the metaphor of rugged adaptive landscapes, is actually nonexistent.

A much more promising avenue is provided by the BDM model and its multidimensional generalizations, unified within the framework of holey adaptive landscapes. This general framework enables the analysis of various types of isolation barriers, including premating, postmating–prezygotic, and postzygotic, as well as different types of selection, including viability selection, fertility selection, and sexual selection. To date, a series of simple models of allopatric and parapatric speciation have been developed and provide a foundation for a quantitative theory of speciation. Hopefully, this theory will be a step toward uniting micro- and macroevolutionary theories in a new evolutionary synthesis (Eldredge 1985; Carroll 2000).

Spatial subdivision is an inherent feature of all species, and the importance of spatial aspects in ecology and evolution is well recognized. With regard to speciation, the two most important consequences of spatial subdivision are isolation by distance and variation in the selection regimes experienced by different subpopulations of the same species. Both of these promote geographic variation that can eventually lead to reproductive isolation. Mathematical models of allopatric and parapatric speciation reviewed in this chapter support these expectations.

Although much more theoretical work remains to be done, some general conclusions regarding the dynamics of allopatric and parapatric speciation can already be drawn:

- One generalization concerns the waiting time to speciation. If the population is strongly subdivided, the sizes of local subpopulations are small and mutation rates are high, so the waiting time to speciation driven by mutation and random genetic drift only can be on the order of a few hundred to a few thousand generations. Yet, in more realistic and general situations, the waiting time to speciation driven by mutation and random drift is expected to be extremely long.
- However, even relatively weak selection (e.g., selection for local adaptation or sexual selection through sexual conflict) that directly or indirectly promotes genetic divergence in the loci (or traits) responsible for reproductive isolation can make rapid speciation plausible. This is true for both allopatric and parapatric speciation. In a sense, this conclusion parallels some well-known facts of molecular evolution: neutral evolution is slow, but adaptive evolution can be extremely rapid.

The waiting time to parapatric speciation is very sensitive to the parameter values assumed in the models: a small change in a parameter value can increase or decrease this time by orders of magnitude. Most of the parameters of the parapatric models (such as the migration rate, intensity of selection for local adaptation, population size, and, probably, the mutation rate) directly depend on the state of the biotic and abiotic environments that a population experiences. This suggests that parapatric speciation can be triggered by environmental change (Eldredge 2002).

In contrast to the waiting time to parapatric speciation, which can be extremely long, the duration of speciation is on the order of one divided by the mutation rate over a subset of the loci that affect reproductive isolation. This is true for a wide range of migration rates, population sizes, intensities of selection for local adaptation, and the number of genetic changes required for reproductive isolation. Given typical parameter values, the duration of speciation is predicted to be on the order of 1000 to 100 000 generations (Gavrilets 2000a).

An important open question concerns the likelihood that natural or sexual selection promotes the evolution of reproductive isolation in allopatric and parapatric scenarios, and thus significantly accelerates speciation. Since direct ecological interactions between diverging groups of organisms in these scenarios are absent (in the allopatric case) or limited (in the parapatric case), less opportunity is expected for selection to promote reproductive isolation. By contrast, in the sympatric case, ecological competition can induce strong frequency-dependent selection and can thus create an opportunity for rapid speciation (e.g., Dieckmann and Doebeli 1999; Kondrashov and Kondrashov 1999; Chapters 3, 4, 5, 7). Which of the speciation modes (i.e., allopatric, parapatric, or sympatric) is more conducive to *rapid* speciation depends on a delicate balance of different factors, with the most important being the strength of selection and the level of migration. In general, for any of the three speciation modes, speciation driven by mutation and random genetic drift will, typically, be slow, whereas speciation driven by selection can be rapid.

7

Adaptive Dynamics of Speciation: Spatial Structure

Michael Doebeli and Ulf Dieckmann

7.1 Introduction

Extant patterns of species abundance are usually considered to be suggestive of allopatric speciation, because even closely related species are often geographically segregated (e.g., Barraclough and Vogler 2000; see Chapters 15, 16, and 17). Even though, in many cases, the ecological abutment between related species does not correspond to any obvious geographic barriers to gene flow, such patterns of geographic segregation are taken as strong indicators that speciation has occurred, either in allopatry or in parapatry. For the latter case it is assumed implicitly that there exists some sort of environmental discontinuity on either side of which different types are favored by selection or have evolved by genetic drift (Turelli *et al.* 2001). Even though gene flow across the environmental discontinuity can actually enhance speciation through the process of reinforcement, in these parapatric scenarios the reasons for speciation are ultimately the same as those in purely allopatric scenarios, that is, divergent evolution in different geographic regions. This has led to a common understanding that allopatric patterns of abundance between closely related species imply past events of allopatric speciation.

Rather than focusing on patterns of species abundance, recent developments in speciation theory focused on the adaptive processes and mechanisms that lead to disruptive selection and subsequent divergence of emerging new lineages in well-mixed, geographically unstructured populations. While this approach, described in Chapters 4 and 5, highlights the importance of frequency-dependent ecological interactions for evolutionary diversification, nonspatial models evidently cannot explain geographic patterns of species abundance. For this it is necessary to account for spatial structure explicitly, including spatial heterogeneity in environmental conditions and spatially localized ecological interactions.

Most previous studies of the role of spatial structure for evolutionary diversification assumed a discrete spatial population subdivision into local habitat patches within which interactions were unstructured spatially (e.g., Chapter 3; Boxes 4.5 and 4.6 in Chapter 4; Day 2000). This chapter describes how the study of evolutionary branching as a model for adaptive speciation can be extended to spatially structured populations that occupy a continuous spatial area. The resultant models of spatial evolutionary branching establish a firm link between ecological processes and geographic patterns of speciation. An abbreviated version of this theory can be found in Doebeli and Dieckmann (2003).

Adhering to established terminology, the models described below are models of parapatric speciation, because interactions between individuals only occur over short spatial distances and the populations are not panmictic. At the same time, however, these models differ crucially from traditional models of parapatric speciation in that the disruptive selective forces responsible for speciation are not imposed externally by the environment, but instead emerge dynamically from local adaptation in conjunction with spatially localized ecological interactions. Specifically, in contrast to the environmental discontinuities assumed, for example, in traditional models for stepped genetic clines in hybrid zones (Barton and Hewitt 1989), we assume that environmental conditions change gradually over space along a linear environmental gradient. If evolution were solely driven by adaptation to this gradient and limited only by gene flow, the establishment of a corresponding phenotypic gradient over space would be expected (Slatkin 1978; Kirkpatrick and Barton 1997; Barton 1999). A crucial perspective that the results described below add to this simple picture is that local adaptation along an environmental gradient has the potential to increase the strength of frequency-dependent selection in the system: if local adaptation leads to a correlation between phenotype and spatial location, and if interactions are spatially localized, then individuals tend to interact relatively more often with other individuals of similar phenotypes. As we show later, this mechanism actually greatly facilitates evolutionary branching in spatially structured populations. Interestingly, this facilitation is most pronounced for environmental gradients of intermediate slope. Moreover, when evolutionary branching occurs, the newly emerging lineages are often spatially segregated, and show a pattern of species abutment. In this way, the models for parapatric speciation studied here link local processes that drive evolutionary diversification to global patterns of species abundance. These results show that, contrary to the predictions of traditional parapatric models, ecological contact is not necessarily a hindrance to speciation, but, instead, can be a prerequisite for speciation.

Section 7.2 briefly reviews traditional approaches to parapatric speciation. We then present individual-based models for evolutionary branching in spatially extended populations, first for clonally reproducing individuals (Section 7.3) and second for sexual populations in which assortative mating must evolve for speciation to occur through evolutionary branching (Section 7.4). Finally, in Section 7.5 we point out a potential link between adaptive speciation in spatially structured populations and the origin of species–area relationships: for a given resource diversity along an environmental gradient, conditions for adaptive speciation are less restrictive in larger spatial areas.

7.2 Classic Models of Parapatric Speciation

Parapatric speciation occurs when an ancestral population splits into divergent descendant lineages that occupy different geographic areas in the ancestral species range under the maintenance of at last some gene flow across the spatial boundaries between the emerging species. As for the theoretical plausibility of this scenario, it is agreed widely that "any mechanism that can produce divergence among allopatric populations can also cause divergence in parapatry" (Turelli *et al.* 2001, p. 337). Parapatric speciation is thus generally envisaged as a process in which the divergence of types occurs in different geographic regions of an ancestral species range, accompanied or followed by the emergence of reproductive isolation between the diverging lineages. Divergence may result from selection or drift, and reproductive isolation between diverging subpopulations may be a pleiotropic by-product of local adaptation, as in allopatric speciation, or it may be an adaptive consequence of reinforcement (see Turelli *et al.* 2001 for a review).

Situations in which divergence is caused by genetic drift are described in detail in Chapter 6. In such cases it is usually assumed that reproductive isolation occurs automatically (through pleiotropic side-effects) as soon as the genetic distance between local populations is large enough (Gavrilets 1999). By contrast, when divergence is caused by local adaptation, reproductive isolation can arise either pleiotropically or through reinforcement (i.e., through the evolution of mating barriers between locally adapted populations driven by selection against hybrids). The latter mechanism (which generates a speciation process that is partially adaptive; see Chapter 19) has received considerable attention in the theoretical literature [Liou and Price 1994; Noor 1995; Kirkpatrick and Servedio 1999; see also the comprehensive review by Turelli *et al.* (2001)]. Even though details of genetic architecture, population structure, hybrid inferiority, and mating systems differ between the various studies, the general conclusion from these studies is that reinforcement is a theoretically plausible evolutionary scenario (Turelli *et al.* 2001).

While the evolution of traits that influence prezygotic isolation in the presence of selection against hybrids is naturally at center stage in studies of parapatric speciation through reinforcement, the ecological reasons as to why hybrid inferiority exists in the first place have received less attention. Typically, it is assumed tacitly that the ecological divergence, which leads to the establishment of different local types and to hybrid inferiority in the contact zone, is caused by externally given discontinuities in the environment (e.g., by the existence of local habitats with different adaptive peaks within a species' range, or by stepped environmental clines). Indeed, reinforcement classically refers to evolutionary processes that unfold upon secondary contact between populations for which ecological divergence has occurred in allopatry. Even though the same process can, in principle, occur during primary contact (i.e., under conditions of continual gene flow), the underlying environmental discontinuities necessary for local adaptation and hybrid inferiority must then be assumed *a priori*.

A notable exception to this conventional pattern are the models by Endler (1977) for parapatric speciation along clines. In these models, the fitness of different genotypes can vary linearly, rather than stepwise, along environmental gradients, and yet stepped genotypic clines can occur. These steps are, however, contingent on the special genetic architecture considered: with one locus and two alleles determining fitness, there simply exists a point along the environmental gradient at which the relative fitness of the two alleles changes sign. It is therefore easy to show that Endler's stepped genotypic clines disappear when more loci or alleles are allowed to affect fitness.

In contrast to these traditional approaches to parapatric speciation, here we focus on the intrinsic ecological mechanisms that can generate divergence and stepped phenotypic clines along continuous environmental gradients. To understand the underlying adaptive processes of divergence, we first discuss clonal models, in which reproductive isolation is not an issue. We then extend these models to sexual populations, in which reinforcement during primary contact can lead to the evolution of assortative mating and hence allow for speciation.

7.3 Evolutionary Branching in Spatially Structured Populations

Whatever the mechanism of adaptive speciation, some form of disruptive selection must be involved. Recent advances in the theory of adaptive dynamics (Metz *et al.* 1996; Geritz *et al.* 1998; Doebeli and Dieckmann 2000) demonstrate that disruptive selection regimes caused by frequency-dependent interactions emerge dynamically during the evolutionary process in generic models for all basic types of ecological interactions, a finding that was foreshadowed in earlier work by Eshel (1983), Christiansen (1991), and Abrams *et al.* (1993a). Such disruptive selection can lead to evolutionary branching, that is, to the splitting of evolving lineages into two phenotypic clusters (Chapters 4 and 5). Here we extend the study of evolutionary branching to spatially structured models for resource competition so as to investigate the effect of localized interactions in geographically extended populations on adaptive speciation along environmental gradients. We first focus on the simpler case of asexual populations to reveal ecological settings that are speciation prone.

Our starting point is the generic Lotka–Volterra models for frequency-dependent competition in spatially unstructured populations, which are introduced in Chapter 5, and which we briefly recall here. In these models, individuals vary with respect to a quantitative trait x, which could be a morphological, behavioral or physiological character. The deterministic dynamics of the density n_x of a population that is monomorphic for trait value x is given by

$$\frac{dn_x}{dt} = rn_x[1 - n_x/K(x)] \,, \tag{7.1a}$$

where r is the intrinsic growth rate of the species, which is assumed to be independent of the phenotype x. The carrying capacity $K(x)$ is the equilibrium density of populations that consist of x individuals and reflects the abundance of resources available to such individuals. We assume $K(x)$ to be of normal form,

$$K(x) = K_0 \exp\left(-\frac{1}{2}(x - x_0)^2/\sigma_{Kx}^2\right) \,. \tag{7.1b}$$

This implies that some intermediate phenotype x_0 has maximal carrying capacity, and that the decline of $K(x)$ to either side of x_0 is measured by the standard deviation σ_{Kx} of the normal distribution z. The equilibrium population density

$\hat{n}_x = K(x)$ of a population monomorphic for x is asymptotically stable. Considering such a resident population at its carrying capacity, the fate of a mutant phenotype x' is determined by its per capita growth rate when rare,

$$f(x', x) = \frac{\mathrm{d}n_{x'}}{n_{x'}\,\mathrm{d}t} = r[1 - a(x', x)K(x)/K(x')] \,. \tag{7.1c}$$

Here $K(x')$ is the carrying capacity of the mutant x', and $a(x', x)$ measures the strength of competition exerted by phenotype x on phenotype x'. This function is assumed to have the form

$$a(x', x) = \exp\left(-\frac{1}{2}(x' - x)^2/\sigma_{ax}^2\right), \tag{7.1d}$$

which implies that competition is strongest between individuals of similar phenotypes, as would occur, for example, when similarly sized individuals compete for similar types of food. Consequently, the total competition that impinges on an individual depends on the phenotypic composition of the population it is part of. In particular, a rare mutant x' pitched against a resident x at carrying capacity $K(x)$ experiences competition from a discounted number of individuals $a(x', x)K(x)$ for an amount of resources that is proportional to $K(x')$, a consideration that immediately allows us to understand the mutant's fitness $f(x', x)$ in Equation (7.1c).

The adaptive dynamics of the trait x is determined by the selection gradient $g(x) = \partial f(x', x)/\partial x'|_{x'=x}$. According to the canonical equation of adaptive dynamics (Dieckmann and Law 1996), the rate of mutation-limited evolutionary change $\mathrm{d}x/\mathrm{d}t$ is proportional to $g(x)$, with the constant of proportionality determined by the mutational process that generates genetic variability. The same basic proportionality also arises in models of quantitative genetics that describe evolution in genetically polymorphic populations (Lande 1979b). Using the functional forms of $K(x)$ and of $a(x', x)$, Equations (7.1b) and (7.1d), it is easy to see that the trait value x_0 that maximizes the carrying capacity is the only trait value for which the selection gradient vanishes, $g(x_0) = 0$ (i.e., the only evolutionary equilibrium in phenotype space). The phenotype x_0 is also a global evolutionary attractor. This is because $\mathrm{d}g/\mathrm{d}x(x_0) < 0$, which implies that for resident trait values x smaller than x_0 selection acts to increase x, and for resident trait values larger than x_0 selection acts to decrease x. Thus, independent of the initial trait value considered, evolutionary trajectories converge toward x_0.

However, x_0 need not be evolutionarily stable, which means a population monomorphic for x_0 may be susceptible to invasion by nearby mutants. Evolutionary stability is determined by the second derivative of the fitness function $f(x', x)$ with respect to x', evaluated at x_0: if $\partial^2 f(x', x)/\partial x'^2|_{x'=x=x_0} < 0$, then x_0 is a fitness maximum, and hence evolutionarily stable. In this case x_0 represents a final stop for the adaptive dynamics.

In contrast, if $\partial^2 f(x', x)/\partial x'^2|_{x'=x=x_0} > 0$, then x_0 is a fitness minimum, and hence an evolutionary branching point (Chapter 4). In this case, the population first evolves to x_0 and then splits into two phenotypic lineages that diverge from x_0 as well as from each other. For the model investigated here it is established

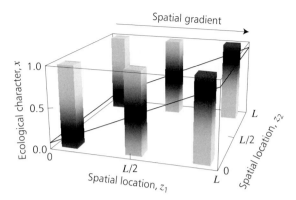

Figure 7.1 Environmental gradient in carrying capacities. The dark shading corresponds to phenotypes that maximize local carrying capacity; these gradually change with spatial location in the z_1 direction, while the z_2 direction is ecologically neutral. At any given location, the carrying capacity decreases with phenotypic distance from the capacity-maximizing phenotype (indicated by diminished darkness).

easily that x_0 is a branching point if $\sigma_{ax} < \sigma_{Kx}$, that is, if around x_0 the strength of competition decreases faster with phenotypic distance than the carrying capacity (Chapter 5).

This analytical theory for the adaptive dynamics of resource competition in well-mixed populations serves as a reference point for the spatially structured models introduced below. It can be derived from underlying individual-based stochastic models (Dieckmann and Law 1996). These individual-based models are described in Chapter 5; we now extend them to spatially structured populations by making the following assumptions. In addition to its phenotype x, each individual is characterized by its spatial location (z_1, z_2) in a square of continuous space with sides of length L. In this spatial arena, resources are distributed such that for each spatial location (z_1, z_2) there is a phenotype x_0 with maximal carrying capacity. We assume that this optimal phenotype varies linearly with one spatial dimension, $x_0(z_1) = \alpha z_1 + x_0(0)$, where α is the slope of the gradient, but is independent of the other spatial dimension z_2 (Figure 7.1). Such a resource gradient in one spatial dimension could, for example, represent an altitudinal temperature or humidity gradient along a mountain side that induces a change in the optimal phenotype with altitude. As in the nonspatial model described above, the carrying capacity K takes the normal form,

$$K = K_0 \exp\left(-\frac{1}{2}[x - x_0(z_1)]^2/\sigma_{Kx}^2\right),\qquad(7.2a)$$

and is thus a function of both phenotype x and spatial location z_1 (Figure 7.1).

As in the well-mixed case, we further assume that competition may be frequency dependent (i.e., the strength of competition between two individuals may depend on their phenotypic distance), so that competition is strongest between individuals with similar phenotypes. In addition, we assume that the strength of competition decreases with spatial distance between individuals. Thus, in our individual-based models the effective population size that determines the death rate of a given individual through competition (Box 7.1) depends on both the absolute number of other individuals in the neighborhood and on their phenotypes. Specifically, the relative strength a of competition between two individuals with phenotypes x and x' and with spatial distance Δz between them is given by a product of two normal functions,

$$a = \frac{1}{2\pi\sigma_{az}^2}\exp\left(-\frac{1}{2}(x-x')^2/\sigma_{ax}^2\right)\exp\left(-\frac{1}{2}\Delta z^2/\sigma_{az}^2\right). \tag{7.2b}$$

The parameters σ_{ax} and σ_{az} thus determine how fast the strength of competition between individuals decreases, respectively, with their phenotypic and spatial distance. We here envisage a situation in which individuals in a given phenotypically monomorphic, but spatially distributed population experience the same total amount of competition for different values of the spatial widths σ_{az} of the interaction kernel. Thus, if σ_{az} is small, spatially very close individuals have a relatively large impact, whereas if σ_{az} is large, the same spatially close individuals have less of an impact and spatially distant individuals become more important in such a way that the total competitive impact is the same. This assumption assures that in the absence of an environmental gradient, equilibrium population sizes of monomorphic populations are independent of σ_{az}. This, in turn, corresponds to the biologically reasonable assumption that the equilibrium population size supported by a given spatial distribution of resources is independent of the value of σ_{az}, which thus only measures the relative impact of spatially near and far individuals, respectively. This is why the normalizing constant $1/(2\pi\sigma_{az}^2)$ is incorporated into the competition kernel a given by Equation (7.2b).

Finally, to describe movement in the spatially structured model we assume that individuals can move through the spatial arena over distances and at rates independent of their phenotypes (Box 7.1). Given that ecological interactions between individuals are localized, populations can become spatially structured if movement occurs rarely or only covers short distances, whereas frequent movement over long distances results in well-mixed and hence spatially unstructured populations. Based on these ecological determinants, the evolutionary dynamics of the quantitative trait x can be investigated. We first do this in asexual populations by allowing for small mutations during birth events (Box 7.1).

As is explained in Box 7.1, a total of 11 parameters are needed to describe the spatially structured asexual populations. However, this complexity can be reduced considerably by considering the relevant limiting cases, and by appropriately rescaling units for phenotype, spatial distance, and time. In the salient limit of large spatial areas, and hence large L, of large local population sizes, and hence large K_0, and of small mutations, and hence small values of $u_{as}\sigma_{as}^2$ (Box 7.1), we are left with seven relevant parameters. Then, by using σ_{ax} as the unit of phenotype and σ_{az} as the unit for space in Equations (7.2a) and (7.2b), we are left with the two dimensionless parameters $\alpha\ \sigma_{az}/\sigma_{Kx}$ and σ_{ax}/σ_{Kx} as determinants of these two equations, respectively. For the remaining parameters m, β, and σ_m/σ_{az} (with the latter arising from taking σ_{az} as the unit of spatial distance), note that by taking the birth rate β as the unit of time, we are left with a single rate parameter m/β. In the limit of small movement distances, the compound parameter $(m/\beta)\,(\sigma_m/\sigma_{az})^2$ $= (\widetilde{\sigma}_m/\sigma_{az})^2$ describes spatial movement (where $\widetilde{\sigma}_m = \sigma_m\sqrt{m/\beta}$ is the expected movement distance during the average lifespan of an individual; see Box 7.1). This can be seen by considering the deterministic approximation to the individual-based model that results from consideration of the limit of infinite local population size (Box 7.2). Taking the limit of small movement distances in this deterministic approximation results in a dynamical system in which movement is described solely by the parameter $(m/\beta)\,(\sigma_m/\sigma_{az})^2$ [see the movement term in Equation (a) and its expansion given by Equation (b) in Box 7.2]. In sum, the three essential parameters for our spatially structured populations are the scaled slope of the environmental gradient $\alpha\,\sigma_{az}/\sigma_{Kx}$, the scaled width of the competition kernel σ_{ax}/σ_{Kx}, and the scaled movement distance $\widetilde{\sigma}_m/\sigma_{az}$. In the following, we refer to the latter simply as "mobility".

To aid with the biological interpretation of these dimensionless parameters, we note the following. If the scaled slope of the environmental gradient $\alpha\,\sigma_{az}/\sigma_{Kx}$ equals 1, then movement of a capacity maximizing phenotype by σ_{az} in the z_1 direction reduces its carrying capacity by $1/e$. If the scaled width of the competition kernel σ_{ax}/σ_{Kx} equals 1, then the phenotypic distance that reduces the strength of competition by a given amount is the same as the phenotypic distance from the capacity-maximizing phenotype that reduces the carrying capacity by the same amount. Finally, if the mobility $\widetilde{\sigma}_m/\sigma_{az}$ equals 1, the expected movement distance during an average lifespan equals σ_{az}, the width of the spatial interaction kernel.

Critical aspects of spatial structure are determined by the steepness of the environmental gradient and the mobility. If the gradient is very shallow, the environment becomes essentially spatially homogeneous. If mobility is large, the population becomes well-mixed and hence spatially unstructured. In either of these cases the system behavior approaches that of the nonspatial model: evolutionary branching occurs if $\sigma_{ax} < \sigma_{Kx}$, with the individuals belonging to the two different phenotypic clusters scattered throughout the spatial arena.

The system's behavior changes rather drastically if both the spatial resource gradient is steep enough and the mobility is low enough. In this case, if evolutionary branching occurs, it is accompanied by spatial segregation of the diverging phenotypic clusters (Figure 7.2). Thus, when frequency-dependent interactions occur under conditions of ecological contact and cause spatial evolutionary branching, the environmental gradient serves to organize the new lineages that emerge from this intrinsically sympatric process into geographic abutment (Figure 7.2). If the process that generates the pattern were not taken into account, the resultant pattern of species abundances would suggest allopatric or parapatric divergence merely driven by local adaptation. But this conclusion would be mistaken: in the absence of frequency-dependent selection, a gradual change in environmental conditions simply results in a gradual change in phenotypic composition (provided, of

Box 7.1 Individual-based and spatially explicit speciation models

Here we briefly detail the setup of our individual-based stochastic models for spatially structured populations.

- *Events.* At each computational step, individuals are assigned birth rates β_i, death rates δ_i, and movement rates m_i, $i = 1, ..., N$, where N is the current population size; these rates are updated after each event. Based on the total rates $\beta_{\text{tot}} = \sum_{i=1}^{N} \beta_i$, $\delta_{\text{tot}} = \sum_{i=1}^{N} \delta_i$, $m_{\text{tot}} = \sum_{i=1}^{N} m_i$, and $e_{\text{tot}} = \beta_{\text{tot}} + \delta_{\text{tot}} + m_{\text{tot}}$, the time that elapses until the next event is drawn from an exponential probability distribution with mean $1/e_{\text{tot}}$, while the type of that event is chosen according to the probabilities $\beta_{\text{tot}}/e_{\text{tot}}$, $\delta_{\text{tot}}/e_{\text{tot}}$, and $m_{\text{tot}}/e_{\text{tot}}$. The affected individual i is chosen with probability $\beta_i/\beta_{\text{tot}}$, $\delta_i/\delta_{\text{tot}}$, or m_i/m_{tot}, and the chosen individual either gives birth to one offspring, dies, or performs a spatial movement, depending on the type of event that is occurring.

- *Traits and genotypes.* In the asexual models, ecological phenotypes $0 \leq x \leq 1$ vary continuously. In the sexual models, phenotypes are determined by up to three sets of diallelic diploid loci with additive effects and free recombination (Chapter 5). The first set of l_1 loci determines the ecological trait x. The second set of l_2 loci determines a mating trait y that varies between -1 and $+1$ and determines mating probabilities, which are based either on the ecological trait or on a neutral marker trait (see *Birth* below). In the latter case, the marker trait also varies between 0 and 1 and is encoded by the third set of l_3 loci.

- *Spatial gradient.* Individuals have a spatial location (z_1, z_2), with $0 \leq z_1, z_2 \leq L$. We denote by $\phi_\sigma(v) = \exp(-\frac{1}{2}v^2/\sigma^2)$ and $\tilde{\phi}_\sigma(v) = \phi_\sigma(v)/\sqrt{2\pi}\sigma$, respectively, a normal function and the corresponding normal probability density with mean 0 and variance σ^2. The carrying capacity for the ecological phenotype x at spatial location (z_1, z_2) is then given by $K = K_0 \phi_{\sigma_{K_x}}(x - x_0(z_1))$, where $x_0(z_1) = \alpha(z_1 - L/2) + L/2$ is the phenotype that maximizes K at location z_1, and $0 \leq \alpha \leq 1$ is the slope of the environmental gradient (Figure 7.1); x_0 thus varies over space in the range $\left[(1 - \alpha)L/2, (1 + \alpha)L/2\right]$.

- *Movement.* Individuals move at a fixed rate $m_i = m$ and undergo displacements Δz in the z_1 and z_2 directions independently drawn from $\tilde{\phi}_{\sigma_m}(\Delta z)$, resulting in an average movement distance σ_m. Boundaries are reflective in the z_1 direction and periodic in the ecologically neutral z_2 direction. Note that at demographic equilibrium, the expected movement distance during the average lifespan of an individual is $\tilde{\sigma}_m = \sigma_m \sqrt{m/\beta}$, where β is the birth rate. The reason for the appearance of β is that at statistical equilibrium the average lifespan of an individual equals the inverse of the average per capita birth rate, and the model assumptions are such that the life times of the individuals vary with the local circumstances, while the per capita birth rate β is constant.

- *Death.* The effective population size experienced by an individual i with phenotype x at location (z_1, z_2) is a weighted sum, $\tilde{N} = \sum \phi_{\sigma_{ax}}(\Delta x) \frac{\phi_{\sigma_{az}}(\Delta z)}{2\pi\sigma_{az}^2}$, where the sum extends over all pairs $(\Delta x, \Delta z)$ of phenotypic and spatial distances between the focal and other individuals. The resultant logistic death rate of individual i is $\delta_i = \tilde{N}/K$.

continued

Box 7.1 *continued*

■ *Birth.* In asexual populations, individuals reproduce at a fixed rate $\beta_i = \beta$. Offspring express the parental phenotype unless a mutation occurs with probability u_{as}, in which case their phenotype x' is chosen according to $\tilde{\phi}_{\sigma_{as}}(x' - x)$. For sexual populations, an individual i slated for reproduction chooses a partner j proportional to phenotype-based mating propensities p_{ij}, depending on its mating character and the partner's phenotypic distance in either ecological or marker character as follows (see Figure 5.2). If mating propensities depend on the ecological trait, then, for an individual with mating trait $y > 0$, mating propensities fall off with a difference Δx in the ecological trait of potential mating partners according to $\phi_{\sigma_+}(\Delta x)$, where the width of the "mating kernel", σ_+, is given by $\sigma_+ = \varepsilon/y^2$, where ε is a parameter that indicates how fast assortativeness increases with increasing y. Thus, individuals with $y > 0$ mate assortatively, with the degree of assortativeness increasing as y approaches the value of 1. Individuals with mating trait $y = 0$ choose partners randomly with respect to phenotype. For individuals with mating trait $y < 0$, mating propensities increase with increasing phenotypic distance Δx of potential mating partners according to the function $1 - \phi_{\sigma_-}(\Delta x)$, where the degree of disassortativeness σ_- is determined by the mating trait y as $\sigma_- = y^2$. The mating propensities of an individual are normalized across all potential partner phenotypes. If mating propensities depend on the marker trait, the distance Δx in the ecological trait is replaced by the distance in the marker trait in the formulas above.

In our spatial models, the location-based component q_{ij} of mating propensities decreases according to $\phi_{\sigma_p}(\Delta z)/(2\pi\sigma_p^2)$ with the spatial distance Δz between potential partners. This induces a cost to the preference for locally rare phenotypes, $\beta_i = \beta N_p/(c + N_p)$, where $N_p = \sum_{j=1, j\neq i}^{N} p_{ij} q_{ij}$ is the number of suitable mating partners locally available to individual i, and c determines the cost's strength. Notice that assortativeness often evolves despite this cost. For sexual populations, only females are modeled. In effect, our models therefore describe hermaphroditic organisms. However, the models also apply to populations with separate sexes if males are assumed to have the same density and frequency distributions as females. After recombination, the offspring genotype is subjected to allelic mutations according to a reversal probability u_s per allele. Offspring undergo an initial movement event from the location of their parent.

■ *Parameters and initial conditions.* Unless otherwise stated: $l = 10$, $L = 1$, $K_0 = 500$, $\sigma_{Kx} = 0.3$, $\alpha = 0.95$, $\sigma_{ax} = 0.9$, $\sigma_{az} = 0.19$, $\beta = 1$, $u_{as} = 0.005$, $\sigma_{as} = 0.05$, $\sigma_p = 0.2$, $c = 10$, $u_s = 0.001$, $m = 5$, $\sigma_m = 0.12$ (i.e., $\tilde{\sigma}_m \approx 0.27$), and $\varepsilon = 0.05$. In the limit of large L and K_0 and small $u_{as}\sigma_{as}^2$, the asexual model has no more than three essential dimensionless parameters (see main text): σ_{ax}/σ_{Kx}, $\tilde{\sigma}_m/\sigma_{az}$, and $\alpha\,\sigma_{az}/\sigma_{Kx}$.

For the simulations reported here, we used the following initial conditions. In the asexual models, K_0 individuals were distributed randomly over space and had the phenotype that was optimal in the center of the environmental gradient. In the sexual models, K_0 individuals were distributed randomly over space, with genotypes assigned randomly assuming allele frequencies of $1/2$ at all loci.

Box 7.2 Deterministic approximation to the individual-based models

Here we assume that population sizes are locally infinite, both with regard to spatial location and to location in phenotype space. Since we assume a one-dimensional environmental gradient, without differentiation in carrying capacities along the other spatial dimension, the spatially structured population is described by a function $n(x, z, t)$, which denotes the density of phenotype x at spatial location z along the gradient and at time t. If, as in Box 7.1, $\phi_\sigma(x) = \exp(-\frac{1}{2}x^2/\sigma^2)$ and $\tilde{\phi}_\sigma(x) = \phi_\sigma(x)/(\sqrt{2\pi}\sigma)$ denote, respectively, a normal function and the corresponding normal density with mean 0 and variance σ^2, the temporal dynamics of the distribution $n(x, z, t)$ can be described by the partial differential equation

$$
\begin{aligned}
\frac{\partial}{\partial t}n(x, z, t) &= (1 - u_{as})\beta \int \tilde{\phi}_{\sigma_m}(z' - z)n(x, z', t)\,\mathrm{d}z' \\
&\quad + u_{as}\beta \iint \tilde{\phi}_{\sigma_{as}}(x' - x)\tilde{\phi}_{\sigma_m}(z' - z)n(x', z', t)\,\mathrm{d}x'\,\mathrm{d}z' \\
&\quad - n(x, z, t)\frac{\tilde{n}}{K(x, z)} \\
&\quad - mn(x, z, t) + m \int \tilde{\phi}_{\sigma_m}(z' - z)n(x, z', t)\,\mathrm{d}z' .
\end{aligned}
\tag{a}
$$

Here $K(x, z)$ is the carrying capacity of phenotype x at spatial location z (other parameters are as in Box 7.1). The first term on the right-hand side is the contribution to change in $n(x, z, t)$ from birth events without mutation at location (x, z'), that is, from parents x at spatial location z' whose offspring move to z. The second term represents the contribution of births to the population density at (x, z) that come from all locations (x', z') through a mutation of size $x - x'$ along the phenotypic axis and a spatial movement of distance $z - z'$ along the spatial axis. The third term represents the rate of death caused by competition from individuals at all locations (x', z'). Similar to Box 7.1, the effective density experienced by individuals at location (x, z) is given by $\tilde{n} = \iint \phi_{\sigma_{ax}}(x' - x)\tilde{\phi}_{\sigma_{az}}(z' - z)n(x', z', t)\,\mathrm{d}x'\,\mathrm{d}z'$, so that the per capita death rate is $\tilde{n}/K(x, z)$. The second-to-last term on the right-hand side of Equation (a) describes movement away from spatial location z, and the last term describes the effect of movement of phenotype x from all other spatial locations z' to spatial location z.

System (a) can be simulated numerically and, for suitable parameter combinations, equilibrium distributions $\hat{n}(x, z)$ are obtained that reflect the formation of spatially segregated phenotypic clusters, and thus represent spatial evolutionary branching. However, both the phenotypic and the spatial segregation obtained in the deterministic system are often less sharp than those seen in the individual-based models. In particular, close to the boundary in parameter space that delineates the region in which branching occurs in the individual-based models (see Figure 7.3), it can happen that branching only occurs in the individual-based model, but not in the deterministic system. This is illustrated in panels (a) and (b) below, for which the same parameter values are used as in Figure 7.3a, with $\sigma_m = 0.184$ and $\sigma_{ax} = 1$. With these parameters, the

continued

Box 7.2 *continued*

deterministic system (left panel) does not branch, in contrast to the individual-based model (right panel). This illustrates that the assumption of locally infinite population sizes made to derive the deterministic approximation is problematic not only for reasons of biological realism, but also because important features of the original model can be lost.

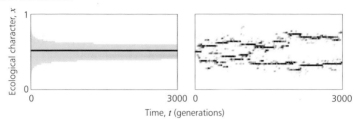

Equation (a) can be simplified further using various approximations. For example, because the distributions $n(x, z, t)$ are continuous, all phenotypes are present at all locations (albeit possibly with very small densities). Therefore, mutations are of lesser importance in the deterministic approximation, and we can simplify Equation (a) by setting $u_{as} = 0$. Furthermore, for small movement kernels σ_m, we can expand the last term on the right-hand side of Equation (a), which describes movement to location z, to obtain the second-order approximation

$$\int \tilde{\phi}_{\sigma_m}(z' - z)n(x, z', t)\,dz' \approx n(x, z, t) + \frac{1}{2}\sigma_m^2 n_{zz}(x, z, t) , \tag{b}$$

where $n_{zz}(x, z, t)$ denotes the second partial derivative of $n(x, z, t)$ with respect to z. It turns out that one cannot take advantage of a similar expansion for small σ_{ax} and σ_{az}, since this renders Equation (a) dynamically unstable. This problem is avoided when σ_{ax} and σ_{az} are not assumed merely to be small, but are assumed to vanish altogether (the phenotypic and spatial components of the interaction kernel are then given by Dirac delta functions). Box 7.3 presents results for yet another limiting case. Numerical analysis of such simplified systems is generally much more convenient than using individual-based stochastic simulations and may thus enable a quicker exploration of the underlying parameter space. In addition, such systems are also more likely to yield analytical insights.

course, that a sufficiently fine-grained set of phenotypes can be coded for genetically). Therefore, frequency-dependent selection is essential for the emergence of stepped phenotypic clines along the linear environmental gradients in our models.

A second and perhaps more important effect of spatial structure is that, for significantly sloped environmental gradients and low mobility, evolutionary branching occurs over a much wider range of parameters than in the nonspatial models (i.e., for values of σ_{ax} that are much larger than σ_{Kx}). The degree to which spatial structure facilitates branching and the abrupt onset of this facilitation as a function of mobility (illustrated in Figure 7.3a) are surprising. If mobility exceeds a certain

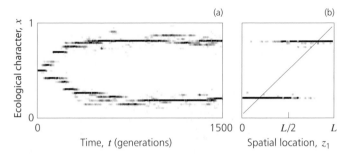

Figure 7.2 Evolutionary dynamics of adaptive divergence in asexual populations. (a) The distribution of phenotypes as a function of time (using the same shading scheme as in Figure 7.1). (b) The frequency distribution of phenotypes as a function of spatial location z_1 at the end of the time series shown in (a). The diagonal line indicates the environmental gradient (see Figure 7.1). Parameters as given in Box 7.1.

threshold value, parameter requirements for branching in the spatial and nonspatial models are almost exactly the same. However, as mobility decreases below this threshold, parameter requirements in the spatial model become suddenly and drastically less restrictive than in the nonspatial model. In fact, if mobility is small enough, evolutionary branching occurs even for effectively infinite σ_{ax} (i.e., even if there is no intrinsic frequency dependence in the competitive interactions). This is illustrated in Figure 7.3b, for which the scaled phenotypic width of the competition function is chosen to be very large. Nevertheless, evolutionary branching occurs once mobility falls below a critical level. Interestingly, this critical level depends on the slope of the environmental gradient, and is highest for environmental gradients of intermediate steepness.

The mechanisms that generate these effects can be illustrated as follows. An environmental gradient initially induces gradual spatial differentiation caused by local adaptation along the gradient. Thus, local adaptation implies a correlation between spatial location and phenotype. When, as assumed here, significant competition occurs only between individuals that are spatially sufficiently close, this correlation decreases the strength of competition between phenotypically distant individuals, and hence increases the degree of frequency dependence in the system. Such gradient-induced frequency dependence can lead to evolutionary branching, even if the phenotypic width of the competition function is very large. The effect tends to be weaker if local adaptation is very incomplete because of gene flow along shallow gradients, or if dissimilar phenotypes are spatially close because of local adaptation along a very steep environmental gradient. Therefore, facilitation of evolutionary branching through gradient-induced frequency dependence is highest for intermediate environmental gradients, as illustrated in Figure 7.3b. In this figure, frequency dependence results entirely from localized interactions between spatially differentiated individuals, and no evolutionary branching at all is expected in the nonspatial model.

The individual-based asexual models are characterized by three essential parameters, so we can use the information provided by Figures 7.3a and 7.3b to

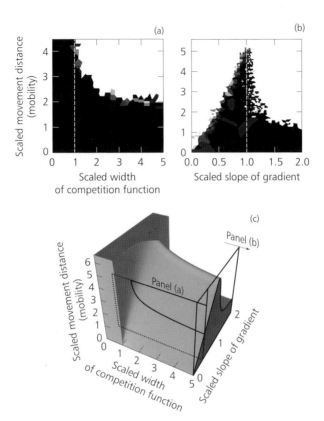

Figure 7.3 Requirements for evolutionary branching in asexual populations. The axes correspond to the model's three dimensionless parameters (see Box 7.1): scaled width of competition function σ_{ax}/σ_{Kx}, scaled movement distance (mobility) $\sqrt{m/\beta}\,\sigma_m/\sigma_{ax}$, and scaled slope of gradient $\alpha\,\sigma_{ax}/\sigma_{Kx}$. (a) and (b) show a subdivision of parameter space into polygons (Voronoi tessellation based on simulation data), shaded according to the recorded time to evolutionary branching: black corresponds to branching within the first 500 generations, white corresponds to no branching after 5000 generations, and shades of gray correspond to branching between generations 500 and 5000 (including multiple branching, which occurs for very low movement distances). (a) Effect of direct frequency dependence. Variation of time until branching with scaled width of competition function and mobility (scaled movement distance) for asexual populations (at $\alpha\,\sigma_{ax}/\sigma_{Kx} = 0.425$). In nonspatial models, only parameter combinations to the left of the dashed line are expected to induce branching. (b) Effect of gradient-induced frequency dependence. Variation of time until branching with scaled slope of gradient and mobility for asexual populations with $\sigma_{ax} \gg \sigma_{Kx}$. In nonspatial models no branching is expected at all. (c) Complete characterization of asexual model. Evolutionary branching occurs for parameters within the shaded block. The positions of the slices shown in panels (a) and (b) are indicated.

represent system behavior schematically in a three-dimensional plot. This characterization of the branching behavior of the system is shown in Figure 7.3c, which has as axes the three dimensionless parameters $\alpha \, \sigma_{az}/\sigma_{Kx}$ (scaled slope of the environmental gradient), σ_{ax}/σ_{Kx} (scaled width of the competition kernel), and $\tilde{\sigma}_m/\sigma_{az}$ (scaled movement distance, that is, mobility). Figures 7.3a and 7.3b are indicated in Figure 7.3c as planar cross sections of the three-dimensional solid that represents the parameter combinations that lead to spatial evolutionary branching. Figure 7.3c was extrapolated from Figures 7.3a and 7.3b using additional simulations to determine the height and the position of the central ridge. Figure 7.3 again makes it clear that evolutionary branching is facilitated greatly in spatially structured populations. The central ridge in Figure 7.3c illustrates that facilitation of evolutionary branching is generally greatest for environmental gradients of intermediate slope. More precisely, evolutionary branching is most likely for parameter combinations for which the scaled slope of the gradient $\alpha \, \sigma_{az}/\sigma_{Kx}$ is approximately equal to 1. This observation may serve as a starting point for empirical tests of our models, for we thus expect evolutionary diversification to be most likely for organisms and environments for which $\alpha = \sigma_{Kx}/\sigma_{az}$. In as much as the quantities α (the steepness of an environmental gradient), σ_{Kx} (the phenotypic width of the carrying capacity function along the gradient), and σ_{az} (the width of the spatial interaction kernel) are measurable in natural populations, this observation could serve as a basis for comparative studies of diversity in different taxa.

An obvious limitation of the analysis presented in Figure 7.3 is that this analysis is based on numerical simulations. It would clearly be very useful to have an analytical theory for the evolutionary dynamics of our spatially structured populations, for such a theory might, for example, allow us to derive analytical criteria for spatial evolutionary branching as a function of parameter values. In fact, by assuming locally infinite population sizes, for which "local" must be understood both spatially and phenotypically, it is possible to derive a deterministic approximation of the individual-based models (Box 7.2). Although the resultant partial differential equation is, again, amenable to numerical investigation only, it would, in principle, allow for a more tractable investigation. However, results from the deterministic approximation differ considerably from those obtained with the individual-based models. In particular, the sharp spatial segregation between lineages that emerge from evolutionary branching, as shown in Figure 7.2, are blurred in the deterministic approximations (Box 7.2). This, in turn, implies that the sharp bifurcation boundaries for evolutionary branching shown in Figure 7.3 are washed out in the deterministic system, so that the critical threshold in parameter values is less clear. Moreover, even when accounting for this artifact, the deterministic system also predicts the boundary's location inaccurately. These quantitative differences may arise from taking the limit of infinite local population sizes in both spatial and phenotypic dimensions (an assumption that appears to be even more difficult to justify biologically than the often used limit of infinite global population size). As a result, the implications of reproductive (and other) pair correlations, of local density fluctuations, and of demographic stochasticity are ignored, all of which have been shown to affect ecological and evolutionary dynamics critically (Dieckmann *et al.* 2000).

It appears to be rather difficult to extract analytical results from such deterministic approximations, but they lend themselves to more tractable numerical analysis and, for example, enable quicker searches of parameter space. A useful application of this approach is explained in Box 7.3, which shows that under certain assumptions spatial evolutionary dynamics can be studied using quite simple deterministic systems. In general, however, the individual-based implementation appears to be inevitable if artifacts caused by biologically unrealistic assumptions are to be avoided. This is particularly true for the investigation of evolutionary dynamics in spatially structured sexual populations with multilocus genetics, which are considered in Section 7.4.

7.4 Extension to Sexual Populations: Parapatric Speciation

To address the question of parapatric speciation, the individual-based spatially structured models can be extended to describe sexual populations in which the quantitative trait x is determined additively by a number of diploid loci (Box 7.1). Here the genetic assumptions are the same as those in Dieckmann and Doebeli (1999) and are described in Box 7.1. The ecological processes remain the same as before, but instead of reproducing clonally, individuals now depend on having suitable partners available within a given spatial neighborhood. The mating system then influences whether evolutionary branching, and hence parapatric speciation, is possible:

- If mating is random with respect to phenotypes, evolutionary branching no longer occurs, regardless of the relative magnitude of the parameters σ_{ax} and σ_{Kx}. Just as in the nonspatial model (Chapter 5), random mating brings about recombination between extreme phenotypes, which prevents the evolution of a phenotypic dichotomy.

- It has been shown before that evolutionary branching in well-mixed sexual populations is possible if the evolution of assortative mating is allowed for (Dieckmann and Doebeli 1999). Here we consider the same two general scenarios for assortative mating as described in Chapter 5. In the first scenario, assortative mating is based on a similarity in the trait x that determines the ecological interactions; in the second scenario assortative mating is based on an ecologically neutral marker trait (in the latter case, a linkage disequilibrium between the marker trait and the ecological trait must evolve for evolutionary branching to occur). As shown in Chapter 5, evolutionary branching in well-mixed sexual populations can occur because of evolution of both types of assortative mating (with the parameter requirements being more restrictive when assortative mating is based on a marker trait). These conclusions essentially carry over to the spatial model studied in this chapter; the resultant speciation processes are illustrated in Figures 7.4a and 7.4b for both types of assortative mating.

Box 7.3 Evolutionary branching in a reaction–diffusion system

Ferenc Mizera and Géza Meszéna

Here we consider the deterministic model for evolution along a spatial gradient given by Mizera and Meszéna (2003). The gradient is similar to those used in the individual-based models of Section 7.3, but incorporates the additional realistic assumption that the maximal carrying capacity K_0 is a function with an intermediate maximum along the gradient. Thus, not only does the phenotype that maximizes the carrying capacity change along the gradient, but the phenotypically maximal carrying capacity itself has a maximum at the center of the spatial axis that defines the gradient. We assume that this unimodality in the phenotypically maximal carrying capacity is given by

$$K_0(z) = K_1 \, \exp\left(-\frac{1}{2}z^2/\sigma_{Kz}^2\right), \tag{a}$$

where σ_{Kz} describes how fast the maximal carrying capacity $K_0(z)$ declines from the optimum K_1 with spatial distance from the center of the gradient at $z = 0$.

In addition, we assume that the slope α of the environmental gradient equals 1, so that we can identify phenotype with spatial location. Thus, the carrying capacity of phenotype x at spatial location z is given by

$$K(x, z) = K_0(z) \, \exp\left(-\frac{1}{2}(x - z)^2/\sigma_{Kx}^2\right). \tag{b}$$

We also assume that there is no intrinsic frequency dependence in the competitive interactions (i.e., we consider the limit $\sigma_{ax} \to \infty$), and that interactions only occur between individuals inhabiting the same spatial location z, but not between individuals from different spatial locations (i.e., we consider the limit $\sigma_{az} \to 0$ for the width of the spatial interaction kernel). Finally, birth, death, and movement are calculated as described in Box 7.1, except we assume that the offspring have the same spatial position as their parent.

The adaptive dynamics of the trait x can be studied using invasion analysis (Chapter 4) based on a deterministic reaction–diffusion equation. If $n(x', z, t)$ denotes the density of a mutant phenotype x' at spatial location z and at time t, the dynamics of the distribution $n(x', z, t)$ is given by

$$\frac{\partial n(x', z, t)}{\partial t} = r\left[1 - \hat{n}(x, z)/K(x', z)\right]n(x', z, t) + D\frac{\partial^2}{\partial z^2}n(x', z, t), \tag{c}$$

where $\hat{n}(x, z)$ is the equilibrium distribution of the resident phenotype x along the gradient, and r is the birth rate. Spatial movement is described by the diffusion coefficient $D = \frac{1}{2}m\sigma_m^2$, where m is the rate of spatial movement and σ_m is the average movement distance.

Using a numerical invasion analysis by determining the growth rate of mutants x' that appear in monomorphic populations of residents x, it can be shown that $x_0 = 0$ is always a convergence-stable singular strategy. This is simply a consequence of the total resource availability being highest at the center $z = 0$ of the spatial axis.

continued

Box 7.3 *continued*

In addition, it can be shown that the evolutionarily singular attractor $x_0 = 0$ is evolutionarily stable if the diffusion coefficient D and the environmental tolerance σ_{K_x} are large. However, when the diffusion coefficient and/or the environmental tolerance become small, the singular strategy becomes evolutionarily unstable, and hence becomes an evolutionary branching point. This is illustrated in panel (a) below, which shows the boundary between evolutionary stability and instability of the singular attractor $x_0 = 0$ in the two-dimensional parameter space given by the environmental tolerance σ_{K_x} and the diffusion coefficient D. (Other parameter values used for the panel below are $\sigma_{K_z} = 0.2$ and $r = 1.0$.)

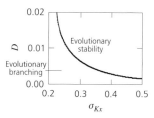

Individual-based simulations of this model confirm that if the singular point is an evolutionary branching point, and if evolutionary dynamics are started from a monomorphic population away from the singular point, the trait x first converges to the singular point, after which it splits into two phenotypic clusters. An example of such evolutionary dynamics is shown in the panels below, in which panel (a) shows the frequency distributions of phenotypes, indicated by shading along the horizontal axis, as a function of time, which runs along the vertical axis. Panel (a) illustrates that after an initial convergence to the singular point $x_0 = 0$, repeated evolutionary branching can occur in this system. Panels (b), (c), and (d) show the density distribution across space of the populations that represent the most abundant phenotype in each of the three branches present at the end of the time series shown in panel (a). (Parameter values for the panels are $D = 6 \ 10^{-5}$, $\sigma_{K_x} = 0.3$, $\sigma_{K_z} = 0.25$, and $r = 0.01$.) Note that the emerging lineages exhibit spatial segregation along the environmental gradient. This again illustrates how environmental gradients in resource availability can generate parapatric patterns of species distributions because of intrinsically sympatric ecological processes.

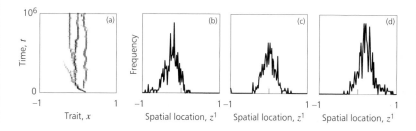

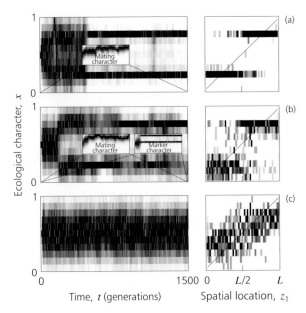

Figure 7.4 Evolutionary dynamics of adaptive divergence in sexual populations. The left panels show the distribution of phenotypes as a function of time (same shading scheme as in Figure 7.1) and the right panels show the corresponding frequency distribution of phenotypes as a function of spatial location z_1. The diagonal lines indicate the environmental gradient (see Figure 7.1). (a) Evolutionary branching with spatial segregation in a sexual population with the same parameter values as in Figure 7.2 and with assortative mating based on ecological similarity. The evolution of the degree of assortativeness is shown as an inset in the left panel (intermediate values of the mating trait correspond to random mating, low values to disassortative mating, and high values to assortative mating). Parameters as given in Box 7.1. (b) Evolutionary branching with spatial segregation in a sexual population with assortative mating based on a marker trait. The evolution of assortativeness and the branching in the marker trait are shown as insets in the left panel (the two marker branches are in linkage disequilibrium with the two branches of the ecological trait). Parameters as given in Box 7.1, except for $\alpha = 0.9$, $\sigma_{ax} = 0.5$, and $\sigma_m = 0.07$. (c) Evolution of a phenotypic gradient in a sexual population with random mating. Parameters as in (b), except that random mating with respect to phenotype was enforced.

Geographic differentiation in the presence of a spatial gradient has also been observed by Case and Taper (2000) in a model for competition between two species (Box 7.4). That model did not, however, address the question of initial diversification in a single population (instead, the existence of two separate species was assumed *a priori*). Indeed, the type of model used by Case and Taper (2000) serves to illustrate that if random mating prevents diversification of a single species that evolves along a linear gradient, then the evolution of a linear phenotypic gradient that tracks the environmental gradient is expected. Thus, when Kirkpatrick and Barton (1997) earlier on applied the same formalism to a single species to study the problem of species ranges, they found that a linear environmental gradient in the optimal phenotype leads to the evolution of a corresponding phenotypic gradient (Box 7.4). The same observation can also be made for our model: in the absence of evolutionary branching, a phenotypic gradient evolves (Figure 7.4c). We can thus conclude that the speciation process described here critically depends on evolutionary branching and on its release through the evolution of assortative mating; both of these processes are triggered by disruptive selection that emerges dynamically from frequency-dependent interactions under conditions of ecological contact. These strictly local interactions can thus ultimately drive the evolutionary diversification of ancestral populations into globally segregated species.

Once the degree of spatial structure caused by low mobility reaches a critical threshold, speciation through spatial evolutionary branching in sexual populations again occurs over a much wider range of parameters than in the corresponding nonspatial sexual populations. This is shown in Figure 7.5 for the case in which assortative mating is based on similarity in the ecological trait. Also, regarding the effect of the gradient's slope – derived in Section 7.3 from the numerical analysis of the spatially extended asexual model (Figure 7.3b) – we find expectations corroborated. As in the asexual case, speciation is especially facilitated for environmental gradients of intermediate slope (Figures 7.5b and 7.5c). As can be seen by comparing Figures 7.3 and 7.5, parameter ranges that allow spatial evolutionary branching in sexual populations are slightly smaller than the corresponding ranges for clonal populations.

Our assumptions that assortative mating originates from mate choice based on phenotypic similarity between partners (either in an ecologically relevant trait or an ecologically neutral marker trait) conform with premating isolation mechanisms investigated in previous analyses of reinforcement (e.g., Liou and Price 1994; Servedio 2000). Other isolation mechanisms, such as mate choice based on absolute preference (Servedio 2000) or on male traits and female preferences (Chapter 5), have yet to be considered in the context of evolutionary branching in spatially structured sexual populations. Compared against these more mechanistic models of assortative mating, the approach taken in this chapter must be interpreted as being more phenomenological. This, incidentally, has the advantage of mimicking many different mechanistic modes of assortativeness, which would otherwise need to be studied in different specific models.

We have thus arrived at a description of parapatric speciation as a dynamic consequence of two intertwined processes. First, spatially localized and frequency-dependent interactions along the environmental gradient can be the driving force of evolutionary divergence. Second, a dynamics akin to reinforcement causes the evolution of assortative mating. Both processes together allow an ancestral sexual population to become organized into phenotypically distinct descendant species that occupy different, spatially segregated regions along the environmental gradient, and thus result in a stepped phenotypic cline. These results provide an intrinsic ecological explanation for parapatric divergence in geographically continuous populations by linking local processes to the resultant global patterns of species abundance.

Box 7.4 Evolution on linear gradients in randomly mating sexual populations

In this box we briefly review the models of Kirkpatrick and Barton (1997) and of Case and Taper (2000) for the evolution of a quantitative trait along a linear environmental gradient. These models show that in a single species with random mating the linear environmental gradient induces a corresponding phenotypic gradient in the quantitative trait, and that interspecific competition can lead to spatial segregation of the competing species along the environmental gradient.

The models are formulated as partial differential equations with diffusive movement of individuals along one spatial dimension z. At spatial location z population growth is logistic, and the local population growth rate decreases with increasing distance between the mean phenotype $\bar{u}(z)$ and an optimum phenotype $x_0(z)$, which varies linearly with z. Thus, populations are only viable at location z if their mean $\bar{u}(z)$ is sufficiently close to the optimum $x_0(z)$. Specifically, the local dynamics at location z are of the form

$$\left(\frac{dn(z)}{dt}\right)_{local} = rn(z)[1 - n(z)/K] - [\bar{u}(z) - x_0(z)]^2/(2\sigma_{sel}^2) \,, \tag{a}$$

where $n(z)$ is the population density at location z, and σ_{sel} measures the strength of stabilizing selection toward x_0. The first term on the right-hand side of Equation (a) describes logistic growth, and the second term describes how growth rates decrease as the distance between the mean and the optimal phenotype at location z increases. According to the model of Lande (1979b), the local evolution of the mean phenotype is assumed to be given by

$$\left(\frac{d\bar{u}(z)}{dt}\right)_{local} = G\sigma_{sel}^{-2}[x_0(z) - \bar{u}(z)] \,, \tag{b}$$

where G is the product of the heritability with the phenotypic variance of the trait and is assumed to be constant over space and time. The role of σ_{sel} is akin to that of σ_{Kx} in our individual-based models. Equations (a) and (b) are complemented by diffusive movement along the z axis, characterized as in Box 7.2 by a diffusion coefficient $D = \frac{1}{2}m\sigma_m^2$, where m is the rate of spatial movement and σ_m is the average movement distance. It is assumed that competition only occurs between individuals that exist at the very same spatial location, and that intrinsic frequency dependence in local interactions is absent. This leads to the following system of partial differential equations to describe population growth and evolution along the environmental gradient,

$$\frac{\partial n}{\partial t} = D\frac{\partial^2 n}{\partial^2 z} + \left(\frac{dn}{dt}\right)_{local} \,, \tag{c}$$

$$\frac{\partial \bar{u}}{\partial t} = D\frac{\partial^2 \bar{u}}{\partial^2 z} + 2D\frac{\partial \ln n}{\partial z}\frac{\partial \bar{u}}{\partial z} + \left(\frac{d\bar{u}}{dt}\right)_{local} \,. \tag{d}$$

The first term on the right-hand side of Equation (c) and the first two terms on the right-hand side of Equation (d) represent the effects of dispersal and concomitant gene flow, while the remaining terms represent the local dynamics described in Equations (a) and (b), respectively.   *continued*

Box 7.4 *continued*

Kirkpatrick and Barton (1997) used this deterministic model to show that if dispersal is large enough, then gene flow swamping the boundary areas of a species range from the center can prevent local adaptation to the optimal phenotype near the boundaries, so that local equilibrium population densities decline to zero toward the species' boundaries. Therefore, gene flow can prevent range expansion. The resultant ecological and evolutionary equilibrium is shown schematically in panel (a) below, which illustrates local adaptation along a linear environmental gradient in the form of a linear phenotypic gradient in the mean trait value. As a result of gene flow from the center of the species range, the phenotypic gradient exhibits increasing distances from the optimal phenotype toward the edges of the spatial area.

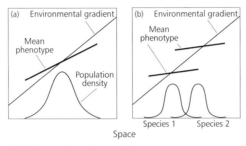

Case and Taper (2000) extended this model to two competing species. One of their main results is that character displacement in the contact zone between the species can lead to even larger degrees of local maladaptation, which thus limits species' ranges even further. This typically results in the two species occupying different spatial areas, and hence leads to spatial segregation along the environmental gradient. A schematic representation of this scenario is shown in panel (b) above, which is the same as panel (a) except that now the spatial arena is occupied by two competing species. Panel (b) illustrates that in this case species ranges are generally smaller because of character displacement as a consequence of interspecific competition in regions of overlap between the two species. Note that, again, the mean phenotype in both species always shows a continuous gradient. The model of Case and Taper (2000) demonstrates that interspecific frequency dependence can induce patterns of spatial segregation between two competing species, a finding in good agreement with the dynamics of our speciation models, in which intraspecific competition is dynamically and gradually transformed into interspecific competition.

7.5   A Note on Species–Area Relationships

The results presented in Sections 7.3 and 7.4 show how spatial structure facilitates parapatric speciation. It is therefore tempting to ask whether these findings also have implications for the general problem of understanding species–area relationships. It is an empirically well-documented fact that species diversity tends to increase with the size of the area over which diversity is sampled, which leads to the

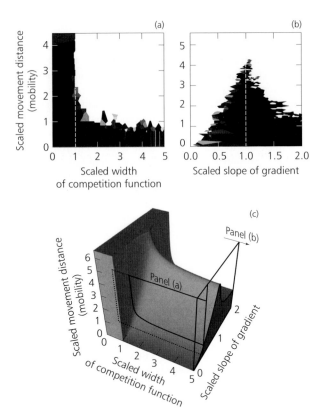

Figure 7.5 Requirements for evolutionary branching in sexual populations in which assortative mating is based on the ecological trait. The basic setup of this figure is the same as that for Figure 7.3. The model's three dimensionless parameters (see Box 7.1) are displayed on all axes. Panels (a) and (b) show a subdivision of parameter space into polygons, shaded according to the recorded time to evolutionary branching: black corresponds to branching within the first 500 generations, white corresponds to no branching after 5000 generations, and shades of gray correspond to branching between generations 500 and 5000. (a) Effect of direct frequency dependence. Variation of time until branching with scaled width of competition function and scaled movement distance (mobility) for asexual populations (at $\alpha \, \sigma_{ax}/\sigma_{Kx} = 0.425$). In nonspatial models only parameter combinations to the left of the dashed line are expected to induce branching. (b) Effect of gradient-induced frequency dependence. Variation of time until branching with scaled slope of gradient and mobility for asexual populations with $\sigma_{ax} \gg \sigma_{Kx}$. In nonspatial models no branching would be expected at all. (c) Behavior of sexual model. Evolutionary branching occurs for parameters within the shaded block. The positions of the slices in panels (a) and (b) are indicated.

characteristic species–area relationships often described by power laws (Rosenzweig 1995). This is one of the most ubiquitous patterns found in ecology, and many alternative explanations have been conjectured to date. The suggested mechanisms are likely to complement, rather than exclude, each other, so that observed species–area relationships presumably are the product of the joint action of several disparate mechanisms.

A classic explanation by MacArthur and Wilson (1967) is based on their "equilibrium model of island biogeography", that is, on the assumption that equilibrium population sizes linearly increase with island size, so that the extinction of species occurs more rarely on larger islands. This perspective is purely ecological and makes no reference to the effect of island area on the rate at which species are being formed, rather than being destroyed. By contrast, it has also been suggested that speciation occurs at higher rates in larger areas. For example, Losos and Schluter (2000) recently argued that the greater species richness of *Anolis* lizards found on larger islands in the Antilles results from higher speciation rates on larger islands, rather than from higher immigration rates from the mainland, or from lower extinction rates. Higher speciation rates in larger areas are, in turn, often attributed to an increased resource and habitat diversity supposedly harbored by larger areas.

Utilizing the spatial speciation model for asexual populations introduced above, here we illustrate that even when there is no difference in resource diversity, larger areas are expected to promote speciation. We suggest that such a relation holds because, according to the speciation mechanism discussed in this chapter, a given characteristic scale of individual mobility enables a higher degree of evolutionary self-structuring along longer gradients, even when these gradients span the same phenotypic range. To test this hypothesis we performed simulations as for Figure 7.3a, but in a spatial arena scaled down to half the area used for Figure 7.3a while retaining the gradient's phenotypic span. To achieve this we increased the slope of the environmental gradient in the smaller spatial arena to $\sqrt{2}$ times the slope of the gradient used for Figure 7.3a. The resultant parameter ranges for evolutionary branching are shown in Figure 7.6 and can be compared to the results for the larger spatial area in two ways:

- Figure 7.6 can be compared directly to Figure 7.3a, that is, to results for a larger spatial area that harbors the same resource diversity (i.e., the same range of phenotypes that locally maximize resource utilization).
- Alternatively, the ranges shown in Figure 7.6 can be compared to those for a larger area with the same slope of the environmental gradient. This implies, however, that the resource optima in the larger area span a larger phenotypic range, so that resource diversity is increased together with the area. In this case, Figure 7.6 should be compared to a corresponding figure for the larger area that is obtained as a cross-section of Figure 7.3c, taken parallel to the x–z plane at a y value (where y is the gradient slope) that is $\sqrt{2}$ times the y value that corresponds to the cross-section representing Figure 7.3a. Note that a cross-section at $\sqrt{2}$ times the gradient slope used for Figure 7.3a would predict larger ranges of parameter values that lead to evolutionary branching than does Figure 7.3a.

Whatever the comparison, Figure 7.6 shows that for low mobility, evolutionary branching still occurs much more readily in the smaller spatial area than in

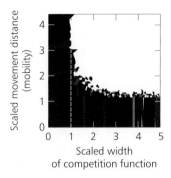

Scaled width
of competition function

Figure 7.6 Requirements for evolutionary branching for asexual populations that occupy half the spatial arena used for Figure 7.3a. As in Figure 7.3a, the panel shows a subdivision of parameter space into polygons, shaded according to the recorded time to evolutionary branching. The panel shows variation of time until branching with scaled width of competition function and with scaled movement distance (mobility). To ensure the same overall resource diversity as in the larger area used for Figure 7.3a, the parameter value for the scaled slope of the environmental gradient was set at α $\sigma_{ax}/\sigma_{Kx} = \sqrt{2}$ 0.425 in the smaller arena. Note that evolutionary branching occurs for a smaller range of parameters than in Figure 7.3a.

spatially unstructured populations. However, comparison either with Figure 7.3a directly, or with the cross-section of Figure 7.3c at $\sqrt{2}$ times the gradient slope used for Figure 7.3a, shows that branching occurs for a substantially smaller range of parameters than in a larger spatial area, irrespective of whether the larger area is assumed to have the same resource diversity or the same gradient slope as the smaller area. This means that there are mobilities that allow for greatly facilitated branching in the larger spatial area, but that are too large to have the same effect in the smaller area. Therefore, in our model parapatric speciation rates are higher in larger areas. Our interpretation of this result is that relatively more localized ecological interactions are more conducive to local adaptive processes of diversification. This should not be confused with mechanisms of isolation by distance, in which local processes at spatially very distant locations lead to diversification. In contrast, in our models divergence requires local ecological contact. Sympatric ecological interactions drive local adaptive diversification, and our results suggest that such processes of adaptive speciation are more likely with more localized interactions.

With regard to the *Anolis* lizards on Caribbean islands, the mechanism described here may be responsible for some aspects of the species–area relationships discussed by Losos and Schluter (2000). Indeed, resource diversity does not appear to be significantly lower on smaller islands in the Antilles (Roughgarden 1995). Yet the large islands of the Greater Antilles typically harbor many species of *Anolis* lizards, while the smaller islands of the lesser Antilles contain at most two species – although morphological variation within these species can be quite large (Malhotra and Thorpe 1997c; Chapter 16). These observations conform with the notion that on the larger islands new species can arise more easily out of existing genetic variation, because evolutionary branching and the concomitant evolution of reproductive isolation is more likely in larger spatial areas.

7.6 Concluding Comments

In this chapter we describe a theory of adaptive parapatric speciation that links ecological processes driving local divergence to global patterns of species abundance. In many traditional models of parapatric speciation it is assumed that some form of geographic differentiation is induced by discontinuities in the external environment, and that speciation is driven by divergent local adaptation or genetic drift in spatially distant locations and is hindered by gene flow because of ecological contact (Gavrilets 1999; Turelli *et al.* 2001; Chapter 6). This applies as much to models of parapatric speciation driven by sexual selection (Box 7.5) as to those driven by natural selection based on ecologically relevant traits. By contrast, the models described in this chapter focus on the adaptive processes that can generate local divergence in spatially extended, but genetically and environmentally continuous, populations. Ecological contact is a critical prerequisite for the operation of this alternative mechanism of parapatric speciation.

Our results show that local ecological contact may, in fact, be the driving force for parapatric speciation. In our models, gene flow is, of course, still a hindrance to local divergence, but the mechanisms that generate local disruptive selection require ecological contact. Local disruptiveness, in turn, selects for assortative mating, which reduces and eventually eliminates gene flow between the emerging species. The latter process is akin to reinforcement, but for the fact that in our models selection for prezygotic isolation emerges dynamically from frequency-dependent ecological interactions, as opposed to being the consequence of secondary contact. Evolutionary branching in spatially structured sexual populations shows how adaptive speciation can result in spatial segregation between the emerging species, and thus suggests an answer to Endler's (1977) old question of how sharp geographic differentiation can evolve in a single, spatially and genetically continuous species despite the presence of gene flow and in the absence of abrupt environmental changes.

The dynamic and continuous unfolding of local processes into a global pattern, as observed in our models, indicates that the classic allopatry–sympatry controversy about speciation may often be ill-posed: what, in the end, results in an allopatric pattern of species abundance can be generated by an intrinsically sympatric evolutionary process. Inferring past processes from extant patterns is always difficult, and to understand speciation it is particularly important to distinguish between the two. This also applies to the study of hybrid zones, which are usually thought of as originating from secondary contact between species that were formed in allopatry (Barton and Hewitt 1989). Our results reinforce the question of whether many of these zones are, in fact, primary and have arisen as stepped phenotypic and genetic clines out of genetically continuous ancestral populations by the processes described in this chapter. That such processes can be an important

---

Box 7.5 Fisherian runaway processes in spatially extended populations

Payne and Krakauer (1997) studied models for parapatric speciation that involve divergent Fisherian runaway processes driven by sexual selection in different regions along a spatial gradient. In these models, two alternative marker traits in males correspond to two alternative preference traits in females. In nonspatial models for homogeneous populations, sexual selection is expected to lead to fixation of either one of the two marker-preference pairs, with the actual outcome depending on the initial conditions.

By contrast, a spatial gradient in viability selection that favors one male marker trait to the left of a given spatial location x_0 and the other male marker trait to the right of x_0 can lead to spatial pattern formation, with one marker-preference pair evolving to the left of x_0 and the other evolving to the right of x_0; this generates a zone of intermediate frequencies with strong linkage disequilibrium between the two spatially segregated species (Lande 1982; Payne and Krakauer 1997). The steeper the environmental gradient, the narrower is this zone, and hence the more likely is complete reproductive isolation and thus speciation.

Payne and Krakauer (1997) also suggested that a spatial gradient in viability selection is not necessary to produce divergent runaway processes in different spatial locations. Instead, they proposed that alternative and spatially segregated marker-preference pairs could also evolve and persist in separate domains for extended periods of time, if dispersal along the spatial axis depends on mating success, such that males who experience less success exhibit higher movement rates. In this case, the spatial location of the interface between the emerging species depends on the initial conditions. Importantly, however, the patterns of spatial segregation between the two marker-preference pairs reported by Payne and Krakauer (1997) turn out to be transient (de Cara and Dieckamnn, unpublished).

In both of these cases of parapatric diversification, sexual selection causes different runaway processes in different spatial locations, through spatial variation either in viability selection or in the initial conditions. The analysis of these models illustrates that in most models of parapatric speciation ecological contact between the incipient species, which leads to genetic mixing, is a hindrance to the emergence of spatially segregated species. The results of Payne and Krakauer (1997) show that a viability gradient can, nevertheless, stabilize divergent Fisherian runaway processes in different spatial domains.

---

agent for generating adaptive splits into abutting sister species may thus provide new perspectives on old problems (Doebeli and Dieckmann 2003; Tautz 2003) – perspectives that are supported by a number of recent empirical studies, such as on intertidal snails (Wilding *et al.* 2001) and on *Anolis* lizards (Thorpe and Richard 2001; Ogden and Thorpe 2002; see Chapter 16).

We conclude that spatial structure can facilitate speciation because local adaptation along an environmental gradient increases the degree of frequency dependence in spatially localized ecological interactions, and hence the likelihood that these interactions generate disruptive selection and evolutionary branching. With local adaptation and sufficiently low levels of mobility, short interaction distances imply that individuals interact more often with other individuals of similar phenotypes, which results in an increased negative frequency dependence of their fitness on similar phenotypes. This is a potentially important mechanism for generating divergence that seems to have been overlooked in past work on parapatric speciation. Interestingly, this mechanism implies that the degree of frequency dependence induced by spatial structure actually decreases for very steep environmental gradients, because in this case very different locally adapted phenotypes occur in close spatial proximity, so that dissimilar phenotypes compete even if their interactions are spatially localized. As a consequence, facilitation of evolutionary branching is most pronounced for environmental gradients of intermediate slope, a result that is fundamentally different from those expected in classic scenarios of parapatric speciation along linear gradients (e.g., Endler 1977; see also Box 7.4). Studying such spatial facilitation of adaptive divergence may even shed new light on the problem of species–area relationships, because this mechanism operates more effectively in larger spatial areas, and thus provides an intrinsic explanation for higher speciation rates in larger areas. In sum, as anticipated in classic speciation theories, geographic structure may, indeed, play an essential role in the generation of diversity, but its importance and the role of spatially gradual environmental change may only be appreciated fully when adaptive processes of divergence through spatially localized and frequency-dependent ecological interactions are taken into account.

Acknowledgments   Michael Doebeli gratefully acknowledges financial support from the National Science and Engineering Council (NSERC) of Canada, and from the James S. McDonnell Foundation, USA.

Ulf Dieckmann gratefully acknowledges financial support from the Austrian Science Fund, from the Austrian Federal Ministry of Education, Science, and Cultural Affairs, and from the European Research Training Network *ModLife* (Modern Life-History Theory and its Application to the Management of Natural Resources), funded through the Human Potential Programme of the European Commission.

Plate 1 Variation among sympatric color morphs of Lake Victoria cichlids (Chapter 8).
Top: *Neochromis omnicaeruleus* at Makobe Island. Letters A to J (females) and K to M
(males) refer to Figure 8.2. A to D is a series of female genotypes that possess the WB
gene (from top: AAM_AM_A, AaM_AM_A, AaM_Am, Aamm), I to F is a series of female genotypes that possess the OB gene (from top: BBM_BM_B, BbM_BM_B, BbM_Bm, Bbmm), E is a
female of genotype aabb- -, J is a rare hybrid between WB and OB, and K to M are male
types $AabbM_AM_A$, aabb- -, and $aaBbM_BM_B$. B and A are X-linked. Bottom: The sympatric Lake Victoria cichlid fish sister species *Pundamilia pundamilia* (top row: left male,
right female) and *Pundamilia nyererei* (bottom row: left male, right female). These species
are reproductively isolated in clear water environments by assortative female mating preferences. Where the water is turbid they hybridize and in very turbid water one slightly
variable species occurs.

Plate 2 Male sticklebacks from Enos Lake, Vancouver Island (Chapter 9).
Top: Limnetic stickleback (*Gasterosteus aculeatus*) mainly feed on zooplankton in the open
water and have a slender body and small mouth. Bottom: Benthics are larger and deeper-
bodied fish that feed on benthic invertebrates inshore and in deeper sediments. Benthic
males in Enos Lake have lost the typical red throat and instead exhibit black nuptial coloration, probably because of a red shift in the available light spectrum in deep water caused
by tannins (Boughman 2002). *Source*: Pictures by E. Cooper.

Plate 3 Charr at the spawning grounds in Lake Thingvallavatn, Iceland (Chapter 10).
Top: A group of large benthivorous charr (*Salvelinus alpinus*) males compete for a female
(arrowed). Bottom: A pair of charr in the spawning act, i.e., at the moment of releasing
gametes. Pictures show an example of the ecological setting inhabited by charr, with a
rugged bottom of broken lava where eggs are deposited between the stones.

Plate 4 Two examples of insects forming host races (Chapter 11).
Top: The fruit fly *Rhagoletis pomonella* (male left, female right) has formed a new host race
by shifting from hawthorns to apples in North America. Bottom: The brown planthopper
Nilaparvata lugens forms host races on rice and related grasses in Asia. *Source*: Top picture
by J.K. Clark, copyright University of California Regents.

Plate 5 Predator–prey interactions among mites (Chapter 12).
Top: A two-spotted spider mite (*Tetranychus urticae*, right) is under attack by a specialist predatory mite (*Phytoseiulus persimilis*, left). Bottom: A soil predatory mite (*Hypoaspis aculeifer*, left) attacking a bulb mite (*Rhizoglyphus robini*, right). *Source*: Pictures by Bert Mans.

Plate 6 Disruptive selection due to different pollinators (Chapter 13).
Hawkmoths (*Hyles lineata* – top) and hummingbirds (*Selasphorus* spp. – bottom) pollinating the red flowers of *Ipomopsis aggregata* (left) and the pale pink flowers of *I. tenuituba* (right). Both pollinators visit flowers of both species in a hybrid zone in Colorado, USA. However, hawkmoths prefer narrow flowers, while hummingbirds prefer wide, red flowers. As a result, disruptive selection on floral phenotype can occur in this hybrid zone, as postulated in some models of plant speciation. Because pollinator preferences are not absolute, however, they would not impose strong reproductive isolation between *I. aggregata* and *I. tenuituba* in sympatry.

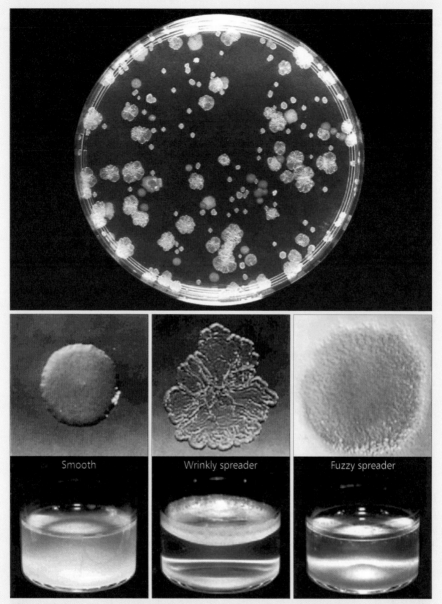

Plate 7 Phenotypic diversity and niche specificity among bacterial colonies (Chapter 14).
Diversity among colony phenotypes within cultures is maintained by frequency-dependent
selection (see Rainey and Travisano 1998). Top: Plate of *Pseudomonas fluorescens* colonies
that have evolved in a spatially heterogeneous environment. Middle: Colony phenotypes of
three distinct evolved morphs. Bottom: Niche specificity of the same three morphs in liquid
culture; the differentially evolved morphs show marked niche segregation. *Source*: Rainey
and Travisano (1998).

Plate 8 Subspecies and species of *Salamandra* in Europe (Chapter 15).
Top: Examples of two subspecies of *S. salamandra* from Germany, *S. s. salamandra* (above) and *S. s. terrestris* (below). Middle left: *S. corsica* from Corsica. Middle right: *S. lanzai*, *S. atra*, and *S. atra aurorae* (from left to right) from the alps. Bottom left: *S. infraimmaculata* from the Near East. Bottom right: *S. algira* from North Africa. *Source*: Pictures by S. Steinfartz.

Plate 9 Sympatric anole species from the Lesser Antilles (Section 16.2).
Top: *Anolis trinitatis*. Bottom: *A. griseus*. *A. trinitatis* is the smaller of the two and both
are widely sympatric on the island of St Vincent. Sympatric pairs always differ in size; it
remains to be shown whether this is because of size assortment or character displacement.

Plate 10 Dewlap color and pattern diversity in Caribbean anole species (Section 16.3). Top left: *Anolis sagrei* is a trunk–ground anole from the Bahamas, but occurs also elsewhere in the Caribbean, including Cuba. Top right: *A. mestrei* is a trunk–ground anole from Cuba. Bottom left: *A. grahami* is a trunk–crown anole from Jamaica. Bottom right: *Chamaelinorops barbouri* is a leaf–litter dwelling species from Hispaniola with no ecological counterparts on other islands. *C. barbouri* falls phylogenetically within *Anolis*.

Plate 11 Giant senecios and giant lobelias in Africa (Chapter 17).
Lobelia deckenii subsp. *deckenii* (left) and *Dendrosenecio kilimanjari* subsp. *cottonii* (right) show convergent evolution: both have large leaf rosettes perched atop wide stems cloaked with a mantle of marcescent foliage. The shown giant lobelia has an emerging inflorescence, which will grow to a height of ca. 1 m at maturity.

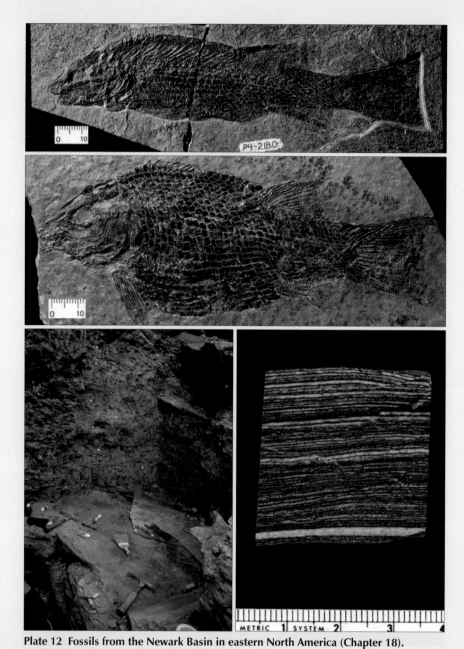

Plate 12 Fossils from the Newark Basin in eastern North America (Chapter 18).
Top: *Semionotus greenwoodi.* Middle: *S. saginatus.* Bottom left: The excavation site at
an Early Jurassic deposit of deep lake sediments of cycle P4. Bottom right: A sample of
lake sediment from the Late Triassic Lockatong Formation showing sedimentary couplets
(varves) that comprise a black organic-rich layer and a white layer rich in calcium carbonate.
Both layers were deposited within the course of a single year.

Part B
Ecological Mechanisms of Speciation

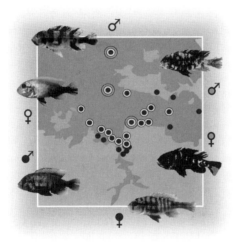

Introduction to Part B

The theoretical studies reviewed in Part A give credence to the expectation that processes of adaptive speciation will be encountered in nature, under a wide range of circumstances. Part B confronts these ideas with reality and explores the extent to which the mechanisms implicated in models of adaptive speciation have been observed:

- What mechanisms of frequency-dependent disruptive selection operate in particular systems?
- What mechanisms for the development of assortative mating are available in principle, and which ones have been recognized operating in actual systems?

To observe these mechanisms in action requires systems with two crucial features: they must be sufficiently accessible in all their parts, and they must be caught at suitable stages of speciation. The latter is only effectively possible when speciation is happening in the wake of recent environmental change.

This combination of prerequisites leads one immediately to think of bacteria as the only group of organisms for which the full processes of speciation can be studied in the laboratory in ways that allow all the subprocesses involved to be discerned (Chapter 14). Somewhat unfortunately, however, bacteria have a rather different kind of sex than higher organisms, and hence are not the ideal test bed for those theories of speciation targeted at the latter. Chapters on higher organisms – reproducing sexually and showing all the intricate behavior that appeals to watchers of nature documentaries – are therefore the primary focus of Part B. Only a few such systems conform to the requirements laid out above: the immediate candidates include fish flocks in young lake systems (Chapters 8, 9, and 10) and insects in the process of host switching, or increased specialization, as a result of recent changes in agricultural landscapes (Chapters 11 and 12). For plants, with their intricate relationships between flowers and pollinators, no studies have yet found systems in the process of adaptive speciation. Nevertheless, the mechanisms that drive such processes are present in principle, although they can only be documented indirectly (Chapter 13). The following chapters impart our taste buds with a healthy mix of the expected and the weird, as only nature can provide.

Part B begins with a discussion of the haplochromine cichlids in the African Great Lakes, well known for their extremely fast adaptive radiation. In Chapter 8, van Alphen, Seehausen, and Galis discuss the mechanisms that are believed to have caused this spectacular diversity. They argue that the primary reason for the speed and extent of the radiation is the potential for fast speciation through sexual selection. Two possible mechanisms are identified: local heterogeneity in ambient lighting that interacts with mate choice based on color patterns, and continual need for rescue from a sex-ratio distorter that operates in a complicated genetic interaction with some color-pattern genes. The resultant high degrees of assortative mating provide a fertile substrate for repeated adaptive radiation driven by resource competition and facilitated by the versatility of the cichlid jaw apparatus.

Whereas the tropics are interesting for their great riches in biodiversity and in mechanisms for niche separation and ensuing speciation processes, the far north has the advantage that it provides an abundance of near replicas of young lakes with relatively simple ecologies. This abundance is explored in Chapters 9 and 10.

Chapter 9 describes one of the cornerstones of empirical research on adaptive divergence and speciation: the three-spined sticklebacks in postglacial lakes of British Columbia. Rundle and Schluter investigate the role of selection and divergent adaptation for the evolution of reproductive isolation between limnetic and benthic species pairs. They envisage a scenario in which a phase of allopatric divergence between lacustrine and marine forms is followed by secondary invasion of the marine form, after which divergence and reproductive isolation evolve in sympatry because of ecological and reproductive character displacement. Evidence for a role of natural selection in this process comes from experiments that show premating isolation to be determined by phenotypic rather than geographic distance. Additional experiments have shown that benthic females in coexistence with limnetics demonstrate a higher degree of preference for benthic males than do solitary benthic females. These findings indicate that disruptive selection on ecological and mating traits was important for the evolution of reproductive isolation in lake sticklebacks, and thus provides evidence of processes thought to occur during adaptive speciation.

In Chapter 10, Snorrason and Skúlason take a wider view on the adaptive speciation of northern lake fish, with an emphasis on charr. Molecular evidence suggests that similar radiations have occurred on various occasions and at more than one locality. Ecologically, these radiations can be seen as genetic assimilations of resource polymorphisms. The great variety in lake types and ages allows replications of equivalent processes with regard to time and ecological conditions, the latter described by habitat diversity and environmental stability. From this, a scenario emerges in which developmental plasticity dominates in unstable environments and is replaced by genetic differentiation in stable environments. How the latter is aided by assortative mating through differentiation in spawning grounds or spawning time turns out to depend on the opportunities offered by the lake topography; where these options are not available, size-dependent mate choice has evolved.

Insects comprise the largest number of described species. In Chapter 11, Bush and Butlin suggest that resource specialization, in particular among phytophagous insects, is a major pathway for adaptive speciation in insects. Reproductive isolation can arise from co-speciation with the host, from allopatric speciation with or without host shift, or from sympatric speciation. The best-studied case of the latter class is sympatric speciation in the fruit fly *Rhagoletis pomonella*. The host shift in this case occurred very recently and reproductive isolation was achieved by assortative mating on the different host plants. The new host races maintain distinct gene pools, in spite of a considerable amount of ongoing gene flow, which indicates strong selection against hybrids. The authors use this example to discuss major general issues of sympatric speciation, such as preference and performance, negative trade-offs, and reinforcement.

Adaptations of phytophagous pests to agricultural plants provide other excellent opportunities to study mechanisms of incipient differentiation and adaptation. In Chapter 12, Egas, Sabelis, Vala, and Lesna highlight aspects of such systems that facilitate adaptive speciation. Herbivorous arthropods display preferences toward hosts, which translate into mate-choice preferences. Learning can play an important role, as it retards stabilizing selection and enhances disruptive selection. Also, infection by *Wolbachia* bacteria, which results in cytoplasmic incompatibilities for certain mating combinations, can ease speciation. Taken together, this complex of factors promotes adaptive speciation in phytophagous pests, which may be one of the explanations for the observed fast emergence of new pest species.

Processes of speciation in plants, and in particular in angiosperms, are influenced by their interaction with pollinators. In Chapter 13, Waser and Campbell discuss various relevant mechanisms. First, flower morphology can adapt to locally available pollinators, a process that could lead to reproductive isolation and character displacement upon secondary contact. Second, when the distribution of pollinator phenotypes acts as a resource-availability spectrum, frequency-dependent competition for pollinators could drive evolutionary branching according to an adaptive speciation scenario. Alternatively, reproductive isolation may arise as a pleiotropic by-product of local plant adaptation, which could occur under conditions of close spatial proximity and would also depend on the behavior of the pollinators. Reciprocal plant-transplantation experiments are expected to better elucidate and discriminate between these mechanisms.

To study experimental evolution in microorganisms may be the only viable approach to observing entire processes of speciation. In Chapter 14, Travisano reviews evolution experiments in bacteria that study the emergence of diversity out of genetically homogeneous ancestral populations. Allopatric divergence is investigated in experiments in which replicate populations are reared separately under identical environmental conditions. This reveals parallel evolution in traits that are highly correlated with fitness, whereas divergence in other traits is driven by genetic drift and historical contingency. Sympatric divergence is observed in experiments in which trade-offs in glucose metabolism result in specialization on either glucose or its breakdown products. Sympatric divergence is also observed when frequency-dependent selection for differential resource utilization leads to the evolution of three novel bacterial types from a single ancestral strain. These results underscore the role of experimental evolution as a promising tool in the study of adaptive diversification.

The chapters in this part highlight that there clearly is more to adaptive speciation than theorists could imagine, even in their wildest dreams.

8

Speciation and Radiation in African Haplochromine Cichlids

Jacques J.M. van Alphen, Ole Seehausen, and Frietson Galis

8.1 Introduction

The explosive radiation of cichlid fishes in the African Great Lakes has intrigued biologists for many decades. These lakes are outstanding, both in species richness and in the composition of their fish fauna. Several of them contain as many or even more fish species than all the rivers and lakes of Europe together (Lowe-McConnell 1987; Kottelat 1997). About 90% of the fish species in each lake belong to a single family, the cichlids (Cichlidae; Teleostei) and are endemic to that lake. Estimates of the phylogenies of these species flocks suggest that the species of Lakes Victoria, Malawi, and Tanganyika have evolved *in situ* (Meyer *et al.* 1990; Lippitsch 1993; Nishida 1997). Even more remarkable, for Lakes Malawi and Victoria the species flocks are derived from one or only a few closely related ancestral species and are all haplochromines. In comparison to the diversity of these lakes, riverine cichlid fish faunas in Africa and South America are considerably less diverse.

The unusually fast ecological radiation of haplochromine cichlids and the exceptionally dense species packing of these fishes demands an explanation. Most lacustrine species flocks of other fish taxa, even other cichlid taxa, are less diverse in ecology and species numbers. The versatility of the pharyngeal jaw apparatus, physiological properties, and their mouth-brooding behavior may all be necessary attributes, but these alone are not sufficient to explain the exceptional diversification of haplochromines. In this chapter, we argue that it is the combination of a number of factors.

Of the three African great lakes, Lake Tanganyika is the oldest and its age has been estimated to be between 9 and 12 My (Cohen *et al.* 1993). Lake Malawi's age has been estimated to be 1–2 My (Fryer and Iles 1972). Lake Victoria is the youngest (0.25–0.75 My; Fryer 1997). Most likely it dried up 200 000 years ago (Martens 1977) and seems to have dried up again in the late Pleistocene, after which it refilled between 15 500 (Beuning *et al.* 1997) and 14 600 years ago (Johnson *et al.* 1996a). Paradoxically, the young Lakes Victoria and Malawi contain more endemic species (between 500 and 1000 each) than the old Lake Tanganyika, with about 250 species. In particular, the Lake Victoria species flock must have been the result of truly explosive speciation, even if the lake did not completely dry out during the late Pleistocene.

One theory about the rapid speciation in Lake Victoria is the isolation of cichlids in marginal lagoons (satellite lakes), when the water level was low. These smaller water bodies were created and reunited with the main lake repeatedly, owing to fluctuations in the lake level. Allopatric speciation during isolation would contribute new species to the flock of the main lake when the water table rose again. New geological evidence (Johnson *et al.* 1996b, 2000; Beuning *et al.* 1997) suggests that Lake Victoria is merely 3.5 times older than Lake Nabugabo, the prime example of a satellite lake, which has persisted in a stable state for about 4000 years. Yet Lake Nabugabo contains only five endemic species. Hence, satellite lakes are unlikely to have contributed much to the diversity of over 500 species that originated in a period of possibly only 15 000 years.

Some species may have recolonized Victoria, after the Pleistocene drought, from Lake Edward, Lake Kivu, or the Malagarasi River. However, these must still have undergone explosive radiation within the Victoria lakebed to account for the large number of species today, because Lake Victoria does not share any species with these ecosystems and only a few of its many genera. Similarly, evolution of the species flock in Lake Malawi was attributed to allopatric speciation. Because Lake Malawi has probably not been split into separate lakes and there are few satellite lakes along its shore, it was suggested that the origin of the flock was by microallopatric speciation in habitat patches within the lake (Fryer and Iles 1972; Ribbink *et al.* 1983). This hypothesis assumes that water-level fluctuations caused periodic losses of suitable habitat and forced populations to colonize newly formed habitats (Arnegard *et al.* 1999).

However, the evidence that habitat stenotopy limits dispersal, a prerequisite for microallopatric speciation, is equivocal. Molecular population genetics suggest low dispersal rates between occupied habitat patches (van Oppen *et al.* 1997; Arnegard *et al.* 1999; but see Danley *et al.* 2000), but artificial reefs are rapidly colonized, with the first immigrants representing the most species-rich genera (McKaye and Gray 1984). Moreover, in both Lakes Victoria and Malawi, even those species that live in continuous habitats have speciated rapidly (Turner 1996; Shaw *et al.* 2000). Microallopatric population differentiation is known to occur within Lake Victoria (Seehausen 1996; Witte *et al.* 1997; Seehausen *et al.* 1998; Bouton *et al.* 1999), which probably has resulted in speciation among littoral cichlids (Seehausen and van Alphen 1999). It is, however, difficult to explain the explosive rate of speciation and ecological diversification as a product of microallopatric speciation alone. The sympatric occurrence of sibling species and color morphs, often on a small number of neighboring islands or even a single island, suggests that at least part of the speciation in Lake Victoria might occur in sympatry. One reason why investigators are skeptical about sympatric speciation is that it is hard to prove (Schilthuizen 2001). This is because it can be demonstrated only on the basis of a painstaking analysis of all the steps in the speciation process and because data on distribution alone do not provide a convincing proof (Seehausen and van Alphen 1999, 2000). In this chapter all the present-day evidence that supports the notion that some of the speciation in the haplochromines occurs in sympatry is presented.

In particular, the evidence for the hypothesis that sexual selection by mate choice for male coloration is the driving force behind the generation and reproductive isolation of color morphs of Lake Victoria haplochromines is reviewed. Note that no postzygotic isolation, in the form of reduced hybrid fertility, or viability between species of these fishes exist. The polygynous breeding system of haplochromines is associated with strong asymmetric investment in parental care, in which the females invest much more than males. Such a breeding system is conducive to sexual selection on male characters. We suggest that sexual selection, together with disruptive selection on feeding and on other characters, may have led to the present species diversity.

A second hypothesis discussed in this chapter is that a gene that causes sex reversal can cause speciation in haplochromines. The presence of this gene has probably resulted in the evolution of some of the mate preferences observed, and could be at the basis of the cascade of events that led to speciation by sexual selection upon coloration.

We discuss the manner in which genetic isolation caused by disruptive sexual selection is accompanied or followed by ecological diversification that results in niche differentiation between species. The striking diversity of feeding niches that characterizes cichlids of Lake Victoria suggests that niche differentiation occurred by rapid specialization for different feeding niches. Finally, we try to indicate why cichlids have shown adaptive radiation more often than other percoid relatives and what allows cichlids to specialize rapidly for different feeding niches.

8.2 Sexual Selection and Speciation in Cichlids

The reproductive behavior of haplochromines, the frequent sexual dimorphism in coloration, the polygynous mating system, and the male display during courtship in haplochromines suggest that sexual selection by female mate choice for male coloration could play an important role in the evolution of these fish.

The breeding system

The large species flocks of Lakes Malawi and Victoria consist of haplochromine cichlids. Haplochromines have colorful and conspicuous males, while the females of many species are cryptically colored. Haplochromines have a polygynous breeding system with female parental care by mouth brooding and, in many species, guarding of young fry. Females with ripe eggs visit males on their territories, where males initiate courtship. A courtship bout begins with the male approach, followed by a lateral display by the male, during which it displays its body and fin coloration by posing in front of the female with an erected dorsal fin. If the female remains or approaches the male, he quivers with the body, with the dorsal fin partly folded and the anal fin stretched out. This is followed by lead swimming to the spawning pit. There the male presents its egg dummies on the anal fin while circling in the pit.

Sexual selection and breeding system

Some researchers have suggested that mating preferences could be instrumental in speciation of cichlids. Kosswig (1947) suggested that mating preferences could lead to the isolation of family groups in cichlids, Fryer and Iles (1972) and Dominey (1984) suggested that microgeographic divergence in courtship might have been important in speciation of monogamous taxa. However, speciation rates in monogamous lacustrine cichlids are not higher than in other lacustrine fish that lack pair bonds and complex courtship. If anything, the evidence suggests the contrary: the most species-rich lineages of cichlids are polygynous. Until recently, studies on the functions of sexual dimorphisms in polygynous cichlids have been few, and few investigators have related these dimorphisms to speciation. McElroy and Kornfield (1990) studied male courtship behavior in Malawi cichlids and found that it was highly conserved among closely related species, and hence unlikely to have played a role in speciation.

Traits on which sexual selection could operate

Males of some cichlids on sand bottoms in Lakes Tanganyika and Malawi form leks and construct crater-shaped spawning sites from sand. McKaye (1991) hypothesized that sexual selection on these extended phenotypes could lead to speciation, but there is little direct evidence in support of this hypothesis. Hert (1989) demonstrated that females of a rock-dwelling Malawi cichlid prefer males with egg dummies on their anal fin to males from which the egg dummies are removed. Goldschmidt (1991) demonstrated that interspecific variation in egg dummy size correlated negatively with light intensity in the habitat. Goldschmidt and de Visser (1990) suggested that sexual selection on egg-dummy morphology could lead to speciation. Again, no direct evidence in support of this hypothesis is available. Male display during courtship suggests that dorsal fin and body coloration also play important roles in mate choice. At least two studies have analyzed hue variation in male cichlids (McElroy *et al.* 1991; Deutsch 1997). In cichlid evolution, the appearance of male nuptial coloration is generally preceded by that of a polygynous mating system (Seehausen 2000). Some studies have found circumstantial evidence to support speciation by selection on male nuptial coloration (Marsh *et al.* 1981; McKaye *et al.* 1982, 1984). However, experimental tests of this hypothesis were not undertaken until recently (Seehausen *et al.* 1997; Knight *et al.* 1998; Seehausen and van Alphen 1998).

Such a test involves a number of steps:

■ Provide evidence that female mate choice is based on male coloration.
■ Provide evidence that across species female preference for male coloration is correlated with male coloration.
■ Show that female preference can act as a genetic barrier between populations that differ in preference and color.
■ Provide evidence that within-species variation exists for male coloration and female preference.

8.3   Sexual Selection in *Pundamilia*

Seehausen and van Alphen (1998) tested some of the assumptions discussed in Section 8.2 using two sibling species of the genus *Pundamilia*. *P. nyererei* and *P. pundamilia* are two anatomically very similar forms that behave as biological species in places where they occur sympatrically, except in places with exceptionally low water transparency where they interbreed (Seehausen *et al.* 1997). They usually show some ecological differences (Seehausen 1997; Seehausen and Bouton 1997). They can be distinguished by their male nuptial coloration. Males of *P. nyererei* are bright crimson dorsally, yellow on the anterior flanks, and their dorsal fin is crimson. Males of *P. pundamilia* are grayish white dorsally and on the flanks, and have a metallic blue dorsal fin (see Plate 1, lower part). Subtle differences in head anatomy help to distinguish the otherwise very similar females.

Evidence for female choice based on male coloration

The behavioral responses of females to males of both species were studied in choice experiments under either white light or monochromatic light:

- Females of both species exhibited strong species assortative mate choice under white light. However, in tests under monochromatic light, when the interspecific differences in male coloration were masked, no preference was observed (Figure 8.1).
- Hybrids between *P. nyererei* and *P. pundamilia* can be obtained easily under no-choice conditions in the aquarium. Intermediate phenotypes resembling F1 or F2 hybrids were not found in most of their common geographic range, but occurred in abundance at several places with exceptionally low water transparency.

In summary, experimental evidence and field observations show that female mate choice based on male coloration prevents interbreeding when the light conditions allow discrimination, but that hybridization occurs when male coloration cannot be assessed by the female. This experiment thus provides evidence that female choice of male coloration can be responsible for reproductive isolation between existing species.

Field evidence from comparisons between communities

Field evidence supports the experimental evidence: Seehausen *et al.* (1997) reasoned that if sexual selection on coloration plays a role in reproductive isolation between species, ambient light conditions could set limits to the effects of sexual selection. Sexual selection can only operate on variation in male coloration and in mate preference where light conditions allow the perception of the color variation. They predicted that the extent of interspecific color variation, the number of species, and the number of color morphs within a species should increase with the width of the ambient light spectrum.

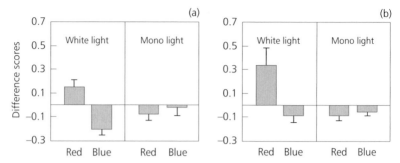

Figure 8.1 Differences in female mate choice behavior toward red and blue males, depending on female ancestry and ambient light conditions. Female responsiveness is measured as the proportion of encounters in which the male display was followed by a female approach (a) or follow (b). The shown mean difference scores and standard errors are based on subtracting female responsiveness to blue males from female responsiveness to red males. Such difference scores were obtained both under white light (left boxes) and under monochromatic light (right boxes), and for females from red fathers (left bars) and from blue fathers (right bars). Each bar is based on the scores from four females, each tested three times.

Data collected at two series of islands to test these hypotheses showed that the ambient light spectrum explained the variation in these parameters better than any of a large number of other ecological variables. The distinctiveness of nuptial-hue difference between sympatric species, the number of sympatric color morphs within a species, and the number of coexisting species within genera increased with the width of the ambient light spectrum. Moreover, the brightness and distinctiveness of male coloration had decreased in one population within the previous ten years, following a decrease in water transparency because of eutrophication. Finally, the frequency of phenotypes that are intermediate between sympatric forms was correlated negatively with water transparency. These observations are all consistent with sexual selection by direct mate choice as the principle factor that keeps closely related species genetically isolated.

Intrapopulation variation in preference and color

To see whether intrapopulation variation in preferences for male nuptial coloration exists in *Pundamilia*, females of one population were tested. In this population the males vary between blue and dull red in the dorsal fin and between blue gray and dull red on the dorsum. Whereas most females consistently and significantly preferred blue over red, some were unselective and others tended to prefer red males. This demonstration that intraspecific variation for male coloration co-occurs with intraspecific variation in female preference (Seehausen 2000) shows that intraspecific disruptive sexual selection on coloration can potentially play a role in the evolution of color polymorphisms. Such color polymorphisms could result in speciation.

8.4   Sexual Selection in *Neochromis omnicaeruleus*

To investigate if and how sexual selection acts on color variation within a species, Seehausen *et al.* (1999) studied the polymorphic species *N. omnicaeruleus*. If morphs would mate selectively, color polymorphisms might represent incipient stages in speciation by disruptive sexual selection. Seehausen *et al.* (1999) tested whether polymorphism has the properties of an incipient stage in speciation, and whether mating preferences can disrupt gene flow in sympatry prior to ecological differentiation, by investigating:

- Ecological correlates of sympatric color variation;
- Genetics of polymorphic color variation;
- Frequency of color phenotypes in the natural population;
- Female and male mate choice in aquarium experiments.

The investigations suggest that *N. omnicaeruleus* represents an original species with blue males and brown females, and two incipient species that are black and orange blotched or black and white blotched.

Morphometric differences

Wild-type *N. omnicaeruleus* differs from its close relative *N. rufocaudalis* by a smaller cheek depth, larger eyes, and a longer mandible. The larger cheek depth and smaller eyes leave more space for the adductor mandibulae muscles. Together with the shorter jaws, this provides *N. rufocaudalis* with more biting power than *N. omnicaeruleus* and makes *N. rufocaudalis* a more specialized algae scraper. While these differences occur between closely related and ecologically very similar species, no morphometric differences were found between color morphs of *N. omnicaeruleus*.

Genetics of color variation

According to Kornfield (1991), haplochromines have XY sex determination with the male as the heterogametic sex. Fish do not have dimorphic sex chromosomes. Nevertheless, in the following the chromosome that carries the main sex-determining genes is called the sex chromosome and the others autosomes.

In nature, nine different phenotypes of females were found and four male phenotypes (see Plate 1, Figure 8.2, and Table 8.1). The most common, "plain"-type (P), has females that are yellow–brown with dark vertical bars and males that are bright metallic blue. The other common types are either "white and black blotched" (WB), or "orange blotched" (OB). Other types are intermediates between these three, as well as almost entirely black and almost entirely orange, and occur in low frequencies.

Color in *N. omnicaeruleus* is associated with sex modifiers. WB and OB are associated with a dominant X-linked sex reversal gene. The presence of a recessive autosomal male determiner (the "male-rescue gene") suppresses the dominance of this gene over the y-allele. This rescue gene is associated or identical with the modifiers of WB and OB coloration. Tight linkage between sex determiners and

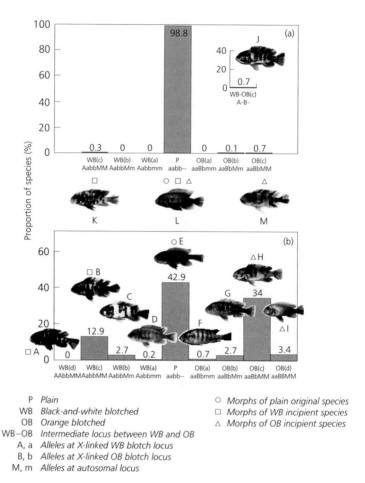

P   Plain
WB   Black-and-white blotched   ○ Morphs of plain original species
OB   Orange blotched   □ Morphs of WB incipient species
WB–OB   Intermediate locus between WB and OB   △ Morphs of OB incipient species
A, a   Alleles at X-linked WB blotch locus
B, b   Alleles at X-linked OB blotch locus
M, m   Alleles at autosomal locus

Figure 8.2 The color polymorphism in *N. omnicaeruleus* (see Plate 1 for the morphs in color, identified by letter). Phenotypes, genotypes, and natural frequencies of the phenotypes: (a) males; (b) females. M in the genotype notations represents two recessive male rescue alleles and additive blotch expressivity modifiers M_A and M_B (see text). The three common P phenotypes in both sexes are those from P parents, WB(c) mothers, and OB(c) mothers. They differ at the autosomal locus and in mating preferences. Female and male morphs with mutually compatible mating preferences are designated by symbols. Information on mating preferences within and between the original species and the WB incipient species is based on the reported mate-choice experiments. That preferences within and between the original species and the OB incipient species parallel those in the WB system is an assumption supported by experiments with a wild OB(c) female that preferred blue over WB males, but did not discriminate between a blue and an OB male (unpublished), and by field underwater observations. Courtship between OB(c) females and P males has been observed on many instances, and a focal observation (20 minutes) of one OB male revealed courtship only with an OB female, although three P, three WB, and two OB females were encountered by that male.

Table 8.1 Possible genotypes and the resultant phenotypes found in nature. The suffixes (a) to (d) indicate increasing blotchiness.

Genotype	aabb	Aabb	AAbb	aaBb	aaBB	AaBb	AaBB	AABb	AABB
Females									
mm	P	WB(a)	–	OB(a)	–	–	–	–	–
mM	P	WB(b)	–	OB(b)	–	–	–	–	–
MM	P	WB(c)	WB(d)	OB(c)	OB(d)	WB/OB(c)	–	–	–
Males									
mm	P	–	–	–	–	–	–	–	–
mM	P	–	–	OB(b)	–	–	–	–	–
MM	P	WB(c)	–	OB(c)	–	–	–	–	–

color genes has been found in other fish as well. OB and/or WB color morphs occur in haplochromines in several African lakes and are usually associated with a strong female determiner (Fryer and Iles 1972; Holzberg 1978; Lande *et al.* 2001). Similar genes have also been observed in the platyfish *Xiphophorus maculatus* (Orzack *et al.* 1980) and in the rice fish *Oryzias latipes* (Wada *et al.* 1998). The evidence for the link of color genes and sex modifiers in *N. omnicaeruleus* is given below.

Breeding experiments in the laboratory showed that WB and OB fish homozygous for the gene that codes for WB or OB are almost black or orange, respectively, while heterozygous fish, obtained from crosses with P fish, have blotched phenotypes with much more white (WB) or black (OB). Crossings between heterozygous WB females and heterozygous WB males showed that the WB gene is sex linked (Table 8.2). The same was found in crossings between heterozygous OB females and heterozygous OB males. Crosses between WB females and OB males yielded no homozygous blotched offspring, which suggests that WB and OB are alleles on different loci. The occurrence of intermediate phenotypes between blotched and plain in crosses between blotched and some plain individuals is consistent with the presence of a locus (M, m) with additive effects on the intensity of expression of WB and OB genes. Thus, the minimum model that can explain the observed color segregation is a three-locus, six-allele model. Two alleles at each of two sex-linked loci determine whether a phenotype is "plain" (aabb), WB (Aabb, AAbb), or OB (aaBb, aaBB). At the modifier locus, M results in a higher intensity of expression of the blotch gene than m.

The absence of males (homozygous for blotched) and the observed recombination fraction ($r = 0.052$) obtained in crosses with blotched males (heterozygous for color and sex) suggest that both blotch loci are X-linked. The third locus (m, M) appears to be autosomal, because crosses indicate that both sexes can be homozygous for either allele and because the recombination fractions ($0.4 < r < 0.6$) between the M locus and the sex chromosome are high.

When at least one parent carried the allele m at the modifier locus, crosses expected to yield blotched males (e.g., males of a P line mated with blotched females) produced significantly skewed sex ratios. Two crosses of homozygous WB

Table 8.2 Results of breeding experiments. First column and first row give the parental phenotypes. Allele A represents the blotch allele at both the WB and the OB locus, which show parallel inheritance, allele M represents M_A and M_B (see text). P, plain morph; B, blotched morphs (a) through (d) represent degrees of blotchiness as affected by the modifier locus with alleles M, m. Mate preferences of all WB-line individuals and all P-line individuals crossed with them were tested prior to the breeding experiments. Observed and expected phenotypic segregations are given as a percentage in each box. The sex effects of the X-linked female determiner and the autosomal male determiner are incorporated. In the first column, third row, the probability for the B(d) phenotype in the cell in the third row of the first column is the sum of all the probabilities for the AAMmxx and the AAMMxx genotypes, which cannot be distinguished visually. Deviations from expectations are tested by χ^2 for color and sex separately, and for the color–sex combination using the recombination fraction of 0.052.

Father/ Mother	Blotched AaMMxy Obs.	Exp.	P of blotched aaMMxy Obs.	Exp.	P of P-line aammxy Obs.	Exp.
Blotched (d) AAMMxx	B(d) 41.50	47.50	B(c) 44.90	50.00	B(b) 100	100
	B(c) 4.70	2.50				
	B(d) –	2.50	B(c) 55.10	50.00		
	B(c) 54.70	47.50				
	Color χ^2=1.53,df=1,p=0.22		Color χ^2=0,p=1		Color χ^2=0,p=1	
	Sex χ^2=0.47,df=1,p=0.69		Sex χ^2=0.51,df=1,p=0.48		Sex χ^2=0,p=1	
	Combχ^2=0.44,df=3,p=0.44					
Blotched (c) AaMMxx	B(d) 23.00	23.70				
	B(c) 25.50	23.70			B(b) 39.50	48.70
	P 0.60	2.60			P 28.90	25.00
	B(d) –	2.60				
	B(c) 26.70	23.70			B(b) 5.30	1.30
	P 24.20	23.70			P 26.30	25.00
	Color χ^2=1.14,df=2,p=0.56				Color χ^2=0.42,df=1,p=0.52	
	Sex χ^2=0.16,df=1,p=0.69				Sex χ^2=0.54,df=1,p=0.46	
	Combχ^2=0.44,df=5,p=0.19				Combχ^2=1.30,df=2,p=0.52	
Blotched (d) AaMmxx	B(d) 21.40	23.70			B(b) 12.50	24.35
	B(c) 21.40	11.85			B(a) 12.50	25.00
	B(b) 17.90	23.08			P 12.50	25.00
	P 3.60	2.60				
	B(d) –	2.60			B(b) 6.30	0.65
	B(c) 17.90	11.85			P 56.20	25.00
	B(b) –	0.62				
	P 17.90	23.70				
	Color χ^2=3.14,df=3,p=0.37				Color χ^2=3.14,df=3,p=0.37	
	Sex χ^2=0.12,df=1,p=0.73				Sex χ^2=0.12,df=1,p=0.73	
	Combχ^2=4.78,df=6,p=0.57				Combχ^2=4.78,df=6,p=0.57	
P aammxx	B(b) 50.90	47.50	P 28.20	50.00	P 54.90	50.00
	P 3.50	2.50	0			
	B(b) 5.30	2.50	P 71.80	50.00	P 45.10	50.00
	P 40.40	47.50				
	Color χ^2=0.86,df=1,p=0.35		Color χ^2=0, p=1		Color χ^2=0, p=1	
	Sex χ^2=0.44,df=1,p=0.51		Sex χ^2=13.50,df=1,p=≤0.05		Sex χ^2=1.38,df=1,p=0.24	
	Combχ^2=1.77,df=3,p=0.62					

females with P males yielded all female clutches. Crosses of heterozygous OB females and P males yielded significantly fewer OB males than expected on the basis of a 0.5 Fisherian sex ratio. This can be explained by a female determining gene (W), tightly linked to both X and a blotched color allele, of which the dominance over Y is complete in WB, but not in OB. P males from blotched parents and blotched males, when crossed with blotched females, produce normal sex ratios, which do not deviate from 0.5. Hence, these must differ genetically from males of the P line. Males from blotched parents must not only possess a gene that enhances blotch expression in their blotched heterozygous offspring, but also a recessive male-rescue gene that counteracts the sex effects of W. These maleness effects and color effects are associated and might be caused by the same locus (Table 8.2).

Phenotype frequencies in nature

A total of 2600 adult specimens were collected between 1990 and 1996 at Makobe island in Lake Victoria. The frequencies of the ten observed phenotypes differ strikingly: P, WB, and OB females and P males are much more abundant than are the others. In particular, it is striking that WB and OB males are extremely rare in nature (Figure 8.2). Many genotypes that can easily be obtained in laboratory breeding are lacking in nature, which indicates that mating is not random. The latter has also been shown in mate-choice experiments.

Mate-choice experiments

The difference between P males of a true breeding plain population P_P and P males that originate from blotched parents (P_{WB} and P_{OB}), as discovered in the experiments on the genetics of coloration, was also reflected in their mate-choice behavior. In female choice/male no-choice trials WB females were courted more often by P_{WB} than by P_P males, and more often by WB males than by P_P males. The WB females were equally responsive to the three kinds of males. P_P females were courted as often by P_P males as by P_{WB} males. They were also courted equally often by WB and P males, but responded to and followed P males more often than WB males. Hence, independent of male courtship activity, P females preferred P males, but WB females did not exhibit preferences. In this no-choice situation, P_P was the only choosy male type. In choice experiments, P_P males also ignored WB females. WB males courted WB females more often than P females. P males from P parents courted P females more often, but P males from WB parents behaved as WB males and courted WB females more often (Figure 8.3).

Summarizing the evidence

Thus, Seehausen *et al.* (1999) have shown that differences at possibly as few as three loci give rise to conspicuous and, in nature, discrete color variation, with three common morphs: "plain", "white-and-black blotched", and "orange blotched". Genetics alone does not explain the relative frequencies of the different morphs in nature. The combination of a high frequency of blotched females with a low frequency of the M allele at the modifier locus in plain morphs in nature is

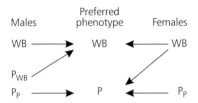

Figure 8.3 Summary of male and female mating preferences in *N. omnicaeruleus*. One arrow indicates a mating preference for either plain colored or white-and-black blotched partners. Two arrows indicate that no preference is exhibited for the two types of partners.

incompatible with random mating. Intermediate phenotypes were distinctly less abundant than expected under random mating. The evidence from the mate-choice experiments is in agreement with the field evidence that mating between color morphs is nonrandom. A sex-determining gene linked to color can explain the evolution of preferences. Mating preferences in color morphs of *N. omnicaeruleus* might have evolved by sex-ratio selection to avoid matings that result in progeny with skewed sex ratios or to avoid fitness loss through the production of unfit YY-homozygotes. For complete interruption of the gene flow between the blotched and plain morphs, preference of blotched females for blotched males may be required. However, the rarity of blotched males in nature selects against the spread of a gene for such a preference and may prevent full speciation.

Modeling speciation in haplochromines with sex reversal

Recent models by Lande *et al.* (2001), based on the genetics and the mate-choice behavior summarized above, show that the combination of sexual selection by mate choice and sex-ratio selection provides a pathway for sympatric speciation. The model considers mate choice by both sexes, as observed by Seehausen *et al.* (1999). Novel color genes can invade against sexual selection because they are initially largely restricted to females (as the male-rescue gene is rare). This causes sexual selection on mating preferences in males for the new female colors. Once the new preference is sufficiently frequent, it, in turn, causes sexual selection by female choice on the initially rare males with new color (and male-rescue genes), leading to sudden complete speciation. The evolution of a novel mating preference in the model is driven by the spread of a dominant sex-determining gene W. This gene can spread in the population when it has some intrinsic selective advantage. This would be so when either the female-biased sex ratio that W produces is selected for by intrademic group selection, or when W has a pleiotropic effect that increases reproductive success to compensate for selection against female-biased sex ratios in panmictic populations. Matings between XY males and WY females will result in YY zygotes.

If YY males are not viable, because of deleterious Y-linked mutations, the establishment of the W gene, followed by the spread of a W-linked new color mutation can result in the above-described sympatric speciation scenario: preference for the new color evolves and a dominant female determiner W together with a recessive suppressor gene M spread to fixation. This suppressor counteracts the effects of W in animals homozygous for M, overcomes the production of inviable YY individuals, and restores the sex ratio.

If YY males are viable, the establishment of the W gene results in a change in the heterogametic sex (XX females and XY males in the parental population, YY males and WY females in the derived population). This is in accordance with the observation that species of the tilapiine cichlid genus *Oreochromis* often differ in the heterogametic sex.

Lande *et al.*'s model explains X-linked sex-reversal, often observed in cichlids, and the strong association between sex reversal and color, as found for the blotch polymorphism in *N. omnicaeruleus* and other haplochromine cichlids.

The model also shows that sex-ratio selection, together with sexual selection for color, can result in sympatric speciation. Full speciation was not found in the empirical study of *N. omnicaeruleus*, possibly because the rarity of blotched males in nature prevents the evolution of preference by blotched females for males of their own kind. The rarity of these males, in contrast to the high frequency of blotched females, can be explained if two opposed selection forces act on the polymorphism. Given that blotched males are produced at normal ratios when blotched females mate with males of the MM genotype, we have to invoke natural selection against them to explain their rarity. Territorial males that perform conspicuous displays to attract females must expose themselves more than females to the abundant avian predators. This natural selection against blotched males might then be balanced by sexual selection.

Last but not least, Lande *et al.*'s model offer an explanation why rock-dwelling haplochromines have undergone more speciation than other cichlids. Rock-dwelling cichlids are territorial habitat specialists with little dispersal. Hence, populations in rocky habitat patches often experience relatively high degrees of genetic isolation (van Oppen *et al.* 1997; Arnegard *et al.* 1999). Such spatially structured populations are more likely to accumulate recessive, deleterious Y-linked mutations that result in low YY-male fitness.

At present we do not know what constitutes the intrinsic selection advantage that allows the W gene to spread initially. The fitness of YY-homozygotes is a subject of ongoing research.

Field evidence for sympatric speciation in haplochromine cichlids

In nature, closely related species are isolated neither by the timing of reproduction nor by the location of reproduction (Seehausen *et al.* 1998). This makes it most likely that direct mate choice is the factor that keeps closely related species genetically isolated and maintains species diversity in the absence of postzygotic barriers to hybridization. Compelling evidence for this is given in the study by Seehausen *et al.* (1997) reviewed above.

The hypothesis that sympatric speciation, caused by disruptive sexual selection, contributed to the origin of haplochromine species diversity, makes two predictions:

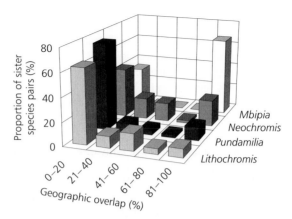

Figure 8.4 The proportions of congeneric species pairs plotted against geographic distribution overlap, for four genera of rock-dwelling Lake Victoria haplochromines.

- The distribution of geographic range overlap between closely related species should have a bimodal frequency with peaks at little range overlap and at large range overlap.
- The proportion of pairs of closely related species that are different colors is higher among sympatric pairs than among allopatric pairs.

An analysis of patterns of geographic distribution of 41 species supported these predictions (Figure 8.4; Seehausen and van Alphen 1999, 2000). Without sympatric speciation, the distribution of range overlap would be expected to have a single hump. However, it shows a bimodal frequency distribution, with peaks at both the fully allopatric and fully sympatric positions.

8.5 Pharyngeal Jaw Versatility and Feeding Diversification

Survival of a new form depends on its ability to respond to ecological demands. Here, the versatile anatomy of cichlids, their phenotypic plasticity, and their capacity to adapt their behavior through learning come into play. Here, we discuss the greater versatility of the cichlid anatomy over that of their percoid ancestors.

Ecological diversification

The genetic isolation of subpopulations by sexual selection, as described for haplochromines, is nonadaptive in an ecological sense, because it results in ecological homologs. Disruptive natural selection is needed for the divergence of young species. The slight, but significant and adaptive, ecological differences that are often observed between closely related species may evolve under disruptive natural selection, which can easily operate after the termination of gene flow by assortative mate choice for males with different body coloration. For lasting coexistence, ultimately both mate choice and natural selection need to be disruptive (Lande and Kirkpatrick 1988).

Most cichlids are opportunistic feeders, and all feed on nutrient-rich food when it is available. Yet, there is ample evidence for ecological radiation in the haplochromines. Species have diversified into specialists on detritus, aufwuchs, zooplankton, phytoplankton, insects, fish, fish scales, fish parasites, gastropods, and so on. Species are further segregated by depth in the water column, by habitat, and by behavior (e.g., different modes are used to eat algae from the rocks). This enables a very large assemblage of coexisting species. Although Lake Victoria haplochromines most likely evolved from a single ancestor over a short evolutionary time scale, they represent an impressive array of ecological adaptations.

Pharyngeal jaw apparatus in cichlids and other labroids

The cichlids belong to the labroid fishes (Perciformes), an extensively diversified group. This diversification has been attributed to the development of a morphologically specialized and functionally versatile pharyngeal jaw apparatus (Fryer and Iles 1972; Greenwood 1974; Liem 1973, 1978, 1979; Yamaoka 1978; Liem and Sanderson 1986; Jensen 1990). The structure and function of the derived pharyngeal jaw apparatus of labroids has been studied extensively (Greenwood 1965; Liem 1973, 1978, 1979; Yamaoka 1978; Hoogerhoud and Barel 1978; Liem and Greenwood 1981; Kaufman and Liem 1982; Liem and Sanderson 1986; Stiassny and Jensen 1987; Galis 1992, 1993a, 1993b; Vandewalle *et al.* 1992), in contrast to that of nonlabroid Perciformes, which is less derived and retains more ancestral conditions [Liem (1970) for Nandidae, Lauder (1983a, 1983b) for Centrarchidae, Wainwright (1989) for Haemulidae, and Vandewalle *et al.* (1992) for Serranidae].

Liem (1973) suggested that the exceptional speciosity of cichlids is a consequence of an evolutionary innovation: the versatile pharyngeal apparatus. Recent support for Liem's hypothesis is provided by a comparison of the pharyngeal jaw apparatus in cichlid fishes with that of their presumed generalized percoid ancestors (Galis and Drucker 1996). The results of this comparison suggest support for a more general hypothesis by Vermeij (1974), namely that speciose taxa are characterized by more independent constructional components than taxa that are less speciose. A large number of independent components increase the number of potential solutions for a particular biomechanical problem. Therefore, body plans with more independent components can be modified and diversified more easily than those with fewer independent elements.

Galis and Drucker (1996) compared the mechanics of the pharyngeal biting of the centrarchids (a less-derived percoid family) with those of cichlids, by developing a mechanical model of pharyngeal biting in centrarchids. This was based on anatomic observations and manipulations in live fishes. The centrarchid model was tested by electrically stimulating individual muscles and observing the resultant movements.

Similar muscle stimulation experiments were carried out with the cichlid fish *Labrochromis ismaeli*, while anatomic studies were carried out with the cichlids *O. niloticus*, *Cichlasoma citrinellum*, and *L. ismaeli*.

The anatomic research revealed that there are three important structural couplings among the elements of the pharyngeal jaw apparatus of centrarchids (Figure 8.5). Two of these are modified in the cichlids:

▪ *Fourth epibranchials – upper pharyngeal jaws.* In the centrarchids, there is an articulation between the lateral face of the upper pharyngeal jaw and the medial surface of epibranchial 4, as has been found in heamulids (Wainwright 1989). This coupling transmits the force exerted by the m. levator posterior and externus 4 on epibranchials 4 to the upper pharyngeal jaws. In the cichlids this coupling is lost. The fourth epibranchials lie against the surface of the upper pharyngeal, but the two elements are capable of independent movement. Unlike in the centrarchids, the m. levator externus of the cichlids runs from the neurocranium to the lower pharyngeal jaws, rather than to epibranchial 4, and it suspends the fused lower pharyngeal jaws in a muscle sling. The force of this muscle therefore acts directly upon the lower pharyngeal jaw, instead of indirectly upon the upper pharyngeal jaw via rotation of the fourth epibranchials.

▪ *Fourth ceratobranchials – lower pharyngeal jaws.* In the Centrarchidae the lower pharyngeal jaws are firmly connected to the fourth epibranchials. This greatly restricts independent movement of the lower pharyngeal jaws and the fourth ceratobranchials. In the Cichlidae, this coupling is replaced by a more flexible connection, which permits movement of the lower pharyngeal jaws with respect to the ceratobranchials.

▪ *Fourth epibranchials – lower pharyngeal jaw(s).* In the centrarchids, there is a strong connection between the fourth epibranchials and the lower pharyngeal jaws, which is both ligamentous and muscular. This direct connection strengthens the second coupling in linking the movements of the fourth epibranchials and the lower pharyngeal jaws. The third coupling has been retained in cichlids.

The anatomic model and its test by electrostimulation of muscles showed that these couplings unite the movements of the upper and lower pharyngeal jaws in centrarchids, and therefore seriously constrain the number of biting movements and, thus, the number of activity patterns of the pharyngeal jaw muscles. In contrast, loss of the first and second couplings in cichlids resulted in independent movement of the upper and lower pharyngeal yaws. This allows the generation of biting forces in many different directions (Galis 1992). In addition to this, the lower pharyngeal jaws in cichlids are fused and a strong bond exists between the upper pharyngeal jaws. This provides a twofold increase in the force that can be applied to the prey (Figures 8.5a, 8.5b, and 8.5c). Thus, the functional flexibility of cichlids is enhanced in comparison with that of more primitive labroid fish like the centrarchids (Figure 8.6). This increase in versatility of the pharyngeal allows efficient handling of many different prey items (Liem 1973; Stiassny and Jensen 1987; Galis and Drucker 1996) and the resultant flexibility allows improved phenotypic response to changing requirements. Like other forms of phenotypic plasticity, it may be genetically assimilated subsequently (Galis and Metz 1998). This decreases the probability of extinction after colonization of a new environment and increases the

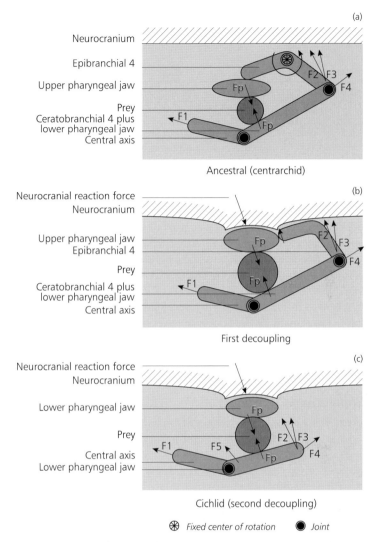

Figure 8.5 Representations of biting with pharyngeal jaws. (a) Presumed centrarchid-like ancestors of cichlids (coupled movement of the upper and lower pharyngeal jaw is caused by F1 and F2; rotation of epibranchial 4, part of the fourth gill arch, causes the upper pharyngeal jaw to move down). (b) Fishes with the hypothesized intermediary state after the decoupling of the upper and lower pharyngeal jaw (there is no rotation and, thus, the upper pharyngeal jaw is not pushed down, but pushed up against the neurocranium; the fourth gill arch and the lower pharyngeal jaw are together pulled up by F1 and F2). (c) Cichlids after the decoupling of the lower pharyngeal jaw and the fourth gill arch (ligamentous and muscular connections have disappeared between the lower pharyngeal jaw and the fourth gill arch and the muscles that produce F1 and F2 are inserting directly on the lower pharyngeal jaw instead of on epibranchial 4).

(a)   (b)   (c)

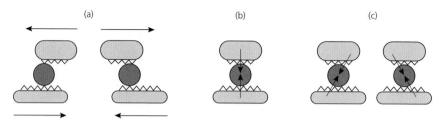

Figure 8.6 Possibilities of movement of the pharyngeal jaw apparatus of a centrarchid-like ancestor of cichlids (a *or* b) and of a cichlid (a *and* b *and* c). Note the much increased versatility of the pharyngeal jaw apparatus after the two decouplings. *Sources*: Galis (1992) and Galis and Drucker (1996).

likelihood of long-term population survival. The enhanced functional flexibility of cichlids not only provides phenotypic plasticity, but also means that only small evolutionary changes are necessary to change feeding specialization. In addition to the versatility of the pharyngeal jaws, the cichlid design enables rapid adaptation to new ecological circumstances by changes in allometric growths of structures in the head (Greenwood 1974, 1981a, 1994). Such changes could possibly be induced by changes in the timing during development, and could be coded for by only a few genes (Gould 1977). Such changes can also be induced phenotypically, as juveniles grow up. This phenotypic plasticity is important for the survival of individuals that have to adapt to a new environment (Galis and Metz 1998; Bouton *et al.* 1999).

8.6   Concluding Comments

The unusually fast ecological radiation of haplochromine cichlids and the exceptionally dense species packing in several lineages of African cichlids raises the question of what is different in these cichlids. Most lacustrine species flocks of other fish taxa, even other cichlid taxa, are less diverse in ecology and species numbers. The versatility of the pharyngeal jaw apparatus, physiological properties, and mouth-brooding behavior may all be necessary attributes, but these alone are not sufficient to explain the exceptionally fast radiation of haplochromines. The evidence presented above indicates that the combination of a number of factors may be required to explain such radiation. The polygynous breeding system of haplochromines is associated with strong asymmetric investment in parental care, in which the females invest much more than males. Such a breeding system is conducive to sexual selection on male characters. It has resulted in the sexually dimorphic breeding coloration, with brightly colored males and often dull females. If water transparency is high, disruptive sexual selection on male coloration can result in genetic isolation between fish that differ in male coloration and female preference for male coloration. Disruptive sexual selection can act either in full sympatry, or maintain genetic isolation of populations that have developed different male coloration in allopatry and have become sympatric secondarily by range expansion. The relative contribution of sympatric and allopatric speciation to the haplochromine species flocks is still unclear. A premating barrier between populations with differently colored males alone is not sufficient to allow long-term coexistence of these populations. However, such a premating barrier prevents the homogenizing effect of recombination in a contiguous population. This facilitates the build-up of ecological differences between the populations by disruptive natural selection. Ecological differences between closely related haplochromines comprise their microdistribution and/or small functional differences in the feeding apparatus (Seehausen and Bouton 1997). Such differences may evolve fast because of the versatile cichlid jaw apparatus and because of the possibly simple genetics that underlie changes in the feeding apparatus.

One additional factor that might drive rapid speciation is a dominant sex-reversal gene that, when linked to color genes, results in sex-ratio selection acting in concert with sexual selection on both male and female coloration. Finally, some aspects of the reproductive biology of haplochromines are important. Haplochromines have parental care by mouth brooding, which frees them from reproductive constraints experienced by free-spawning fish. Such constraints can result from insufficient oxygen supply for eggs and larvae and/or the necessity to migrate to shallow water or rivers for spawning. The mouth-brooding behavior of haplochromines makes them relatively independent of the spawning substrate (Fryer and Iles 1972), which can be a scarce commodity on the soft muddy bottoms that prevail at greater depths in lakes.

Other fish have formed species flocks in sympatry, such as the Lake Tana barbs (Nagelkerke and Sibbing 1996) and the tilapiine cichlid species from Lake Barombi-Mbo (Schliewen *et al.* 1994). Yet, the haplochromines of Lakes Malawi and Victoria stand out by their species numbers (one or two orders of magnitude more than the flocks in Lakes Tana and Barombi-Mbo). The haplochromine flocks also stand out in comparison with tilapiine cichlids, co-occurring with them in Lakes Malawi and Victoria. Why have these *Oreochromis* species not radiated into large species flocks? They share almost all the attributes that make the haplochromines prone to rapid radiation. This suggests that something present in the genetic make-up of haplochromines, but absent in *Oreochromis*, predisposes haplochromines to rapid speciation. Possibly this is the presence of a color-linked sex-reversal gene, as found in the blotched morphs of haplochromines; possibly it is a genetic mechanism that allows species repeatedly to split into red–yellow and blue species pairs without exhausting the genetic variation for coloration in the daughter species. Detailed genetic studies are needed to solve this riddle. This is yet another reason to continue the study of haplochromines as a model for speciation. The lack of postmating barrier between species, the high rate of speciation, and many of their other biological attributes allow the processes that may operate in a wide variety of sexually dimorphic animals to be studied with relative ease.

9

Natural Selection and Ecological Speciation in Sticklebacks

Howard D. Rundle and Dolph Schluter

9.1 Introduction

The idea that selection may be fundamental to the origin of species dates back at least to the synthesis of modern evolutionary theory, being present in the writings of Fisher (1930), Muller (1942), and Dobzhansky (1951). Nevertheless, despite almost three-quarters of a century since the idea was first proposed, little progress has been made in testing the role of selection in speciation (Coyne 1992; Schluter 1996a, 2001; Futuyma 1998). As the various chapters of this volume attest, however, the topic is enjoying a resurgence of interest (see also Schluter 2001). The purpose of this chapter is to summarize tests for the role of selection in speciation in a natural system: the sympatric populations of limnetic and benthic threespine sticklebacks that inhabit postglacial lakes in British Columbia, Canada (Plate 2). Our investigation specifically addresses the role of divergent selection between environments and niches in the origin of reproductive isolation, and the role of reinforcement in its completion. Throughout, we adopt Mayr's (1942) biological species concept, in which species are groups of actually or potentially interbreeding natural populations that are isolated reproductively from other such groups. We broaden this definition slightly to recognize that imperfect reproductive isolation can exist between species that nevertheless maintain their distinctiveness in nature (Rundle *et al.* 2001).

We focus on the question of "ecological speciation", a term that encompasses various speciation scenarios in which divergent natural selection between niches or environments is ultimately responsible for the evolution of reproductive isolation (Box 9.1). Ecological speciation differs from adaptive speciation as defined in Chapter 1 in two ways (Boxes 1.1 and 19.1). First, ecological speciation can also occur in allopatry (Box 9.1) and, second, ecological speciation does not include cases in which disruptive selection and subsequent divergences result entirely from sexual selection. Ecological speciation includes cases in which reproductive isolation evolves wholly as an indirect by-product of adaptation to alternative resources and environments, as well as several mechanisms in which natural selection directly favors the evolution of reproductive isolation (Box 9.1). Later we attempt to determine whether, in the stickleback speciation process, direct natural selection has played a role. Speciation by sexual selection is also ecological speciation if divergent natural selection between environments drives divergence in mate preferences, leading to reproductive isolation (Schluter 2000).

Box 9.1 Defining ecological speciation

Ecological speciation occurs when reproductive isolation evolves ultimately as a consequence of divergent natural selection between niches and environments (Schluter 2000, 2001). Reproductive isolation may evolve indirectly as a by-product of adaptive divergence of other traits, or selection may directly favor the evolution of reproductive isolation. Ecological speciation is adaptive speciation (as defined in Chapter 1) if it is driven by frequency-dependent ecological interactions. Ecological speciation does not, however, encompass speciation by purely sexual selection, in the absence of divergent natural selection. Sexual selection is a part of ecological speciation only if divergent mate preferences are ultimately the outcome of divergent natural selection between environments, but not otherwise (summarized in Schluter 2000; Boughman 2002).

Indirect (by-product). In this form, reproductive isolation evolves indirectly as a by-product of adaptation to alternative environments or niches. Environmental differences lead to divergent natural selection on phenotypic traits (morphology, physiology, or behavior) and the resultant divergence in phenotype may bring about reproductive isolation (premating and/or postmating) as a side-effect, both in sympatry and allopatry (see Box 9.2). Divergent selection may arise from external differences between the two environments or niches occupied by the two populations (the "environment" in a narrow sense). Divergent selection may also arise from frequency-dependent interactions between individuals of the two populations, such as competition or predation (the "environment" in a wider sense). For instance, ecological character displacement caused by competition for shared resources in sympatry can cause ecological traits to diverge, which may produce some reproductive isolation as a by-product.

Direct. In this process, selection directly favors the evolution of reproductive isolation. It generally occurs when two populations are sympatric and individuals that mate heterospecifically have reduced fitness. Natural selection then directly favors individuals that discriminate against heterospecific mates, which leads to stronger premating isolation. Reduced fitness of individuals that mate heterospecifically can arise in two ways. First, if hybrid offspring have reduced fitness (i.e., partial postmating isolation exists), selection favors those individuals that hybridize less. This is the hypothesis of reinforcement (Dobzhansky 1940; Blair 1955). Second, heterospecific matings may impose fitness costs on the individuals (because of parasites, courtship costs, increased predation risk, etc.), and so those individuals that mate within their own population have a higher fitness. In contrast to reinforcement, here the fitness cost is borne by the mating individual and is not determined by the viability or fertility of its offspring.

Processes of speciation that do not involve divergent selection between environments are regarded as "nonecological", because chance events dominate the initiation of the evolution of reproductive isolation. Mechanisms include genetic drift, founder events, and fixation of alternative advantageous alleles in allopatric populations that experience similar selection pressures. Speciation initiated by any of these alternative mechanisms may be completed by reinforcement in a second, sympatric

continued

Box 9.1 *continued*

phase. For this reason, demonstration of reinforcement does not by itself distinguish ecological from nonecological speciation. Tests of ecological speciation must focus instead on the role of selection at other stages of speciation, particularly early on in the process. Genetic drift, founder events, and alternative mutations may lead to divergent sexual selection independently of environment, which should also be regarded as a nonecological process of speciation.

Ecological speciation may occur in a variety of geographic settings: allopatric, sympatric, and between. A useful general framework considers the process that leads to two coexisting species potentially to have two stages (Box 9.2). In the first stage, reproductive isolation begins to evolve between two allopatric populations as a by-product of adaptation to their unique ecological environments or niches. The second stage occurs after secondary contact: selection in sympatry strengthens reproductive isolation and completes the speciation process. Within this framework, completely allopatric and completely sympatric speciation are extremes in which one or the other stage is absent. Among the goals of speciation research is to determine how much reproductive isolation is built during each stage, and by what mechanisms.

At present there are two kinds of evidence for ecological speciation. First, laboratory experiments using *Drosophila* demonstrated the feasibility of the model, both in allopatry (Kilias *et al.* 1980; Dodd 1989) and in sympatry (Rice and Salt 1988, 1990). For example, premating isolation evolved between laboratory populations that adapted to different environments, but not between populations that independently adapted to the same environment (Figure 9.1). Such results indicate that the model is feasible, but they do not indicate the importance of ecological speciation in nature. Second, a few results from nature are consistent with the hypothesis of ecological speciation. At least three examples are known in which traits with adaptive significance also form the basis of the reproductive isolation between species: Darwin's finches in the Galapagos Islands (Boag and Grant 1981; Ratcliffe and Grant 1983; Price *et al.* 1984), yellow monkey flower (*Mimulus guttatus*) living on copper-contaminated and uncontaminated soils (Macnair and Christie 1983; Christie and Macnair 1984), and sympatric threespine sticklebacks (Nagel and Schluter 1998). This implies that reproductive isolation somehow followed adaptive divergence in traits.

Here we review further evidence for ecological speciation in wild sticklebacks. Measurements of the strength of premating isolation among populations are consistent with a dual role of selection during two phases of ecological speciation. Initial premating barriers probably arose between allopatric populations as a by-product of adaptation to different environments, and later these were strengthened in sympatry by direct selection against heterospecific mating. Our results

Box 9.2 The geographic context of ecological speciation

Much attention has been given to the geographic context of speciation; the majority of this work focused on the possibility of fully sympatric speciation versus the necessity of an allopatric phase. The hypothesis of ecological speciation is general and makes no claim as to the geographic context of speciation. An all-purpose geographic scenario for the ecological process that leads to two coexisting species is for reproductive isolation to begin during an allopatric phase and to be completed in sympatry, as illustrated below.

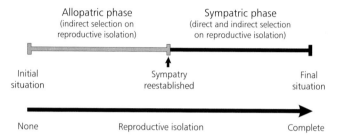

Reproductive isolation is initially absent on the left of the figure above, at the start of the speciation process, and evolves to completion on the right. The vertical arrow indicates the point at which sympatry between the two populations is reestablished, and may occur anywhere along the time line (shown here at the center). Allopatric and fully sympatric speciation represent the two extremes of this general scenario, whereby one or the other of the phases is missing and reproductive isolation evolves completely in either allopatry (complete reproductive isolation achieved before sympatry is established) or sympatry (no allopatric phase precedes the evolution of reproductive isolation). The mechanisms of divergent selection depend to some extent on geography. Premating and postmating isolation may evolve as by-products of adaptation to different environments or niches in either phase, but frequency-dependent interactions are added as an agent of divergent selection in the sympatric phase. Direct selection on premating isolation occurs in the sympatric phase, either via reduced hybrid fitness (i.e., reinforcement) or fitness costs incurred during heterospecific encounters. The sympatric phase corresponds to adaptive speciation processes as defined in Chapter 1 (also see Box 19.1).

Evidence suggests that both phases were present in the origin of sympatric benthic and limnetic sticklebacks (Box 9.3). The amount of reproductive isolation that evolved during each phase remains uncertain, however. The placement of the division that separates allopatric and sympatric phases, and the precise mechanisms of selection involved, are the subjects of continuing research.

clearly demonstrate ecological speciation and reveal that at least part of the isolation evolved as a result of direct selection (adaptive speciation, *sensu* Dieckmann *et al.*, Chapter 1). However, the precise amount of premating isolation that evolved during each phase is uncertain and remains to be determined.

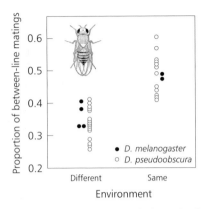

Figure 9.1 Proportion of matings that occur between independently evolved lines of *Drosophila*, derived from a common ancestor, as a function of the similarity of their environments. *Sources*: Open circles *D. pseudoobscura*, Dodd (1989); filled circles *D. melanogaster*, Kilias *et al.* (1980).

9.2 Natural History of the Sympatric Sticklebacks

Marine threespine sticklebacks (*Gasterosteus aculeatus*) colonized freshwater lakes and rivers along the coast of British Columbia after the retreat of the Pleistocene glaciers 10 000–12 000 years ago, and gave rise to permanent freshwater populations (McPhail 1993). While most lakes or rivers contain a single population, species pairs evolved in a few small, low-elevation lakes (Figure 9.2). In every case, one species of each pair, the benthic, is a larger, deeper bodied fish (Plate 2), with fewer, shorter gill rakers, that feeds on invertebrates in the littoral zone of the lake. The other species, the limnetic, is a smaller, more fusiform fish (Plate 2), with longer, more numerous gill rakers, that feeds primarily on zooplankton in the open water of the lake (McPhail 1984, 1992, 1994; Schluter and McPhail 1992). Phenotypic differences between limnetics and benthics are genetic and persist over multiple generations in a common laboratory environment (McPhail 1984, 1992; Hatfield 1997). Both comparative (Schluter and McPhail 1992) and direct experimental evidence (Schluter 1994) indicate that these differences are the result of divergent natural selection caused, in part, by frequency-dependent resource competition. Within each lake, limnetics and benthics constitute biological species. They are isolated reproductively and behaviorally (Ridgway and McPhail 1984; Nagel and Schluter 1998), and hybrids, which are intermediate in phenotype, suffer a reduced fitness as a result of ecological mechanisms (Schluter 1995; Hatfield and Schluter 1999). Sexual selection is also likely to act against hybrid males (Vamosi and Schluter 1999).

The geographic context of limnetic–benthic speciation has been contentious, but the present evidence favors a "double-invasion" scenario rather than fully sympatric speciation (Box 9.3). Under this scenario, originally proposed by McPhail (1993), the limnetic–benthic pair in each lake is the result of two separate invasions

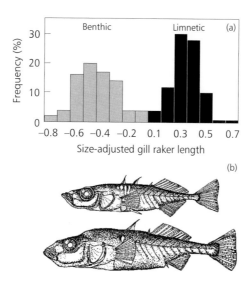

Figure 9.2 (a) Frequency distribution of size-adjusted gill raker lengths for Paxton Lake benthic (gray) and limnetic (black) sticklebacks. Gill rakers are protuberances from the gill arch that are thought to function during feeding to sieve particles of food or to direct the movement of water through the oral cavity. Plankton-feeding fish tend to have more numerous, longer gill rakers (see Schluter and McPhail 1993). *Source*: Schluter and McPhail (1992). (b) Limnetic (above) and benthic (below) sticklebacks from Paxton Lake, British Columbia. *Source*: Schluter (1993).

by the marine stickleback into freshwater, in which the second, more recent, invader evolved into the modern limnetic. Speciation thus involved an initial stage of allopatry during which the first invader was adapting to freshwater prior to the second invasion of the marine form. The species have persisted subsequently, despite at least some gene flow between them.

While the geographic context of speciation has been debated, genetic evidence indicates that the phenotypic similarity of benthics and of limnetics from three different lakes (Paxton, Priest, and Enos) is the result of parallel evolution and not shared ancestry. The evidence includes unique assemblages of mitochondrial DNA (mtDNA) that characterize the species pairs from each of these lakes (Taylor *et al.* 1997; Taylor and McPhail 1999) and a phylogenetic and genetic distance analysis of six nuclear microsatellite loci (Box 9.4; Taylor and McPhail 2000). For example, each lake is dominated by mtDNA haplotypes not found in any other lakes, each of which is a small number of base-pair transitions away from common marine haplotypes (Taylor and McPhail 1999).

In Section 9.3 we turn our attention to mechanisms of speciation, and describe two separate tests of ecological speciation in our attempt to understand the evolution of reproductive isolation from beginning to end. The first takes advantage of the independent evolution of the limnetics and benthics from these three lakes.

Box 9.3 Double invasion and stickleback speciation

The geographic context of speciation of the limnetic–benthic pairs, whether they arose within each lake via sympatric speciation or instead had an allopatric phase, is unclear and not all the evidence points to one conclusion. The weight of evidence, however, favors the double-invasion scenario of McPhail (Schluter and McPhail 1992; McPhail 1993) as depicted below (figure modified from Taylor *et al.* 1997). In this scenario each coexisting benthic and limnetic pair results from two separate invasions by the marine threespine stickleback (*Gasterosteus aculeatus*) into fresh-water after the retreat of the glaciers at the end of the Pleistocene period (< 13 000 years ago).

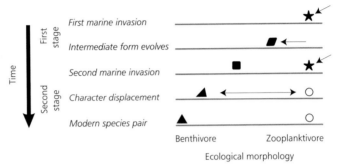

The sequence began with the first invasion of lakes by the marine species, which then evolved into an intermediate phenotype characteristic of most small, single-species lakes today (Schluter and McPhail 1992). The marine form invaded a second time after a secondary rise in sea level a few thousand years later. Ecological character displacement, driven by frequency-dependent resource competition, increased the ecological and phenotypic divergence in sympatry (Schluter 1994). The first invader gave rise to the present day benthic species, while the second invader remained a plankton specialist like its marine ancestor, and has evolved into the present-day limnetic species. The double-invasion scenario can be accommodated within the framework of adaptive dynamics, as outlined in Box 9.5.

The evidence in favor of double invasion is as follows:

■ Allozyme frequencies from two lakes indicate that the limnetic species is similar to the present-day marine species (Nei's $D_N \approx 0.02$; McPhail 1984, 1992), whereas benthics are more distant (Nei's $D_N \approx 0.07$). However, the distances in all cases are small.

■ Like the marine species, limnetics from the two lakes tested could successfully complete development in 28 parts per thousand (p.p.t.) of sea water, from egg to hatchling, whereas the success rate of benthics from the same lakes was poor (Kassen *et al.* 1995). This supports the idea of two invasions spaced apart in time, with the most recent invasion leading to the modern limnetic species (assuming that salinity tolerance has decayed with time since the colonization of freshwater).   *continued*

Box 9.3 *continued*

- Analysis of six microsatellite loci showed that allele frequencies in the limnetics are more similar to those of the present-day marine species than to allele frequencies of the benthics (Taylor and McPhail 2000), although the differences are again small. In addition, the microsatellite allele frequencies fit the unconstrained maximum-likelihood phylogeny (Box 9.4) significantly better than a tree in which the limnetic and benthic species from each lake are constrained to be sister species, as would be the case if they had resulted from sympatric speciation (Taylor and McPhail 2000).
- There are morphological indications of recent hybridization between the limnetic species in Emily Lake (the lake of lowest elevation in the drainage that includes the Priest Lake species pair) and the marine species that breeds in the stream that drains Emily Lake to the ocean (McPhail, personal communication). This marine population is indicated as Marine 4 in Box 9.4.

In contrast to these indications of double invasion, restriction fragment length polymorphism (RFLP) analysis of mtDNA is more consistent with sympatric speciation: dominant haplotypes within each lake occur in both species, which suggests they are sister species (Taylor and McPhail 1999). However, we suggest that these results show a recent mtDNA gene flow between sympatric species (Taylor *et al.* 1997; Taylor and McPhail 2000).

9.3   Parallel Speciation of Limnetics and Benthics

One mechanism of ecological speciation occurs when reproductive isolation evolves as a by-product of adaptation to different environments (Box 9.1). When replicate, closely related populations independently adapt to their environments, a remarkable pattern termed parallel speciation may result (Schluter and Nagel 1995). Parallel speciation is a special case of the phenomenon of parallel evolution, a form of homoplasy in which a similar trait evolves repeatedly in closely related, independently evolving lineages. Parallel evolution provides strong evidence for natural selection in the evolution of the trait, because genetic drift is unlikely to produce predictable changes in independent lineages in correlation with the environment (Clarke 1975; Endler 1986; Schluter and Nagel 1995).

Parallel speciation is a form of parallel evolution in which the traits that evolve predictably in correlation with the environment also affect reproductive isolation. When reproductive isolation evolves as a by-product of adaptation to the environment in multiple, independent populations, the predicted outcomes are twofold:

- Reproductive compatibility between populations that evolved independently in similar environments.
- Reproductive isolation between populations that evolved independently in different environments.

Box 9.4 Independent parallel evolution

Currently, three drainages in the region (Paxton Lake, Priest Lake, and Enos Lake) contain a pair of limnetic and benthic species (a fourth pair in Hadley Lake drainage was recently extirpated by introduced catfish). How many origins of limnetic and benthic populations have led to this pattern? One possibility is that each species arose exactly once, and that the pair subsequently colonized other drainages together. The other possibility is that each species pair arose independently from the others, and acquired their similarities by parallel evolution under common natural selection pressures. The weight of evidence favors the latter possibility.

The first line of evidence comes from mtDNA. Each of the drainages is dominated by haplotypes that are not found in any other drainage nor the sea (Taylor *et al.* 1997; Taylor and McPhail 1999). The majority of these are distinguished from common marine haplotypes by a single restriction site; most of the remaining unique haplotypes are a single site away from other haplotypes in the same drainage. In no case are unique haplotypes from two different drainages distinguished by as little as one restriction site. In other words, haplotypes in each drainage trace their origins to the sea, not to other lakes. Unfortunately, mtDNA gene flow between limnetics and benthics within each drainage (Box 9.3) prevents us from teasing apart their separate histories in this way.

The second line of evidence is from analyses of allelic variation at six microsatellite loci (Taylor and McPhail 2000). When the limnetics and benthics are considered apart from other populations, almost none of the total genetic variation among individuals is partitioned between the classes "limnetic" and "benthic" (2.4–4.4%, which is not significantly different from zero). This variance component would be larger if limnetics and benthics arose exactly once. In addition, the maximum-likelihood phylogeny of all populations tested, shown below (unrooted tree modified from Taylor and McPhail 2000), suggests a polyphyletic origin of both limnetics and benthics. Despite uncertainty in most of the groupings (numbers below indicate bootstrap support levels greater than 50%), this tree fits the data significantly better than those that constrain either limnetic or benthic populations to be monophyletic (Taylor and McPhail 2000).

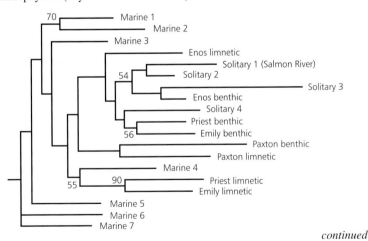

continued

Box 9.4 *continued*

The phylogeny illustrates the greater overall similarity of marines to limnetics than to benthics (Box 9.3). The four solitary populations tested, from different lakes and streams, tend to cluster with the benthics, which suggests that most were formed at the time of the first invasion rather than the second. Measurements of salinity tolerance bear this out for the one solitary population tested (Kassen *et al.* 1995). The marine samples are very similar to one another in microsatellite allele frequencies, as would be expected from high levels of gene flow among sites. Monophyly (with the exception of Marine 4) of the freshwater populations is also consistent with high levels of gene flow among marine populations and does not require a single marine population to be the common ancestor of all freshwater populations.

As in parallel evolution generally, the repeated evolution of similar mechanisms of premating isolation in independent populations that inhabit similar environments strongly implies that natural selection was the cause (Schluter and Nagel 1995). Reproductive compatibility is predicted by selection regime rather than by geographic proximity or phylogenetic history. Despite its significance to our understanding of ecological speciation in nature, prior to our work no conclusive tests of parallel speciation were found in the literature.

We took advantage of the independent evolution of limnetics and benthics in Paxton, Priest, and Enos Lakes to test two specific predictions of parallel speciation (Rundle *et al.* 2000):

- Populations of the same "ecomorph" from different lakes (e.g., limnetics from Paxton, Priest, and Enos Lakes) should be compatible reproductively, despite the known isolation between limnetics and benthics within a lake (Nagel and Schluter 1998).
- Limnetics and benthics from different lakes (e.g., Enos limnetics and Paxton benthics) should be isolated reproductively, even though they have not encountered one another before.

These predictions were tested by measuring the strength of premating isolation between various combinations of populations. The strength of premating isolation was measured in no-choice mating trials conducted in the laboratory using wild-caught individuals. 753 mating trials were performed. Details of the methodology, population combinations, and sample sizes can be found in Rundle *et al.* (2000). Statistical analyses treated populations of females (not individual females) as replicates and corrected for phylogeny. Strong reproductive isolation between limnetics and benthics within each lake was used as a benchmark for other comparisons (Figure 9.3, comparison A). Females spawned significantly more often with males of their own population (38%) than with males of the other species within the same lake (15%); note that this mating frequency between sympatric species is artificially higher in our no-choice laboratory setting than in the wild, where hybridization rarely if ever occurs.

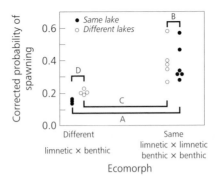

Figure 9.3 Population mean probabilities of spawning between pairs of populations as a function of ecomorph. Comparison A represents the test for reproductive isolation between limnetics and benthics within a lake. Comparisons B and C represent the two tests of parallel speciation. The strength of reproductive isolation between limnetics and benthics from the same or different lakes is compared in D. Our statistical tests employed a conservative paired *t*-test that treated each population of females as a replicate and corrected for phylogeny, and the comparisons shown here represent the nature of the tests, but do not depict the actual analyses performed. *Source*: Rundle *et al.* (2000).

Both predictions of parallel speciation were met. First, reproductive isolation was lacking among populations of the same ecomorph from different lakes (Figure 9.3, comparison B). Females mated just as readily with males of the same ecomorph from a different lake (39%) as with males from their own population (38%). Second, reproductive isolation was present between limnetics and benthics from different lakes (Figure 9.3, comparison C). Females mated significantly more often with males of their own ecomorph from a different lake (39%) than with males of the other ecomorph from a different lake (20%). Interestingly, the probability of spawning between limnetics and benthics from different lakes (20%) was slightly higher than that between limnetics and benthics from the same lake (15%), a difference that approached statistical significance ($t_5 = 2.36$, $P = 0.065$; Figure 9.3, comparison D). This last result is the first hint that premating isolation between limnetics and benthics was strengthened in sympatry.

The parallel evolution of premating isolation in correlation with the environment provides strong evidence for ecological speciation in nature. In a little over 10 000 years populations of sticklebacks descended from the same marine ancestor, but that inhabited different environments, have evolved strong premating isolation. Remarkably, during this same period of time no reproductive isolation has arisen between populations that adapted independently to similar environments. Similar results have been observed in a number of other systems (Funk 1998; McPeek and Wellborn 1998; McKinnon *et al.*, unpublished; see Morrell 1999), which suggests that parallel speciation may not be a rare occurrence. Our results strongly implicate natural selection in speciation and suggest that reproductive isolation between limnetics and benthics has evolved as a by-product of adaptation to different habitats. The by-product mechanism can occur both in allopatry and in sympatry (Box 9.2), but the degree to which this isolation evolved during each of the two phases remains to be determined.

Next we consider the possibility that some portion of the isolation results from selection that directly strengthens premating isolation in sympatry (e.g., adaptive speciation via reinforcement), and that this occurred in parallel in each lake. As noted in Figure 9.3 (comparison D), premating isolation between limnetics and benthics may be slightly stronger when the two populations are from the same lake than when the two populations are from different lakes, suggesting a role for selection in sympatry. We now turn our attention to this second stage of ecological speciation.

9.4 Premating Isolation Strengthened in Sympatry

To test whether selection strengthened premating isolation between sympatric limnetics and benthics during a second stage of ecological speciation, we tested for reproductive character displacement. Reproductive character displacement is the pattern of greater premating isolation between two taxa in areas of sympatry rather than allopatry (Brown and Wilson 1956; Howard 1993). Its presence suggests that a greater degree of reproductive isolation may have evolved in sympatry. Indeed, if the second stage of ecological speciation commonly involves the strengthening of premating isolation in sympatry, reproductive character displacement is the predicted outcome. Comparison D in the previous section (Figure 9.3) is not a sufficient test of reproductive character displacement because all populations were sympatric with some other population. Reproductive character displacement is a pattern that involves a comparison of populations from areas of sympatry versus allopatry, so a stronger test requires that solitary (i.e., allopatric) populations be included.

A number of mechanisms of ecological speciation can produce reproductive character displacement, two of which also fall under the definition of "adaptive speciation". Most attention has been paid to reinforcement, defined as the process whereby premating isolation is strengthened in sympatry as an adaptive response to reduced hybrid fitness (Dobzhansky 1940; Blair 1955). Other processes include direct selection in sympatry for premating isolation that results from fitness costs borne by individuals that mate heterospecifically, and a nonadaptive process termed "biased extinction". These alternative mechanisms are described below. While reproductive character displacement was once considered a rare phenomenon (e.g., Littlejohn 1981; Phelan and Baker 1987), more recent work suggests otherwise (Coyne and Orr 1989, 1997; Howard 1993; Noor 1995, 1997; Sætre et al. 1997; Higgie et al. 2000).

We focused on reproductive character displacement of benthic female mate preferences (Rundle and Schluter 1998). Our study used females from three populations: the benthic from Priest Lake (a two-species lake) and two solitary populations, one from Beaver Lake and the other from the Salmon River. The Salmon River population is related closely to the Priest Lake benthic (Box 9.4), whereas

Box 9.5 Double invasion and frequency-dependent selection

H.D. Rundle, D. Schluter, U. Dieckmann, J.A.J. Metz, and M. Doebeli

The double-invasion scenario described in Box 9.3, which includes an initial allopatric phase and a later sympatric phase, requires that selection be frequency dependent. That is, the fitness landscape of the second invader is dependent on the presence and the phenotype of the first invader. Without frequency dependence, the two forms (depicted by a square and a star in Box 9.3) would not have persisted as competitors in sympatry following the second invasion. Instead, one of the two morphs would have been driven to extinction by the other. Moreover, ecological character displacement can only occur under frequency-dependent selection: in its absence, the second invader is forced to pursue the same evolutionary course as did the first one initially. Below we use the adaptive dynamics framework to depict the possible role of frequency dependence in a simple scenario. We consider a single trait axis along which the two stickleback phenotypes are differentiated. A low value along this axis represents the zooplanktivore phenotype. Greater values indicate more benthic-feeding phenotypes.

    The linear pairwise invasibility plot in panel (a) below represents one scenario of events that might have led from a single marine stickleback ancestor to two sympatric species within a lake. The signs $(+, -)$ indicate, for each value of the resident phenotype, the invader phenotypes that can $(+)$ and cannot $(-)$ invade when there is competition for resources. The crossed lines delimit regions of different sign.

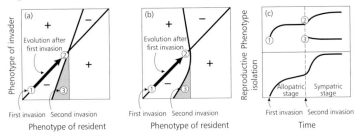

The hypothetical sequence of events begins at (1) when the marine form, a zooplanktivore, first invades the lake. Subsequently, adaptation to the new environment [arrow from (1) to (2)] causes this first invader to evolve toward an intermediate phenotype that represents an evolutionary branching point (intersection of the two thick lines). A lake with a single intermediate population at the branching point can be invaded by nearby phenotypes larger or smaller (i.e., the population experiences disruptive selection). The area shaded in gray indicates the trait values of zooplanktivore-like phenotypes permitted to invade once the resident population is close to the branching point. In the scenario in panel (a), the marine form invades a second time (3) before the first population has reached the branching point, which

continued

Box 9.5 *continued*

results in two phenotypes within the lake. Although the new marine invader and the evolved lake form now reside on the same side of the branching point, they can coexist and continue to evolve. Ecological character displacement, driven by frequency-dependent competition for resources, causes further divergence between the two populations until they eventually reach an evolutionarily stable combination of phenotypes [top section of panel (c)].

Under this simple scenario, competitive exclusion precludes a second invasion until the phenotype of the first invader approaches the branching point. Insufficient reproductive isolation may also preclude a second invasion that is too soon after the first. Thus, the timing of the second invasion is dependent not only on favorable geologic events that allow the movement of individuals from the sea into the lake, but also on the evolution of the resident within the lake. As shown in the bottom section of panel (c), partial premating isolation is assumed to build prior to the second invasion (the allopatric phase) as a by-product of phenotypic divergence. After the second invasion (the sympatric phase), natural selection directly on premating isolation (i.e., reinforcement) or on correlated traits might further strengthen premating isolation.

Alternative invasion scenarios that involve more complicated forms of interaction between phenotypes are possible, and lead to nonlinear pairwise invasibilities as in panel (b) above. For example, the first invader might evolve not to a branching point, but instead to an evolutionarily stable attractor that permits no invasion by nearby phenotypes [intersection of the two thick lines in panel (b)]. In this case the intermediate population is under stabilizing selection, not disruptive selection. Such stabilizing selection might occur if zooplankton and benthic resources in the lake peak at different times in the season, such that a generalist intermediate phenotype can exploit both of them in sequence when more specialized zooplanktivore or benthic phenotypes cannot.

In this scenario, the line that delimits $(+)$ and $(-)$ regions in the lower part of the plot is curved to the left. In this case a second invasion is possible if the invading zooplanktivore morph is sufficiently different in phenotype from the resident morph, as the latter population evolves toward an intermediate phenotype. This might happen if a portion of the zooplankton resource base cannot be exploited by an intermediate phenotype. Following a successful invasion, ecological character displacement and reinforcement might cause further divergence between the two populations, as in the previous scenario.

We do not know which of the two above scenarios, intermediate branching point or intermediate stable attractor, best describes the double invasion process in sticklebacks. Distinguishing them would require determining whether the intermediate form in single-species lakes most often experiences stabilizing or disruptive selection. A series of future experiments could also compare the fitnesses of different invading phenotypes in the presence of alternative resident phenotypes.

the phylogenetic affinities of the Beaver Lake population are unknown. We compared the probability of spawning of these females with limnetic and benthic males from Paxton Lake, another two-species lake. Reproductive character displacement of benthic female mate preferences would be indicated if Priest benthic females show a more marked difference in their propensities to mate with limnetic and benthic males than do females from solitary populations. To minimize lake effects on mating behavior, none of the female populations came from the same lake as the males.

Reproductive character displacement is mainly of interest if inferences can be made as to the process that caused it. With this in mind, our test for reproductive character displacement accounted for ecological character displacement that may strengthen premating isolation in sympatry as a by-product (Box 9.1). Divergent selection created by competition for resources may cause ecological characters to diverge and, if these characters also affect mate choice, reproductive character displacement may evolve as a side-effect. Indeed, both comparative (Schluter and McPhail 1992) and direct experimental evidence (Schluter 1994, 1996a) indicate that ecological character displacement has occurred between limnetics and benthics. We were especially concerned about displacement of body size, because this trait affects the probability of interspecific mating (Nagel and Schluter 1998). For this reason we compared the mate preferences of benthic females only with solitary populations of females that closely resembled true benthics in morphology. As the resemblance was not perfect, we also corrected statistically for body-size differences in the analysis. A total of 239 no-choice mating trials were conducted in the laboratory. Details of the methodology, populations used, and sample sizes are given in Rundle and Schluter (1998).

Reproductive character displacement of benthic female mate preference was clear (Figure 9.4). Priest benthic females spawned much more readily with benthic males than with limnetic males, whereas this difference was small in females from both solitary populations. This pattern remained after statistically correcting for differences in the body size of the females from the three populations.

To distinguish the various processes that can produce reproductive character displacement is difficult. The first task is to rule out mechanisms in which selection plays no part, such as biased extinction. Under biased extinction, incipient species come together in multiple, independent localities, but the populations persist as separate species only in the sites in which the initial level of premating isolation is sufficiently strong. Cases in which isolation is weak result in a fusion of the populations or the extinction of one by reproductive interference (Butlin 1987, 1989). Sympatric pairs that survive are thus a biased subset of all possible combinations; they include only the pairs in which premating isolation was strong enough to permit coexistence. Under this mechanism, premating isolation does not evolve in sympatry. However, speciation may still be regarded as ecological (but not adaptive) if selection causes the initial isolation to evolve in allopatry. In this case, only the first stage of ecological speciation occurs and speciation is entirely allopatric.

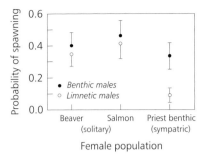

Female population

Figure 9.4 Mean probability of spawning (±1 SE) for various combinations of male species and female population. The pattern of reproductive character displacement is indicated by the lower probability of spawning with heterospecific males exhibited by the populations of sympatric as opposed to solitary females. In some instances, error bars are shown in only one direction for clarity. *Source*: Rundle and Schluter (1998).

Biased extinction predicts that a range of levels of premating isolation should exist among randomly paired solitary populations, and the most extreme solitary pairs should exhibit virtually complete isolation, such as that found between existing sympatric species. With only two solitary populations investigated thus far, we are unable to test rigorously this prediction. However, it is striking that relatively little isolation occurs in both of these populations, chosen because they are very benthic-like in morphology, and at least one (Salmon River) is phylogenetically close to the benthic population tested. For this reason, we view biased extinction as insufficient on its own. Nevertheless, our results imply that some variation in reproductive isolation exists among solitary populations, because the isolation detected in the two solitary populations tested seems too weak to prevent fusion with limnetics. However, no-choice mating trials may be too insensitive to detect the variation present, and may underestimate levels of premating isolation (e.g., isolation between sympatric limnetics and benthics is weaker in our trials than in the wild).

The next task is to identify which ecological mechanism is responsible for reproductive character displacement. These include direct and indirect selection for increased premating isolation in sympatry (Box 9.1). We examine each of these in turn.

Reproductive character displacement can evolve indirectly when unique selective pressures created by interactions between the two species in sympatry (e.g., competition, predation) strengthen premating isolation as a by-product (Box 9.1). For instance, as noted above, competition for resources may cause ecological character displacement, which in turn may cause reproductive character displacement as a by-product. By choosing solitary females that closely resembled benthic females in morphology, we minimized this possibility in our study and thus view this mechanism as unlikely. However, it is conceivable that additional unmeasured ecological characters that diverged in sympatry are responsible for the increased reproductive isolation, and that our sampling of solitary populations did not control for such confounding factors.

Alternatively, reproductive character displacement is the outcome of direct natural selection on premating isolation (Box 9.1). This occurs if individuals that avoid heterospecific mating have higher fitness for one of two reasons. First, if partial or complete postmating isolation exists, individuals that avoid heterospecific encounters tend to have more fit (i.e., nonhybrid) offspring. This is the hypothesis of *reinforcement*. Reinforcement requires a number of preconditions, which include that hybridization has occurred in the past and that hybrids are selected against (Butlin 1989; Howard 1993). These preconditions appear to be met in the sticklebacks (Rundle and Schluter 1998; Hatfield and Schluter 1999; Taylor and McPhail 1999, 2000). The second mechanism that results in higher fitness occurs when individuals that avoid heterospecific matings avoid the costs incurred as a result of these encounters. In this "reinforcement-like" process (Servedio 2001), the cost to heterospecific encounters is not manifest through a reduced fitness of the individual's offspring, but rather by a reduced fitness of the individuals themselves. For instance, a benthic female that spawns with a limnetic male may be exposed to new and harmful parasites, while the limnetic male exposes the eggs already in his nest to egg predation by the benthic female (Foster 1994). As yet no attempts have been made to measure such costs of heterospecific encounters in these sticklebacks.

In conclusion, the combined data are consistent with the hypothesis that premating isolation is strengthened directly by selection in sympatry. This occurs as an adaptive response to either reduced hybrid fitness (reinforcement) or to costs imposed by heterospecific encounters.

9.5 Concluding Comments

Selection has long been thought to play a fundamental role in the formation of new species, but only recently has evidence from nature begun to accumulate in support of this (Schluter 1996a, 2000, 2001). The hypothesis of ecological speciation, under which reproductive isolation ultimately evolves as a result of divergent selection between environments or niches, gains strong support from the species pairs of threespine sticklebacks that inhabit postglacial lakes in British Columbia, Canada. Our results show that reproductive isolation has evolved in correlation with the environment, and that reproductive character displacement of the mate preferences of benthic females has occurred. This suggests that selection has had a dual role that corresponds to two stages in the evolution of premating isolation:

- Premating isolation evolved initially as a by-product of adaptation to different environments during the first, allopatric stage.
- Later, premating isolation was strengthened directly by selection during a second, sympatric stage.

Nevertheless, we do not yet fully understand the roles of these two mechanisms in the evolution of the premating isolation. The results shown in Figure 9.3 indicate that most premating isolation is explained by a common environment, with only a small component explained by a shared lake (i.e., sympatry; comparison D). It is tempting to conclude that the majority of the reproductive isolation thus evolved as a by-product of adaptation to alternative environments, with only a minor role for direct selection in sympatry. An alternative hypothesis is that these same results represent parallel responses to direct selection on premating isolation in sympatry. For instance, if reinforcement occurs by strengthening the dependence of mate choice on some key trait (e.g., body size), and this trait evolves in parallel in different lakes, parallel reinforcement may result. Such a process is consistent with the lack of premating isolation observed between limnetic males and solitary, benthic-like females (Figure 9.4). However, it seems unlikely that reinforcement could occur in parallel in independent lakes, unless it acted to strengthen mate preferences that had already evolved in parallel as a by-product of adaptation to the environment. Additionally, some premating isolation must have been present at secondary contact to prevent fusion of the two populations because postzygotic isolation, while present, is too weak to prevent fusion on its own (Schluter 1995; Hatfield and Schluter 1999). To determine the relative roles of by-product and direct selection in sympatry requires further studies of the mate preferences of solitary populations that differ in phenotype to establish how much isolation evolved during the allopatric phase. Conclusions as to the relative contribution of mechanisms of adaptive speciation await this data.

We need also to consider that mechanisms other than reinforcement may have caused the enhanced divergence of mate preferences in sympatry. Reproductive character displacement may also occur as a result of fitness costs to individuals that court or spawn with members of the other species. To distinguish between these alternatives remains an interesting possibility that is yet to be explored.

Finally, we have not yet identified all the traits that underlie the parallel mate preferences, although body size is a strong candidate (Borland 1986; Nagel and Schluter 1998). The probability of spawning between ecomorphs within a lake is strongly size dependent in no-choice mating trials, in which hybridization occurs only between the largest individuals of the smallest species (limnetics) and the smallest individuals of the larger species (benthics; Nagel and Schluter 1998). Body size is probably an adaptation to foraging in one or other of the two lake environments (Schluter 1993). However, body size is correlated with a number of other characters that need to be separated to understand their individual roles in causing the isolation.

Regardless of the precise mechanism or the exact traits, we have shown that reproductive isolation evolved in parallel under a common selective regime; this resulted in the repeated evolution of the "same" species in different lakes. Selection leading to premating reproductive isolation probably occurred during two phases, the first in allopatry and the second in sympatry. This combination of roles, in which premating isolation evolves both as a by-product of adaptation and later as a direct adaptive response, may be general and help to explain the high rates of phenotypic divergence and speciation that characterize some of the spectacular adaptive radiations discussed elsewhere in this volume.

10

Adaptive Speciation in Northern Freshwater Fishes

Sigurður S. Snorrason and Skúli Skúlason

10.1 Introduction

Fish constitute the most species-rich group of the vertebrates. They have adapted to highly diverse habitats in the sea and fresh waters of the world and display great phenotypic variability. Some fish groups have gone through extremely rapid local radiations with clear connections to ecological factors. Such extensive radiations are, for example, seen in various groups of cichlids in the African rift lakes (e.g., Echelle and Kornfield 1984; Meyer 1993; van Alphen *et al.* in Chapter 8). Radiations in those vast lakes have been termed intralacustrine, which emphasizes that the relevant speciation events in the monophyletic species flocks are not necessarily sympatric (Smith and Todd 1984; Meyer 1993). A convincing case for sympatric speciation is seen in studies of cichlids in Cameroon crater lakes (Schliewen *et al.* 1994). In many cases, the evolution of diversity is seen to occur within populations of what are still considered single species, in which so-called varieties or morphs have emerged as a result of adapting to different habitats and food resources (e.g., Robinson and Wilson 1994; Skúlason and Smith 1995; Smith and Skúlason 1996). From a purely ecological perspective, these phenomena can be viewed as resource polymorphisms (Skúlason and Smith 1995). From an evolutionary perspective, they can be seen as multiple instances of incipient speciation. Thus, resource polymorphisms provide important opportunities to study the role of ecological factors in the evolution of new species (Schluter 1996b, 1998, 2001).

In the wake of the most recent glacial epoch, extensive new fresh water habitats were formed in the Northern Hemisphere. These systems can be viewed as a multitude of evolutionary theaters in which we can find the early acts of speciation. The variation in the degree of phenotypic segregation seen in northern polymorphic fish can be considered as point samples from a multitude of evolutionary trajectories started not earlier than 10 000 years ago; trajectories that in many cases, depending on the circumstances, appear to move toward the formation of new species (Snorrason *et al.* 1994a; Skúlason *et al.* 1999).

With regard to fish species, these systems may still be considered as in a colonization phase. The systems are usually species poor, so that invaders are presented with a diversity of uncontested habitats and food resources. As reflected in many studies of northern freshwater fish, this has often led to character release in the form of resource polymorphisms (Robinson and Wilson 1994; Skúlason and Smith 1995). Furthermore, many of these studies have revealed repeated patterns of ecological specializations within and across fish species (Skúlason *et al.* 1999; Robinson and Schluter 2000).

Recent advances in analyses of variation in neutral genetic markers, along with more detailed ecological and experimental data, have facilitated tests to dissect the causal relationships, whereby ecological and genetic factors and historical constraints interact in the evolution of northern freshwater fish. On the one hand, a clear pattern is seen in how phenotypic differences among sympatric forms correlate with the availability, number, and discreteness of habitat and food resources (e.g., Hindar and Jonsson 1982; Malmquist *et al.* 1992; Robinson *et al.* 1993). On the other hand, genetic studies have shown that in many species (e.g., salmonids, sticklebacks, sunfish, smelt, and whitefish) the evolution of sympatric morphs has occurred locally, following postglacial invasions of common ancestors (see the discussion in Section 10.4). The repeated monophyletic origin of coexisting morphs provides unique opportunities to study the effect of ecological, developmental, and genetic factors in diversifying evolution (Smith and Todd 1984; West-Eberhard 1986, 1989; Rice 1987; Diehl and Bush 1989; Wilson 1989, 1992; Rice and Hostert 1993; Bush 1994; Skúlason and Smith 1995; Smith and Skúlason 1996; Skúlason *et al.* 1999; Robinson and Schluter 2000).

The striking similarities in the patterns of diversity seen in different groups of northern freshwater fish strongly suggest the operation of common evolutionary mechanisms. However, the differences among species, and among localities within species, indicate that the degree of segregation currently realized (see, e.g., Skúlason *et al.* 1999) depends on variations in the environmental parameters and on the species-specific constraints that shape the segregation process.

In this chapter, the focus is on speciation in relation to resource polymorphisms. Resource polymorphisms represent diversifying evolution on a fine scale that produces distinct ecological groups. Below the following issues are treated:

- Ecological factors that promote morph formation and shape selection forces are emphasized, especially in the early steps toward divergence.
- Various mechanisms that facilitate assortative mating of different morphs, and thus keep them genetically separated and promote their genetic divergence, are discussed.
- The first two points are linked to the phenotypic variability in representative groups of fish, and some underlying mechanisms of diversification are discussed.
- The above points are integrated and an evolutionary framework proposed to encompass the process that results in polymorphism and adaptive speciation and, at the same time, highlight the present variation among systems.
- Finally, these ideas are discussed in relation to new models of sympatric speciation.

10.2 Ecological Factors that Promote Diversification

In general, resource polymorphism can originate and be maintained by density- and frequency-dependent disruptive selection (Rice 1984b; Wilson 1989; Wood

and Foote 1990; Pfennig 1992; Hori 1993; Smith 1993). The most important ecological factors proposed as facilitators of resource polymorphism are resource diversity combined with inter- and intraspecific competition (Skúlason *et al.* 1989a;
Schluter and McPhail 1993; Robinson and Wilson 1994; Snorrason *et al.* 1994a;
Skúlason and Smith 1995; Smith and Skúlason 1996).

Northern freshwater systems are extensive and offer numerous, often discrete,
resources for fish. Many lakes have a sizable open water (pelagic or limnetic) habitat and one or more types of benthic habitat, and in such lakes limnetic and benthic
morphs of fish often coexist [reviews in Schluter and McPhail (1993) and Robinson
and Wilson (1994)]. Habitat diversity and complexity can be further influenced by
several factors. Lakes in volcanic areas, such as Iceland, offer new dimensions to
benthic habitats because of the highly complex and open lava substrate (Snorrason
et al. 1994a; Skúlason *et al.* 1999; see Plate 3). Similarly, soft-bottom habitats in
lakes offer two distinct conditions depending on whether they are bare or vegetated
(Robinson *et al.* 1993). The stability (predictability) of habitats is also important.
In general, it is to be expected that the stability of alternative resources promotes
diversification and morph formation.

It is of fundamental importance to recognize that recently formed postglacial
freshwaters are in many ways "immature" systems, in the sense that they are
still sparsely colonized. For invading fish this means that interspecific competition is low or absent, which allows access to a wider array of resources (Schluter
and McPhail 1993; Robinson and Wilson 1994; Snorrason *et al.* 1994a; Skúlason
and Smith 1995; Smith and Skúlason 1996). Subsequent to invasion, population
expansion leads to an increase in intraspecific competition for the preferred resources, which forces diversification through utilization of alternative resources.
It is easy to envisage how the combination of the above factors (stable resources,
relaxed interspecific competition, and intense intraspecific competition) can lead
to character release and the evolution of increased resource-related phenotypic diversity. Depending on the genetic and developmental mechanisms and the mating
system, this variation may become discontinuous, and different, discrete forms can
appear that are competitively superior in utilizing their particular resource. Subsequently, intermorph resource-competition can lead to character displacement,
which further reinforces phenotypic differences among morphs (Robinson and
Wilson 1994).

Risk of predation can play an important role in the diversification process. The
presence of a fish predator forces fish to seek refugia and food in sheltered habitats. In lakes that have more than one habitat of this nature available, this can
promote local adaptations to the respective resources. This is especially relevant
for relatively small fish, such as the threespine stickleback, which is suitable prey
for several larger fish species and, furthermore, is able to utilize complex benthic
habitats better than some larger species of fish. This may well be the process that
underlies the evolution of sympatric stickleback morphs in Iceland, which occupy
either littoral lava substrate or vegetated soft-bottom habitats (Doucette 2001). Potentially, predation risk can influence the way competition drives diversification,

such as through modification of habitat use and feeding behavior. Such interactions certainly deserve further attention.

10.3   Factors that Facilitate Assortative Mating

Whether resource polymorphisms are consolidated by genetic divergence of morphs ultimately depends on mechanisms that lead to morph-specific assortative mating. Therefore, it is important to understand how ecological opportunities (external factors) and characteristics of the organisms (internal factors) can influence – promote or constrain – the formation of reproductive barriers. The evolution of prezygotic isolation among morphs can be influenced by broad life-history traits that characterize genera or even larger taxonomic assemblages. Fishes of the subfamily Salmonidae (salmon, trout, and charr), many of which constitute pioneering species of northern postglacial freshwaters, are localized benthic spawners (Balon 1985). They lay relatively large eggs, which they protect by depositing them in gravel nests, so called redds, which are subsequently covered. If the stones are too large to be moved, the eggs are simply deposited between stones and sink into the sub-benthic spaces (see Plate 3). The gravel or stone beds also serve as a suitable nursery habitat for the young, as they provide both food and shelter from predators. The redds must be situated in places where the oxygen demand of the developing embryos is assured, usually through water movement, such as in rivers and streams or, in the case of lakes, near shores or in spring water sites (Crozier and Ferguson 1986; Skúlason *et al.* 1989a; Ferguson and Hynes 1995; Jonsson and Skúlason 2000). In each lake or river, these stringent requirements necessarily dictate the possibilities for suitable spawning grounds. In some lakes there is but a single suitable site, whereas other lakes harbor many sites, which can differ in various ecological characteristics. Another important trait of salmonids is the precise homing of sexually mature fish to natal spawning grounds. In situations with several distinct spawning sites, homing reduces or even cuts off the gene flow among sites, and thus increases the likelihood of morph divergence through differential adaptations.

Another class of separating mechanisms involves divergence in the timing of reproduction, which can lead to complete reproductive isolation (allochrony; Smith and Todd 1984). Most often salmonids spawn in the autumn, a timing tied in with seasonal variation in temperature and food availability (e.g., for first-feeding juveniles in spring). This means that the development of embryos takes place at winter temperatures. Some lakes offer spawning and nursery grounds that are less affected by season, such as areas with a high flow of stable groundwater at constant low temperature. In several such cases morphs of arctic charr have been observed to breed much earlier in the year (Brenner 1980; Skúlason *et al.* 1989a; Klemetsen *et al.* 1997), which should isolate them reproductively from other morphs that reproduce according to the common schedule. Such differences in breeding time may be a direct result of seasonal differences in the availability of food preferred by particular morphs, mediated through constraints on annual patterns of gonad production (Skúlason *et al.* 1989a). Such scenarios, in which the interplay between external and internal factors leads to intralacustrine spawning allopatry and/or spawning allochrony, are believed to be of key importance in morph formation among lake-spawning fish (Smith and Todd 1984).

In those cases for which the gene flow between morphs becomes impeded by spawning allopatry or allochrony, the situation is comparable to the classic allopatric scenario. In the extreme case where there is essentially no gene flow, (potential) mating incompatibilities can only build-up through differential mutation and genetic drift, or as a by-product of differential adaptive processes that operate within alternative morphs (see, e.g., Turelli et al. 2001). The former tends to be a slow process. However, in the case of divergent selection incompatibilities can arise much more quickly (Rice and Hostert 1993; Reznick et al. 1997; Hendry 2001; Schluter 2001). To evaluate the possible relevance of such mechanisms, consideration must be given to the low age of northern freshwater systems and to the possibility that these environments may have undergone changes (e.g., fluctuations in water level), so that the divergent populations may have passed through different phases of spawning allopatry and sympatry.

In less extreme cases, some intermorph gene flow will still occur. In these cases, at least initially, the genetic integrity of the morphs can be secured by disruptive selection only. For instance, studies on two sympatric forms of the Pacific salmon Oncorhynchus nerka, which interbreed, strongly suggest that genetic segregation is based primarily on reduced fitness of hybrids (Wood and Foote 1990, 1996). Such disruptive selection can accelerate the build-up of prezygotic barriers that involve traits important in mate choice, such as color patterns, and size and shape of body parts (see Section 10.5). The likelihood of such reinforcement processes being effective can depend heavily on intrinsic factors (e.g., how mate choice relates to various phenotypic attributes). For example, sympatric morphs of salmonids often differ in size at sexual maturity (see Section 10.4), which provides a ready substrate for behavioral isolation through size-specific assortative mating (Sigurjónsdóttir and Gunnarsson 1989). Behavioral reproductive isolation among populations and closely related species of fish is likely to be common (e.g., Dominey 1984), and a number of well-studied cases have shown how sexual selection based on selective mate choice promotes and maintains reproductive isolation among sympatric resource-based morphs and species (Rundle and Schluter 1998; Seehausen et al. 1998; see also modeling work in Lande et al. 2001).

10.4 Nature and Basis of Phenotypic Variation

Fish biologists (e.g., Svärdson 1979; McPhail 1984) have, for a long time, been interested in polymorphism among northern freshwater fish and its potential importance in the study of the processes of divergence and speciation, but it is only in the past two decades that these systems have been brought to the attention of a wider audience of evolutionary biologists (e.g., Robinson and Wilson 1994; Skúlason and Smith 1995). In this section, first a brief description of the nature of this variation is given. Next, the proximate basis of polymorphisms (genetics and/or environment) and how this ties in with ecological variables are discussed. Finally,

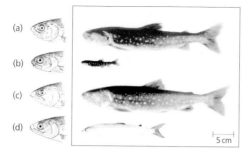

Figure 10.1 Four morphs of arctic charr (*Salvelinus alpinus*) from Thingvallavatn, Iceland. (a) Large benthivorous charr (33 cm), (b) small benthivorous charr (8.5 cm), (c) piscivorous charr (35 cm), and (d) planktivorous charr (19 cm). All these individuals are sexually mature. Note the differences in head morphology, with the planktivorous and piscivorous morphs displaying the ancestral characteristics (pointed snout and terminal mouth), and the benthivores displaying a derived, paedomorphic condition (blunt snout and subterminal mouth).

a hypothesis is given as to how polymorphism could arise through changes in developmental regulation which, if true, would explain the fast speeds by which polymorphic systems appear to have evolved.

Morphs, or incipient species, of northern freshwater fishes generally differ phenotypically in a broad suit of characters, which range from behavior and life-history characteristics to morphology. Typically, this variation is connected to habitat use and feeding; repeated patterns of this type are seen among populations within species (e.g., in different lakes) and across species, which indicates evolutionary parallelism and suggests the operation of common evolutionary mechanisms (Skúlason and Smith 1995). A common dichotomy seen among several species are broad adaptations connected to pelagic (limnetic) life as opposed to living in benthic habitats [Taylor and Bentzen 1993; Smith and Skúlason 1996 (Table 1 therein); Robinson and Schluter 2000]. Typically, differences in body morphology among sympatric morphs are seen in head, jaw, and gill-raker structures associated either with benthic or more limnetic feeding behavior, and/or structures associated with swimming patterns, such as fin size and position, length, and streamlining of the caudal area (Snorrason *et al.* 1994a; Jonsson and Skúlason 2000). An extreme case of morphological diversification is seen in the four morphs of arctic charr in Thingvallavatn, a rift lake situated in the neovolcanic zone of Iceland (Figure 10.1). In this case two benthic morphs display specialized benthic morphology with blunt heads, subterminal position of the mouth, and large pectoral fins (Snorrason *et al.* 1994a). This trophic specialization is further reflected in shapes and relative sizes of bone elements of the skull, in particular those that are closely related to the function of the feeding apparatus, such as some of the tooth-bearing bones (Ingimarsson and Snorrason, unpublished).

Most often, sympatric morphs of northern freshwater fish show clear-cut differences in diets, a differentiation that reflects morph-specific utilization of habitats

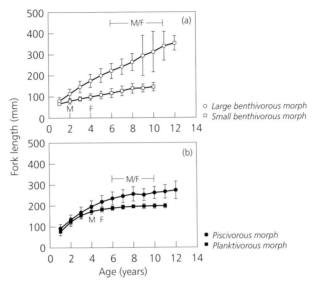

Figure 10.2 Growth curves of the four arctic charr morphs from Thingvallavatn, Iceland. Fork length is the length from the tip of the snout to the mid-marginal cleft of the caudal fin. Vertical bars indicate standard deviations. (a) Large benthivorous morph (open circles) and small benthivorous morph (open squares); (b) piscivorous morph (filled circles) and planktivorous morph (filled squares). M and F indicate the age at which the majority of males and females attain sexual maturity. *Source*: Snorrason *et al.* (1994a).

(e.g., Larson 1976; Hindar and Jonsson 1982; Bourke *et al.* 1997). These studies provide strong, indirect evidence of differences in foraging behavior and also reflect phenotypic constraints on behavior. Direct, experimental evidence shows that feeding behavior can differ markedly among morphs, as with regard to prey type [threespine stickleback (Doucette 2001); arctic charr (wild-caught fish, Malmquist 1992; naive, laboratory-reared fish, Skúlason *et al.* 1993)] and habitat heterogeneity (Mikheev *et al.* 1996). In some cases "morphs" have been defined on the basis of distinct alternative feeding tactics (e.g., McLaughlin *et al.* 1999). Direct field observations of young brook charr showed two distinct foraging tactics, sitting-and-waiting and actively searching; individuals differed in morphology according to which tactic they used (McLaughlin *et al.* 1999; McLaughlin 2001).

Sympatric morphs often differ in life history, such as in growth patterns, age and size at maturity, and various other reproductive characters (e.g., Taylor and Bentzen 1993; Griffiths 1994; Wood and Foote 1996; Jonsson and Skúlason 2000). In extreme cases of morph diversification such variation is extensive, as in Thingvallavatn, where variation in size at sexual maturity among the four morphs of arctic charr almost encompasses the entire interpopulation variation for the species (Figure 10.2; Jonsson *et al.* 1988).

Phenotypic differences among individuals are based on a combination of genetic, environmental, and maternal effects (e.g., egg volume and chemistry). The relative importance of these effects varies within and among species, and different traits are differentially affected. When a high proportion of the phenotypic variation in a trait or species arises through the variable responses of developmental processes to any environmental variation, the trait or species is said to be phenotypically plastic (Schlichting and Pigliucci 1998). In general, fish tend to be phenotypically plastic for many characters (e.g., Liem and Kaufman 1984; Stearns and Crandall 1984; Meyer 1987, 1990; Noakes *et al.* 1989; Thorpe 1989; Wainwright *et al.* 1991; Wimberger 1991, 1992, 1994; Travis 1995). For example, Liem and Kaufman (1984) showed that two morphs of the cichlid species *Cichlasoma minckleyi* differed dramatically in their ability to switch between prey types. In some cases sympatric morphs may result primarily from plasticity (Nordeng 1983; Meyer 1987; Robinson and Wilson 1996). However, several common-garden and family breeding experiments have shown that many phenotypic attributes that differ within and among morphs and populations have a strong genetic basis (Gjedrem 1983; Thorpe *et al.* 1983; Sutterlin and Maclean 1984; Nilson 1990; Silverstein and Hershberger 1992; Crandell and Gall 1993). While most of these studies involve complex polygenic traits, some cases of resource polymorphism are based on variation at a single locus, an elegant example being the left- or right-handedness of the mouth in scale-eating cichlids maintained by frequency-dependent selection (Hori 1993).

Arctic charr are extremely plastic in behavior, growth, and maturation (Nordeng 1983; Vrijenhoek *et al.* 1987; Brown *et al.* 1992; Jobling *et al.* 1993). In rearing and transplantation experiments, Nordeng (1983) demonstrated that progeny from intramorph crosses of three coexisting life-history forms of arctic charr in Norway all yielded the three adults forms; indeed, the occasional individual represented all three forms at different times in its life. A common-garden rearing experiment of the progeny of two morphs of arctic charr in Vangsvatnet, Norway (which differed mainly in growth pattern, adult size, and body color, and to a lesser extent in trophic morphology), showed that adult size differences had a small genetic component, which indicates considerable phenotypic plasticity in these characters. Coloration was a function of size, but differences in trophic morphology had a clear genetic component (Hindar and Jonsson 1993). An experiment that involved two sympatric size-morphs from Stora Rösjön, Sweden, showed much clearer genetic contribution to both growth and maturation patterns than did both of the above rearing studies (Svedäng 1990). Rearing experiments in the laboratory with the progeny of the four morphs from Thingvallavatn showed a clear genetic component to differences in growth, age at first maturity, trophic morphology, body color, and foraging behavior (Skúlason *et al.* 1989b, 1993, 1996; Eiríksson *et al.* 1999). A study on laboratory-reared progeny of benthic and pelagic morphs from Loch Rannoch in Scotland showed that substantial differences in social behavior had a genetic basis (Mikheev *et al.* 1996).

A comparison of the available data on growth, maturity, and behavior for wild- and laboratory-reared morphs of arctic charr shows that morphs and/or their progeny respond differently to variable environmental conditions, that is, the morphs have different norms of reaction (Mikheev *et al.* 1996; Skúlason *et al.* 1996; Ruiz *et al.*, unpublished). For example, small benthivorous charr in Thingvallavatn appear to be less plastic in foraging behavior than planktivorous charr (Malmquist 1992; Skúlason *et al.* 1993, 1996), which ties in neatly with the temporal stability of resources in their respective preferred habitats (Malmquist *et al.* 1992; Snorrason *et al.* 1992, 1994b). Phenotypic plasticity can also be related to variable feeding opportunities. Rearing experiments that involve sympatric limnetic and benthic forms of threespine stickleback from British Columbia showed that a greater plasticity in trophic morphology of the limnetic form, compared to the benthic form, was associated with a more variable diet (Day *et al.* 1994).

It is of paramount importance to unravel the developmental basis of the morphological and life-history variation among sympatric morphs, since this can illuminate critical steps in its evolution. Many of these differences may result from heterochronic shifts in developmental events (Strauss 1990). It is clear, for example, that the timing of ossification in crucial regions of a skeletal element can greatly constrain or even dictate its shape.

Among salmonids, the limnetic morphotype (pointed snout, terminal mouth, fusiform body) constitutes the most usual morphology, and the benthic morphotype (blunt snout, subterminal mouth, stocky body; see Figure 10.1) is considered a paedomorph (Skúlason *et al.* 1989b; Alekseyev 1995). In other words, morphs with benthic features retain embryonic and juvenile characteristics in the adult phenotype. In Arctic charr, this appears to relate to the differential timing and/or speed of skull ossification in late embryonic stages (Eiríksson 1999; Eiríksson *et al.* 1999; Eiríksson *et al.*, in press). Some benthic morphs also retain juvenile coloration in the adult stage (Hindar and Jonsson 1982; Snorrason *et al.* 1994a). The developmental simplicity of intermorph differences implied by the heterochrony hypothesis may help to explain the speed and repeatability of their evolution.

Unfortunately, the genetic and molecular basis of such developmental shifts is still poorly understood. Important advances are now being made by employing the development of genome-wide linkage maps of quantitative-trait loci. Recent studies on limnetic and benthic species of threespine stickleback from Lake Priest, British Columbia, show how changes in various skeletal structures are controlled by genetic factors that map to independent chromosomal regions. Notably, these studies indicated genetic linkage or pleiotropy between certain functionally related traits, such as lengths of pelvic and dorsal spines (Peichel *et al.* 2001).

10.5 Ecological Determinants of Diversity Patterns

For some closely related sympatric species of northern fish, good evidence now suggests that they have evolved locally from morphs or ecological forms that became reproductively isolated through some form of disruptive selection (McPhail 1994; Smith and Todd 1984; Skúlason and Smith 1995; Smith and Skúlason

1996; Schluter 1996b; Lu and Bernatchez 1999; Robinson and Schluter 2000). Inevitably, this has raised the question whether those morphs or ecological forms initially originated in sympatry, before the onset of reproductive isolation. Although some studies indicate the allopatric origin of morphs or species (Bernatchez and Dodson 1990; Ferguson and Taggart 1991; Ferguson and Hynes 1995), a growing number of molecular genetic studies suggest that sympatric morphs of the same species have repeatedly evolved locally and do not represent separate invasions of allopatric populations (e.g., Hindar *et al.* 1986; Foote *et al.* 1989; Bodaly *et al.* 1992; Taylor and Bentzen 1993; Hindar 1994; Gíslason *et al.* 1999).

It may be assumed that the fish species that invaded postglacial freshwater systems were cold resistant and very flexible in terms of development and behavior (Box 10.1), and thus were able to cope with adverse and unpredictable environments (Power 2002). We have hypothesized that the presently known variation in the status of polymorphism among some pioneering fish species of northern postglacial lakes reflects a continuum of outcomes in which at one end are found unstable systems that still harbor monomorphic, but highly flexible, populations, while at the other end are found genetically segregated morphs, which , through adaptation to special niches, have lost some of the original plasticity (Snorrason *et al.* 1994a; Skúlason *et al.* 1999). This scenario assumes that mechanisms of flexibility are costly and that, as ecosystems stabilize and become more predictable, flexibility will be lost whereas specialization will increase. The extent to which individual polymorphic systems or even morphs within systems have advanced in this sense depends on many factors, such as the age of the system and the stability and discreteness of suitable niches. On a comparative basis, some predictions of such a scenario can be tested. More direct tests of the above ideas involve comparisons of morphs within systems under experimental conditions. Two morphs of threespine stickleback from Thingvallavatn have been described recently (Kristjánsson *et al.* 2002a); the one that lives in a very stable habitat (lava rubble with high influx of cold spring water) was found to be less flexible in foraging trials than a morph that lives in a more variable vegetated habitat (Doucette 2001; see also Day *et al.* 1994).

Some authors have hypothesized that phenotypic plasticity can, in some cases, be crucial in setting the stage for sympatric ecological divergence (West-Eberhard 1989). Given a lake with various vacant niches for fish, it is easy to envisage how a trophically plastic invader could immediately utilize various niches. If alternative niches are ecologically stable and/or predictable, this would quickly set the scene for ecological divergence. This may look counterintuitive, since the invader might seem to do very well by staying plastic. However, the cost of plasticity (e.g., when an ability to utilize two niches in some way constrains the scope for improvement in the alternative niches) and the availability of mutations that increase or decrease (Kawecki 1997) the efficiency with which one of the alternative niches is exploited, open possibilities for ecological divergence. In particular, mutations that lead to heterochronies are prime candidates for effecting a fast morph divergence.

Box 10.1 The role of foraging behavior in the formation of sympatric morphs

Flexible foraging behavior in fish can be of major adaptive importance, especially in environments in which resources fluctuate in some important way (Dill 1983; Noakes 1989). Skúlason et al. (1999) have hypothesized that flexibility (plasticity) in foraging behavior and mobility plays a crucial role in the early stages of morph segregation, before life-history and morphological differences evolve. This view is supported by studies of the recently emerged young of the brook charr (*Salvelinus fontinalis*), which inhabits still-water pools along the sides of streams. Either they are sedentary and eat crustaceans from the lower portion of the water column or they are more active and eat insects from the upper portion of the water column. Studies have shown that this divergent foraging behavior reflects short-term divergent selection brought about by intraspecific competition in the presence of an alternative food source (McLaughlin et al. 1999). It is thus reasonable to suggest that a population of fish that invaded a fishless lake (with diverse resources) at the end of the most recent glaciation could, driven by increased population size, very quickly occupy the different resources, primarily on the basis of behavioral differences.

Variation in early foraging and social behavior can have long-term life-history consequences for fish (e.g., Metcalfe 1993). Thus, early behavior can determine growth patterns, which are closely correlated with other developmental events at later stages, such as "decisions" to migrate, shift niches, or become sexually mature (Thorpe et al. 1992; Metcalfe 1993; Forseth et al. 1994; Skúlason et al. 1996). Thus, variation in early behavior can have immediate consequences for habitat and food selection, with possible long-term consequences for growth patterns, ontogenetic niche shifts. and maturation patterns (Forseth et al. 1994; Snorrason et al. 1994a; Skúlason et al. 1996). Similarly, body morphology can be influenced by behavior, for instance plastic trophic structures in fish can respond to different diets (Meyer 1987; Wimberger 1991, 1992, 1994; Adams et al. 2003).

A comparative study of lakes that host two sympatric size-morphs of arctic charr illustrates well how flexibility in feeding behavior can have cascading effects on differences in growth patterns in arctic charr (Griffiths 1994). It was demonstrated that the morph that attained larger adult body size usually had a broader diet selection (a characteristic expected in the original invaders), while the diet of the morph with smaller adult body size was more restricted. In other words, the small morph was ecologically more specialized. This is further supported in a study of two morphs, which differed greatly in adult size, in a small lake in Iceland; here, differences in flexibility in diet selection between morphs were evident in young fish and independent of body size (Jonsson and Skúlason 2000).

The Thingvallavatn charr morphs (see Figure 10.1) represent an advanced case of morph specialization (Snorrason et al. 1994a) and, as such, can be used in an attempt to describe and explain the evolution of a polymorphic system. The postglacial history of the lake has been described in detail (Sæmundsson 1992). The lake was most likely isolated from potentially invading fish very soon after it became established in its present form by a lava dam some 9130 ($\pm$260) years ago. We suggest that, after the postglacial invasion of arctic charr to Thingvallavatn, the evolution of morphs occurred in the early stages, primarily as a consequence of ancestral phenotypic plasticity in behavior, life history, and (to a lesser extent) morphology (Box 10.1). As time passed and the global stability of the lake increased (i.e., the inflow of glacial water ceased), isostatic movements decreased, the present outlet was established, and the lake developed its characteristics as a spring-fed system (Sæmundsson 1992). After this, the stability of niches in the lake increased, the ecological segregation of morphs became stable, and their phenotypic differences more clear, a process consolidated by hard-wired genetic modifications of developmental programs that influence behavior, life history, and morphology (Skúlason *et al.* 1989a, 1993, 1996).

Both the continuum from great flexibility to considerable specialization that exists in arctic charr in different lakes (Skúlason *et al.* 1992; Snorrason *et al.* 1994a) and its relationship with environmental and genetic effects on development (Nordeng 1983; Svedäng 1990; Skúlason *et al.* 1989b, 1993, 1996; Hindar and Jonsson 1993) indicate that an evolutionary process similar to that suggested above for Thingvallavatn may be ongoing in many other polymorphic systems of freshwater fish in the Northern Hemisphere. However, flexibility and phenotypic plasticity continue to be favored in less predictable environments.

The degree of genetic divergence between sympatric morphs is highly variable, both within and among species. In some cases, gene flow may be unimpeded – in which case the polymorphism is presumably environmentally induced through an appropriate system of reaction norms (Nordeng 1983) or because of segregating loci in a simple genetic polymorphism (Hori 1993) – while in other cases, sympatric forms are partially (Hindar *et al.* 1986; Magnusson and Ferguson 1987; Foote *et al.* 1989; Danzmann *et al.* 1991; Bodaly *et al.* 1992; Hindar 1994; Volpe and Ferguson 1996; Gíslason *et al.* 1999) or completely reproductively isolated (Ferguson and Taggart 1991; Hartley 1992; Hartley *et al.* 1992; McPhail 1994; Gíslason *et al.* 1999). The degree of gene flow in partially isolated morphs depends on ecological and topographic conditions in individual systems, and by their geological age (see Section 10.3).

As already emphasized, successful divergence ultimately depends on the build-up of reproductive isolation, a process that should be accelerated in lakes that offer opportunities for temporal and/or spatial spawning segregation or opportunities for isolation because of variable mate choice (see Section 10.3). Irrespective of morph origin (sympatric or allopatric), there is now growing consensus that polymorphic systems with moderate gene flow are of key importance in the consolidation of mating barriers (Rice and Hostert 1993; Smith and Skúlason 1996; Schluter 2001; Via 2001; Wu 2001). This rests on the logical expectation that gene flow will bring on disruptive selection and thus accelerate the build-up of mating barriers. The outcome inevitably depends on the strength of the selection, as it must thwart the mixing effect of recombination (Maynard Smith 1966; Felsenstein 1981). The speed of the process (if it takes off) also depends on the strength of selection and the genetic mechanisms involved, such as whether these are direct effects on "mating" genes or only indirect effects (via linkage or pleiotropy) of genes that affect ecological traits on intermorph matings (e.g., Kondrashov *et al.* 1998), as well as the number of genes under selection (Rice and Hostert 1993). Recent empirical evidence from sockeye salmon indicates that the two ecotypes (commonly found in Western North America), one breeding in streams and the other along lake beaches (Wood 1995), may form very quickly in sympatry. In a population recently introduced into Lake Washington, Washington, USA, evidence was found of reproductive isolation arising between such ecomorphs within as few as 13 generations (Hendry 2001; Hendry *et al.* 2000; see also Kristjánsson *et al.* 2002b).

Assuming the above scheme of ecologically driven sympatric divergence, coupling of relevant phenotypic differentiation and genotypic divergence would be expected. This is supported by recent studies that show the level of genetic isolation among sympatric morphs of Icelandic arctic charr and North America whitefish across lakes to be, indeed, positively correlated with the magnitude of morphological differences among the respective morphs (Gíslason *et al.* 1999; Lu and Bernatchez 1999; see Figures 10.3a and 10.3b). For one small mountain lake in Iceland, genetic analyses showed that two arctic charr morphs are completely isolated reproductively (Gíslason *et al.* 1999), and should at this time be regarded as different biological species (Mayr 1963; Wu 2001).

10.6 Concluding Comments

For ecologists, the adaptive speciation models pioneered by Rosenzweig (1978), Bengtsson (1979), Gibbons (1979), Pimm (1979), Seger (1985a), Wilson and Turelli (1986), and Johnson *et al.* (1996b), and see also Chapters 3 and 4, are very attractive, especially in the way ecological dynamics continually shape the fitness landscapes (Dieckmann and Law 1996; Metz *et al.* 1996; Geritz *et al.* 1997, 1998; Chapters 4 and 5). In this section we examine how the above information and ideas of divergence in northern freshwater fish conform to the adaptive speciation concept, which for operational purposes can be described as "speciation driven by disruptive selection under conditions of at least minimal ecological contact between the diverging lines" (see Box 1.1 in Chapter 1). For reasons of transparency, the focus here is on the diversification in arctic charr (see summary framework in Box 10.2). Some apparent contrasts between our conceptual model and the mathematical model of Dieckmann and Doebeli (1999; Chapter 5) are highlighted below.

At the end of the most recent glaciation, arctic charr invaded many newly formed lakes (it may be assumed that in many cases it was the first pioneer among fish). Where conditions were favorable, the charr quickly underwent a population explosion, which led to intense intraspecific competition for resources, accompanied by phenotypic shifts or character release immediately realized through phenotypic plasticity in behavior and life history. Both of these steps are well documented for recently introduced populations of arctic charr (Nordeng 1983; Skúlason *et al.* 1999). After these initial steps and given a minimum degree of habitat diversity, phenotypic plasticity probably, in the early stages, played a crucial role in facilitating rapid and extensive diversification of phenotypes, which

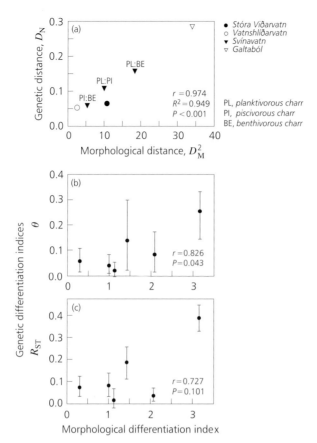

Figure 10.3 Relationship of morphological differentiation versus genetic differentiation for morphs in several lakes. (a) Relationship between Mahalanobis distance (D_M^2) and Nei's genetic distance (D_N) for morphs of arctic charr in four Icelandic lakes. *Source*: Gíslason (1998). (b, c) Relationship between morphological differentiation index (eigenvalues) and genetic differentiation indices (θ, R_{ST}) for dwarf and normal whitefish (*Coregonous* sp.) ecotypes (morphs) in six lakes in eastern Canada and northern Maine. Also given are correlations for six microsatellite loci combined. *Source*: Lu and Bernatchez (1999).

resulted in the formation of distinct resource morphs, especially relating to behavior and life-history characteristics (Nordeng 1983; Vrijenhoek *et al.* 1987).

Polymorphisms engendered by plastic responses in principle provide a fertile base for subsequent genetic assimilation (Waddington 1953a), especially in systems where habitats and other resources have attained long-term stability (predictability). The question at this point is how does reproductive isolation among morphs arise? The literature on arctic charr resource morphs suggests at least two possibilities:

Box 10.2 Framework for the evolution of different morphs of arctic charr

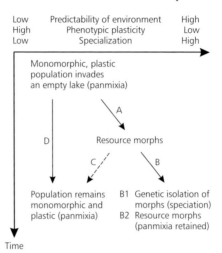

Above is a graphic presentation of a framework for the evolution of different morphs of arctic charr and speciation. A basic assumption of the framework is that, after an initial phase of instability, many freshwater systems stabilized, in the sense that habitats and food resources for fish became more predictable, opening the possibility for adaptive specialization and morph formation (arrows A and B). Whether resource polymorphism of this kind can lead to reproductive isolation (B1) depends on the interactions of ecological factors (e.g., static and dynamic characteristics of resources, and the scope for variation in spawning sites and/or time) and intrinsic factors (e.g., genetic variation in a relevant ecological trait, scope for developmental modifications, and homing to spawning grounds; see Sections 10.2 and 10.3). In stable systems, in which assortative mating does not become established (e.g., because of interference with other species and/or restricted scope for variable spawning sites or spawning times), genetic polymorphisms can be maintained by frequency-dependent selection (B2). If ecosystems revert to a state of unpredictability, polymorphic systems will revert to plastic monomorphy (arrow C). In systems that remain unstable and unpredictable, populations remain plastic and monomorphic (arrow D). Early stages of polymorphism are expected to involve behavioral and life-history traits; morphological traits become more prominent later.

▓ The diversification phase described above can be accompanied by the evolution of reproductive isolation among forms when resource-based features of the respective ecological and/or phenotypic segregation correlate with factors that promote temporal and/or behavioral segregation in spawning. In lakes in which opportunities for spawning are diverse and correlated with discrete, predictable (in a spatial and temporal sense) resources, arctic charr morphs can develop different seasonal spawning times and locations (see Sections 10.2 and 10.3).

■ Along similar lines, certain, discrete differences between resource morphs can promote isolation directly based on mating behavior (see Section 10.2). Thus, size-assortative mating may follow the appearance of differently sized life-history morphs so commonly seen in arctic charr.

Recent evidence suggests these processes can progress very quickly in the wild (Hendry *et al.* 2000). Once a state is reached in which gene flow between morphs has become severely restricted, the stage is open for further specialization of the morphs and ultimately, given time enough, the formation of new species. The advancement of this process depends heavily on the stability of the utilized niches. Also, loss of phenotypic plasticity is expected in the characters that show the highest degree of specialization, as seen when arctic charr polymorphism is compared across lakes (Skúlason *et al.* 1999; see Section 10.5).

While the evolutionary process of sympatric divergence in arctic charr laid out here conforms clearly with the essence of the adaptive speciation concept (i.e., selectively driven speciation under conditions of ecological contact between the diverging lines) some apparent contrasts with the theoretical structure of adaptive speciation models must be highlighted [see Dieckmann and Doebeli (1999), and Chapters 4 and 5]. First, in contrast to the unimodal resource spectrum assumed in the mathematical model, clearly ecological situations better described as multimodal are being dealt with here. Recent applications of adaptive speciation models to situations of "ontogenetic niche shifts" are very promising in this respect (Claessen and Dieckmann 2002; Box 10.3). Second, the role of plasticity in the early phases is emphasized here (which may be somewhat of a bias, as some of the northern pioneers are very plastic). In a unimodal situation, a high degree of plasticity apparently has the effect of increasing the spread of the competition curve, and thus making branching less likely. However, plasticity always comes at a cost. In particular, fine tuning a plastic response is a much more involved process than doing so for a genetically fixed one. Under heavy competition, fine tuning matters greatly. If this fine tuning takes the form of better adapting the morphology to either of two different behaviorally chosen roles, the system is close again to the simple Dieckmann–Doebeli scenario.

The idea that plasticity could jump-start divergence by immediately exposing invaders to disruptive selection as well as causing assortative mating suggests various avenues of future research. "Field experiments" (planned or incidental), in which species are introduced into a new, multipotential (e.g., offering alternative habitats) locality (see, e.g., Hendry *et al.* 2000) are of obvious interest, especially with respect to novel habitats and diets, as well as potential spawning areas and nursery habitats. Laboratory experiments that test the scope of plasticity and address its function are also important (e.g., Meyer 1987; Adams *et al.* 2003). The proposed trade-off between plasticity and specialization calls for comparisons of plasticity among populations along a gradient from low to high environmental stability and/or predictability and lake age. Parallel to such research, the genetics of plasticity and the candidate target genes of disruptive selection must be studied.

Box 10.3 Adaptive speciation through ontogenetic niche shifts
David Claessen and Ulf Dieckmann

This box introduces a simple mechanistic model to demonstrate how adaptive speciation in size-structured populations can result from the interplay between ontogenetic niche shifts (Werner and Gilliam 1984) and environmental feedback (Heino *et al.* 1997). Arctic char, for example, undergo an ontogenetic niche shift from benthivory to planktivory at body sizes of 10 to 17 cm (Snorrason *et al.* 1994a; Langeland and L'Abee-Lund 1998).

Motivated by this and other systems, the model (Claessen and Dieckmann 2002) is based on the assumption that individuals predominately exploit one type of prey (prey 1) while their body size l is small, $l < x$, and a second type of prey (prey 2) once they have grown to larger sizes, $l > x$. This niche shift is assumed to be irreversible and is determined by an individual's genotype x, referred to as its switch size.

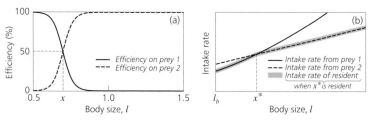

Individuals are born at size l_b. Their intake rates follow a size-dependent type-II functional response, determined by handling times and attack rates (Mittelbach 1981; Persson and Greenberg 1990). Handling times decrease with body size according to an allometric function. Attack rates are modeled as the product of a maximum feasible attack rate, increasing allometrically with body size, and the sigmoidal efficiency functions depicted in panel (a). Allocation of ingested energy follows the κ-rule (Kooijman and Metz 1984): a fraction $1 - \kappa$ of the total energy intake is invested in reproduction; the remaining energy is used first to meet metabolic costs (proportional to body volume) and then to invest in somatic growth. As a result of mutation, there is a small possibility that the genotype of individual offspring will differ slightly from that of the parents. The two prey populations are unstructured and are described by semi-chemostat dynamics (Persson *et al.* 1998).

Environmental feedback arises from the impact of consumption on the density of the two prey populations. If the niche shift occurs at a very large switch size x, hardly any individuals survive to exploit the second niche. Prey 1 is then overexploited while prey 2 is underexploited. If, by contrast, the ontogenetic niche shift occurs very early in life, prey 1 is abundant whereas prey 2 is depleted.

Panel (c) shows a stochastic realization that corresponds to the latter situation. The population starts out from a switch size $x < l_b$ and thus essentially only

continued

Box 10.3 *continued*

exploits prey 2, leaving prey 1 at high abundance. For the chosen parameter values, the switch size x^* is both convergence stable and evolutionarily unstable. As illustrated in the figure, this means that directional selection first brings an almost monomorphic population into the vicinity of this switch size; however, once x^* has been reached, selection turns disruptive and two different ecomorphs emerge.

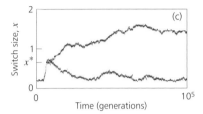

The mechanism that underlies this somewhat counterintuitive process can be understood by considering panel (b). The size-dependent intake rates from the two prey types depend on their exploitation and thus on the switch size that is resident in the population. For a given resident, adopting the switch size at the intersection of these curves either maximizes or minimizes the total intake. When x^* is resident, the resultant intake rates intersect at $l = x^*$. Convergence toward x^* is driven by environmental feedback. Perturbing the resident switch size a little away from x^* explains how this works: a resident that shifts exploitation at a slightly smaller size, $x < x^*$, underexploits prey 1 and overexploits prey 2. This means that, in panel (b), the continuous curve for prey 1 moves down while the dashed curve for prey 2 moves up. If such a resident now gives rise to a mutant that shifts later, this mutant profits from prey 1 more than the resident itself; it can therefore invade. The opposite is true for perturbations with $x > x^*$. This ensures evolutionary convergence toward x^*.

By considering the fate of mutants in the environment set by the resident x^*, we can see that x^* is not evolutionarily stable. Compared to the resident, a mutant that switches at a smaller size profits more from an increased intake rate from prey 2 than it loses on prey 1. Conversely, a mutant that switches at a larger size profits more from an increased intake rate from prey 1 than it loses on prey 2. This shows that x^* is a fitness minimum and thus gives rise to disruptive selection. Since it is also convergence stable, x^* is an evolutionary branching point (Chapter 4). These conclusions are valid, if, as for panel (c), the intake rate increases faster with body size for prey 1 than for prey 2. In the opposite case, x^* is convergence stable and evolutionarily stable: no evolutionary branching is then expected.

To assess whether the eco-evolutionary mechanism proposed here can explain observed resource polymorphisms and the potentially ensuing processes of adaptive speciation, estimates of how intake rates in two alternative niches scale with body size will be of critical importance.

Rapid advances in molecular genetics are now opening new avenues in the analysis of adaptive speciation. Genome-wide linkage maps are currently being constructed to assess the genetic linkage of quantitative traits that are potential key players in divergence processes (Peichel *et al.* 2001; Via and Hawthorne 2002). Applying these techniques to the diverse systems of polymorphism in fresh water fishes of the "northern postglacial theater" is bound to pay off (Rogers *et al.* 2001). It is crucial also that such programs are used to integrate the various subdisciplines involved, whether they are defined by "end products" of phenotypic expressions (physiology, behavior, morphology), genetic architecture (linkage of target genes, sex linkage, etc.), or by developmental expression mechanisms, such as the structure of relevant, developmental expression cascades and their evolutionary potential.

Acknowledgments   We thank Bjarni Kristjánsson and Lisa Doucette for allowing us to use their unpublished work on stickleback morphs in Thingvallavatn.

11

Sympatric Speciation in Insects

Guy L. Bush and Roger K. Butlin

11.1 Insect Diversity, Body Size, Specialization, and Speciation

Insects are among the most abundant multicellular organisms on earth. Sampling in the tropics and subtropics suggests that there are 10 to 30 million species of insects (Erwin 1982). As a group, invertebrate insects, therefore, represent an inordinately large percentage of the world's fauna, far outnumbering their larger and more conspicuous vertebrate cousins, such as fish (about 24 000), amphibians (about 4000), reptiles (7907), birds (9808), and mammals (4629). In fact, the number of species in many families of Heteroptera, Diptera, Coleoptera, and Hymenoptera outnumber the species in most classes of vertebrates.

Why are there so many insect species? One apparent reason is body size (Bush 1993). Their small size allows insects to subdivide habitats and specialize on resources that large animals are unable to exploit. Frequently, much of the life cycle occurs exclusively on the resource, which is particularly important since it induces assortative mating. Specialization permits a habitat occupied by a single vertebrate species to support a much greater number of small, often closely related, invertebrate species that can coexist in close sympatry with minimal competition. Vertebrate sister species usually have similar ecological needs, and are seldom sympatric because they competitively exclude one another. In contrast, it is estimated that the majority of insects, over 70% of which either feed on plants or are parasites and parasitoids, are highly host specific (Jaenike 1990). Sister species may sometimes even feed on different parts of the same host.

The occurrence of specialization, however, is not uniform throughout the Insecta. Specialists occur most commonly in the more advanced Neoptera insect orders, such as the Paraneoptera and Endopterygota. Highly species-rich genera of monophagous and stenophagous plant and animal parasites and parasitoids are characteristic of these insects. In contrast, specialists or species-rich genera in the ancestrally apterous Archaeognatha, Monura, and Thysanura are found rarely. Nor are they common in the Paleoptera (Ephemeroptera and Odonata) and basal Pterygota (Plecoptera, Dictyoptera, Grylloblattodea, Dermaptera, Orthoptera, and Phasmatodea), which are represented mostly by relatively larger bodied polyphagous, omnivorous, and insectivorous orders (Naumann 1991).

Two possible sources of diversity among specialist species can be envisaged: generalists may give rise to multiple descendant specialists, or specialist species might multiply by shifts from existing to new hosts. Kawecki (1998) said that "Divergence of an initially generalist population into specialized subpopulations that show preference for different habitats or resources has been considered the most likely route to sympatric speciation." This is true from the theoretical point of view. Models derived from Maynard Smith (1966), Seger (1985a), including Kawecki's own model, and Doebeli (1996a), for example, consider the division of one population that exploits two hosts into two separate monophagous populations. Progress toward the evolution of reproductively isolated specialist populations from a generalist ancestor has been observed under laboratory conditions in ingenious experiments by Rice (1987) with *Drosophila melanogaster*. A case for such a pathway was made by Gibbons (1979) for divergence in ovipositor length between species of ichneumonid parasitoid wasps of the genus *Megarhyssa*. There is also clear evidence that *D. sechellia*, which specializes on the toxic fruit of *Morinda citrifolia*, is descended from the generalist *D. simulans* (Jones 1998). However, relatively few models consider host shifts (Johnson *et al.* 1996b), although this actually appears to be a dominant mode of speciation in insect specialists (see below).

Allopatric speciation is likely to be the most prevalent mode of divergence among insect generalists. As closely related sister species have very similar ecological resources and mate recognition systems, they seldom live sympatrically. Only when sister species have evolved sufficiently strong pre- or, in many cases, postmating reproductive isolation and different ecological requirements can two sister species coexist without either hybridizing or competitively excluding one another. If they do hybridize, they may form a relatively stable hybrid zone (Butlin 1990) or occasionally generate a new parthenogenetic species (Bullini 1994).

Competition is less likely to prevent co-existence between closely related sister species of insect specialists (Bush and Smith 1997). Recently diverged insect specialists each spend most of their life history on different hosts, which reduces the chances of encountering one another when feeding or searching for mates. Interspecific competition from unrelated insects, if present, may exert a much stronger competitive effect on adaptation. Mating is also usually restricted to a different host or habitat for each species, and host preference is generally a major component of their mate-recognition system.

As a result of these differences in biological traits responsible for specialization, the process of speciation in insect specialists can differ from the speciation of insect generalists. Speciation in specialists potentially can occur in at least five ways:

- By cospeciation;
- In allopatry on the same resource;
- In allopatry accompanied by a resource shift;
- In sympatry accompanied by a resource shift;
- By interspecific hybridization.

By sympatry we mean that populations occur within the same region, such that gene flow is obstructed as a result of the biology of the insects rather than because of physical barriers that prevent their meeting. Parapatry is an intermediate situation in which ranges abut, and so only part of each population is in contact. Most commonly, the resource is part of a host, such as a fruit, and shifts occur between the same parts of different host species. We concentrate on examples of this type. However, the resource shift could be between parts of the same host. For example, in *Blepharoneura* (Tephritidae: Diptera) one species infests male flowers while its sympatric sister species uses female flowers of the same host plant, *Gurania costaricensis* (Cucurbitaceae; see Condon and Steck 1997). Similarly, two sister taxa of the human body louse *Pediculus humanus* (Pediculidae: Anoplura; see Levene and Dobzhansky 1959), regarded as species by some (Chirov and Ozerova 1997), coexist on the same individual. *P. h. corporis*, specialized in a variety of ways to life on the body, lays its eggs only on cloth threads, while the head louse *P. h. capitis* is adapted to an existence on the head and glues its eggs on hair. Interspecific hybridization is not considered here, as there is insufficient evidence that it is responsible for speciation in insects, except in the origin of parthenogenetic taxa (Bullini 1994).

11.2 Cospeciation

Cospeciation occurs when an ancestral association between species simultaneously splits into descendant associations. It most often involves a coordinated process of allopatric speciation between a host and its specialized phytophagous, commensal, or parasitic insect. Although originally thought to be common, it is clear that cospeciation, or co-cladogenesis, is relatively rare (Godfray 1994). To qualify as cospeciation, the insects must be associated tightly with their hosts from the time at which the ancestral host originated (Mitter *et al.* 1991; Menken and Roessingh 1998). As the association between host and commensal or parasite has been continuous throughout their evolutionary history, the two cospeciation clades should be of similar age.

A clear example of cospeciation involves ectoparasitic lice and their hosts. A phylogenetic study using mitochondrial cytochrome oxidase I found clear evidence of cospeciation between most of the 15 allopatric pocket gophers (Geomyidae) and their associated lice (Mallophaga; see Hafner *et al.* 1994; Page and Hafner 1996). However, even in this example a single case of host switching has occurred. Indeed, host switching may often be the rule in other lice and their hosts. Little evidence of cospeciation has been found between swiftlets (Apodiformes: Apodidae) and their *Dennyus* lice (Phthiraptera: Menoponidae; see Clayton *et al.* 1996). Host switching is so common in some groups of lice that Barker (1994) suggests the axiom that lice and their hosts invariably cospeciate be abandoned. Whether or not cospeciation occurs in lice may depend on the degree of allopatry between host species. If hosts remain allopatric, as do most pocket gophers, there is little opportunity for host shifting. When host species become sympatric, lice may replace one another after bouts of interspecific competition and extinction (Barker 1994).

Cospeciation is the rule in the association between figs and their fig-wasp pollinators (Machado *et al.* 1996). An in-depth phylogenetic analysis by Weiblen (2001), which combines molecular and morphological data, finds clear evidence of extensive cospeciation between dioecious New Guinea figs (*Ficus* subg. *Ficus*, Moraceae) and their highly specialized fig-wasp pollinators (Hymenoptera: Agaonidae) accompanied by only a few host switches. There is strong evidence of correlated evolution between fig style and wasp ovipositor lengths, as well as of other changes in the fig-breeding systems that foster a stable mutualism. This study on figs and their pollinators indicates a tight one-on-one specificity and life-cycle interdependence that should promote and sustain cospeciation. As mating occurs within the fig, inbreeding results in a loss of genetic variation. Extreme specialization and loss of genetic variation limits the ability of pollinators to shift host. These results are consistent with the view that figs and their wasp pollinators cospeciated allopatrically, although pollinator-driven sympatric speciation cannot be ruled out (Weiblen 2002).

11.3 Allopatric Speciation on the Same Host

Although the assumption is often made that host specialists have speciated allopatrically without a host shift, few detailed studies document this mode of speciation in insects. One such example is speciation in tephritid flies that belong to the *Rhagoletis suavis* species group (Bush and Smith 1998). The six species in this group feed as larvae exclusively on the husks of walnuts (*Juglans* spp.). The husk contains large quantities of juglone, a phenolic compound toxic to many organisms. This compound may serve as a dietary requirement (Bush and Smith 1998), much as cactus alkaloids are required for reproduction by cactus-infesting *Drosophila* (Heed and Kircher 1965). Such compounds appear to impose a serious biological constraint on the ability of these specialists to shift to new hosts.

Taxa of the *R. suavis* species group are allopatric or, when parapatric, they are allochronically and altitudinally isolated. Cospeciation can be ruled out because all *Juglans* species used as hosts evolved at least 35 million years before the first ancestor of the *R. suavis* group colonized walnuts about 2–5 million years ago. Ample fossil evidence shows that the distribution of walnut species shifted over the millennia, which provided opportunities for populations that infest walnut to diverge in isolation. As noted below, a shift to a new host in the absence of geographic isolation accompanies speciation events in other *Rhagoletis* species groups in which the rate of speciation is 2–3 times faster than that in the *R. suavis* group.

Allopatric speciation has also occurred in the Hawaiian Island *Drosophila*. At least 22 inter-island colonization events resulted in speciation without a major shift in feeding habits (White 1968). Indeed, island archipelagos provide the best examples of allopatric speciation in insects, because it is possible to establish patterns of colonization and rates of divergence with some degree of accuracy (White 1968; Roderick and Gillespie 1998). However, most of the 700 species of Hawaiian *Drosophila* result from intra-island rather than inter-island speciation events. Many of these speciation events involved shifts in habitats and use of host-plant resource (Kambysellis *et al.* 1995).

Another well-studied example of insect specialists that speciate in allopatry without host shifts involves *Heliconius* butterflies, although parapatric speciation cannot be excluded (Mallet *et al.* 1998). *H. erato* and its close relatives are all specialized on *Passiflora* species in the subgenus *Plectostemma*. The evolution of alternative mimetic color patterns and/or environmental associations, now associated with strong assortative mating (McMillan *et al.* 1997), apparently drive speciation.

11.4   Allopatric Speciation with a Host Shift

A shift to a new host accompanies the majority of speciation events that involve plant feeders, parasites, and parasitoids (Strong *et al.* 1984; Mitter *et al.* 1991), although it is important to note that evolutionary changes in host association are not necessarily connected to speciation (Singer *et al.* 1992). There is still disagreement over when, where, and how speciation by a host shift occurs. Three broad possibilities must be distinguished: sympatry, allopatry without a restriction in population size (vicariance), and allopatry with a period of reduced population size (Table 11.1, the peripatric model). To explain the origin of the many observed sympatric sister species of phytophagous, parasitic, and parasitoid insects, Mayr (1947, 1963), Futuyma and Mayer (1980), and other authors maintain that it must occur allopatrically. They base this view on the assumption that such insects generally have several alternative hosts. Speciation occurs when a population of a widely distributed species becomes geographically isolated on only one of its normal hosts. Over time, the isolated deme specializes on the host and eventually speciates. The new species may then re-establish sympatry with the parental species. Although speciation by this allopatric process undoubtedly can take place in nature, its frequency is not documented and thus it is uncertain how important it has been in generating new species. It is likely to be a relatively slow mode of speciation, and is therefore unlikely to account for all of the millions of highly specialized sympatric sister species of plant-feeding insects that appear to evolve rapidly. However, small peripatric isolates may evolve more rapidly or there may be strong selection for specialization in some populations, so that a clear distinction from sympatric host shifts is difficult.

An association between phylogeography, based on allozymes, and host-plant use in *Larinus* weevils on *Onopordum* and *Cynara* thistles in the Mediterranean region is suggestive of allopatric divergence (Briese *et al.* 1996). Since there are no major differences in heterozygosity, this appears to have happened without population bottlenecks. Futuyma *et al.* (1995), Funk (1998), and Knowles *et al.* (1999) propose a possible example of peripatric speciation. On the basis of limited mitochondrial DNA sequence data and population sampling, they conclude from their phylogeographic analysis that founder events and peripatric speciation (Mayr 1963) were involved in host shifts and speciation in the beetle genus *Ophraella* (Coleoptera; Chrysomelidae; see Knowles *et al.* 1999). They focused their analysis on the origin of *O. bilineata* whose distribution broadly overlaps that of its putative parental species, *O. communa*. However, their evidence that *O. bilineata* went through a genetic bottleneck is not convincing. *O. communa* and *O. bilineata* have equivalent levels of genetic variation that Knowles *et al.* (1999)

Table 11.1 Contrasting expectations from allopatric and nonallopatric speciation.

	Allopatric		Nonallopatric		Comments
	Vicariance	Peripatric	Sympatric	Parapatric	
Ranges of sister species at time of origin	Nonoverlapping	One species with restricted peripheral range	Broadly sympatric	Abutting	Secondary range overlap is needed to establish species status
Host association	Same or different	Same or different	Different	Different	
Rate of speciation	Slow	Fast	Very fast	Very fast	
Postzygotic isolation	Through genetic incompatibility	Through genetic incompatibility	Through host adaptation only	Through host adaptation only	
Morphological divergence	Likely	Likely	Unlikely	Unlikely	Assumes that morphological divergence accumulates slowly. The opposite pattern is true for traits directly involved in resource use
Parallel speciation	Unlikely	Possible	Likely	Likely	Parallel speciation is the independent origin of reproductive isolation in separate populations in response to the same environmental conditions (Schluter 1996a)
Phylogenetic relationships	Paraphyly unlikely, gene trees congruent	Paraphyly likely, gene trees may be incongruent because of lineage sorting[a]	Paraphyly likely, gene trees may be incongruent because of incomplete barrier to gene exchange	Paraphyly likely, gene trees may be incongruent because of introgression near boundary	Here, paraphyly means that populations of the ancestral species may be more distantly related to one another than the descendant species is to the population from which it was derived. This is reflected in paraphyletic relationships in at least some gene trees

[a] See Box 15.3.

believe resulted from postspeciation hybridization. However, they did not rule out incomplete lineage sorting (see Box 15.3) to account for the observed shared polymorphisms. Nor does the present broadly parapatric overlap of *O. bilineata* with *O. communa* and the paraphyletic origin of *O. bilineata* from *O. communa* support their view that *O. bilineata* speciated peripatrically (Table 11.1). Their data are equally compatible with the hypothesis that *O. bilineata* speciated as a result of a sympatric or parapatric host shift with substantial gene flow during the early stages of divergence and host-race formation, or indeed with allopatric speciation in a large population. The potential roles of population bottlenecks and divergence through selection on a new host have been confounded in this argument. They may interact, but allopatric host shifts do not require bottlenecks.

In contrast, similar phylogeographic analyses now provide good examples of paraphyletic patterns of host-plant utilization in which a derived host race or species has arisen within a phylogenetic cluster of a widely distributed and geographically differentiated species (Brown *et al.* 1997; Bush and Smith 1998; Dobler and Farrell 1999). In these examples, there is no evidence that host shifts occurred allopatrically, but it remains difficult to infer sympatric divergence without additional evidence because paraphyly can result from colonization of a new region by individuals from one part of a widespread species (Table 11.1). A possible example here is *D. simulans*, which is paraphyletic to the island endemic *D. sechellia*, a specialist on a novel host (Solignac and Monnerot 1986).

Barraclough and Vogler (2000) have attempted to overcome this problem by comparing range overlap between species pairs and time since speciation, and using the relationship to infer the extent of sympatry at the time of speciation. Their analysis suggests that speciation was predominantly allopatric in the taxa studied, which included *Rhagoletis* fruitflies and *Flexamia* leafhoppers. However, they did not include closely related sympatric taxa that are good candidates for sympatric host shifts, such as *R. mendax* and *R. zephyria*. These were omitted because the phylogenetic hypothesis on which the analysis was based failed to separate them from their ancestral species, in this case *R. pomonella* (for the marker used, *R. mendax* and *R. zephyria* sequences are embedded within the clade of *R. pomonella* sequences.

11.5 Sympatric Speciation with a Host Shift

Sympatric speciation first involves the colonization of a new host and the development of a biologically distinct host race (Box 11.1). Although it is not yet demonstrated that host races can originate peripatrically, there is greater opportunity for host races to develop sympatrically within a large, widely distributed species with abundant variation. In such species, the testing of new alleles and allelic combinations is continuous on potentially new hosts until the right combination allows successful colonization. This adaptive process is less likely to prove successful when a few individuals colonize a new patch of habitat.

As a working definition, we recognize host races as populations that specialize on alternative hosts and differ genetically from one another in host preference and

Box 11.1 Stages of host-race formation

1. A new host relatively free of competitors becomes available for colonization because of changes in the host or insect range, or because of changes in host ecology or physiology.
2. Mutation and/or recombination generate individuals able to utilize the new host from within the parental species (if appropriate genotypes are not already available).
3. A number of males and females that exhibit a genetically based preference for and/or ability to survive on the new host, colonize the new host over successive generations and establish a population. Low efficiency of host utilization may be compensated by low competition.
4. Assortative mating occurs among the earliest colonists of the new host (mating occurs on the preferred host). This reduces gene flow between the original and new host-associated populations and allows adaptive gene combinations to be maintained in each population.
5. During the course of adaptation to the new host, a genetically distinct host race evolves as host-associated differences in fitness and host fidelity increase over time between the host-associated populations.
6. For loci not involved in host-associated adaptations, genetic similarity between the original species and the new host race is maintained by continued low levels of gene flow, particularly during the early stages of host-race formation.
7. Speciation is completed by reinforcement or because continued divergence incidentally reduces gene exchange to zero.

host fidelity, but still exchange genes (see Berlocher and Feder 2002). Current gene exchange can, in principle, be distinguished from incomplete lineage sorting on the basis that recent mutations or parallel geographic variations in neutral markers are shared. Host shifts appear to occur most commonly following the introduction of a new host that is relatively free of competing insect specialists (Box 11.1). In the case of phytophagous insects, hosts shifts are sometimes restricted to related plant species, but in others, shifts are to unrelated plants. The latter usually involve host shifts between chemically similar host plants (Strong *et al.* 1984; Becerra 1997).

New sympatric host races are usually established not by a few, but by many individuals over many generations, and because they remain open to gene exchange at many loci, they are likely to share similar levels of genetic variation (Vouidibio *et al.* 1989; Abe 1991; Stanhope *et al.* 1992, 1993; Bush 1992; Guldemond and Dixon 1994; Mackenzie and Guldemond 1994; Mackenzie 1996; Feder 1998). In the few cases examined closely, the new host races appear to differ genetically at loci responsible for adaptation to the new host, such as those that contribute to phenological changes in *R. pomonella*, and for mate recognition (Feder 1998). Host races remain distinct at these key adaptive loci and at loci closely linked to them, but low levels of gene exchange may continue for other loci (Feder *et al.* 1994; Feder 1998). There is no convincing evidence that recently evolved sympatric or parapatric host races or monophagous species experienced genetic bottlenecks in their past, as expected from some modes of allopatric divergence.

We see sympatric speciation as a progressive process, because adaptation to the new host initially occurs without cessation of gene exchange. Whether complete reproductive isolation evolves rapidly or over a protracted period depends on the intensity of divergent selection, and the degree of host preference and assortative mating (Johnson *et al.* 1996b). As a population adapts in response to strong divergent selection on a new host, it becomes progressively more isolated from the parent species. The criteria to establish exactly when a host race becomes a species are yet to be established and may be impossible to determine because gene exchange becomes harder to detect as it becomes more and more limited. Mallet (1995) suggests that two sympatric populations are species when they maintain distinct genetic clusters. Although this definition is very useful, recently established host races, such as the apple race of *R. pomonella* (see Plate 4), would fall into the species category (if one bases a decision on the divergent group of loci), which may be premature (Feder 1998). In the absence of definitive criteria that specify when a host race establishes an evolutionary lineage independent from its parent species, we will regard a recently established genetically distinct population on a new host as a host race not a species.

Examples of host races established sympatrically or parapatrically on plants cover a diverse range of insects (Diehl and Bush 1984; Craig *et al.* 1993; Mopper 1996; Secord and Kareiva 1996). A few well-studied examples are discussed here, while some of the many other possible cases are outlined in Table 11.2 (and see Berlocher and Feder 2002).

Walsh (1864) was the first to propose a model of sympatric host-race formation and speciation to explain the origin of many observed sympatric and closely related species of phytophagous host-specific insects. The fruit fly *R. pomonella* (Tephritidae), which Walsh described, infests fruits of North American haws (*Crataegus* spp: Rosaceae). In 1860, larvae were found attacking the fruits of introduced apples (*Malus pumila*: Rosaceae) in the Hudson River valley of New York (Bush 1969; Bush *et al.* 1989). The fly quickly spread throughout the range of apples in northeastern North America and established a host race that differs genetically from the parental sympatric hawthorn-infesting populations (Feder *et al.* 1988, 1997a; McPheron *et al.* 1988). These sympatric races, which continue to hybridize at low levels (Feder 1998; Feder *et al.* 1988; Feder *et al.* 1997b), differ in host preference, habitat specific mating preference, eclosion time, and host-associated fitness trade-offs for developmental rates and larval survival (Feder *et al.* 1988; Feder and Filchack 1999; Filchack *et al.* 1999). Strong, divergent natural selection and adaptive trade-offs for phenological attributes and host preferences, and the propensity of these host specialists to mate assortatively maintain their racial distinctions (Bush 1994; Feder 1998; Johnson *et al.* 1996b).

Miyatake and Shimizu (1999) demonstrated experimentally in another tephritid fly, *Bactrocera cucurbitae*, how such phenological shifts can promote race formation. Selection for short- and long-development periods resulted in an indirect

Table 11.2 Host shifts and sympatric speciation in insects.

Order	Family/Genus/Complex	Host plant	Comments	References
Hemiptera	*Uroleucon* (aphids) 44 species in subgenus *Uroleucon*	Asteraceae	Mitochondrial DNA phylogeography suggests rapid radiation of host-specific species following colonization of North America	Moran *et al.* 1999
	Auchenorrhyncha (treehoppers, leafhoppers, and planthoppers)	Many taxa	Many highly host-specific species. Sexual signals are transmitted through the host-plant substrate	Claridge 1985
	Enchenopa	A variety of tree species	Monophagous and extremely philopatric. Life cycle closely linked to host-plant phenology. Transfer experiments show how chance colonization of a new host can generate substantial reproductive isolation	Wood *et al.* 1999
	Nilaparvata lugens	Host races on *Oryza* and *Leersia* grasses	Host-associated performance and oviposition preference each controlled by a few genes (see text). Substrate transmitted sexual signals controlled by more loci and probably diverged by drift. Evidence for two independent sympatric host shifts.	Butlin 1996; Sezer and Butlin 1998a, 1998b
	Erythroneura (>500 species)	Trees, shrubs, and vines	Many sympatric, putative sibling species specialized on different, usually related, hosts	Ross 1958
	Dalbulus	*Tripsacum* and *Zea*	Two independent host shifts onto cultivated corn in the past 5000–8000 years. Substrate-transmitted signals more distinct between sympatric than between allopatric species	Nault 1985

continued

Table 11.2 *continued*

Order	Family/Genus/Complex	Host plant	Comments	References
	Miridae (most species-rich family of Heteroptera), e.g., *Lopidea*		Phylogeographic analysis suggests at least 25% of species diverged sympatrically	Asquith 1993
	Orthotylinae, *Sarona* (40 species in Hawaii)	17 genera in 14 families	Inter-island colonists retain ancestral hosts, but up to 70% of within-island speciation involves a host shift	Asquith 1995
Lepidoptera	"Butterflies"		>50% of species are monophagous Host shifts common during speciation	Jansen 1988 Janz and Nylin 1998
	Zeiraphera diniana (larch budmoth)	Host races on *Larix* and *Pinus*	High vagility suggests allopatric divergence is unlikely. Mating is on host plant, host odors as well as pheromones involved in mate finding	Emelianov *et al.* 1995
	Spodoptera frugiperda	Races on corn and rice/forage grasses	Broad sympatry possible because of strong premating isolation	Prowell 1998
	Lycaenidae (40% of all butterfly species)		Specific associations with both host plants and ant mutualists. Host switching common in evolutionary history	Pierce 1987; Pratt *et al.* 1994
	Yponomeuta, Laspeyresia, Hedylepta, Ostrinia, Greya		Predominantly monophagous species with ecological and behavioral attributes likely to promote sympatric speciation	Bess 1974; Phillips and Barnes 1975; Harrison and Vawter 1977; Brown *et al.* 1997; Menken and Roessingh 1998

continued

Table 11.2 *continued*

Order	Family/Genus/Complex	Host plant	Comments	References
Coleoptera	>150000 phytophagous species, mainly Chrysomelidae and Curculionidae			Farrell 1998
	Epilachna viginitioctomaculata complex		Host shifts associated with sympatric host-race and species formation	Katakura 1997
	Phyllotreta	Brassicaceae	Host-associated divergence and possible sympatric speciation	Verdyck 1998
	Blepharia	*Bursera*	Molecular phylogeny shows pattern of host shifts associated with host-plant chemistry rather than phylogeny	Becerra 1997
	Haltica, Diabrotica, Lochmaea, Chrysochus, Oreina		High degree of host specificity and frequent range overlap suggest sympatric divergence	Phillips and Barnes 1975; Krysan *et al.* 1989; Kreslavskiy and Mikheyev 1994; Dobler *et al.* 1996; Dobler and Farrell 1999
Hymenoptera	Insect parasitoids: possibly as much as 10% of all metazoan species		Highly host specific. Haplodiploidy, sibling mating often on the host, and associations with microorganisms essential for disarming host immune systems may all promote sympatric divergence and speciation	Askew 1968, 1971; Godfray 1994; Godfray and Waage 1988

continued

Table 11.2 *continued*

Order	Family/Genus/Complex	Host plant	Comments	References
Hymenoptera	*Megarhyssa, Andricus*, Aphidiinae, Euphorinae		Case studies suggesting sympatric divergence	Gibbons 1979; Shaw *et al.* 1988; Tremblay and Pennacchio 1988; Ramadevan and Deakin 1990; Abe 1991
	Ants: *Solenopsis invicta*		Gene flow restricted between monogynous and polygynous social forms	Shoemaker and Ross 1996
	Leptothorax		Molecular phylogenies show that social parasites are monophyletic and not related to hosts. May speciate by host shifting	Baur *et al.* 1995
Diptera	*Rhagoletis pomonella* group		See text	Bush and Smith 1997, 1998; Berlocher 2000
	Blepharoneura	Cucurbitaceae	Not just highly host-plant specific, but also specialized on plant parts. Speciation apparently associated with shift of host species rather than resource	Condon and Steck 1997
	Other phytophagous and parasitic families		Species rich with many host specialists, often mating on the host. Sympatric speciation likely to be common	

(pleiotropic) phenological shift in mating time. This difference in the time of mating caused significant reproductive isolation in mate-choice tests between the two populations. When coupled with genetic variation in other life-history traits, such as host preference, host-induced pleiotropic shifts in mating time can increase reproductive isolation substantially in insects that mate on their hosts.

Biological studies on the gall-forming tephritid fly *Eurosta solidaginis*, which infests the goldenrod *Solidago altissima* (Asteraceae), revealed the presence of a sympatric host race using *S. gigantea* (Waring *et al.* 1990; Craig *et al.* 1997; Itami *et al.* 1998). Although the two races are interfertile, they differ slightly in emergence time and prefer to mate and oviposit on their respective host plants. Each host race also survives significantly better on its own host, and F1 and backcross progeny do less well than the parental races on their host plants. A phylogeographic analysis (Brown *et al.* 1996) established that the host race on *S. gigantea* is derived from eastern populations of the widespread *S. altissima* race.

In Java, central Sumatra, and peninsular Malaysia, *Epilachna vigintioctomaculata* (Coccinellidae), whose normal hosts are species of *Solanum*, has established a population on the legume *Centrosema pubescens*, introduced from South America about 200 years ago (Nishida *et al.* 1997). The two host-associated populations are interfertile, but differ in size and host preference. In many respects, these host races resemble the stage of divergence present between the apple and hawthorn host races of *R. pomonella* (Feder *et al.* 1988). Similar genetic studies on the *E. vigintioctomaculata* may help clarify the status of the host-associated populations.

Host-race formation in the soapberry bug *Jadera haematoloma* (Rhopalidae: Hemiptera) has evolved rapidly as the result of independent sympatric or parapatric host shifts from native to introduced Sapindaceae (Carroll and Boyd 1992; Carroll *et al.* 1997). This bug occurs throughout the range of its native hosts, the soapberry tree (*Sapindus saponaria* var. *drummondii*) in the south-central region of the United States, the serjania vine (*Serjania brachycarpa*) in south Texas, and the perennial balloon vine (*Cardiospermum corindum*) in southern Florida. Within the past 20–100 years morphologically distinct taxa have independently colonized and adapted to the "flat-podded" golden-rain tree (*Koelreuteria elegans*) from *C. corindum* in Florida, and to the "round-podded" golden-rain tree (*K. paniculata*) and the heartseeded vine (*C. halicacabum*) from *S. saponaria* var. *drummondii* in south-central United States. Nymphs and adults feed by inserting their long slender tubular beak through the fruit wall to seeds that are located within the host fruits at varying distances from the fruit perimeter. Although the races are interfertile, they differ greatly in beak length, which is highly correlated to fruit size and shape.

Many host races exist among the 4000 species of aphids, 99% of which use one or a few hosts. As in other insect specialists, a host shift in aphids can initiate assortative mating and a substantial escape from gene flow. This allows individuals that speciate on different hosts to diverge in response to selection (Akimoto 1990; Guldemond and Mackenzie 1994; Mackenzie 1996; Via 1999).

Aphids manifest several unique sets of biological traits that predispose them to rapid host-race formation on introduced plants. Many aphids have various forms of cyclic reproduction. A round of sexual reproduction may follow several parthenogenetic (thelytokous) generations on the same host plant. If one or more asexually reproducing females with a genetically modified host preference shift to a new host plant, they can generate a large number of individuals with a similar preference for the same host. Where mating occurs on this host, males and females with similar genetic host preferences continue to feed on the new host during the reproductive stage of the cycle. A large number of their progeny are thus genetically predisposed to exploit the same host the following year for several rounds of asexual reproduction. More complicated life cycles, such as host alternation, may facilitate, modify, or impede host-race formation and speciation (Guldemond and Mackenzie 1994). For example, specialization on different hosts in the asexual summer populations may be opposed by mating on a common winter host.

11.6   Conditions Needed for Sympatric Shifts

These examples, and many others (Table 11.2), provide empirical support for the argument that sympatric speciation by host shifts is not just possible, but is a major route to speciation in specialist phytophagous insects. However, it is difficult to exclude alternative explanations completely in specific cases and, therefore, it is important to demonstrate that host shifts are also theoretically plausible. Numerous difficulties with the evolution of novel host associations in the face of gene flow have been raised, but there are now either empirical or theoretical grounds to reject all of them. Here, we briefly review the key issues (see also Via 2001, 2002).

Problem of crossing the gap

Host shifts are unlikely because an insect population that is well adapted to its current host is not expected to contain suitable genetic variation to utilize on an alternative host. The necessary variation needs to be maintained in the host population, because a rare mutation is unlikely to coincide with a rare colonization event. Futuyma *et al.* (1995) describe this as a genetic constraint on host shifts. Much evidence shows that host shifts are more common between hosts with similar chemical defenses (e.g., in butterflies; Janz and Nylin 1998), which indicates that some constraints do operate. However, if alternative hosts are too similar, oligophagy may be a more likely outcome than speciation.

This argument can be countered in two ways, either with a "two-phase" model, or with a broader view of host-associated adaptation. The two-phase model is derived from Turner's (1981) arguments about the evolution of mimicry. In the case of a host shift, mutations of small effect are likely to reduce fitness on the current host without providing a significant ability to use the new host. Mutations of a large enough effect to "jump the gap" and confer high efficiency of utilization of the new host may be rare, but mutations with moderate ability to use the new host may be more common. If these genotypes also gain from greatly reduced competition (or predation or parasitism) on the new host, they may have high fitness despite low efficiency. Once a population is established on the new host, selection favors modifiers that increase the efficiency of novel resource exploitation. This is the second phase. Overall, the process is expected to leave a genetic signature when host utilization is controlled by one or a few major genes, responsible for the initial jump, with epistatic modifiers responsible for the fine-tuning of performance. Evidence for this type of genetic architecture occurs in the host races of the brown planthopper *Nilaparvata lugens* (see Plate 4) on rice and a related grass *Leersia hexandra* (Sezer and Butlin 1998b). Host-associated performance is controlled by a small number of loci with evidence for epistatic interactions.

The picture envisaged in the constraint argument is of two distinct resource types with a gap that has to be crossed to move from one to the other. This may be realistic for host plants that use different defensive compounds, but adaptation to a host has many more dimensions than simply overcoming defense. This is clear in *Rhagoletis*, for example, in which emergence timing, as opposed to distinct host chemistry, seems to have the greatest influence on host-associated fitness (Feder and Filchack 1999). Emergence timing is a quantitative trait that is likely to exhibit genetic variation within the parental population on hawthorn. Genotypes at the early extreme of this distribution are most likely to colonize apples and have the highest fitness on apples. Rather than crossing a gap, the host shift simply involves response to directional selection.

Problem of maintaining multiple-niche polymorphism

Sympatric host shifts face the same difficulty as other modes of sympatric speciation. A resource-use polymorphism must be maintained for long enough, and with sufficiently strong selection against intermediate genotypes, for reproductive isolation to evolve.

Broadly, there are two reasons why this argument no longer carries much weight:

- Multiple-niche polymorphisms can be stabilized by density- and frequency-dependent selection (Udovic 1980; Wilson and Turelli 1986; see Chapter 3). When the population on one host is at low density, its fitness tends to be higher because negative interactions between individuals are reduced. There has been much debate about whether phytophagous insects compete for resources sufficiently for this to be an important effect, but the evidence now suggests that it can be in many instances (Denno *et al.* 1995). Competition for food may also be supplemented or replaced by the effects of escape from predators or parasites.
- Host choice can also stabilize multiple-niche polymorphism, because it increases the probability that individuals will live in the habitat type to which they are well adapted (provided choice and performance are correlated, see below; de Meeûs *et al.* 1993; Johnson *et al.* 1996b). Host choice, usually by ovipositing females, is a near universal feature of specialist insect species.

Problem of negative trade-offs

Sympatric speciation requires that an increase in fitness on one host be accompanied by a decrease on the alternative host. If this is not true, selection favors genotypes that perform well on both hosts: a specialist evolves into a generalist rather than splitting into two host-specific species. The expectation is that antagonistic pleiotropy must underlie this trade-off between performances on different hosts, otherwise recombination could create successful generalist genotypes.

Empirically, a negative genetic correlation is expected between performances on different hosts, but this has rarely been observed (Via 1990, 2002). Estimation of genetic correlations is difficult for most organisms, but in aphids it is possible to exploit the alternation of sexual and asexual generations to design more powerful experiments. It is significant, therefore, that Mackenzie (1996) found a strong negative correlation in performance on nasturtium and broad bean among 77 clones of *Aphis fabae*. Since this correlation was maintained after allowing recombination, it does appear to arise from antagonistic pleiotropy.

Generally, studies of trade-offs have concentrated on host-associated performance measured by survival, growth rate, or adult size. However, this is only one aspect of fitness and trade-offs might occur in other fitness components. Phenological matching between *R. pomonella* and its hosts (Feder 1998; Feder and Filchack 1999) is a good example in which there is necessarily a negative correlation between fitness on the two hosts across different emergence and diapause times. Interactions with predators or parasites that find their prey by searching host plants may also contribute significantly to host-associated fitness differences and contribute to trade-offs. For example, *R. pomonella* on apples is attacked by only one parasitoid species, while larvae on the ancestral host are attacked by two (Feder *et al.* 1995). Also, the agromyzid fly *Liriomyza helianthi* suffers 17% less parasitism when moved experimentally to a novel host (Gratton and Welter 1998).

Recent theory actually questions whether negative pleiotropy is necessary at all. In the "independent adaptation" model by Kawecki (1998), specialization evolves because of a coevolutionary "arms race" between insect and host. Specialists can "keep up" with the evolution of their hosts' defenses, while generalists are unable to track the independent changes that occur in different hosts. This effect selects for host choice and therefore isolation without the need for negative pleiotropy. A possible example of the independent genetic basis of adaptation is a Y-linked locus in the flea beetle *Phyllotreta nemorum* that influences performance on one host only (Nielsen 1999).

Problem of simultaneous evolution of preference and performance

A mutation that increases performance on a novel host is not at a selective advantage unless it is in an individual on the novel host. Conversely, a preference for a novel host is not advantageous unless it is associated with an ability to exploit the host. Since the two types of mutation are very unlikely to occur together, a successful host shift is highly improbable. Even if variation in both characters is present in a population, a genetic correlation is needed for a host shift to be achieved.

This argument predicts that host shifts would be most likely if loci existed with pleiotropic effects on both performance and host preference. Roessingh *et al.* (1999) suggest that the same chemosensory cues might be used in oviposition preference and as feeding stimulants, which would make pleiotropic mutations a possibility. Via and Hawthorne (2002) found a close linkage between preference and performance loci in crosses between host races of *Acyrthosiphon pisum* (Homoptera: Aphididae). However, Sezer and Butlin (1998a, 1998b) found no association between genes for host-associated performance and oviposition preference in crosses between host races of the brown planthopper *N. lugens*. Within populations, the evidence for genetic correlations between preference and performance is generally equivocal (Via 1990).

Empirical work on the relationship between preference and performance suggests a way out of this problem. Host ranges of insects may be limited by specialization in either preference or performance, so that only a single change is needed for a host switch. For example, *Liriomyza* leaf miners can survive on hosts that are not currently used (Gratton and Welter 1998), so only a change in preference is needed. In bruchid weevils of the genus *Stator* significant numbers of eggs are laid on unsuitable hosts. Only a change in performance is required for a host shift (Johnson and Siemens 1991).

In the treehopper *Enchenopa binotata*, Wood *et al.* (1999) have used experimental shifts between hosts to show that the tendency to stay, mate, and oviposit on the individual plant on which it is reared is sufficient to generate significant isolation after an initial chance transfer. Hosts are not equivalent, but there is significant genotype–host interaction for fitness components. Therefore, host fidelity provides an opportunity for adaptive divergence and speciation.

Problem of reinforcement

Reinforcement is the evolution of increased reproductive isolation in response to selection against hybrid or intermediate genotypes. Theory suggests that it is opposed by gene flow and recombination (Butlin 1989), although recent models and examples suggest that these barriers are not so great as had been supposed (Butlin and Tregenza 1997; Noor 1999; Servedio 2000). In the context of sympatric speciation, this very important step is required for progress toward speciation after the establishment of a multiple-niche polymorphism. In a recent model by Dieckmann and Doebeli (1999), strong disruptive selection at an "evolutionary branching point", at which a pattern of disruptive selection is generated and stabilized by density-dependent selection, can result in linkage disequilibrium between performance loci and mating loci despite interbreeding and recombination. This disequilibrium leads to assortative mating and hence speciation. A similar outcome was found in the rather different model of Kondrashov and Kondrashov (1999).

The situation envisaged in these models might be considered equivalent to the stage in host-associated speciation at which differently adapted host races still exchange genes. However, two reasons indicate that it is different: competition may not be a major source of selection in phytophagous insect specialists (as noted on page 244, first bullet item), and host preference may act to generate linkage disequilibrium that facilitates the evolution of nonhost-associated assortative mating, with no need for strong selection against intermediates (Johnson *et al.* 1996b).

A role for reinforcement in host-associated speciation has been suggested for *Cryptomyzus* aphids (Guldemond and Mackenzie 1994). In this genus, species with different summer hosts share the same winter host, on which mating takes place. Divergence in the diurnal pattern of pheromone signaling could be a result of reinforcement. However, in other cases differential adaptation of host races apparently persists in the face of gene flow without any divergence in mating behavior (as in *Rhagoletis*; Feder 1998). Since the rate of gene exchange is already low because of host fidelity (about 6%; Feder *et al.* 1994), the advantage resulting from behavioral changes that would cause more complete assortative mating is also small. Speciation is only completed when host preference is coupled with some form of assortative mating independent of host choice (Johnson *et al.* 1996b). The evolution of such nonhabitat assortative mating is a form of reinforcement that is greatly facilitated by the linkage disequilibrium generated by habitat-dependent assortment. It is driven by selection against hybridization, not competition, because the former tends to break down advantageous associations between host selection and fitness alleles. In *N. lugens* host races, Butlin (1996) suggested that host fidelity reduced gene flow to the point at which mating signals and preferences of host races diverged by drift. Eventually, this completes the speciation process, just as it would in allopatric populations.

11.7 Concluding Comments

In this chapter, we have tried to show the following for specialist phytophagous insects:

▪ There are many species-rich taxa of host specialists in which sister species that use different hosts are frequently sympatric.

▪ There are well-supported examples of sympatric host races maintained by a combination of adaptation to alternative resources and restricted gene exchange because of mating on the host.

▪ All of the theoretical objections that have been raised against the process of sympatric speciation via host specialization have proved to be surmountable based on either new theory or empirical evidence.

One must conclude that this route to speciation is at least possible.

However, there is evidence for allopatric divergence that leads to speciation in generalist phytophagous species (e.g., Butlin 1998) and, possibly, in specialists for which speciation is driven by something other than a host or habitat shift. No examples of allopatric host shifts that lead to speciation are firmly established. On balance, the evidence suggests that sympatric divergence is the major source of diversity in specialist phytophagous and parasitoid insects, which are the most species-rich groups of living organisms. However, there is a real difficulty in distinguishing among alternative modes of speciation after the process is complete (Table 11.1), and thus considerable uncertainty remains about their relative prevalence. This is the main challenge for the future. To provide answers will require the development of more discriminating predictions from the alternative models and new empirical approaches.

12

Adaptive Speciation in Agricultural Pests

Martijn Egas, Maurice W. Sabelis, Filipa Vala, and Iza Lesna

12.1 Introduction

Agricultural crops provide an ideal environment for adaptive speciation of pest species. They represent recently colonized habitats, harbor small incipient pest populations, and form environments in which not all the resources are already occupied [together referred to as ecological opportunity, see Schluter (2000)]. These characteristics exactly meet the conditions predicted by Dieckmann and Doebeli (1999) to favor species that split into specialists by the process of evolutionary branching (see Chapters 4 and 5). Additionally, in agricultural systems the ecological environment is constant relative to the natural world, which also favors adaptive speciation. Moreover, the economic importance of agricultural crops warrants extensive research into pest species, which increases the probability that adaptive speciation will be documented. It is therefore not surprising that the best examples of ecological speciation in sympatry involve pest species [e.g., the apple maggot fly *Rhagoletis pomonella* (Feder 1998), the pea aphid *Acyrthosiphon pisum* (Via 1999), and the two-spotted spider mite *Tetranychus urticae* (Gotoh *et al.* 1993)].

Agriculture provides a diversity of crops, and plant breeding creates a unique level of heterogeneity in both resistance and palatability within a crop species. In this world, pest species may evolve to become specialists (feeding on one or a few plant species, or even genotypes) or generalists, depending on fitness trade-offs (Levins 1962; Egas *et al.*, in press; Egas *et al.*, unpublished). Such trade-offs may be found in food conversion efficiency, detoxification, phenology, and defenses against or escape from natural enemies. Given the fitness trade-off and the availability of resources, an ideal, omniscient pest organism knows where to forage to maximize its fitness. Real pest organisms, however, are not omniscient and need to sample the environment (Stephens and Krebs 1986). In the process of sampling, individuals acquire experience, which may affect (foraging) behavior – this is what defines learning in this context.

It is long recognized that learning can influence evolution (Box 12.1). When learning adaptively (i.e., leading to improved reproductive performance), foragers distribute themselves over the resources in a fashion closer to the optimal distribution (Pulliam 1981). Adaptive learning favors host-plant specialization and host-race formation (Egas *et al.*, in press; Egas *et al.*, unpublished). Moreover, genetically based differences in host preference and performance may be reinforced by learning, which creates a correlation between host use and fitness on different

Box 12.1 The guiding and hiding effects of learning on evolution

Baldwin (1896) was one of the first to propose a role for learning in evolution by natural selection, together with Morgan (1896) and Osborn (1896). Learning is said to "guide evolution" (Hinton and Nowlan 1987; Maynard Smith 1987) when it promotes the speed of adaptation. This is commonly called the Baldwin effect (e.g., Robinson and Dukas 1999). However, learning may also have the effect of masking fitness differences among individuals: the "hiding effect" of learning (Papaj 1993; Mayley 1997; see also Jaenike and Papaj 1992).

The general conditions under which the guiding or hiding effects of learning on evolution can be expected are still unclear. Moreover, there are many different definitions of learning, ranging from neurological descriptions that involve changes in the input-to-output mapping of neural networks, through behavioral changes caused by accumulated experience (our operating definition in this chapter), to definitions in the field of artificial intelligence, such as "any environmentally driven phenotypic change that increases an individuals's survival chances (fitness)" (Mayley 1997; see also Hinton and Nowlan 1987). The third definition already includes the fitness effect of learning, and would be interpreted by most biologists as a definition of phenotypic plasticity. This shows the overlap in definitions when behavior is considered as part of the phenotype (as, e.g., in the study of host-plant preference), and that learning can thus be said to affect evolution by either changing the fitness of individuals with a certain phenotype (through alterations in their behavior), or by changing the phenotype of individuals (concomitantly altering their fitness). In the following paragraphs we give examples of these two possibilities. For simplicity, we assume that fitness is neither frequency nor density dependent (or, at least, hardly so).

We illustrate the effects of learning through behavioral change with an example based on the impact of foraging behavior on the evolution of specialization (Egas et al., in press; Egas et al., unpublished). Consider a consumer that exploits two resources, constrained by a fitness trade-off. This means that when an individual's fitness on resource 1 is high, its fitness on resource 2 is low. We use an individual's degree of specialization on resource 1 as the phenotypic variable x. We consider selection regimes that are either disruptive (panel a) or stabilizing (panel b), and that learning affects the foraging behavior of consumers.

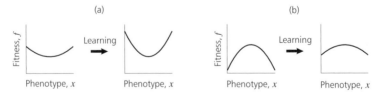

In the absence of learning (left graphs in a and b), consumers are assumed to harvest resources in proportion to their relative densities; whereas with learning (right graphs in a and b), consumers are assumed to learn to spend more time on the resource that yields higher fitness (such learning is called adaptive). When selection is

continued

Box 12.1 *continued*

disruptive (panel a), learning to forage on the resource that yields higher fitness can increase the fitness differences between individuals. By contrast, when selection is stabilizing (panel b), the same effect of learning reduces fitness differences among phenotypes, and thereby reduces the speed of adaptation. Both effects arise because learning does not greatly alter the distribution of generalists over the two resources, since generalists are equally well adapted to either resource. For specialists, however, learning leads to a higher utilization of the resource on which they specialize, and hence to higher fitness. From this example, it is not difficult to see that learning may also change the selection regime itself (for instance, from stabilizing to disruptive; Egas *et al.*, unpublished).

For an analogous example of the effects of learning through phenotypic plasticity, we keep the framework of consumers that exploit two resources. However, instead of learning to forage, individuals are now assumed to be plastic in their performance on the resources, such that individuals of any phenotype x can adapt their phenotype up to an amount Δx to increase their fitness. This plasticity effect is similar to induced performance in herbivorous arthropods, as discussed in Section 12.3. It is easy to see that, when selection is disruptive (panel a), this kind of learning again accelerates evolution, whereas in a regime of stabilizing selection (panel b) evolution is slowed down: the fitness landscape that applies to a situation with such learning from one without learning is derived by a transformation of the horizontal axis. In panel a, for instance, phenotypes x on the far left of the central generalist phenotype change to phenotypes $x - \Delta x$, whereas on the far right phenotypes change to $x + \Delta x$; phenotypes closer to the generalist can be expected to exhibit less plasticity.

Based on these examples, we anticipate that adaptive learning is more likely to guide evolution than to hide selection in regimes of disruptive selection, and *vice versa* in regimes of stabilizing selection. From the considerations above, we can also appreciate that under directional selection the effects of learning may depend on the curvature of the fitness landscape in the region of the phenotypes considered. However, more research is certainly needed before we can assess the generality of such statements.

hosts, and again facilitates the evolution of host specialization and host-race formation (Jaenike and Papaj 1992).

By deciding where to forage and how long to stay, individuals become nonrandomly distributed over the resources; thus, their populations exhibit spatial structure. This, in turn, affects the mating structure of the pest population, and thereby reinforces genetic divergence when mate choice is a by-product of (micro)habitat choice. More generally, mate choice may depend on the environment (context-dependent mate choice), as has been shown for several organisms (Qvarnström 2001). Thus, flexible behavior may contribute to prezygotic reproductive isolation.

An important factor in postzygotic reproductive isolation, at least in arthropod species, is the presence of the microbe *Wolbachia*. These bacteria, related to

Rickettsia, live inside the cells of the infected host, and are vertically transmitted through the eggs. *Wolbachia* is renowned for manipulating the reproductive mechanisms of its host – causing, among other effects, parthenogenesis in parasitoid wasps and reproductive incompatibilities in a large number of arthropod species [see the review in Stouthamer *et al.* (1999)]. For these reasons, *Wolbachia* has been implicated in speciation (Hurst and Schilthuizen 1998; Werren 1998).

In this chapter, we address, for population differentiation and speciation in agricultural pests, the consequences of the following effects:

- Adaptive learning;
- Adaptive mate choice;
- *Wolbachia*-induced reproductive incompatibility.

We start with a review of the evidence for rapid evolutionary change of arthropod herbivores in agricultural settings.

12.2 Crops as Ecological Niches

Agricultural crops occupy vast areas, and are waiting to be colonized by herbivores. The herbivores may either stem from the ancestral plants in the area of origin of the crop or from related wild plants endemic in the area where the crop is grown. It is hard to predict whether specialists or generalists will dominate among the first invaders of a new crop. On the one hand, most phytophagous arthropods are specialists (Futuyma and Gould 1979; Chapman 1982; Strong *et al.* 1984), which may increase the likelihood of specialists being among the first invaders. On the other hand, generalists form part of the many host-associated communities of phytophagous arthropods. Empirical observations, however, reveal a pattern whereby generalists invade first, after which the community becomes more specialized the longer the plant coexists with the community of phytophagous arthropods. For example, examination of the literature of 23 annual crop plant species and the associated 498 species of phytophagous arthropods in Japan, which covers several centuries, showed that 83% of the invading phytophages had a more generalized host range than the median of the native community and that the proportion of specialists increased by 6.6% per century after introduction (Andow and Imura 1994).

Agricultural crops also harbor a unique genetic diversity in cultivars through plant-breeding programs and patterns of seed distribution, but also through the traditional development of local crop varieties. Although we are only just beginning to understand how coevolution drives diversity in phenotypes of plant resistance and insect virulence in natural insect–plant communities (Berenbaum and Zangerl 1998; Kareiva 1999), there is much more evidence for arthropod pests in agriculture (Diehl and Bush 1984; Rausher 2001). For example, biotypes in the Hessian fly, a cecidomyiid that attacks wheat in Europe and North America, have been shown to possess virulence loci that match resistance loci in the plant in a one-to-one relation. Another example is the brown planthopper, a notorious pest of rice in Asia. Rice varieties greatly vary in their susceptibility to planthoppers and several "major" resistance genes have been identified. Resistance-breaking mechanisms in the planthoppers inherit polygenically and it is possible to select new biotypes from individuals of another biotype. In Sri Lanka, planthopper populations have adapted locally to domestic rice varieties or wild rice (that they thrive better on the variety from which they were collected demonstrates this). In the laboratory, mate choice appears to be random, but in the field behavioral preferences for particular rice varieties may act as a partial barrier to interbreeding.

The extent to which the biotypes are isolated reproductively is often not known, so it may be unclear as to whether they represent host-adapted races or not (Diehl and Bush 1984; see also Chapter 11). An illustrative example derives from extensive studies on two-spotted spider mites that feed on various crops in European greenhouses (Helle and Sabelis 1985). Greenhouses represent nonseasonal ecological islands in a seasonal environment. The selection regime within the greenhouse is dramatically different, which has led to novel adaptations, such as the absence of diapause (Helle 1962), resistance to pesticides (Helle and Overmeer 1973), and suppressed long-distance dispersal (Margolies 1995). The expectation is that each greenhouse population originated from a nearby local population, and that the greenhouse restricts gene flow with the open field. This is suggested by population genetic analyses, using allozymes and molecular markers, which show that greenhouse populations are differentiated genetically from their respective source populations nearby (Tsagkarakou *et al.* 1998). Reproductive incompatibilities between greenhouse strains have been reported (Helle and Pieterse 1965), and occur – albeit at a lower frequency – among local populations (<10 km apart) under natural conditions [dune areas of Belgium and The Netherlands (de Boer 1980, 1985)]. Increasing evidence now suggests that such incompatibilities can arise from infection with *Wolbachia* endosymbionts (Breeuwer 1997; Vala *et al.* 2000; discussed in Section 12.5). Unfortunately, host-plant selection and mate choice are little studied, but one example shows that greenhouses harbor lines of two-spotted spider mites adapted to tomato plants (lower mortality after colonization, higher growth rates) that differ in host-plant choice and are isolated reproductively from strains collected from greenhouse cucumber through mate choice and postzygotic isolation (Gotoh *et al.* 1993). Spider mite systems fulfill all the necessary conditions for adaptive speciation (Crozier 1985; Fry 1992; Navajas 1998), although experimental evidence of differentiation in sympatry is still lacking.

Together, these examples show that local adaptations rapidly evolve in agricultural environments and that this process is possibly associated with the evolution of prezygotic isolation (e.g., host preference and mate choice) and postzygotic isolation (e.g., *Wolbachia*-mediated incompatibility) among local populations.

12.3 Adaptive Learning of Host Preference

In herbivorous arthropods, the guiding effect of learning on evolution (see Box 12.1) may facilitate the evolution of host-plant specialization and host-race formation (Jaenike and Papaj 1992; Egas *et al.*, in press; Egas *et al.*, unpublished). Specialization and host-race formation are observed repeatedly in agricultural pests, including the apple maggot fly *R. pomonella* (Feder 1998; Filchack *et al.* 2000), the pea aphid *A. pisum* (Via 1999; Caillaud and Via 2000; Via *et al.* 2000), and the two-spotted spider mite *T. urticae* (Fry 1992; Gotoh *et al.* 1993). Is there a possible role for learning in the speciation process of these organisms? In general, the effects of experience are almost always found when looked for in herbivorous arthropods (Papaj and Prokopy 1989; Bernays 1995). Below, we provide examples of flexible (experience induced) and fixed (not induced by experience) preference and performance for these three species.

Absence of induced preference and performance, pea aphids

Several recent papers give evidence for sympatric speciation in pea aphids on red clover and alfalfa (Via 1999; Caillaud and Via 2000; Via *et al.* 2000). Sympatric populations are highly specialized ecologically and gene flow between them is low. Recently, Caillaud and Via (2000) reported on the aphid behavior that underlies acceptance of the two host plants. Their results show that the aphids rapidly assess both host plants and reject the alternative host, based on chemosensory cues (i.e., before any feeding had been initiated). There appears to be no role for experience in this acceptance behavior, although aphids do perform a series of additional short penetrations at different locations on the plant before finally leaving the plant. Via (1991) studied the effects of experience with one of the two hosts on induced performance on that host, using replicates of two specialized clones of pea aphids. Experience on the alternative crop had no significant effect on the relative performance of these two clones, which suggests that only genetically based differences occur in host-plant performance among pea aphid clones.

Hence, there is no evidence of experience-induced preference and performance in the pea aphid. Such effects of experience might have been anticipated, as they have been found in other species of aphids (e.g., Douglas 1997; Ramírez and Niemeyer 2000). Instead, the divergence in host use may, in this case, be attributed to differences in chemosensory perception: the host-adapted races have diverged in neuroreceptor coding for stimulants and/or deterrents that determine the behavioral acceptance of a plant as host (see Menken and Roessingh 1998).

Induced preference and fixed performance, apple maggot flies

Effects of experience on host selection and oviposition behavior are relatively well studied in the apple maggot fly (e.g., Prokopy *et al.* 1982, 1988, 1994; Papaj and Prokopy 1986, 1988, 1989). Finding host fruit is influenced by the fruit color experienced previously, whereas oviposition behavior is affected by fruit size and surface chemistry. Increasing experience with apple or hawthorn fruit increases the retention time within the host trees, and reduces the propensity to oviposit in the alternative host fruit. Effects of experience on performance have not been studied, but are not expected to play a role, since the maggots do not leave the fruit until pupation.

The apple maggot fly exhibits induced preference and fixed performance, which suggests that preference and performance are not likely to become correlated through learning. However, these observations have not been coupled at the individual level. Since natural selection acts most strongly at this level, we need to know whether individuals can learn to prefer the host on which their offspring perform best.

Induced preference and performance, spider mites

Although learning is well-documented in arthropods, the fitness effects of learning are not (Papaj and Prokopy 1989; Dukas and Bernays 2000). Recently, Egas and Sabelis (2001) demonstrated fitness effects of host-preference learning in the two-spotted spider mite (Plate 5). Individual mites from two races with different hosts (Gotoh *et al.* 1993) were subjected three times to a choice between their respective host plants, tomato and cucumber, and then subjected to a performance test on each. During the three consecutive choice tests, mites of both strains learned to prefer cucumber over tomato. The performance test showed that cucumber enabled higher oviposition, higher survival, and faster development than tomato. The same two races were also shown to learn food quality in choice tests, which again resulted in fitness benefits (Egas *et al.* 2003). Food consisted of cucumber plants with three different degrees of feeding damage – undamaged (no mites), mildly damaged, and heavily damaged (infested by one of the two mite races). Again, during three consecutive choice tests, mites of both strains learned to prefer less damaged cucumber, and had a higher oviposition rate on food with a lower degree of damage.

In herbivorous arthropods, preference may be induced when feeding on a host plant [see the review in Szentesi and Jermy (1990)]. Clearly, the induction of a greater preference for a host plant is beneficial when the performance on that host plant is increased simultaneously. The association of an improved physiological state with the smell and/or taste of the host plant may explain the induction of preference as a result of increased performance. Although preference induction is widely observed, induction of performance had not been studied in this context. Recently, however, Agrawal *et al.* (2002) and Magowski *et al.* (2003) provided evidence of induced preference, as well as induced performance, on tomato plants in various strains of *T. urticae*, and showed considerable variation in these two traits among strains. In these cases, the spider mites switched on detoxification mechanisms, which allowed them to cope better with the toxins incurred from feeding on tomato plants. Consequently, the mites were able to feed more on the tomato plants, which resulted in an induced preference. Inhibiting part of the detoxification induction mechanisms prevented both induced performance and induced preference (Agrawal *et al.* 2002).

Adaptive learning and speciation

In the pea aphid, no evidence indicates that learning affects host use. Indeed, with a strong preference–performance correlation within clones and a stark difference in host-plant specialization between clones of the two different host-adapted races, there is no need for behavioral plasticity in host use. Of course, we do not know whether learning may have played a role in the early stages of divergence in this species. The other two examples do show the effects of experience on preference and/or performance, and in our view exemplify the situation as it may appear in the early stages of the divergent process of speciation. How, then, may learning affect sympatric evolution toward new species?

In the apple maggot fly, induced preference for the host fruit may contribute to disruptive selection (see also Papaj and Prokopy 1989). The difference in host phenology of the apple and hawthorn (Feder 1998; Filchack *et al.* 2000; see also Chapter 11) means that flies that emerge early from diapause learn to select the early host (apple), and the later flies learn to select the later host (hawthorn). This sets up a correlation between timing of the diapause and utilization of the most abundant host at that time, which results in a disruptive selection on performance characters for one host or the other, correlated with the timing of diapause. The role of learning in this scenario is to change the distribution of the flies from being proportional to relative host density to being skewed toward the more abundant host, and thereby enhances disruptive selection.

The life history of the two-spotted spider mite is different from that of the apple maggot fly, because adult females need to feed on the host plant to oviposit (as do pea aphids; see also Via *et al.* 2000). Hence, they can and do test the food internally (i.e., physiologically) before leaving their offspring on it. In contrast, an apple maggot female necessarily uses indirect cues to assess the value of various fruits for her offspring. Again, learning contributes to a positive correlation between the distribution of individuals over host plants and their fitness on these plants (Pulliam 1981; Kacelnik and Krebs 1985). Moreover, the different life history of spider mites increases this positive correlation further – an effect shown to promote disruptive selection for host-plant specialization (Egas *et al.*, in press; Egas *et al.*, unpublished).

12.4 Adaptive Mate Choice

Once local adaptations evolve, gene flow between biotypes may become restricted through the evolution of assortative mating, which ultimately leads to host-race formation and speciation. In its simplest form, assortative mating among individuals with new adaptations for a host plant can be a by-product of their host-plant choice (*de facto* assortative mating). Mate choice may also be directed to traits that correspond or are even associated with adaptations to the novel host. The latter type of mate choice may, in theory, arise from learning novel mating signals produced by individuals in the process of acquiring adaptations to the new host plant. Indeed, learning has been implicated in mate choice of songbirds (Irwin and Price 1999; Ten Cate and Vos 1999; Ten Cate 2000), but evidence for pest species is lacking. In pest species, however, there is increasing evidence for the female's ability to discriminate between males that confer "good genes" to their offspring. This involves complex forms of mate choice, depending on the quality of the mate (condition-dependent mate choice) or on the particular environment in which the offspring is produced (context-dependent mate choice). Below, we discuss examples that illustrate the three mechanisms of mate choice.

Mate choice as a by-product of food choice, pea aphids

The pea-aphid system discussed in the previous section is an example of *de facto* assortative mating (Caillaud and Via 2000). Aphids of the two host-adapted races have a strong preference for their respective host plants, and mating takes place on the host plant. This is the prime reason for the lack of hybrids in the field (Via 1999; Via *et al.* 2000).

It is long recognized that *de facto* assortative mating may be more widespread when host-plant characters are imprinted in juveniles (Hopkins 1917; Thorpe 1945). The phenomenon of food imprinting or induction of food preference is called Hopkins' host-selection principle. It refers to the increased preference of adult individuals for oviposition on the host plant that was experienced when juvenile. However, there is hardly any evidence that juvenile experience survives the pupal stage [although there is proof that memory transfer from the larval stage to the adult is, in principle, possible; see the review in Szentesi and Jermy (1990)]. Nevertheless, individuals may also be imprinted in early adulthood, when they are still on the host plant that they successfully grew up on (Jaenike 1983). Although very interesting, we know virtually nothing about the occurrence of this neo-Hopkins' host-selection principle (see also Schoonhoven *et al.* 1998).

Condition-dependent mate choice, spider mites

Assortative mate preferences in host-adaptive races of spider mites have been shown to exist. Gotoh *et al.* (1993) demonstrated that males from a tomato-adapted line prefer females of the same line over females of a cucumber-adapted line. Moreover, recent experiments demonstrated a remarkable ability of female spider mites to avoid mating with males with whom they would otherwise produce inviable offspring (Vala *et al.*, in press). Matings between male spider mites infected with a bacterial endosymbiont (*Wolbachia*) and uninfected female spider mites are incompatible (see Section 12.5). When given a choice, female spider mites avoid *Wolbachia*-infected males and prefer to mate with uninfected males from the same isofemale line. Clearly, this mate choice increases their reproductive success and may therefore be called adaptive. How the females perceive whether a male is infected or not is still unclear. Thus, females can somehow perceive the presence of a microorganism that greatly affects the viability of their offspring, yet does not do any measurable physical harm to the male carrying it. This is the first demonstration that the *Wolbachia* infection can affect the prezygotic (rather than postzygotic) reproductive isolation of its host. These results are also very exciting given the vast literature on females of various species that discriminate mating partners based on the load of parasites that are harmful to their current host (Hamilton and Zuk 1982; Møller 1994).

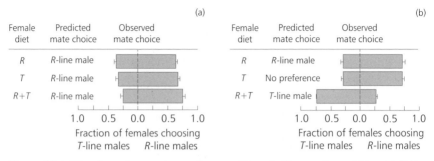

Figure 12.1 Diet-dependent mating preference of female *Hypoaspis aculeifer*: (a) *T*-line females, (b) *R*-line females. Females were put on one of three diets (prey *R* alone, prey *T* alone, or a mix of *R* and *T*), and then offered one *R*-line male and one *T*-line male in a choice test. Mate-choice prediction is based on the relative growth rates of pure-line offspring and hybrid offspring on the diet the female experienced. *Source*: Lesna and Sabelis (1999).

Context-dependent mate choice, soil predatory mites

The benefit to spider mite females of mate choice is dependent on the infection status of the male, but not on the environment they live in. Do females in different environments also prefer those males that improve the performance of their offspring in that environment? A striking example comes from a study on the soil predatory mite *Hypoaspis aculeifer* (Lesna and Sabelis 1999), an important natural enemy of copra mites (*Tyrophagus putrescentiae*, hereafter called prey *T*) and bulb mites (*Rhizoglyphus robini*, hereafter called prey *R*; see Plate 5), two species that feed on fungi and below-ground plant parts (bulbs, roots). A sample of predatory mites collected from a small site (0.25 m²) was first reared in the laboratory on copra mites and then subjected to selection for prey preference. After four generations of selection two lines were obtained, one with a preference for copra mites (*T*-line) and one with a preference for bulb mites (*R*-line). Cross-breeding studies revealed that the preference was inherited as if determined by a single gene (or group of linked genes). Coexistence of these predator genotypes could not be explained by niche differentiation: the *R*-line outperformed the *T*-line irrespective of whether they were reared on bulb mites alone or copra mites alone. However, on a mixture of the two prey species the roles were reversed, because the *R*-line performed very badly (negative growth). Strikingly, the hybrids between the two preference lines outperformed the *T*-line on a diet of prey *T* and the *R*-line on the prey mixture. Only when the diet consisted exclusively of prey *R* was hybrid performance intermediate between the two parental lines. Subsequent tests on mate choice revealed that females of the *T*-line always preferred to mate with males of the other line (Figure 12.1a), which makes sense because hybrids between the two lines performed better (although not significantly so on the prey mix). Females of the *R*-line, however, preferred to mate assortatively, except when fed on a diet of the two prey species together (Figure 12.1b). This switch from assortative to disassortative mating is also understandable, because hybrids between the two lines

Figure 12.2 *Wolbachia* bacteria in an egg cell of *T. urticae*. DNA was stained with DNA-specific fluorescent stain and photographed under a fluorescence microscope (magnification 400). In the center of the picture are the $2N = 6$ chromosomes of the fertilized egg cell. Scattered throughout the cell are *Wolbachia* bacteria, visible as small dots.

performed worse than the parental R-line on a diet of bulb mites alone and equally well on a diet of copra mites alone, but the hybrids performed much better on a diet of the two prey species together. These results show that females can switch their mate choice depending on the performance of the offspring in the current environment (Lesna and Sabelis 1999). This phenomenon is therefore an example of context-dependent mate choice (Qvarnström 2001).

Adaptive mate choice and speciation

When mate choice is a by-product of food choice, sympatric divergence readily occurs through the evolution of specialization and/or food preference (Felsenstein 1981; Rice 1987; Diehl and Bush 1989). Clearly, flexible adaptive mating behaviors (e.g., condition-dependent and context-dependent mate choice) create even more opportunity for reproductive isolation than *de facto* assortative mating (see also Diehl and Bush 1989). Simple genetics underlying food choice (as found for prey preference in the soil predatory mite) and mate choice make adaptive speciation more likely (Dieckmann and Doebeli 1999).

12.5 Symbiont-induced Reproductive Incompatibility

Wolbachia bacteria are obligate endosymbionts that are vertically transmitted from mother to offspring (Figure 12.2). They infect a large number of nematode and arthropod hosts and may induce several reproductive alterations in their hosts [see the review by Stouthamer *et al.* (1999)]. These suggest a role for *Wolbachia* in speciation (Hurst and Schilthuizen 1998; Werren 1998), which led Coyne (1992) to dub this "infectious speciation".

The most enigmatic as well as widespread effect of *Wolbachia* on host reproduction is cytoplasmic incompatibility (CI; Box 12.2), in which *Wolbachia*-infected males are reproductively incompatible with uninfected females, or with

Box 12.2 *Wolbachia* and cytoplasmic incompatibility

Although the molecular details are still unknown, it is hypothesized that the induction of cytoplasmic incompatibility (CI) results from the "modification" by the symbiont of sperm in an infected male. When in a fertilized egg, the modified paternal chromosomes fail to segregate properly, unless bacteria of the same strain are present in the cytoplasm of the egg – so they may rescue the paternal chromosomes (Stouthamer *et al.* 1999). Failure of paternal chromosomes to segregate properly results in either a complete haploid or an aneuploid embryo (Callaini *et al.* 1997).

Levels of CI observed in nature and in the laboratory range from complete (in the laboratory) to undetectable (indeed, strains are known in which infected males do not induce CI, but infected females do rescue sperm chromosomes from males with a different *Wolbachia* infection). Vertical transmission of the endosymbiont can be far from complete (e.g., in fruitflies and in spider mites), which leads to continuous production of some uninfected offspring. When two host-adapted strains are infected with different *Wolbachia* variants, interspecific crosses usually result in bidirectional incompatibility. It is this feature of the *Wolbachia*–host interaction (which leads to *de facto* assortative mating in the host) that first implied a role for *Wolbachia* in speciation processes (Breeuwer and Werren 1990; Coyne 1992).

The figure below illustrates that when a *Wolbachia*-infected male (dark gray) mates with an uninfected female (light gray), no offspring is produced. Apparently, *Wolbachia* in the male "modified" the sperm chromosomes. All three other possible crosses are fully compatible, including the cross of an infected male and an infected female: *Wolbachia* in the egg "rescue" the sperm chromosomes.

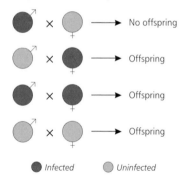

females infected with a different strain of these bacteria (e.g., see Breeuwer and Werren 1990; Stouthamer *et al.* 1999). It was soon recognized that a host species with two different *Wolbachia* strains that are reciprocally, bidirectionally incompatible would instantaneously provide effective reproductive isolation. Recently, Bordenstein *et al.* (2001) provided the first empirical support for this scenario. In two closely related allopatric species of parasitoids, high levels of bidirectional CI preclude the frequent occurrence of hybrids in the laboratory, but crosses with uninfected (cured) individuals of both species show an absence of hybrid breakdown.

Arguments against *Wolbachia*-induced adaptive speciation

Werren (1998; see also Hurst and Schilthuizen 1998) suggested that the role of *Wolbachia* is primarily in allopatric speciation, and not in sympatric speciation. The scenario is analogous to the genetic Muller–Dobzhanski model. A species, fixed for a specific *Wolbachia* strain, becomes divided into two allopatric populations. In each population, a new *Wolbachia* strain replaces the ancestral *Wolbachia* strain, aided by unidirectional incompatibility with it. When the two populations meet again, the chances are high that the two new *Wolbachia* strains are bidirectionally incompatible. (Alternatively, the ancestral state could be uninfected, and the two daughter populations acquired different *Wolbachia* infections independently.)

Werren (1998) and Hurst and Schilthuizen (1998) raise several arguments against a role for *Wolbachia* in adaptive speciation. First, Werren (1998) assumes that it requires near-simultaneous acquisition of different (bidirectionally incompatible) *Wolbachia* strains in subpopulations. Second, Hurst and Schilthuizen (1998) suggest that bidirectional incompatibility may not continue long enough for a permanent split in the host, because *Wolbachia* is lost through coevolution of *Wolbachia* and the host toward lower CI penetrance, or through a double infection from horizontal transfer.

A scenario for *Wolbachia*-induced adaptive speciation

We disagree with these views, because they assume that the drive toward speciation depends solely on *Wolbachia*. Instead, we propose that the coevolution of host and *Wolbachia* during the process of evolutionary branching (see Chapters 4 and 5) may lead to bidirectionally incompatible *Wolbachia* strains in the two host subpopulations. The following example shows the role *Wolbachia* can play in adaptive speciation by maintaining co-adapted gene complexes within its host. Consider a situation in which evolution in an uninfected host population leads to a trait value that would enable evolutionary branching provided some degree of assortative mating could be achieved (Figure 12.3a). At this crucial stage in an infected host population (Figure 12.3b, continuous curve), a *Wolbachia* mutation (dashed curve) that is incompatible with the resident *Wolbachia* strain may help the host to undergo branching, when it occurs in a host female with an extreme phenotype. At the same time, branching of the host population allows coexistence of two *Wolbachia* strains. We assume that part of the population is uninfected and/or bidirectional CI is not complete, otherwise the female with the mutant *Wolbachia* unavoidably mates with an incompatible male and does not reproduce. Our view also holds when a rare migrant-host individual with a different, incompatible *Wolbachia* strain invades the population, instead of a rare mutant *Wolbachia*.

Coupling bidirectional incompatibility to evolutionary branching eliminates two of the arguments against *Wolbachia*-assisted adaptive speciation. First, the assumption of near-simultaneous invasion of different *Wolbachia* strains is not necessary in our scenario. Second, in this situation, selection for lower CI penetrance is counterbalanced by selection against unfit hybrids. Horizontal transfer

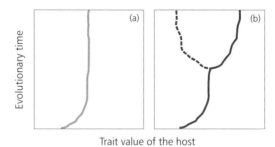

Trait value of the host

Figure 12.3 The proposed scenario for *Wolbachia*-assisted adaptive speciation. (a) Evolution in an uninfected host population leads to the trait value for which evolutionary branching could occur if hybrid offspring could be avoided. (b) In an infected population (with the ancestral *Wolbachia* incompatibility type, continuous curve), branching can occur when a bidirectionally incompatible *Wolbachia* infection (dashed curve) arises.

may still destroy the bidirectional incompatibility when it creates a doubly infected host. Horizontal transfer is possible in parasitoid wasps (Huigens *et al.* 2000), but may not last long when the transfer is from one species to the other (Heath *et al.* 1999). Moreover, horizontal transfer has not been demonstrated for phytophagous species.

The role of *Wolbachia* in speciation may be even larger under unidirectional CI, that is, when reproductive isolation in the other direction is maintained by another mechanism (Giordano *et al.* 1997; Rokas 2000), such as mating preference of infected females for infected males (Shoemaker *et al.* 1999). However, we think that this role is restricted to situations in which some degree of reproductive isolation is already present, not in a fully interbreeding species. In the latter case, we predict the opposite, as reported for the two-spotted spider mite (Vala *et al.*, in press; explained in Section 12.4): uninfected females should prefer uninfected males over infected males to avoid CI, whereas infected females need not have a preference.

12.6   Concluding Comments

In this chapter, we argue that agricultural crops provide ideal conditions for host-adapted race formation among pest species. Hence, agricultural pests provide an excellent opportunity to study speciation processes and test speciation theory. Much can be learned from studies focused on the three issues we addressed:

- We need to know how well herbivorous arthropods detect their individual adaptation to various host plants and translate this experience into a preference for the host plant that yields the highest fitness (Egas and Sabelis 2001). Through its effect on the distribution of foraging herbivores, adaptive learning of host-plant preferences may well promote genetic divergence of host races (as explained in Box 12.1).
- The consequences of host preferences for the mating structure of the population need to be investigated. Research on mate choice suggests that females appear

well aware of the differences among males and choose adaptively with whom to mate, depending on environmental context (Lesna and Sabelis 1999). Such adaptive mate choice will strongly promote adaptive speciation.

▪ There is every reason to study the role of symbiont-induced incompatibilities in keeping coadapted gene complexes intact. In this sense, there are definite possibilities for adaptive speciation aided by *Wolbachia*-induced incompatibility. As soon as such incompatibilities arise, however, the pest insect will not remain an "innocent bystander": it may profit by avoiding incompatible partners (Vala *et al.*, in press). This leads to selection for assortative mating among uninfected individuals, but probably also among individuals infected with the same *Wolbachia* strain when various incompatible strains are present.

Taken together, premating barriers caused by adaptive learning of host-plant quality and adaptive mate choice, as well as postmating barriers through symbiont-induced incompatibilities, may well play a major role in the emergence of novel host races among arthropod pests in agriculture.

Acknowledgments   We thank David Claessen, Arne Janssen, and Steph Menken for comments on an earlier version of this chapter.

13

Ecological Speciation in Flowering Plants

Nickolas M. Waser and Diane R. Campbell

13.1 Introduction

Speciation refers to divergence of phenotype, and the evolution of reproductive barriers between populations that exhibit the different phenotypes. Either of these components of speciation may involve adaptation by natural selection. In this context several terms have been coined (see Chapter 1, in particular Box 1.1). Any speciation in which reproductive isolation follows directly or indirectly from adaptation to ecological conditions has been termed *ecological speciation* (e.g., Hatfield and Schluter 1999). In contrast, this volume defines the term *adaptive speciation* as lineage splitting in sympatry (or parapatry) through a process of frequency-dependent selection. This latter definition thus includes some, but not all, forms of ecological speciation and vice versa.

In this chapter we adopt the broader term ecological speciation, as we review the forms it may take in flowering plants, or angiosperms, and explore whether there is evidence for adaptive speciation more specifically. Most of the about 300 000 angiosperm species rely on animals for pollination, and the dramatic evolutionary radiation of the angiosperms over the past 100 million years is roughly contemporaneous with radiation in several animal taxa from which modern pollinators are drawn. As a result, pollination is thought by many investigators to be a critical ecological factor in speciation (Baker 1963; Crepet 1984; Eriksson and Bremer 1992; Grimaldi 1999; but see Ricklefs and Renner 2000). We outline several scenarios for pollinator-mediated speciation, paying special attention to the unusual role of pollinators as external agents of gene transfer. As both selection and gene flow depend on pollinator responses to floral traits, genetic changes in these traits may simultaneously generate phenotypic divergence and reproductive isolation. Alternatively, divergence of floral phenotypes may be decoupled from selection on traits that prevent successful crosses. A different likely mechanism of speciation is adaptation of plants to local environmental conditions. Finally, selection for floral and reproductive divergence may stem from pollination and other environmental sources combined. For each of these kinds of mechanism we present the available evidence and suggest avenues for future research.

Limited space restricts our attention here to divergence of two incipient species via adaptation to external biotic or abiotic conditions. This constitutes only part of angiosperm speciation. We do not discuss divergence caused by random genetic drift alone [see Levin (2000) for some putative examples]. Nor do we discuss speciation that results from the formation of stabilized diploid hybrids (Gallez and

Gottlieb 1982; Arnold *et al.* 1990; Rieseberg 1995, 1997) or changes in ploidy (Grant 1981; Soltis and Soltis 1993). Intra- and intergenomic conflicts may underlie other speciation events (see Haig and Westoby 1989; Gavrilets 2000b), but these foster adaptation and counteradaptation within the genic environment rather than an external environment.

13.2 Ecological Speciation Driven by Animal Pollinators

Mutualistic associations between plants and pollinators might drive ecological speciation in allopatry, parapatry, or sympatry. As we discuss below, similar mechanisms may generate reproductive isolation regardless of whether the incipient species are in ecological contact, based on the dual role of pollinators as agents of gene transfer and selection. Indeed, there is little direct evidence to date for pollinator-mediated speciation in the absence of spatial separation.

The basic scenario of *pollinator-mediated* divergent selection is especially transparent in the case of allopatry. Populations in geographic isolation encounter pollinators distinct from the ancestral ones (e.g., hummingbirds instead of large bees), and these pollinators select for different floral traits, and so produce phenotypic divergence. For example, the classic concept of pollination syndromes (Grant and Grant 1965; Faegri and van der Pijl 1979) predicts evolution of a deep, tube-shaped flower in a hummingbird-pollinated population (or "pollination race"), versus a broad bell-shaped flower in a bee-pollinated population. As gene flow between species also depends on pollinator behavior, a strong specialization for different pollinators also produces strong or complete reproductive isolation if the incipient species come into secondary contact (Grant 1949). The unusual dual role of pollinators (as agents of selection and movers of genes) thus suggests that speciation may sometimes proceed via a "single-variation" model (*sensu* Rice and Hostert 1993), with the same set of genes producing reproductive isolation as a pleiotropic by-product of phenotypic divergence (Box 13.1).

Initial differences in floral phenotype might often arise by mutations of large rather than small phenotypic effect (Gottlieb 1984; Bradshaw *et al.* 1995), and thus accelerate a "niche shift" to the use of a new pollinator resource. If this process occurs in local isolates, as argued by Levin (2000), then pollination races and ultimately the new species that result may be polyphyletic until lineage sorting (see Box 15.3) and extinction are complete (Levin 1993, 2000; Rieseberg and Brouillet 1994). When the incipient species come into secondary contact, it is possible for reproductive isolation imposed by pollinators to be reinforced by *postpollination* events, if negative assortative mating lowers fitness by generating phenotypes poorly adapted to either pollinator. For example, heterospecific pollen may fail to reach ovules or be outcompeted by conspecific pollen (e.g., Kölreuter 1763; Darwin 1876; Darlington and Mather 1949, p. 253; Paterniani 1969; Rieseberg 1995; although see Mather 1947; Alarcón and Campbell 2000). Reinforcement by a one-allele substitution is essentially automatic if appropriate genetic variation arises (e.g., Crosby 1970; Waser 1993; Waser and Price 1993), and even reinforcement by a two-allele substitution is likely, if reproductive isolation is strong already (Box 13.1; Felsenstein 1981; Dieckmann and Doebeli 1999; Chapter 5).

Box 13.1 Pleiotropy and linkage disequilibrium in ecological speciation

Phenotypic differentiation and reproductive isolation evolve in concert if they are pleiotropic effects of the same genes [figure below, panel (a)]. This "single-variation" model does not necessarily mean only one gene is involved, but rather that the same suite of genes produces phenotypic differentiation and reproductive isolation. For example, imagine a change in floral morphology that corresponds to a shift in major pollinators, with pollinator fidelity strong enough to eliminate pollen exchange between the derivative and parent species. A similar situation arises if separate genes code for phenotype and for mating specificity, so long as positive assortative mating is enhanced in all populations by substitution of the same allele [M in panel (b)]:

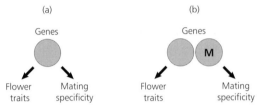

This is the "one-allele" substitution of Felsenstein (1981). In contrast, assortative mating may require substitution of alternative alleles, M and m, in different populations:

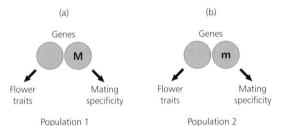

This "two-allele" substitution (Chapters 3 and 5) requires a statistical association between alleles at mating loci and at loci that code for phenotype. Such a linkage disequilibrium will arise and be maintained only with strong selection against hybrids, and genetic drift (or another mechanism such as partial flower constancy of pollinators) sufficient to generate an initial association (Felsenstein 1981; Dieckmann and Doebeli 1999).

In spite of its logical simplicity, this scenario has enjoyed little direct support to date. Many studies describe strong pollinator-mediated selection on floral phenotypes (e.g., Campbell 1989; Johnston 1991; Andersson and Widén 1993; Galen 1996), which illustrates the potential for rapid evolutionary change. However, the usual focus is on selection *within* a species rather than divergence *between* incipient species, so we have only indirect support for the scenario. We now need to extend the studies of selection not only to related species, but also to larger spatial scales within species that are appropriate for exploring divergence in allopatry or parapatry – efforts that will require unusual collaboration and resources to succeed. Examples of such an approach include Levin's and colleagues' (Levin and Kerster 1967; Levin and Schaal 1970) demonstration that *Phlox* species exhibit character displacement in flower color, and reduced hybridization by butterflies upon secondary contact; and preliminary evidence (Peter E. Scott, personal communication) that flower morphology and color, and flowering phenology, vary geographically in ocotillo (*Fouquieria splendens*) in response to changes in pollination environment (e.g., large bees versus resident hummingbirds versus transient migratory hummingbirds).

Another reason for the scanty support for this scenario may be that species interactions are not always as tidy as it tacitly assumes. Many pollination interactions are generalized and opportunistic (e.g., Herrera 1996; Waser *et al.* 1996; Memmott 1999). In such cases, one-on-one coevolution of plants and pollinators is unlikely, and instead new pollination races may adapt to *mixtures* of pollinator species that differ *quantitatively*, not qualitatively, from those of parents (Grant and Grant 1965), with weak-to-moderate reproductive isolation on secondary contact (Waser 1998, 2001).

The scenario of pollinator-mediated speciation in allopatry can be extended to situations in which ongoing gene exchange counteracts divergence (i.e., to parapatry and sympatry). If different pollinator types exhibit sufficiently narrow and invariant preferences, they might select disruptively on floral phenotype even within a single population, which would lead to phenotypic divergence and, pleiotropically, to reproductive isolation. The amount of isolation that ensues depends on how often pollinators fly between flowers of the same rather than different phenotype, and on how much pollen is transferred, two aspects historically labeled "ethological" and "mechanical" isolation, respectively (Grant 1949). If some isolation via pollinators is in place, it may again be reinforced by postpollination events.

As a consequence of the dual role of pollinators, sympatric speciation, like allopatric speciation, can proceed via a single-variation model (see Box 13.1). In this case, the usual difficulty imposed by sympatric speciation of generating linkage disequilibrium between mating loci and a fitness locus is circumvented. Whether sympatric speciation of this form really does occur, however, depends critically on the *strength* of disruptive selection and reproductive isolation produced by preferences of different pollinators for alternative phenotypes.

Once again, there are few telling studies. Mather (1947) inferred (without direct observation) that pollinator behavior explained the rarity of F1 hybrids between the snapdragons *Antirrhinum majus*, *A. glutinosa*, and *A. orontium* placed in mixtures. More recent studies have observed pollinator visitation directly in mixed-species arrays. For example, Fulton and Hodges (1999) reported strong fidelity of hummingbirds to the columbine *Aquilegia formosa*, and of hawkmoths to *Aq. pubescens*. However, unless preferences are absolute (and they rarely are, at least in temperate systems; Wilson and Thomson 1996), it cannot be assumed that

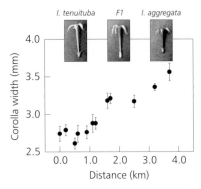

Figure 13.1 Several floral traits show clinal variation across a 4 km wide hybrid zone of *Ipomopsis*. Plants of *I. aggregata* have wider corolla tubes than those of *I. tenuituba*, and populations along the hybrid zone show a gradual transition in this trait. F1 hybrids are intermediate in corolla width, length, and flower color. Bars indicate standard errors.

divergent preferences will always result in the lowest overall visitation to intermediate phenotypes. With the inclusion of natural or experimentally produced hybrids, or selected mutants, it is possible to generate a wider range of floral phenotypes, including intermediates, and to estimate selection more precisely than when only parental species are presented to pollinators. For example, when arrays of hybrid monkeyflowers (*Mimulus*) that express a wide range of floral traits are exposed to hummingbirds and bumblebees, selection favors elements of the phenotypes of the parental species (Sutherland and Vickery 1993; Vickery 1995; Schemske and Bradshaw 1999), but intermediate phenotypes also received substantial visits (Vickery 1992). Similarly, Ippolito and Holtsford (1999) documented disruptive selection by the combined visitation of hawkmoths and hummingbirds to arrays of F2 and F3 hybrids between two species of Brazilian *Nicotiana*. In contrast, Wesslingh and Arnold (2000) found that F1 *Iris* hybrids were visited more frequently than the two parental species. In none of these cases is it yet clear that selection is strong and consistent enough to drive complete divergence of the phenotype, much less reproductive isolation. Inconsistent selection is not a trivial issue. For example, Campbell *et al.* (1997) also documented disruptive selection on flower width in hybrid populations of *Ipomopsis* (Figure 13.1 and Box 13.2), but such selection occurred only in rare years when hawkmoths were present along with the more reliable hummingbirds (Plate 6).

More study certainly is needed of the strength and consistency of disruptive selection mediated by pollinators. This should be followed by investigations of whether differences in the selected traits also confer some reproductive isolation, and whether other traits (including postpollination traits noted earlier) also contribute. Few of the studies described above reported the frequency of interspecific pollinator flights in addition to the overall rate of visitation; those that did found weak-to-moderate, rather than strong-to-absolute, ethological isolation (Campbell *et al.* 1998, 2002; Wesslingh and Arnold 2000).

Box 13.2 Pollinator-mediated disruptive selection

Several quantitative traits of the flowers of *I. aggregata* are under phenotypic selection during the pollination phase of the life cycle (Campbell 1989), and heritabilities and genetic correlations suggest that this could cause rapid evolution (Campbell 1996). In some cases the mechanism of selection is known to involve pollinators. For example, experiments show that hummingbirds select for wider floral tubes (fused corollas), which they probe deeply, thereby removing more pollen per visit (Campbell *et al.* 1991, 1996).

In some parts of the western USA, *I. aggregata* grows in contact with the congener *I. tenuituba* and the two form hybrids with intermediate floral phenotypes. In one such hybrid zone, Campbell *et al.* (1997) used multivariate selection analysis to detect disruptive selection on corolla width within single populations. Here, the component of fitness is rate of visitation to flowers by hummingbirds [broad-tails (*Selasphorus platycercus*) and rufous (*S. rufus*)], and hawkmoths (*Hyles lineata*), which prefer broad and narrow corollas, respectively (Plate 6). The quadratic relationship for visitation (relative to the mean) as a function of corolla width and length, at a value of flower color equal to the average for natural hybrids, is:

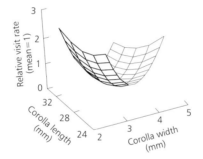

This selection gradient occurs only in rare years when hawkmoths appear in appreciable numbers; in most years pollination is predominantly by hummingbirds. However, this example illustrates how pollinator-mediated disruptive selection may operate in some systems, and so lead to speciation.

Another promising experimental approach is the genetic or phenotypic manipulation of a trait already shown to differ between closely-related species, to see whether elimination of the difference increases interspecific pollination. As is true in some animals (Chapter 16), color patterns often distinguish flowers of different species, and are amenable to manipulation (Waser and Price 1985b). When Meléndez-Ackerman and Campbell (1998) painted flowers of *I. tenuituba* and hybrids red, to match flowers of the congener *I. aggregata*, more of the resultant seeds were hybrid than in a control population with the species differences intact, which suggests that color differences confer some reproductive isolation. A similar approach could be taken with other traits, using phenotypic manipulation or plants known to differ at major loci for floral traits.

Box 13.3 Adaptive dynamics of pollinator-mediated speciation

Disruptive or divergent pollinator-mediated selection on floral phenotypes seems to require, at first glance, two distinct modes of pollinator phenotype (e.g., two pollinators that differ markedly in the lengths of their tongues, proboscises, or beaks). Adaptive dynamics models, however, highlight the existence of an alternative mechanism (Metz *et al.* 1996; Dieckmann and Doebeli 1999). In this mechanism, which also has a unimodal and continuous distribution of pollinator phenotypes (e.g., describing the relative frequencies of pollinators as a function of their tongue length), disruptive selection can result from frequency-dependent competition for pollination among plants.

In this scenario each plant phenotype corresponds to a pollinator phenotype to which it is best adapted. The pollination resources available to a particular plant phenotype are therefore proportional to the frequency of the corresponding pollinator. Models for adaptive dynamics driven by frequency-dependent competition for a unimodal resource (in the present case represented by the pollinators) are then directly applicable (Chapter 5).

These models predict the following. First, the mean phenotype in the parental plant species evolves to occupy the peak of the pollinator distribution, as illustrated by the arrow in panel (a) below:

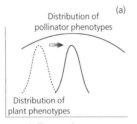 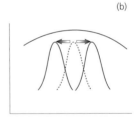

(a)   (b)

Distribution of pollinator phenotypes

Distribution of plant phenotypes

Pollinator phenotype   Pollinator phenotype

Once the mean phenotype in the plant population reaches this point, rare phenotypes on either side of the mean phenotype attract visits by pollinators from more marginal ranges of the pollinator distribution and thereby experience less competition for visits by pollinators. For disruptive selection to occur, this advantage to rare phenotypes must outweigh the disadvantage that results from the smaller abundance of pollinators away from the peak of their distribution and the potential disadvantage of finding other plants to mate with. If these disadvantages are overcome, evolutionary branching in the plant population can bring about two daughter species, as illustrated in panel (b) above.

The pollinator distribution in this scenario corresponds to the resource distribution used in the general models of evolutionary branching driven by intraspecific competition. Also, note the congruence with MacArthur's (1969) "Q-minimization" model, which predicts an optimal match of resource utilization and resource functions for a set of species.

The adaptive dynamics approach may thus offer a resolution to the apparent paradox, observed by Ollerton (1996), that pollinator-mediated plant speciation seems to require a multimodal "pollination space", whereas such a setting seems

continued

Box 13.3 *continued*

unlikely to exist in nature given the generalization and opportunism in plant-pollinator interactions (see Herrera 1996; Waser *et al.* 1996). The scenario described above assumes that, on the time scale of plant speciation, little or no evolutionary change occurs in the distribution of pollinator phenotypes. This assumption may apply to plants that rely on many pollinator species (see Memmott 1999).

In addition, adaptive dynamics models can be constructed for more complex scenarios that lead to the cospeciation of tightly associated plants and pollinators – a process that may have happened, for instance, in some figs and fig wasps (Herre *et al.* 1996).

An example for such a model was provided by Doebeli and Dieckmann (2000). Based on a Lotka–Volterra system for two mutualistically interacting species, individuals in each species are characterized by a quantitative trait that determines the distribution of mutualistic benefit given to individuals of the other species. In this coevolutionary scenario, plants are viewed as resources for the pollinators, and vice versa. The allocation of this benefit depends on the phenotypes of the interacting individuals in such a way that the level of support increases with phenotypic similarity. In addition to determining the strength of the mutualistic interaction, the quantitative trait also determines the amount of external resources available to each individual. This imposes a stabilizing effect on the quantitative traits, with a different optimal trait value in each of the two species. Thus, while the mutualistic interaction is frequency dependent and selects for phenotypic similarity, the stabilizing selection imposed by the distribution of external resources selects for specific trait values that are different in the two species.

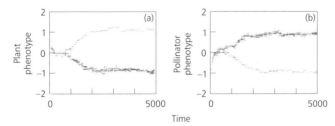

If the conflict between these different sources of selection is large enough, coevolution can result in concurrent evolutionary branching in both species [cospeciation; see panels (a) and (b) above]. These results suggest that evolutionary diversification can also arise when plants and pollinators are specialized to each other, as long as additional external resources are critical and provide the necessary selection for distinct quantitative phenotypes.

The "adaptive dynamics" framework (Chapters 4, 5, and 7) suggests a variant on pollinator-mediated ecological speciation in sympatry or parapatry (Box 13.3). According to this model, distinct types of pollinators, or quantitatively distinct mixtures of the same pollinator types, are not required for the initial pollinator-mediated divergence in floral phenotype. Even with a distribution of pollinator phenotype that is unimodal rather than bi- or multimodal, rapid evolution of floral phenotype and splitting into multiple phenotypic classes might be observed. However, to date no evidence indicates this mechanism of adaptive speciation in angiosperms.

13.3 Adaptation and Speciation in Different Environments

Another potential route to speciation involves adaptation to different environments (other than animal pollinators), with reproductive isolation following as a pleiotropic by-product. Environment differences may be abiotic, such as in soil moisture, chemistry, or solar insolation (e.g., McNeilly and Antonovics 1968; Davies and Snaydon 1976; Lovett Doust 1981; Galen and Stanton 1991), or biotic, involving interactions with soil fungi, herbivores, competitors, and diseases (e.g., Turkington and Harper 1979; Parker 1985). A special case arises when abiotic conditions lead to a paucity of animal pollinators, which may select for traits such as predominant self-pollination or greatly modified pollen, flowers, or inflorescences to achieve wind pollination, any of which will enhance reproductive isolation between the original species and its derivative (e.g., Berry and Calvo 1989; Weller et al. 1998).

A famous example of genetic divergence that involves soil chemistry and competition is the adaptation of plant populations to soils with high concentrations of salts of heavy metals. On tailings (spoils) from British lead and zinc mines (dating in some cases to Roman times), plants are tolerant to the metals, whereas genotypes from adjacent pasture are unable to grow. Conversely, tolerant genotypes are overshadowed and excluded from the pasture by faster-growing intolerant genotypes (McNeilly and Antonovics 1968; Hickey and McNeilly 1975). The boundary between tailings and pasture is abrupt, as is that between genotypic classes. In some species, such as the grasses *Agrostis tenuis* and *Anthoxanthum odoratum*, tolerant and intolerant races differ in seasonal time of flowering, and so are partially isolated reproductively at a prepollination stage. Flowering phenology can evolve rapidly under selection (e.g., Stanford et al. 1962; Carey 1983), so this difference might represent adaptive reinforcement of postzygotic reproductive isolation (hybrid inviability). However, more direct causes of different flowering phenology cannot be ruled out (Antonovics 1968; McNeilly and Antonovics 1968), such as a difference in the timing of appropriate moisture conditions of the soil in pasture versus tailings. In a similar case that involves *Mimulus* species endemic to serpentine soils, Macnair and Gardner (1998) proposed that nutrient-poor, dry soils select for early flowering and self-fertilization. This has the effect of reducing gene flow between species, but it does not constitute adaptive reinforcement.

Adaptation of plants to local abiotic and biotic conditions is reported on various spatial scales across a range from close parapatry to allopatry (e.g., Bradshaw 1960; Jain and Bradshaw 1966; Davies and Snaydon 1976; Turkington and Harper 1979; Waser and Price 1985a; Bennington and McGraw 1995; Mehrhoff and Turkington 1996; Ronsheim 1997; for a historical review see Langlet 1971). Examples of ecological and phenotypic radiation on oceanic islands and habitat islands provide especially compelling cases of apparent adaptive radiation in strict allopatry (Levin 2000; Chapter 17). Thus, it appears possible that ecological speciation via this route is common. What is poorly understood is how reproductive isolation evolves. There is evidence for partial pre- and postzygotic barriers, both for short distances [e.g., the mine tailings example and Waser *et al.* (2000)] and on larger spatial scales within species (e.g., Kruckeberg 1957; Grant and Grant 1960; Hughes and Vickery 1974; Vickery 1978). In some cases these barriers are known to be direct pleiotropic effects of the genes involved in local adaptation (Macnair and Christie 1983; Galen and Stanton 1991; Stanton *et al.* 1994).

13.4 Combined Speciation Mechanisms

Pollinators are assumed to be central to one mode of ecological speciation and, as outlined in Section 13.2, this mechanism of *pollinator-mediated divergent selection* is likely, although evidence is scanty. In addition, pollinators might play a more subtle, indirect role by facilitating the mechanism of adaptation to local environments described in Section 13.3 (Waser 1998, 2001) and thus speciation in local isolates (Levin 2000). The scenario for *pollinator-facilitated local adaptation* is as follows. By their area-restricted foraging, itself a result of natural selection for efficiency in food harvest, pollinators move most pollen relatively short distances (e.g., Levin and Kerster 1974; Campbell and Waser 1989; Karron *et al.* 1995). The resultant restriction of pollen dispersal, coupled with restricted seed dispersal (the situation for most plant species), allows adaptation of conspecific plants in different populations to different local environments. In other words, gene flow between populations is insufficient to counterbalance local adaptation. Thus, pollinator behavior helps to foster phenotypic and reproductive distinctiveness.

This scenario remains mostly unexplored. We do not know, for example, how closely the scale of local selection matches restrictions on gene flow. Part of the uncertainty arises from the difficulty in estimating the long-distance tail of gene flow. New methods of paternity exclusion (e.g., Devlin and Ellstrand 1990) reveal surprising amounts of pollen flow in excess of 1 km in some species (5–24% of paternity in examples cited by Levin 2000), more than anticipated from direct studies of pollinator behavior or pollen dispersal from a source (Campbell 1991a). It remains unclear, however, where exactly this long-distance pollen comes from, how fit offspring sired by such pollen are (but see Waser and Price 1994; Waser *et al.* 2000), and, most importantly, how such gene flow interacts with locally variable selection to determine the final patterns of genetic differentiation.

Other combinations of mechanisms also seem likely. In particular, divergent selection in parapatry or allopatry (or disruptive selection in sympatry) might be mediated by a combination of pollinators and other environmental features. For example, divergent floral phenotypes in *Polemonium viscosum* along elevational gradients may represent different equilibria between opposing selection on flower size related to water loss from petals, attraction of nectar-robbing ants, and interaction with pollinator faunas that comprise different mixtures of bumblebees and flies (Galen 1983, 1996; Galen *et al.* 1999; Galen and Cuba 2001). Brody (1997)

Table 13.1 Selection on floral phenotype by combinations of pollinators and other environmental factors. In some cases the evidence for selection as tabulated here is circumstantial; in others it is more firmly established. In no cases are the mechanisms of selection fully worked out, and neither is net selection shown to vary from site to site, which would lead to divergent selection in parapatry or allopatry. Nonetheless, examples such as these point toward possible cases of floral divergence under the influence of multiple factors.

Species	Traits	Selection	References
Polemonium viscosum	Flower size, corolla flare	Preference by bumble bees, pollen transfer efficiency, water loss, damage from nectar robbers	Galen 1996; Galen *et al.* 1999; Galen and Cuba 2001
Phlox drummondii	Flower color	Pollinator preference, pleiotropic effect on survival and flower production	Levin 1972; Levin and Brack 1995
Delphinium nelsonii	Flower color, nectar guides	Pollinator preference, background color, environment	Waser and Price 1985b
Mimulus cardinalis and *M. lewisii*	Flower color and shape, leaf morphology	Pollinator preference, survival	Hiesey *et al.* 1971
Ipomoea purpurea	Flower color	Preference by bumble bees, pleiotropic effect on plant size and seed size via herbivore resistance	Rausher and Fry 1993; Fineblum and Rausher 1997
Clarkia gracilis	Petal color spotting	Preference by bees, differential survival	Jones 1996
Ipomopsis aggregata	Flower morphology, flowering phenology	Preference of pollinators and predispersal seed predators	Campbell 1991b; Brody 1997; Irwin and Brody 1999

and Irwin and Brody (1999) pointed out the potential for selection on floral phenotype jointly mediated by pollinators and by plant enemies, such as seed predators and nectar robbers. Flowering time is a trait that may be selected by pollinators, but also by other environmental conditions (e.g., Ehrendorfer 1953; McNeilly and Antonovics 1968; Waser 1978). Other possible cases of interaction between various mechanisms of selection and various targets of selection have been identified (Table 13.1), but usually not fully explored. Examples are likely to be common, but to detect them requires investigators not to focus too narrowly on only one aspect of plant biology, such as pollination or vegetative physiology, in isolation of the others.

One approach to study simultaneously pollinators and other potential sources of divergent or disruptive selection makes use of reciprocal transplants across species boundaries. Reciprocal transplants have a long history in botany (Clausen *et al.* 1940; Langlet 1971), but have been applied only recently to mechanisms of speciation. They do not necessarily reveal the nature of selection that *generates* the original species differences, but can elucidate selection that *maintains* differences. The hope is that by observing species or subspecies that are closely related we can learn about the selection responsible for early stages of species differentiation (Harrison 1993). For example, consider an experimental reciprocal transplant across the zone of contact between two related plant species. If selection always favors the parental species, with hybrids less fit in both parental habitats, divergent selection in parapatry is implicated (ecological speciation model of Hatfield and Schluter 1999; Chapter 9). Whether such selection derives from pollinators, from other environmental factors, or from some mixture could then be distinguished by comparing selection at each site that arises from variation in pollination success with selection at other life-history stages. In sagebrush (*Artemisia*), each parental species survived better than the other parent or hybrids in its own home habitat, whereas hybrids survived better than either parent in a hybrid zone between the parental habitats (Wang *et al.* 1997). In contrast, F1 and F2 hybrids between two species of *Iris* survived as well as both parental species, regardless of habitat, once rhizomes had been established (Emms and Arnold 1997). In the *Iris* system, reproductive isolation appears to result from neither pollinator behavior (Wesslingh and Arnold 2000) nor physical features of the environment, but largely from postpollination events that limit the formation of F1 hybrids (Arnold *et al.* 1993; Carney *et al.* 1994).

To date, reciprocal transplant experiments generally have been conducted only over a small portion of the life history, but ultimately it will be necessary to extend them to allow direct comparison of pollination success of hybrids with other fitness components. In an *Ipomopsis* hybrid zone, differences in both pollination and vegetative survival contribute to fitness differences between the two parental species and F1 hybrids (Campbell and Waser 2001). *I. aggregata* appears to have higher fitness at low elevation, largely because its floral traits make it attractive to hummingbirds (Campbell *et al.* 1997), whereas *I. tenuituba* has higher fitness at high elevation largely because of a survival advantage. Reproductive isolation is not strongly developed, but F1 hybrids may suffer because pollinators move pollen more effectively from *I. aggregata* to *I. tenuituba* (Campbell *et al.* 1998, 2002), and hybrid seeds with the latter species as maternal parent survive poorly in parental habitats (Campbell and Waser 2001). Thus, a complex mixture of pollinator-mediated selection and other mechanisms is involved in the dynamics of this hybrid zone, and may have contributed to the original differentiation of the parental species.

In summary, a promising approach is to combine reciprocal transplant studies across multiple habitats with selection analysis (Nagy 1997), or with manipulation of specific traits, to determine if traits are selected divergently, and if so by pollinators, other features of the environment, or both.

Box 13.4 Hot research topics, old and new

The last major full treatment of angiosperm speciation is by Grant (1981). The major sections of his book consider the nature of plant species, divergence of species, secondary genetic fusion of species, derived genetic systems (such as polyploidy), and the evolution of hybrid (= species) complexes. Grant lists the following primary mechanisms of divergence of species:

- Geographical speciation;
- Quantum speciation;
- Sympatric speciation;
- Reproductive isolation as a by-product of divergence;
- Chromosomal repatterning.

We imagine that a modern update would include these same topics in some form. However, our ideal update would also stress some new issues, including:

- Studies of pollinator-mediated disruptive selection on various scales, and its exact mechanisms, including explicit attention focused on the behavior of the animals in choosing flowers and its root causes, such as cognitive constraint (e.g., see Chittka *et al.* 1999). These studies should be quantitative and coupled with investigations of reproductive isolation to determine whether the same floral traits under disruptive or divergent selection also confer reproductive isolation, as implied in the classic scenario.
- Assuming that the mechanisms of reproductive isolation are clearly determined, the investigation of their genetic basis [e.g., via quantitative trait locus (QTL) mapping; Bradshaw *et al.* 1995], with the proviso that the basis of species' differences may not fully reflect the basis of isolation mechanisms that constituted the actual speciation event (Harrison 1991). New molecular methods for the construction of isogenic lines should increase the range of characters that can be manipulated to identify the effects on reproductive isolation.
- The continued marriage of molecular systematics and ecology with the mapping of interactions, such as pollination and herbivory, onto family trees for the plants to infer evolutionary history and lability (see Armbruster 1996; Weller and Sakai 1999).
- Renewed focus on the plant phenotype, with emphasis on the integration of physiology and external morphology, and on the study of pleiotropic linkages between vegetative and reproductive traits (see Fineblum and Rausher 1997). Transplant studies with hybrids could be used to explore the relative importance of selection on vegetative versus floral traits. In addition, new morphometric approaches can be used to characterize the floral phenotype and to explore phenotypic selection (Herrera 1993).
- Study of the sizes of phenotypic changes that appear spontaneously by mutation in plants, and of the role of these in phenotypic divergence under natural selection. As Darwin stressed, *Natura non facit saltum*; but she does take steps of various sizes, and there is renewed discussion about the importance of "macromutations" (e.g., Gottlieb 1984; Doebley 1993).   *continued*

Box 13.4 *continued*

- Exploration of how reproductive isolation is manifest on a hierarchy of scales of spatial or genetic separation, with the goal of reconstructing the temporal sequence of evolution of isolating mechanisms (Coyne and Orr 1997). The spatial context could be incorporated by performing hybrid crosses over a wide range of spatial scales (Edmands 1999), and by combining phylogenetic information with biogeographic distributions (Schaal *et al.* 1998) to reveal the history of genetic exchange.
- A discussion of various forms of *intra- and intergenotypic conflict*, and how these might result in reinforcement of reproductive isolation between populations that diverge in parapatry or allopatry (e.g., Haig and Westoby 1989; Martienssen 1996; Grossniklaus *et al.* 1998; Rice 1998; Gavrilets 2000b). In particular, might such conflicts, and the selection "arms race" that they engender, turn out to be a major source of genetic incompatibilities between individuals from different populations, and if so on what spatial scales and under what circumstances?

13.5 Concluding Comments

In this chapter we are limited to a précis of ecological speciation. An updated treatment of all of angiosperm speciation, including ecological speciation, is overdue. However, a good argument can be made that this update must wait for the results of a renewed empirical attack. There is no lack of research topics at hand (Box 13.4), and new techniques promise answers to old and hitherto intractable questions.

In spite of empirical gaps, our conclusion is that ecological speciation in angiosperms is likely, via several mechanisms, but that no direct evidence is at hand for adaptive speciation as defined in this volume (Chapter 1). Given large values of phenotypic selection coefficients measured under natural conditions, and high heritabilities of phenotypic traits of flowers (Jain and Bradshaw 1966; Antonovics 1971; Bradshaw 1972; Davies and Snaydon 1976; Lovett Doust 1981; Campbell 1989, 1996; Armbruster 1990; Johnston 1991; Conner *et al.* 1996), speciation may also be quite rapid. Finally, divergence in sympatry is not ruled out. The task for the coming decades is to establish the relative importance of different mechanisms and spatial contexts.

In the study of speciation, flowering plants are worthy subjects in their own right. They are also of value as model systems. For example, individual plants "stay put to be counted" (Harper 1977) and so they have proved to be ideal subjects for transplant studies, which are impossible with most animals (Futuyma and Shapiro 1995), and this advantage remains to be fully exploited. While *exact* mechanisms of speciation in plants can no more be extended to other taxa than the reverse (e.g., Mitchell *et al.* 1993), the lessons learned from angiosperms promise to shed much light, however oblique, on ecological speciation in general.

14

Experiments on Adaptation and Divergence in Bacterial Populations

Michael Travisano

14.1 Introduction

The basis of biological diversity continues to be the subject of intense study. Numerous causes of biodiversity have been identified, including adaptation, spatial heterogeneity, ecological interactions, genetic drift, and phylogenetic history. However, the general importance of these and other possible causes has proved difficult to quantify. Much of the difficulty arises from the historical nature of evolution. Evolutionary change can take many generations, which severely limits studies that examine the dynamics of diversification. Compounding this problem are limits to true replication in studies of extant diversity and an inability to examine directly the past selective conditions, both of which again result from the historical nature of evolution. While several approaches have been successful in overcoming some of these difficulties (e.g., Schluter 1996a), general conclusions have been hard to obtain.

A complementary approach to examining extant populations (Endler 1986) is to perform experimental evolution under controlled laboratory conditions (Dykhuizen 1990; Lenski 1995). By performing natural selection experiments in the laboratory, both the selective conditions and the initial genetic composition of the evolving populations can be controlled. Thus, difficulties that arise from limited replication and uncertain selective history are circumvented and evolutionary responses can be attributed directly to the known conditions under which the experiment was performed.

Experimental evolution is particularly powerful when coupled with organisms of short generation times, such as bacteria. With appropriately chosen bacterial species, several thousand generations of evolution can be accomplished during a single year. In addition, ancestral and descendant genotypes can be stored in a viable, but nonevolving, state at $-80°C$. Viable storage allows simultaneous fitness and phenotypic assays of ancestral and descendant genotypes under uniform controlled conditions. With a microbe-based approach, evolutionary responses to specific conditions can be observed over the course of an experiment and subsequently can be examined in great detail. Other benefits to using bacteria for selection studies include the large population sizes, ability to initiate replicate isogeneic cultures, ease of culture, ease of replication, and extensive background information on bacterial physiology, biochemistry, and genetics.

A potential complication in using bacteria for evolutionary studies of speciation involves the definition of bacterial species. Speciation in bacteria generally cannot be assessed by tests for reproductive isolation, except for those species that undergo frequent natural recombination (e.g., *Bacillus subtilis* and *Haemophilus influenzae*), and there is no general consensus for the definition of a bacterial species. Prior to the advent of molecular phylogenetics, bacterial species were defined on the basis of several traits, such as cell morphology, cell-wall structure, resource-utilization profiles, virus sensitivity, and habitat. While systematic studies based on DNA sequence have revolutionized the study of bacterial evolution, dividing them into two broad and phylogenetically unrelated groups (Archaea and Eubacteria), species definitions have remained difficult. A striking, but typical, example is provided by the "species" *Escherichia coli*. Different environmental isolates may vary by as much as 20% in genome size and may be either commensal or pathogenic. Attempts to define new species are not typically made in selection studies with bacteria; instead, phenotypic and ecological divergence is assayed. The focus of bacterial experimental evolution has been on the divergence of ecologically relevant traits, rather than on reproductive isolation.

This chapter reviews several bacterial evolution experiments that address adaptation and the evolution of diversity. To date, bacterial experiments have been performed that correspond to allopatric and sympatric speciation. (Experiments that correspond to parapatric speciation have yet to be carried out, and while they are likely to be fruitful, further speculation on their potential value is beyond the scope of this chapter.) Section 14.2 focuses on adaptation and divergence of isolated replicate populations (i.e., on allopatric divergence). Adaptation and diversification within a population (i.e., sympatric diversification) is considered in Section 14.3. In both sections, adaptation and divergence are observed to occur readily, but only under the appropriate environmental conditions. The chapter concludes with a discussion on future directions.

14.2 Allopatric Divergence

The distinguishing feature of allopatry is the potential for divergent evolution among distinct isolated populations. Divergence among isolated populations potentially could have a variety of causes, such as differences in selective conditions, neutral genetic drift, and differences in the initial genetic composition among populations. A less obvious possible cause is adaptation in the absence of the above factors. Divergence by adaptation to the same selective conditions could occur if allopatric populations can fix different sets of adaptive mutations. In contrast, divergence by genetic drift occurs by fixation of alternative alleles of little selective value.

The repeatability of adaptive evolution

The propensity for divergence among populations during adaptation to a uniform environment was examined in a series of selection experiments with *E. coli* (Lenski *et al.* 1991). In the initial study (which is still ongoing), a single genotype was used

---

Box 14.1 Propagation of experimental bacterial populations

The basis of microbial selection experiments is the propagation of replicate populations under controlled environmental conditions. There are two broadly defined propagation schemes, serial culture and continuous culture. In serial culture, a portion of each population is periodically transferred to fresh medium. Thus, the size of the population varies over time, often by 100-fold or more. Serial culture is most commonly performed using liquid cultures that are shaken to ensure spatial homogeneity, but unshaken liquid cultures or even solid medium can also be used (Korona *et al.* 1994). In contrast, continuous-culture systems have a continuous inflow of nutrients into a culture vessel and outflow of cells and waste products. Typically, the rate of nutrient inflow (and waste outflow) is kept constant over time so that the size of a population remains constant, although pulsed experiments can also be performed (blurring the distinction between serial and continuous culture). A wide variety of different continuous-culture systems are possible, of which the most common is the chemostat (Kubitschek 1970). Genotypes must grow as quickly as the outflow rate so that they are not washed out and lost from the system. Under chemostat conditions, bacterial growth reduces the concentration of the limiting nutrient to an equilibrium level.

Various media are used in microbial selection experiments. A minimal medium contains only the minimal nutrients required for growth. A glucose-minimal medium contains glucose as the sole available carbon and energy source in addition to the essential minerals, sodium, phosphates, a nitrogen source, etc. The maximum population size for the bacteria in minimal media is most often limited by one of the nutrients (e.g., glucose, nitrogen, etc.), although growth in minimal media with abundant resources can be inhibited by the production of waste metabolites. A rich medium contains additional nutrients that, while not required for growth, generally increase the growth rate. Population sizes in rich media can be limited by both exhaustion of resources and production of waste products.

---

to found 12 replicate populations, each of which was selected for $>20\,000$ generations under the same environmental conditions. The populations were sampled periodically and examined for divergence in a number of traits: fitness in the selected environment (Lenski *et al.* 1991; Lenski and Travisano 1994), life-history traits (Vasi *et al.* 1994), cell size (Lenski and Travisano 1994; Elena *et al.* 1996), and performance in novel environments (Travisano *et al.* 1995a; Travisano and Lenski 1996; Cooper and Lenski 2000; see reviews by Lenski *et al.* 1998; Travisano and Rainey 2000).

The selective conditions are daily serial batch culture: every 24 hours a 100 μl aliquot (about 5 million individuals) of each culture is transferred to 10 ml of fresh medium (Box 14.1). This propagation scheme ensures that the bacteria go through about 6.64 ($= \log_2 100$) generations each day prior to transfer. The propagation scheme also ensures that the bacteria experience a "feast and famine" environment with an abundance of initial resources that are exhausted as the population grows. Maximum population density is limited by the amount of carbohydrate, 25 μg/ml

---

Box 14.2 Determination of relative fitness

Fitness in microbial populations can be determined on a head-to-head basis, with two or more genotypes in direct competition in a common environment. Each of the competing genotypes must be distinguishable by a marker, generally differences in nutrient use, or by antibiotic and virus resistance. Most desirable is a marker that has no fitness cost in the selected environment and only has phenotypic effects when exposed to the appropriate assay conditions. Arabinose utilization has been the marker of choice for the populations initiated by Lenski *et al.* (1991), as it has no detectable fitness effects in most minimal media (Travisano and Lenski 1996), but it is an easily scored trait. Genotypes capable of using arabinose as a carbon source (Ara^+) form white colonies on tetrazolium arabinose agar plates, while Ara^- genotypes form red colonies.

To perform a fitness assay, genotypes to be competed are added into a common competition culture environment that is assayed for the initial density of all competitors. The competitors are allowed to grow for an appropriate period of time, which depends upon the particular goal of the fitness assay, after which they are assayed again to determine the final density of all competitors. Fitness of one genotype relative to another is determined by the ratio of Malthusian parameters. The Malthusian parameter r_i for a genotype i is estimated as

$$r_i = \ln(n_i(T)/n_i(0))/T \,, \tag{a}$$

where $n_i(T)$ and $n_i(0)$ are the final and initial densities, respectively, of genotype i. Thus, the relative fitness of genotype i to genotype j is

$$w_{ij} = r_i/r_j \,. \tag{b}$$

Occasionally, the difference in Malthusian parameters is calculated rather than the ratio, since often the Malthusian parameters differ by a large amount, which makes the ratio very sensitive to sampling error. This difference in Malthusian parameters is the selection rate constant

$$s_{ij} = r_i - r_j \,. \tag{c}$$

---

glucose, supplemented to the medium. Under these conditions the rate of fixation of deleterious mutations by genetic drift via bottlenecks is low, since the minimum census population size is about 5 million, and reaches around 500 million every day.

Both parallel and divergent evolution were observed across the 12 replicate populations over the course of 20 000 generations. Highly statistically significant genetic variation for fitness was detected among the populations over most of the experiment (Lenski *et al.* 1991), but the magnitude of divergence was small (Lenski and Travisano 1994). At 10 000 generations, the standard deviation among populations for fitness was only about 0.05, while there was a 48% average fitness improvement relative to the common ancestor (procedures for measuring fitness are described in Box 14.2). Primarily parallel evolution was also observed for

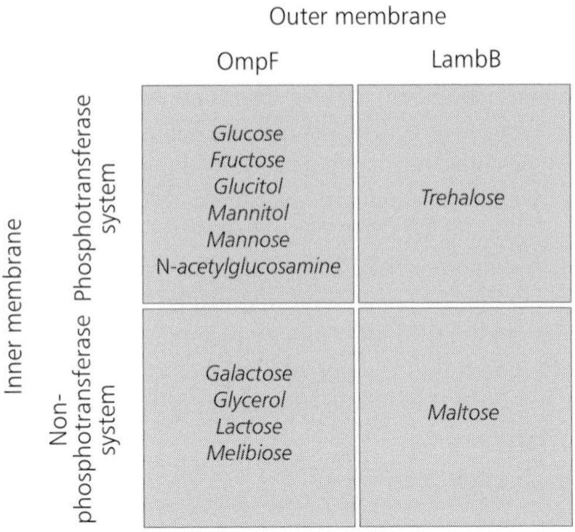

Outer membrane

OmpF   LambB

		OmpF	LambB
Inner membrane	Phosphotransferase system	Glucose Fructose Glucitol Mannitol Mannose N-acetylglucosamine	Trehalose
	Non-phosphotransferase system	Galactose Glycerol Lactose Melibiose	Maltose

Figure 14.1 Classification of the nutrients used to assay adaptation and divergence of the 12 evolved populations after 2000 generations. All 12 nutrients fall into four groups based on mechanisms of nutrient uptake through the inner and outer membranes of *E. coli*. Like glucose, fructose, glucitol, mannitol, mannose, and *N*-acetylglucosamine (NAG) are brought through the outer membrane via the OmpF protein and through the inner membrane via the phosphotransferase system (PTS). Galactose, glycerol, lactose, and melibiose enter *E. coli* through the outer membrane via the protein OmpF (like glucose), but not via the PTS across the inner membrane. In contrast, trehalose enters the cell through the outer membrane via LambB protein and the PTS. Maltose enters the cell by neither the OmpF nor the PTS.

life-history traits tightly associated with fitness after 2000 generations of selection (Vasi *et al.* 1994). Using a demographic model of population growth (Stewart and Levin 1973), maximal growth rate and lag time (the time required to initiate growth upon transfer to fresh medium) were predicted to be strong determinants of fitness. Subsequent experimental work demonstrated adaptation in both traits (faster growth rate and shorter lag times), but with little statistically significant genetic variation.

However, divergence of fitness was observed in some novel nutrient environments among the 2000-generation populations (Travisano *et al.* 1995a; Travisano and Lenski 1996). Divergence was observed in those novel environments in which glucose was substituted with a carbohydrate whose physiological mechanisms of uptake by *E. coli* differ from those of glucose (Figure 14.1). For example, populations ranged from about 30% less fit than the common ancestor to about 20% more fit when assayed in a medium supplemented with maltose, a carbohydrate that shares no mechanisms of nutrient transport with glucose (Figure 14.2). To quantify the correlated responses to selection in the novel environments, a parallel adaptation index *PAI* can be calculated for the selected populations,

$$PAI = p_f / V_G , \qquad (14.1)$$

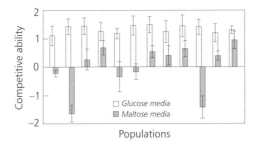

Figure 14.2 Competitive ability of the 12 replicate *E. coli* populations in glucose (white) and in maltose (gray) media after 2000 generations of selection in glucose medium. Competitive ability calculated by the difference in Malthusian parameters (derived minus ancestor), measured in head-to-head competition. Values greater than zero indicate an improved competitive ability of the derived population. Confidence intervals are 95% based upon *t*-distribution with $n - 1$ degrees of freedom. *Source*: Travisano *et al.* (1995a).

where p_f is the proportion of populations with higher fitness and can be determined by counting the number of populations that have a fitness greater than the ancestor and dividing that number by the number of evolving populations, 12 in this case. V_G is the genetic variation of fitness among a set of genotypes that can be determined by performing replicate fitness assays of the genotypes in blocks, and then performing a two-factor random effects analysis of variance (ANOVA) with a block as one factor and the genotype as the other. V_G among the genotypes is equal to the difference in group and error mean-square estimates divided by the number of replicate blocks (Sokal and Rohlf 1981).

The *PAI* is a measure of parallel adaptation, as it increases with the proportion of populations that have adapted (have greater p_f) and declines as populations increasingly differ from one another (have greater V_G). A general improvement in fitness, but with little or no divergence, was observed in novel environments in which the provided carbohydrate shared a common uptake physiology with glucose (Figure 14.3). In contrast, little fitness improvement and substantial genetic divergence was observed in nutrient environments in which uptake of the limiting carbohydrate differed from that of glucose. This outcome strongly suggests that glucose uptake was an important target of selection during the evolution of these populations, but that evolution of the glucose uptake system varied across populations. There appears to have been strong selection for an improved glucose uptake, but different physiological mechanisms to improve this uptake were fixed in each population. After 2000 generations, the 12 populations can be grouped into no less than six phenotypic classes based upon their performance in novel nutrient environments (Figure 14.4).

Divergence thus occurred rapidly, without initial variation either among or within populations and without differences in environmental conditions. In principle, divergence under such circumstances could have arisen through some combination of two processes:

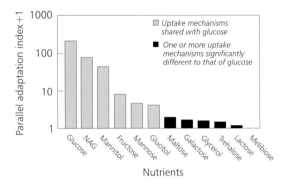

Figure 14.3 Parallel adaptation of the 12 replicate *E. coli* populations in the selected glucose environment and novel nutrient environments. Parallel evolution was observed in nutrients that shared uptake mechanisms with glucose (gray), but not in those whose uptake differed significantly in one or more mechanisms (black). *Source*: Travisano and Lenski (1996).

- Fixation of different adaptive mutations could have resulted in the observed divergence if the adaptive mutations engendered different pleiotropic fitness effects in novel environments.
- Alternatively, divergence could have been caused by the accumulation of different mutations that have no fitness effect in the selected environment, but have different effects in novel environments.

Since no recombination occurred among genotypes within populations over the course of the experiment, the second scenario encompasses both that of linkage hitchhiking and genetic drift. In examining the possible role of genetic drift, it is important to recall Kimura's result that the fixation of neutral mutations is independent of population size (Kimura 1983). Nevertheless, discrimination could not be made between the two possible causes based solely on fitness in novel environments.

However, pleiotropy can be shown to have been the predominant cause of divergence by a genomic analysis of nutrient uptake. All the nutrients examined have nutrient-specific uptake pathways, some of which share specific physiological uptake mechanisms with glucose uptake. For mutation accumulation to be a tenable hypothesis for divergence, divergence should be most extreme in nutrients that have the largest target for mutations to accumulate. Hence, the *PAI* should be correlated negatively with the number of nucleotides that code for nutrient-specific uptake. The correlation coefficients for the populations assayed in the five carbohydrates transported similarly to glucose, the six carbohydrates that differ in carbohydrate transport, and all eleven novel carbohydrates are 0.47, 0.02, and –0.32, respectively (Figure 14.5). None are significantly different from zero ($P < 0.2$).

Divergence among the *E. coli* lineages was the result of adaptation, but did not require niche specialization. However, the propensity for divergence varied among traits. The absence of variation for fitness and traits tightly associated with fitness

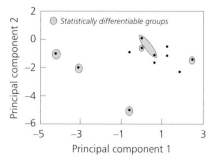

Figure 14.4 Divergence of the 12 replicate *E. coli* populations. Based upon correlated responses to selection in novel nutrient environments, the populations can be divided into no fewer than six genetic groupings by multiple Bonferroni-corrected *t*-tests with a joint confidence level of $P < 0.05$. Statistically differentiable groups are shown enclosed by ovals on this graph of the first two axes of a principal components analysis. *Sources*: Travisano (1993) and Travisano and Lenski (1996).

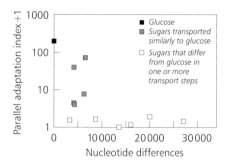

Figure 14.5 Plot of parallel adaptation index versus the number of nucleotides coding for genes required for nutrient-specific uptake and catabolism distinct from that for glucose. The black square represents the values for glucose, the gray squares represent the five sugars transported in a manner similar to glucose, and the white squares refer to those sugars that differ in one or more transport steps from that of glucose. *Source*: Travisano and Lenski (1996).

in the selected environment suggest strong limits on the potential for allopatry to generate divergence in traits that are direct targets of selection. For traits less tightly associated with fitness in the selected environment, the rapid divergence across populations suggests that diversity can readily evolve in allopatry despite uniform selection. This is consistent with the substantial literature on heritability of life-history traits, as it is known that heritability tends to be higher in traits that are less tightly correlated with fitness (Ridley 1996).

Effects of chance, history, and adaptation

Studies on trait evolution focus on at least one of three factors: chance, phylogenetic history, and adaptation. The effects of chance are usually discussed with

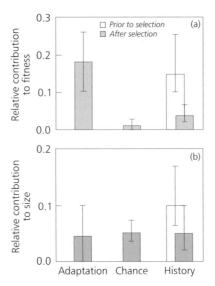

Figure 14.6 Relative effects of chance, adaptation, and history on (a) fitness and (b) cell size. The white bars represent the values prior to selection and the gray bars those after selection. Confidence intervals encompass 95%. *Source*: Travisano *et al.* (1995b).

regard to selectively neutral mutations (e.g., Kimura 1983), but chance effects can be non-neutral, most clearly on the appearance and initial persistence of a new mutation. The effects of phylogenetic history on trait evolution are contentious (Gould 1989; Conway-Morris 1998). Closely related species generally have similar attributes, but it is often unclear how much of the similarity results from a common phylogenetic history and how much from similar selective conditions. Gould and Lewontin (1979) described the difficulties in assessing the role of adaptation on trait evolution. Chief among these is the inability to assess accurately the selective conditions under which the trait evolved and difficulty in performing an appropriately sensitive experiment under the current selective conditions.

Using the replicate populations after 2000 generations of selection in glucose, Travisano *et al.* (1995b) performed a subsequent evolution experiment to assess empirically the relative effects of chance, history, and adaptation on trait evolution. Single colonies from each of the 12 selected populations were used to initiate three replicate cultures, for a total of 36 populations. All 36 populations were propagated for 1000 generations in maltose-minimal medium, an environment in which there was an initial genetic variation both in fitness (Figure 14.2) and cell size (see below). The effects of chance were assayed as divergence among replicate lines that had the same immediate ancestor. Historical effects were measured as persistent differences among the 12 groups. Adaptation was assessed as a net mean change over the 1000 generations across all 36 evolving populations.

For fitness, adaptation to the new selective environment was the dominant factor and most of the historical variation among the 12 groups was lost (Figure 14.6a).

The effect of chance on fitness in the selective environment was small and not statistically significant, consistent with previous results. Clearly, historical differences and chance had only minor effects relative to adaptation of the trait on which natural selection acts. Even so, historical differences did persist, and those differences arose within the first 2000 generations of selection. This suggests that small fitness differences among allopatric populations can evolve quickly and persist in various environments, even if all the populations experience the same sequence of different environments.

Chance diversification and historical differences had a much larger effect on the evolution of cell size (Figure 14.6b). Cell sizes were measured electronically (disruption of an electric potential across a 20 micron aperture through which cells were drawn) for all genotypes. Chance contributed significantly to the variation in cell size across populations, and most of the historical contribution to cell size was maintained over the 1000 generations of selection. In contrast, there was no statistically significant effect of adaptation on cell size. The difference in outcomes between fitness and cell size is probably because of differences in their selective importance. Selective differences among genotypes were not correlated with cell size in this experiment and other experiments (Lenski and Travisano 1994). In contrast, strong selection on cell size would probably have reduced the effect of chance and history on this trait.

The relative roles of chance, history, and adaptation have been difficult to address. Gould addressed it by a "gedanken" experiment of replaying the tape of life on earth (Gould 1989), essentially identical to having replicate allopatric "earths". The outcome of Gould's thought experiment was that, "Any replay of the tape would lead evolution down a pathway radically different from the road actually taken. ... Alter any early event ever so slightly and without apparent importance at the time, and evolution cascades into a radically different channel." Different lineages, with different morphologies, would thrive or perish during each restarting, because of the effects of random chance factors. The accumulation of chance events would result in historical contingencies that limit the range of subsequent adaptation. The above results with bacterial lineages support some of Gould's conclusions. Chance and history can have major effects on diversity and adaptation across populations, particularly for traits most commonly examined, such as those that affect morphology. Even so, diversity for fitness and tightly associated traits is likely to be much more limited.

Genotype-by-environment interactions

One caveat on the relative ease of diversification in allopatry under the same selective conditions is that it may vary for particular combinations of ancestral stock and selective environment. Such divergence requires multiple possible avenues for evolutionary improvement, and genotype and environment combinations may differ in the limitations on adaptive evolution. This was shown in a second selection experiment using the same ancestral *E. coli* genotype as used above, but in which the replicate populations were selected in maltose-limited medium, rather than in

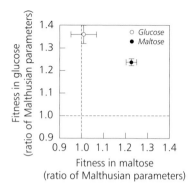

Figure 14.7 Mean fitness of E. coli populations evolved in glucose (open circle) or maltose (filled circle) medium in both nutrient environments. Fitness calculated as the ratio of Malthusian parameters (derived/ancestral), measured in head-to-head competition. Values greater than one indicate improved competitive ability of the derived population. Confidence intervals are 95% based upon t-distribution with $n - 1$ degrees of freedom. *Source*: Travisano (1997).

glucose-limited medium (Travisano 1997). The populations improved in fitness in the selected environment, with little genetic variation for fitness, and also had high fitness in a novel environment (glucose-limited medium) with little genetic variation. Thus, as observed previously, when assayed in their respective selective environments, both sets of populations performed similarly. However, when assayed in the reciprocal novel environments, completely different outcomes were observed (Figure 14.7). The contrasting results suggest that, while divergence readily evolved during selection in glucose-limited medium, adaptation in maltose-limited medium was far more constrained, at least as measured by fitness in a novel environment. Hence, the propensity for adaptive divergence is likely to be highly environmentally dependent and to be sensitive to subtle genetic differences.

14.3 Sympatric Divergence

In sexually reproducing species, sympatric speciation occurs when a single breeding population divides into two or more breeding populations without full geographic isolation. The driving force in sympatric speciation is generally assumed to be ecological specialization. Since definitions of bacterial speciation are problematic, ecological specialization of distinct clonal lineages is a useful criterion by which to assess divergence in sympatry.

Resource partitioning

Resource partitioning is almost the paradigm for ecological specialization, as most examples of divergence and adaptive radiation involve resource partitioning. Darwin's finches have different mouth forms, corresponding to the food source. So do Hawaiian honeycreepers, cichlids, and sticklebacks. The potential for resource partitioning to drive divergence appears high, but how resources are partitioned is less clear.

Multiple resources. Trade-offs in resource use are invoked frequently in studies of resource partitioning, but the actual demonstration of trade-offs is often difficult. To examine the role of specialization in resource partitioning, Dykhuizen and Davies (1980) assayed a genetically engineered set of *E. coli* genotypes that varied in their resource specialization and competitive ability for coexistence. All genotypes were derived from an efficient maltose specialist (ES) genotype that had the lactose operon deleted, which rendered it incapable of utilizing lactose as a nutrient source. Other genotypes, an efficient resource generalist (EG), an inefficient maltose specialist (IS), and an inefficient resource generalist (IG) were constructed by reintroduction of a lactose operon and/or selection for a streptomycin-resistant mutant. Resistance to streptomycin incurs a fitness cost in the absence of the antibiotic, which makes resistance genotypes "inefficient". Using these genotypes and evolved derivatives, exclusion or coexistence was observed depending upon the availability of two carbon sources, maltose and lactose.

Fitness and coexistence assays were performed in chemostats, and not in serial culture as was done in the experiments described above (Section 14.2). In medium that contained only maltose as the energy source, the ES genotype was the superior competitor, most likely because of a mutation that improved fitness inadvertently acquired during the genetic engineering (Dykhuizen, personal communication). In medium that contained mixtures of lactose and maltose, coexistence of a generalist and a specialist genotype was observed within a range of lactose-to-maltose mixtures. The exact coexistence equilibrium frequency was dependent on the nutrient ratio and the specific genotypes. Coexistence did not occur at extreme ratios of lactose-to-maltose, as expected from theoretical models of bacterial growth (Stewart and Levin 1973).

However, at low lactose concentrations, mutants arose that rescued lactose-utilizing populations from extinction. Such mutants were derived from the generalist population, but expressed the lactose operon even in the absence of lactose (constitutive expression). Constitutive expression allowed lactose-utilizing genotypes to persist at lactose concentrations too low to keep the lactose operon induced. In the absence of lactose and maltose specialist competitors in a lactose–maltose mixed medium, generalist populations (IG or EG) became polymorphic because of the appearance of lactose constitutive mutants. Coexistence was only maintained at low input concentrations of lactose; at higher concentrations the lactose constitutive mutants excluded the generalists. The coexistence at lower lactose concentrations indicates that the IG and EG genotypes are more efficient users of maltose and that a pleiotropic effect of constitutive lactose operon expression is a fitness reduction during growth on maltose. Such trade-offs through pleiotropy are essential for resource specialization and diversification.

A "single" resource. Coexistence of multiple resource specialists is discussed in the previous section, and based upon those results it might appear difficult for divergence and coexistence to occur if there is only a single limiting resource. Presumably, a single genotype (species) will be best fitted to exploiting the resource and it will exclude all others. However, it is not uncommon for prey species to support multiple consumers, including predators, scavengers, and detritivores. Single resources often, although not always, consist of multiple component parts that allow for at least the possibility of specialization. While this is self-evident for many prey species (e.g., wildebeest), it is also true for sugars such as glucose.

Chemostats that initially contain a single *E. coli* genotype and have a constant supply of glucose as the sole limiting carbon source are invaded repeatedly by multiple coexisting genotypes in as little as 100 generations (Helling *et al.* 1987). Three different phenotypes can evolve and are distinguished by various phenotypic characteristics, including colony size on agar plates. In a detailed study of one chemostat culture, each of the three derived genotypes was superior to the common ancestor for growth on glucose (Rosenzweig *et al.* 1994). However, the derived isolates persisted in coexistence when cultured together, in any pairwise combination or all together, under the conditions in which they were selected. The derived genotype that was the best competitor for glucose, determined by affinity studies, incompletely oxidized glucose and excreted acetate and glycerol. Each of the other coexisting genotypes was the superior competitor for one of the excreted metabolites. The acetate specialist expressed an enzyme involved in acetate utilization, acetyl CoA synthetase, at much higher levels than the glucose specialist. Examining acetate specialists from replicate chemostat experiments showed that all the acetate specialists isolated had mutations in the regulatory region of the acetyl CoA synthetase locus that resulted in semiconstitutive overexpression of the enzyme (Treves *et al.* 1998). The genetic mechanism of adaptation involved in glycerol specialization is not known.

Adaptation for utilization of glucose resulted in divergence and specialization to glucose breakdown products. Trade-offs in glucose metabolism are likely to have been the cause of specialization, although the causes of such trade-offs are presently a matter for speculation. Acetate and glycerol utilization can be partially inhibited by glucose metabolism, and glucose uptake may be enhanced by excretion of metabolites. Regardless of the physiological mechanism, it is intuitively plausible that a trade-off in uptake rates of primary and secondary nutrients can result in frequency-dependent selection, and hence that the fate of a mutant strain depends on the environment generated by resident strains: if the resident is very efficient on the primary resource, it produces many secondary metabolites, so that it may pay a mutant to be efficient on these secondary resources at the expense of primary efficiency; on the other hand, if the resident is not very efficient on the primary resource, then metabolites are not very abundant, and it may simply pay to be more efficient on the primary resource. Box 14.3 contains a simple model that illustrates how this type of frequency dependence can generate evolutionary branching into a glucose specialist and a crossfeeder specializing on secondary metabolites. In the experiments reported here, evolution of uptake rates proceeded through a small number of mutational steps. Indeed, some distinct genetic differences between strains that constitute a glucose–acetate crossfeeding polymorphism have been suggested (Rosenzweig *et al.* 1994). It is nevertheless reasonable to assume that uptake rates can vary almost continuously, because the metabolic pathways involved are rather complicated and can be affected in many different ways (Helling *et al.* 1987). Moreover, even if single mutations have relatively large effects on uptake rates, detecting an evolutionary branching point in the invasibility analysis presented in Box 14.3 is important because it implies (Chapter 4) that there exist pairs of strains that can coexist in their uptake efficiencies on primary and secondary resources while diffusing.

Spatial heterogeneity

Following resource specialization, spatial heterogeneity has been the most often suggested factor for promoting and maintaining diversity. Distinction is usually made between fine and coarse variation in scale, although it has been difficult to determine the effects of spatial heterogeneity on diversity at any scale (Hedrick 1986). Island biogeography theory has been relatively successful in predicting diversity over different spatial scales (MacArthur and Wilson 1967), but it is less useful in identifying the specific causes and conditions that result in diversification. How spatial heterogeneity affects adaptation and divergence has been difficult to address because of limitations in the ability to make causal associations of extant genetic variation to selectively important environmental factors.

Rainey and Travisano (1998) tested the importance of spatial heterogeneity in adaptation and diversification of replicate populations of the bacterium *Pseudomonas fluorescens* strain SBW25. In 6 ml static broth cultures, microcosms that contained a single ancestral morph rapidly diversified, in as little as three days, into three morphologically distinguishable niche specialists (see Plate 7). In contrast, no diversification was observed during growth of the ancestral morph in a vigorously shaken homogeneous culture over the same period of time. Morphological differences were apparent as differences in colony morphology. The three morphs were denoted smooth (SM), wrinkly spreader (WS), and fuzzy spreader (FS), based upon their respective morphologies, with the ancestor being a SM colony. Within each derived morph class (WS and FS), morphological variation was evident both within and across replicate cultures. When placed in isolation and allowed to grow for one day, each morph had spatially distinct growth; WS morphs grew at the air–broth interface, SM morphs grew in the broth phase, and FS morphs accumulated on the culture vessel bottom (see Plate 7). All phenotypes were stable over at least 100 generations under a variety of nonselective conditions and differed in virus susceptibility, traits commonly used to distinguish different genotypes.

Heterogeneous conditions developed in the static broth cultures because of the growth of the bacteria. As SBW25 grew, it removed oxygen and nutrients from the environment and released metabolic waste products. SBW25 is an obligate aerobe and grows best at the air–broth interface. In static culture, the WS morph formed a mat over the top of the broth, and so maximized its exposure to the air.

Box 14.3 A model for the evolution of crossfeeding polymorphisms
Michael Doebeli

This box describes how gradual evolution of uptake efficiencies for a primary resource and a secondary metabolite can give rise to a crossfeeding polymorphism via
evolutionary branching into resource specialists. To model the evolution of crossfeeding polymorphisms in continuous culture, first consider ecological chemostat
models based on Michaelis–Menten kinetics,

$$
\begin{aligned}
\frac{dn}{dt} &= \frac{x_g g}{k_g + g} n - Dn \, , \\
\frac{dg}{dt} &= -\frac{1}{Y_g} \frac{x_g g}{k_g + g} n - Dg + Dg_0 \, ,
\end{aligned}
\tag{a}
$$

where n is the population density of the bacteria in the chemostat, and g is the
concentration of the primary nutrient glucose. The parameters D and g_0 describe
the rate of dilution (outflow) in the chemostat and the glucose concentration in the
reservoir, respectively. The other parameters characterize the particular strain of
bacteria whose ecological dynamics are given by Equation (a): x_g is the maximal
growth rate, k_g is the concentration of glucose needed to attain a growth rate of
$x_g/2$, and Y_g is the "yield" [see, e.g., Edelstein-Keshet (1988) for a detailed account of Michaelis–Menten dynamics]. To model crossfeeding on secondary waste
products, the dynamics of a metabolite need to be included (e.g., acetate, which
is produced during glucose metabolism at a rate proportional to the rate of glucose
consumption). This secondary metabolite is also consumed by the bacteria, but with
different demographic parameters x_a, k_a, and Y_a. With a denoting the concentration
of acetate, the dynamics of bacteria, glucose, and acetate are described by

$$
\begin{aligned}
\frac{dn}{dt} &= \frac{x_g g}{k_g + g} n + \frac{x_a a}{k_a + a} n - Dn \, , \\
\frac{dg}{dt} &= -\frac{1}{Y_g} \frac{x_g g}{k_g + g} n - Dg + Dg_0 \, , \\
\frac{da}{dt} &= -\frac{1}{Y_a} \frac{x_a a}{k_a + a} n - Da + \frac{Y_{a/g}}{Y_g} \frac{x_g g}{k_g + g} n \, ,
\end{aligned}
\tag{b}
$$

where the parameter $Y_{a/g}$ determines the rate at which acetate is produced during
glucose metabolism.

We have to specify which of the parameters represent evolving traits and, for
simplicity, we assume that all the parameters are fixed externally, except for the
maximal growth rates x_g and x_a, which are assumed to be constrained by some
function $x_a = \phi(x_g)$, with $d\phi/dx_g < 0$ for all x_g. Such a constraint generates
frequency dependence, because specialization on glucose (high x_g) not only leads
to higher rates of glucose consumption and therefore to higher rates of acetate production, but it also implies a low value of x_a, and thus makes the corresponding
strain a weak competitor for the acetate produced. This, in turn, paves the way for
the invasion of strains with high x_a (and correspondingly low x_g).

continued

Box 14.3 *continued*

More specifically, the adaptive dynamics of the trait x_g can be studied using pairwise invasibility plots (Chapter 4). These are calculated numerically by first letting Equations (b) equilibrate for a given resident strain x_g, which results in equilibrium values $\hat{n}(x_g)$, $\hat{g}(x_g)$, and $\hat{a}(x_g)$, and then testing whether the per capita growth rate $f(x'_g, x_g)$ of a rare mutant x'_g in the environment determined by the resident x_g is positive, that is, whether

$$f(x'_g, x_g) = \frac{x'_g \hat{g}(x_g)}{k_g + \hat{g}(x_g)} + \frac{\phi(x'_g)\hat{a}(x_g)}{k_a + \hat{a}(x_g)} - D > 0 . \qquad (c)$$

The pairwise invasibility plot for the trait x_g based on the constraint function $x_a = \phi(x_g) = 3\exp(-x_g/5)$, with parameter values $Y_g = k_g = Y_a = k_a = 1$, $D = 1$, $g_0 = 10$, and $Y_{a/g} = 1.5$, is shown below.

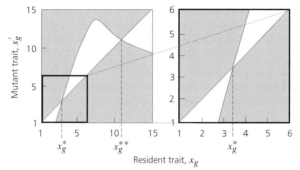

In this pairwise invasibility plot those resident–mutant pairs (x_g, x'_g) for which $f(x'_g, x_g) > 0$ are marked gray. The left panel shows that there are two evolutionarily singular trait values, x_g^* and x_g^{**}. As explained in Chapter 4, these values represent equilibria for the monomorphic adaptive dynamics, because at these values the selection gradient $\partial f(x'_g, x_g)/\partial x'_g \big|_{x'_g = x_g}$ vanishes. From the sign configuration shown in the plot, we see that x_g^* is an evolutionary attractor, while x_g^{**} is an evolutionary repellor. Moreover, the vertical line through x_g^* lies in a gray area (see enlargement on the right), which implies that once the population has evolved to the attractor x_g^*, every nearby mutant x'_g can invade. Therefore, x_g^* is an evolutionary branching point, and after having converged evolutionarily to x_g^* the population is expected to split into two phenotypic branches: a glucose specialist with high x_g and correspondingly low x_a, and a strain with low x_g and correspondingly high x_a, which represents a crossfeeder that specializes on the waste products of the glucose specialist.

The model thus shows how a trade-off in maximal growth rates on primary and secondary resources can lead to the gradual evolution of crossfeeding polymorphisms from a single ancestral bacterial strain. Pairwise invasibility plots that exhibit evolutionary branching points (as in the figure above) imply that there are points off the diagonal in the resident–mutant plane corresponding to pairs of strains that can coexist (Figure 4.2). Therefore, the existence of a branching point in the above analysis implies the evolutionary potential for a crossfeeding polymorphism, even if mutations have large effects on resource uptake efficiencies.

The SM morph probably had less access to oxygen in the broth phase, but no competition from WS morphs, which tend to fall to the bottom if the culture is perturbed. The selective benefit of occupying the culture bottom by the FS morphs was less apparent. The alteration of environmental conditions by bacterial growth and subsequent localization of phenotypes suggested that evolutionary branching had occurred through frequency-dependent interactions in populations with spatial structure (Chapter 7).

Although the initial experiment demonstrated the importance of spatial heterogeneity for the evolution of diversity, it was not sufficient to indicate the persistence of diversity. Transient variation within bacterial cultures can occur through a succession of adaptive mutations, and the length of the initial study (ten days) was too short to exclude this possibility. To test the persistence of the diversity, and the role of spatial heterogeneity in the persistence, ten replicate cultures were initiated and incubated for one week without shaking. As expected, diversity readily increased within the cultures. The cultures were then propagated, each culture being used to initiate two new cultures, for a total of 20 cultures. One set of ten cultures was again incubated without shaking, but the other was moved into a shaking incubator. After one week, the cultures incubated with shaking had lost almost 50% of their initial diversity, while the static cultures remained diverse. A second week of incubation resulted in an additional 50% loss in diversity in the shaken cultures, with the SM morph predominating in every culture. Continued high diversity was observed in the static cultures. Clearly, diversity persisted in the heterogeneous cultures, but was rapidly lost under homogeneous conditions, which demonstrates that the variation within the heterogeneous cultures was unlikely to have resulted from the incorporation of alternative equally advantageous mutations in different locations in the culture.

A third experiment was carried out to determine the role of adaptation in the static culture diversification. The previous experiments were strongly suggestive of populations altering their environment (environmental feedback, see Chapter 4) leading to the emergence and coexistence of multiple populations. However, the spatial localization of SM, WS, and FS growth was purely correlative. Likewise, the necessity of a heterogeneous environment for evolution and maintenance suggested adaptive speciation, but could potentially have indicated merely neutral divergence and a lack of selective differences in the heterogeneous environment. To show adaptation directly, 24 independently evolved and isolated WS and FS morphs were collected and competed in pairwise competitions with one another and with the ancestral SM morph. Competitions were performed over a period of one week in 6 ml static broth microcosms. The expectation was that if adaptation was underlying the diversification, then stabilizing frequency-dependent selection would be observed, so that genotypes have a selective advantage when rare. Thus, it should be tested whether the invasion fitness (Chapter 4) of rare genotypes in a given resident was positive. The results of such experiments are summarized in Figure 14.8, in which directions and weights of arrows, respectively, represent the inferred signs and relative magnitudes of invasion fitness. The intransitivity of

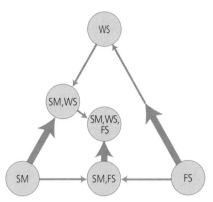

Figure 14.8 Results of multiple invasibility experiments among the smooth morph (SM), the wrinkly spreader morph (WS), and the fuzzy spreader morph (FS) of *Pseudomonas fluorescens* in static broth cultures. The six arrows parallel to the sides of the triangle document the outcome of pairwise invasibility experiments: when an arrow points from a vertex X towards a vertex Y, type Y can invade into a pure culture of type X, whereas an arrow in the opposite direction means that it cannot do so. The two central arrows, as well as the relative weight of arrows, are based on what happens if one lets a culture evolve from an initial SM state: shortly after all three morphs occur together (often after only 24 hours), a higher abundance of the WS morph is correlated with a lower abundance of the FS morph. These experiments suggest that the FS morph can only coexist with WS if SM is also present and that WS invades considerably faster than either FS or SM when they get the chance.

the arrow pattern underscores that no overriding fitness measure exists that could capture all eco-evolutionary interactions, which implies that we are dealing with a complicated pattern of frequency-dependent selection. An interesting open question is whether, in experiments in which types evolved *de novo*, such evolution went through a number of intermediaries, or whether the final types were the result of single (macro-)mutations. Some observations indicate that intermediate types are sometimes involved; evolution, however, proceeded so fast that it was not feasible to sample and analyze these types in any detail. So while the simplifying assumptions of small mutational steps and rates, often invoked in adaptive dynamics theory (Chapter 4), do not apply, the documented invasibility relations (Figure 14.8) are of the type required to bring about the two new types, WS and FS, through two subsequent processes of evolutionary branching.

More complex situations

Successful explanations of the causes of adaptation and diversification often identify a single "critical" factor essential for the observed phenomena. Numerous different factors have been identified, such as population size, mutation rate, niche availability, epistasis, pleiotropy, and temporal variation. In part, the assortment of factors simply obscures the underlying processes, such as evolutionary branching. However, the variety of factors also suggests that adaptation and diversification

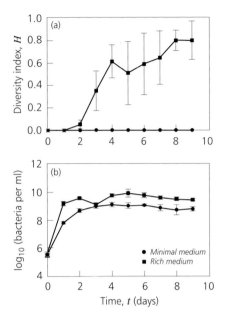

Figure 14.9 SBW25 cultures propagated in rich medium (squares) or minimal medium (circles): (a) diversity and (b) population size. Diversity by the Shannon–Weaver index H. Confidence intervals are 95% based upon t-distribution with $n - 1$ degrees of freedom. *Source*: Travisano and Rainey (2000).

have multiple interacting dependences that are difficult to disentangle. Through the ability to rerun evolution, microbial systems allow the importance of multiple factors on adaptation and diversification to be investigated (see Section 14.2, subsection "Genotype-by-environment interaction").

Travisano and Rainey (2000) examined the importance of multiple environmental factors in adaptation and divergence by perturbing the culture conditions that give rise to diversity in static heterogeneous broth cultures of SBW25. The culture medium provided in the experiments that compared homogeneous versus heterogeneous environments was rich in both mixed short-length proteins (peptone) and glycerol, which possibly allowed resource specialization to evolve. This was investigated by altering the medium so that it contained glycerol, peptone, or both. No diversification was observed in static cultures that contained medium without peptone, which indicates that a complex resource environment was essential for diversification to evolve (Figure 14.9). The robust growth, with or without peptone, indicated that population size was not an important factor in diversification. Moreover, diversification after one week was observed to be correlated linearly with available peptone, which showed that peptone availability limited diversity, and not total available carbon or energy (Figure 14.10). Clearly, spatial heterogeneity alone was insufficient to generate diversity and WS morphs were resource specialists.

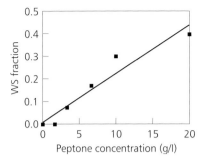

Figure 14.10 Effect of peptone concentration on the percentage of WS in a population after seven days of static incubation. Three replicate microcosms were initiated per peptone concentration. The continuous line indicates the best linear fit to the data (slope = 0.021, $P < 0.005$, five degrees of freedom). *Source*: Travisano and Rainey (2000).

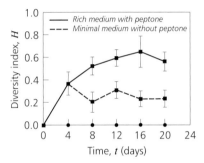

Figure 14.11 Effect of nutrient richness on the maintenance of morphological diversity. Eight replicate populations founded from the ancestral smooth morph were allowed to evolve in nutrient-rich medium. After four days, each microcosm was used to found two fresh sets of eight replicate cultures; one set was incubated in a rich medium with peptone (continuous line) and the other in minimal medium without peptone (dashed line). No diversity evolved in cultures propagated entirely in minimal medium (filled circles). Diversity calculated by the Shannon–Weaver index H. Values shown are means and 95% confidence intervals based on a t-distribution with seven degrees of freedom. *Source*: Travisano and Rainey (2000), and unpublished data.

However, diversity persisted once present. Static cultures that contained rich medium (peptone and glycerol) became diverse after four days of static culture. Cultures maintained under these conditions remained highly diverse. Cultures switched to a minimal medium (glycerol alone) lost diversity, but not completely (Figure 14.11). Subsequent pairwise fitness assays between WS and SM morphs (FS morphs were rare and not studied in this experiment) were performed as described previously, each competitor was added to the appropriate media, with one morph initially common and the other initially rare. Although the morphs occupied different portions of the habitat, the overall population size of the culture

Table 14.1 Competitive relationships of niche specialists in minimal and rich media.

		Common morph			
		Minimal medium[a]		Rich medium[b]	
		SM	WS	SM	WS
Rare	SM	–	1.501 ± 0.105	–	1.564 ± 0.367
Morph	WS	1.074 ± 0.098	–	2.860 ± 0.567	–

Values indicate mean relative fitness and 95% confidence intervals of the rare morph when it invades a population of the common morph at an initial ration of about 1:100. A fitness larger than 1 indicates the ability of the rare morph to invade.
[a] Confidence intervals based on a t-distribution with 23 degrees of freedom.
[b] Confidence intervals based on a t-distribution with 11 degrees of freedom.

was limited so that competition assays were a valid procedure to determine frequency dependence. Frequency-dependent selection was observed to be important in maintaining diversity in both rich and minimal media (Table 14.1), although the equilibrium frequency of WS morphs in minimal medium was about 100-fold less than that in rich medium. The reason for the apparent inability of WS to evolve in the absence of peptone is not known, but there are at least three possibilities:

■ Evolution to the WS morph may involve multiple mutational steps, some of which are deleterious in minimal medium.
■ A minimum number of WS individuals may be required to attach to the culture vessel at the broth surface.
■ An ecologically equivalent genotype may evolve, but does not have an easily differentiable phenotype.

For all three possibilities, initial selection in a rich medium alters the subsequent evolution of the system, but the mechanism differs: genetic, ecological interaction, or multiple adaptive phenotypes.

14.4   Concluding Comments

Adaptation and divergence are essential processes in the evolution of biological diversity. The importance of adaptation and divergence in the generation of biodiversity has long been recognized and at its most basic level predates Darwin. Even so, elucidation of the mechanisms that underlie adaptive speciation is far from complete. A number of substantial successes have been made in understanding diversification. The ability to quantify diversity at the protein and molecular genetic levels has enabled not only quantification of the great amount of genetic diversity in many populations (Lewontin 1974; Avise 1994), but also it has brought vast improvements in inferring the history of diversity (Fitch and Margoliash 1967; Kaplan *et al.* 1991; Kuhner and Felsenstein 1994). Similarly, numerous adaptations have been delineated in many organisms at the phenotypic (Endler 1986), physiological (Huey and Bennett 1987), biochemical, and molecular genetic (Shyue *et al.* 1995) levels. Nevertheless, given this wealth of information it is surprising that the quantitative effects of the specific ecological factors (e.g., competition, facilitation) that cause the evolution of biological diversity remain unknown.

The primary difficulty in assessing the causes of biological diversity has been the inability to observe its evolution directly. Knowledge of prior ecological conditions and the previous genetic structure of populations is therefore imprecise. Such details are necessary to test the theory of adaptive dynamics, which focuses on quantitative variation in ecologically relevant traits. Adaptive speciation, the branching of one morphospecies into two under conditions of at least minimal ecological contact (see Section 14.1), is particularly difficult to test as it requires the specific knowledge of the selective conditions over evolutionary time.

Microbial selection experiments are one approach to test adaptive dynamics theory and adaptive speciation. The ability to observe evolution in action over many generations allows the direct observation of the conditions under which diversity evolves. Moreover, the ability to maintain nonevolving genotypes in suspended animation and also to alter environmental conditions permits the direct experimental investigation of the causes of biological diversity. The studies described herein address the two most important longstanding topics in evolution, adaptation and divergence, and provide an experimental perspective that would otherwise be difficult or impossible to obtain.

Diversity among isolated allopatric populations can evolve rapidly for ecologically relevant traits, without differences in environmental conditions or preexisting genetic variation among or within populations. This rapid evolution indicates that caution must be exercised in assigning adaptive explanations to differences among populations or species. Caution must also be exercised in assessing the role of prior phylogenetic differences on subsequent diversity. The diversity of selectively important traits is more likely to be affected by adaptation than historical contingency or chance. However, many paleontologically important traits may not be under such strong selection and so their diversity may be subject to the effects of phylogenetic history and chance. The effects of chance are likely to be idiosyncratic, depending upon both the genotype of the organism and the selective conditions. These conclusions, based upon results with allopatric populations, demonstrate that evolutionary branching is not the only possible mechanism for the evolution of ecologically relevant diversity.

Support for adaptive speciation and evolutionary branching has been seen in studies that demonstrate sympatric speciation. A single ancestral bacterial genotype can give rise to multiple niche specialists. Diversity evolves in the presence of multiple exogenously supplied resources, but partitioning of a single resource can occur as well. Spatial heterogeneity can also provide ecological opportunity and thus lead to niche specialization, although spatial heterogeneity is not sufficient alone to generate diversity. In all cases, the maintenance of diversity is frequency dependent.

Several avenues for future research with microbes to examine the causes of biological diversity are potentially rewarding. The most obvious direction lies in explicit tests of adaptive speciation theory, none of which have been performed to date, since the theory was developed after the work described herein had been carried out. The theory makes specific predictions on the dynamics of phenotypic change that could be tested. Additionally, the potential for branching can be examined by using multiple initial genotypes that differ in resource use. A second possible avenue for future research is in pre-existing species interactions. The majority of selection studies with microbes that have been performed to date focused on the adaptation and/or divergence of a single genotype. While there are exceptions (e.g., Morin 1999), the role of species interactions remains relatively unexamined. One difficulty in multispecies studies will be the increasing complexity of the experimental system. For example, to determine assembly rules in a ten-member community is difficult even under controlled laboratory conditions. However, such experiments can be performed, and thus enable the build-up of complexity that is a hallmark of natural systems.

Acknowledgments   I thank R. Lenski and P. Rainey for collaborations and stimulating discussions on the issues raised.

Part C
Patterns of Speciation

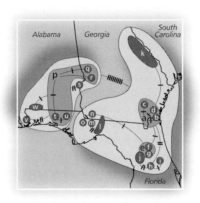

Introduction to Part C

Part A presents the theory of adaptive speciation and Part B describes natural systems for which speciation mechanisms can be assessed. Part C goes one step further and looks at the distribution patterns of populations and species that have the potential to shed light on the underlying mechanisms. At first sight it may seem counterintuitive to examine phylogeographic patterns as evidence for adaptive speciation under sympatric conditions, since such patterns are usually taken to prove the prevalence of allopatric speciation mechanisms. However, when interpreting observed distribution patterns it is necessary to account for the dynamic nature of the speciation process: as shown in Chapter 7, adaptive speciation along an environmental gradient can lead to species abutment. It is also critical to pay attention to the appropriate time scales: most phylogeographic studies deal with time horizons of millions of generations after the initial splits, while the theory of adaptive speciation suggests that the split of populations may occur within thousands of generations. The ecological and spatial conditions under which the split actually occurred are very likely to be eradicated after such long divergence times.

There is a further argument for the study of distribution patterns. Although speciation can occur relatively fast, it is still too slow to be replicated easily in the laboratory, at least for sexually reproducing populations. Similarly, experimental interference in natural settings is problematic, since such action is likely to alter the prevailing ecological conditions and thus disturb those processes that drive the split. This can only be avoided by studying natural experiments of recent speciation that started some time ago and are now waiting to be analyzed by comparing model predictions with observed patterns. While phylogenetic history and time frames of population divisions can be studied by molecular techniques, trends in morphology and population distribution have to be assessed through paleontological analysis.

Part C features four chapters that deal with different aspects of phylogeographic patterns and their analysis. The first chapter highlights the connection between speciation pattern and process, with particular emphasis on the relevant time frames (Chapter 15); the other three chapters focus, respectively, on examples in animals, plants, and paleontology (Chapters 16 to 18). Example systems were chosen so as to discuss the relevant patterns in conjunction with insights into ecological settings. It must be stressed, however, that research on these systems has, so far, not been carried out to test adaptive speciation theory. We therefore much appreciate that all these authors discuss their findings in the light of adaptive speciation theory, although clear distinctions between alternative speciation scenarios cannot be drawn until more detailed data have been collected.

Chapter 15 by Tautz discusses a framework in which to apply molecular techniques to the study of phylogeographic patterns. A four-phase model is suggested to characterize the processes of population subdivision that result from adaptive speciation; each such phase is defined by a combination of morphological and molecular characteristics. This analysis highlights that differential predictions for different speciation scenarios arise for the earliest phases of subdivision, which should therefore become a preferred target of analysis in future studies. Distinctions between the four phases are illustrated for specific natural examples and a point is made that even the phylogeographic patterns found for late phases of population subdivision may sometimes be explained more easily by adaptive speciation.

An excellent example of differential adaptation and speciation are the Anolis lizards on Caribbean islands. Chapter 16 by Thorpe and Losos discusses these systems. Between them, Thorpe, Malhotra, Stenson, and Reardon describe the situations found on the small islands of the Lesser Antilles. Few within-island speciation events have occurred, but often very different ecotypes of the same species have adapted to different habitats. Translocation experiments and common-garden experiments show that these ecotypes are determined genetically and do not result from phenotypic plasticity. Some of the adaptations correlate with molecular differentiation, while ongoing gene flow is found for other pairs of types. Losos examines the situation on the larger islands of the Greater Antilles, which is characterized by evidence for many within-island speciation events. Intriguingly, highly visible signals for assortative mating have evolved in these lizard species (which include differently colored dewlaps, as well as display behaviors) that are not found on the Lesser Antilles. A joint conclusion section reflects ongoing discussions by the experts. While some discrepancies in interpretation may result from different approaches and criteria, it is clear that the Anolis system lends itself to tests of adaptive speciation scenarios that will have to be carried out in the future.

Stunning patterns of convergent adaptation and speciation in plants have been observed for giant senecios and lobelias on African mountains; these are described by Knox in Chapter 17. High mountains scattered across Central and East Africa have been compared to "islands in the sky". Molecular and morphological analysis of the colonization and speciation history has revealed that more than half of the speciation events occurred on individual mountains. More than a dozen species and subspecies are known for each taxon – these must have evolved within only one million years in the senecios and within only a few million years in the lobelias, which suggests that differential adaptation and speciation can be fast, even for perennial plants. Different species of the same taxon abut each other in altitudinal steps, a spatial pattern reminiscent of that described in Chapter 7.

The application of paleontological methods to infer the history of evolutionary diversification is described by McCune in Chapter 18. Straddling the Triassic–Jurassic boundary, the Newark rift lakes in eastern North America underwent repeated cycles of dessication and refilling over a period of 20 000 years. This resulted in a fossil record with an exquisite time resolution down to single years and provides replicas of adaptive radiation in Semionotid fish. Semionotids were encased by heavy scales that fossilized well, and the record reveals a large variety of body shapes, suggestive of ecological diversification. Studies of the lake with the best fossil record showed that the radiation was extremely fast initially, comparable to that of Lake Victoria cichlids, and slowed down substantially later on. For the initial radiation phase, some fairly wild variations in the dorsal ridge scales have also been documented. Both patterns are best explained by a scenario of adaptive sympatric speciation.

Chapters in this part look at adaptive speciation with different experimental and conceptual approaches and emphasize different and sometimes conflicting points of view. It becomes clear that in future studies data acquisition and analysis must be geared toward testing alternative speciation scenarios, much more than is possible with the data currently available. Analyses of the phylogeographic patterns shaped by natural experiments promise a high potential in the study of dynamic processes and ecological mechanisms that lead to population subdivision.

15

Phylogeography and Patterns of Incipient Speciation

Diethard Tautz

15.1 Introduction

The term "phylogeography" was introduced by Avise (Avise *et al.* 1987) to refer to the principles and processes that govern the geographic distributions of genealogical lineages, including those at the intraspecific level (Avise 1994). Since then, the use of DNA markers to study phylogeography has become very popular, with an increasing flood of studies on a diverse range of taxa. The data from these studies are interpreted traditionally within the framework of limited dispersal (isolation by distance) or past vicariance events (isolation by geographic barriers), considered against the geographic and geologic history of the area under study. The resultant patterns are then discussed in the context of passive divergence and are taken as evidence for allopatric speciation scenarios. However, the argument presented here is that at least some of these patterns could also be seen in a different light and might provide evidence for sympatric modes of speciation.

The crucial point in this context is that the consequences of assortative mating should be considered for the generation of patchy distribution patterns, as well as for the maintenance of borders between the patches. The evolution of assortative mating is generally an integral part of sympatric speciation models, both for the ecologically driven ones (Rice 1987; Doebeli 1996a; Johnson *et al.* 1996b; Dieckmann and Doebeli 1999; Kondrashov and Kondrashov 1999) and for those based on sexual selection (Turner and Burrows 1995; Higashi *et al.* 1999). In general, assortative mating amounts to an avoidance of hybridization under contact conditions. Thus, if phylogeographic patterns of very closely related populations are found in direct contact, a previously evolved assortative mating behavior would explain why they have not become genetically mixed, even if geologic or geographic conditions would have allowed this. Evidently, it still has to be explained why they have become geographically separated as well. A possible explanation for this would be that assortativeness in itself leads to spatial segregation (Box 15.1).

The general aim of this chapter is to discuss data sets from phylogeographic settings that suggest that the spatial separation of populations is maintained actively via mechanisms that could originally have evolved in sympatry, rather than being the passive by-product of old differential adaptations in allopatry. However, to evaluate such patterns properly, it is necessary first to discuss the time frames within which different effects can be observed. There are some differential predictions in the adaptive speciation models that contrast with those in allopatric

Box 15.1 Assortative mating and spatial coexistence

Ulf Dieckmann

Coexistence in ecological communities is limited by the diversity of underlying resources and by the similarity of the species involved (Gause 1934; Hardin 1960; Hutchinson 1961; MacArthur and Levins 1967; MacArthur 1970; Armstrong and McGehee 1980; Yodzis 1989; Grover 1997). Such constraints restrict the *number* and *similarity* of incipient species that originate from adaptive radiation. Temporal fluctuations (e.g., Huisman and Weissing 1999; see also Lundberg *et al.* 1999 and references therein) and spatial structure (e.g., Pacala and Tilman 1994; Law and Dieckmann 2000) have been shown to allow a greater number of species to coexist.

Here we illustrate that spatial structure, in conjunction with mate recognition and local assortative mating, can also overcome limiting similarity; this allows ecologically equivalent species to coexist, albeit in spatial segregation. For this book and chapter there are two important implications:

■ Speciation by sexual selection without ecological differentiation of the resultant species can lead to their long-term coexistence in segregated ranges. Flocks of ecological sister species are thus not necessarily ephemeral.

■ When the spatial segregation of populations is promoted by processes associated with assortative mating, phase-1 and phase-2 species pairs, as described in Section 15.2, may already display distribution patterns that could be mistaken as evidence for their allopatric origin (Section 7.3).

To underpin these claims, we introduce a simple reaction–diffusion model to describe the dynamics of two ecological sister species along a one-dimensional spatial coordinate z (varying between 0 and 1, with reflective boundary conditions). Density regulation is of logistic type and acts on mortality, with carrying capacity density $K(z) = 1$. Birth rates are $b_1(z)$, $b_2(z)$; death rates at carrying capacity are $d_1(z)$, $d_2(z)$; and the coefficients of diffusive movement are $m_1(z)$, $m_2(z)$. The dynamics of local densities $n_1(z)$ and $n_2(z)$ are then described by

$$\frac{\partial}{\partial t} n_i(z) = b_i(z) n_i(z) - d_i(z) \frac{n_1(z) + n_2(z)}{K(z)} n_i(z) + \frac{\partial}{\partial z}\left[m_i(z) \frac{\partial}{\partial z} n_i(z) \right] , \qquad \text{(a)}$$

for $i = 1, 2$. The last term on the right-hand side is the standard expression for describing diffusion with spatially varying coefficients (Okubo 1980). No environmental heterogeneity is considered. In particular, we do not allow for the two species being differentially adapted to specific locations of their habitat (Lande 1982; Payne and Krakauer 1997), since this would render them ecologically nonequivalent. Importantly, the two sister species are assumed to be isolated reproductively by mating fully assortatively.

Assortative mating usually depends on differential behavior towards conspecifics and heterospecifics. To highlight the resultant effects, we consider four instances of this model:

continued

Box 15.1 *continued*

▦ *No coexistence with density-independent vital rates.* When all vital rates are independent of local densities, varying the relative abundance of the two sister species has no effect. The demographic and environmental stochasticity (not considered in the deterministic model above) results in one of them being ousted.

▦ *Coexistence with fecundity depending on the density of conspecifics.* If no suitable mates are locally available, the fecundity of a sexual species is bound to decrease, which causes an Allee effect. This stabilizes spatial segregation and coexistence, as shown in panel (a) below, in which continuous curves depict $n_1(z)$ and dashed curves $n_2(z)$ at different moments in time. Parameters: $d_i = 1$, $m_i = 0.001$, and $b_i(z) = \sqrt{n_i(z)}$.

▦ *Coexistence with mortality depending on the density of heterospecifics.* If interspecific interference is particularly vigorous (e.g., through territorial fighting), a high density of heterospecifics can induce extra mortality. With suitable initial conditions, this mechanism leads to spatial segregation and stable coexistence, as illustrated in panel (b) below. Notice that the dynamics of the two ecological sister species in isolation remain identical, and that the differential interference behavior considered here occurs only in response to differentiated mating phenotypes, retaining the species' ecological equivalence. Parameters: $b_i = 1$, $m_i = 0.001$, and $d_i(z) = 1 + n_{\bar{i}}(z)$ with $\tilde{1} = 2$, $\tilde{2} = 1$.

▦ *Coexistence with mobility depending on the density of heterospecifics.* If a high density of heterospecifics induces increased mobility (a simple strategy for returning to the conspecific range and thus for finding mates), the heterospecific range is escaped rapidly. Panel (c) below shows the resultant pattern of segregated coexistence. Parameters: $b_i = 1$, $d_i = 1$, and $m_i(z) = 0.001 + 0.01/\{1 + \exp(50[0.9 - n_{\bar{i}}(z)])\}$.

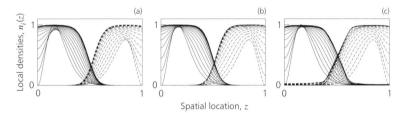

In nature, the three coexistence-enabling mechanisms described above are expected often to act simultaneously. For illustrative purposes, we have assumed reproductive isolation between the ecological sister species. More intriguing questions arise as to how the underlying assortativeness occurs evolutionarily and how the spatial dynamics of incipient species behave under intermediate degrees of assortativeness. These issues are touched on in Chapter 7, but in general still await systematic analysis.

models. These can be analyzed in natural settings and could serve as differentiating criteria to assess whether a sympatric or an allopatric model is more likely to apply. However, to do this, it is usually necessary to carry out a refined marker analysis on natural populations, which is not yet available for most cases. In the following, first the molecular markers that can be used to obtain such a high-resolution analysis of population structure are discussed and then a four-phase model for the patterns to be expected after the initial split of populations is developed. This serves to discuss the differential predictions for the different speciation scenarios.

15.2   Molecules, Morphology, and Time Frames

Most of the molecular markers employed in population genetic studies (Box 15.2) are expected to evolve neutrally and should therefore only be subject to drift mechanisms. This implies that their evolution is determined only by the primary mutation rate, which can be considered to be constant over time and for which reasonable estimates exist for the different markers (Box 15.2). Thus, they can be used to estimate the time frames for population subdivisions.

The situation is different for morphological markers. Most of these are very unlikely to be neutral and are therefore subject to positive or negative selection. Thus, their evolutionary rate depends on selection coefficients that can vary widely, contingent on the ecological setting and its dynamic change. This explains why morphology and molecules normally do not evolve in parallel. In other words, neutral molecular change can occur without any morphological change and, conversely, morphological change can occur in the virtual absence of molecular change.

The differential behavior of these two types of markers can be used to make specific predictions about the patterns that are to be expected after an initial split of populations. The concept of adaptive speciation implies that population subdivisions arise on the basis of selection, rather than neutral drift. Thus, even with small selection coefficients, they are likely to occur within relatively short time frames, probably measuring thousands of generations at most (Doebeli 1996a; Dieckmann and Doebeli 1999; Section 5.2). On this basis, one can propose the following generalized phases of further differentiation after the initial splitting has occurred under sympatric conditions (Figure 15.1).

■ *Phase 1.* This phase is characterized by a response to disruptive selection on traits that allow for the utilization of alternative niches. The divergence that results in these traits is coupled to an increasing degree of assortative mating. In a natural population, two differently adapted types that mate assortatively would be expected to emerge. At this early stage, the expectation is also that most of the polymorphic alleles between the populations would still be shared, because the populations would not have experienced a significant bottleneck, and gene flow may still occur, at least with respect to genes and chromosomal regions that are not involved directly in the differential adaptation. This phase may last less than 100 generations and may thus be detectable only in special circumstances.

Box 15.2 Molecular markers used in population studies

Allozymes. Allozymes are enzymes from the basic metabolic pathway that can be separated on gels and visualized with enzyme-specific staining reactions within the gel. This makes their analysis convenient and many samples can be screened at relatively low costs. Alleles differ through amino-acid substitutions, which change the separation behavior within the gels. It is often assumed that these mutations are neutral, as long as they do not affect the activity of the enzyme. However, one cannot be certain about this, and there are examples for positive selection on particular alleles. Mutation rates are relatively low, compared with those found for other neutral markers.

Mitochondrial DNA. The mitochondrial genome replicates separately from the nuclear genome and is inherited maternally. Thus, no recombination occurs between different alleles (which are called haplotypes), so mitochondrial DNA is particularly useful to trace old lineages. Usually, variations are analyzed by first amplifying subregions of the genome by polymerase chain reaction (PCR) and subsequently sequencing these pieces. The most variable region in the mitochondrial genome is that involved in controlling DNA replication, the so-called "control region" or "D-loop" (this name reflects that unidirectional replication starts in this region and leads to a displacement of the nonreplicated strand, which results in a D-like appearance in the electron microscope). Many positions within the D-loop are considered to evolve neutrally, because they do not code for a protein. Mutation rates range from 0.5% to 5% per million years, depending on the length and structure of the region.

Microsatellites. Microsatellites consist of repetitions of very short nucleotide motifs. Motif length may lie between one and six base pairs, although di- and trinucleotide motifs are those that are most frequently employed in nonhuman studies. The motifs are repeated dozens of times, and different alleles differ with respect to the number of repeats. The individual loci can be amplified by PCR with primers that flank the repeat regions, and the allele lengths can be scored directly on high-resolution gels. Since microsatellites occur mostly in noncoding regions, they are considered to evolve in a neutral fashion, although it cannot be ruled out that occasional loci might respond to positive or negative selection. Mutation rates at microsatellite loci can vary considerably, depending on the motif and total repeat lengths. Loci with many repeats tend to be more mutable than those with few. Mutation rates range between 10^{-4} to 10^{-5} per generation and are thus higher by a factor of at least 10^4 than point-mutation rates in the nuclear genome.

▓ *Phase 2.* Within 100 to 1000 generations, the morphotypes and the assortative mating tendency should become more pronounced and reach the final state. The resultant strong reduction in gene flow means that the neutral alleles in the two subgroups increasingly become subject to independent drift, which produces different relative frequencies of the alleles. This phase can be inferred reliably by study of the degree of genetic subdivision with highly variable molecular markers, such as microsatellites. A number of suitable test statistics enable the

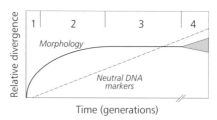

Figure 15.1 Relative divergence of morphological and molecular markers, after an initial split of a population according to the adaptive speciation scenario. Morphological change (continuous curve) is expected to be initially fast, until a new stable state has been attained. This state may then be retained for some time, until further changes are possible (symbolized by the gray triangle). By contrast, molecular change is expected to occur in a clock-like manner (i.e., in linear dependence on the time since the split; dashed line). The numbers across the top refer to the four phases discussed in the text.

significance of allele frequency differences and of degrees of genetic subdivision to be assessed.

▪ *Phase 3.* Within 1000 to 10 000 generations, no further change with respect to morphotype differentiation or assortative mating pattern is expected. However, by now significant molecular differences will have built up. While only allele-frequency changes occur in phase 2, phase 3 is characterized by fixation and lineage sorting (see Box 15.3) of neutral alleles. Furthermore, new mutations will be found that are a single mutational step away from pre-existing alleles and that can be used as diagnostic markers. Molecular phylogenetic reconstruction methods become applicable at this stage.

▪ *Phase 4.* After still longer separation times, of up to millions of generations after the initial split, prediction of the further evolution of the involved adaptive characters is no longer possible. Additional adaptations might occur, but relative stasis with respect to the initial adaptations is also possible. However, there will now be a clear molecular distinction with respect to allele types and frequencies. Many population-specific alleles will have evolved, and differ by multiple mutational steps from alleles that existed previously. The accumulation of many mutations may also have led to postzygotic isolation, and species status will be generally acknowledged.

These four phases, in their strict sense, can only be used to describe the adaptive speciation process. However, they may also be more generally useful as guidelines for the classification of natural populations, to ensure comparability between different studies. The status of molecular-marker differentiation, in particular, is an objectively measurable parameter that can help to assess the time scales within which a given population subdivision has taken place.

Phase 1 would inevitably have to occur under sympatric conditions, since the disruptive selection that is envisaged here would be caused by intraspecific competition among the optimally adapted genotypes. At first glance, there appears to be no obvious reason why sympatry should not also be retained throughout the other phases. However, there is a second component in the adaptive speciation concept that could lead to a more dynamic behavior with respect to spatial segregation of the incipient species, namely assortative mating. The issue of spatial segregation under the adaptive-speciation mode is discussed in Chapter 7, where it is shown that an environmental gradient not only facilitates the splitting of lineages, but can also entail spatial patterns of species abutment. In the following, whether such spatial sorting could also be caused directly by assortative mating behavior, even in the absence of ecological gradients, is discussed.

15.3 Assortative Mating and Patterns of Subdivision

It has long been recognized that sympatric speciation in sexual populations can only work in conjunction with assortative mating. The evolution of alleles to enable choice of mates is usually built into the models. In the older models this required direct linkage with the alleles that conferred ecological advantage, or high selection coefficients to maintain an association of nonlinked alleles. In the model of Dieckmann and Doebeli (1999; Section 5.2), however, it was shown that a linkage disequilibrium can be reached under realistic conditions within the framework of the adaptive-speciation mode. In the sexual selection models of sympatric speciation, it is usually a preference of females for a male character that is modeled (Turner and Burrows 1995; Higashi *et al.* 1999). This also leads to assortative mating, although only with respect to phenotypic markers and preferences.

Thus, the parameters that lead to assortative mating are relatively general, and therefore the expectation is that abundant evidence should be found for this phenomenon. This is, indeed, the case – assortative mating is a rather general phenomenon in natural populations and has been studied in many contexts (Bridle and Ritchie 2001). However, the dynamic long-term consequences of assortative mating for the structure of populations are usually not further considered. Still, it is clear that adhesion properties among independent units should lead to spatial segregation. This is well known in a different context, cell biology. Individual cells can carry different types of adhesion molecules at their surfaces that ensure that cells in a given tissue retain their contacts among each other. Many of these adhesion molecules show homophilic association (i.e., they connect to cells that carry the same type of adhesion molecule, but not to cells that carry different types). If cells that harbor two different types of adhesion molecules on their surface are mixed, active spatial segregation of these cells is observed, which eventually results in the formation of different clusters (Steinberg 1996).

The question is, of course, whether such a simple analogy can be applied to whole organisms. However, a well-established ecological principle can be used to arrive at a similar analogy at the level of organisms. It is known that two species that utilize one and the same ecological resource cannot coexist in the same area (Abrams 1983). If we take a purely sexual selection model for sympatric speciation, this would produce two noninterbreeding populations that still have the same ecological requirements. This would be an unstable situation and should result in spatial segregation into two nonoverlapping areas. Thus, at least for the purely

Box 15.3 Lineage sorting and population splitting

Population subdivisions and restrictions in gene flow between populations are usually traced on the basis of neutral markers (see Box 15.2). However, at the time of subdivision and shortly afterward, the resultant subpopulations still harbor essentially the same set of neutral alleles. Allelic differentiation occurs only after genetic drift processes start to become effective. Depending on the size of the evolving population, it may take a substantial time before fixed differences between the two subpopulations appear among the sampled neutral loci. For maternally inherited mitochondrial lineages, it has been estimated (Neigel and Avise 1986) that it takes on average $4N_e$ generations until the fixation of lineages in each subpopulation is essentially complete; N_e is the effective size of the considered subpopulation. For alleles of nuclear-encoded loci in sexually reproducing species, such fixation takes four times longer, because of diploidy and biparental inheritance (reviewed in Edwards and Beerli 2000). If the subpopulations are sampled before fixation is complete, they will be found to share alleles or lineages that arose long before the split of the population occurred. This effect is called incomplete lineage sorting and applies within the time window that lasts from the restriction of gene flow until the fixation of differences in neutral markers. Incomplete lineage sorting has important implications for the interpretation of gene flow and phylogenetic patterns during and after speciation events:

- The main problem is that it is very difficult to distinguish between those alleles that are shared because of recent common ancestry and those that are exchanged between populations through ongoing gene flow. For example, even if the true amount of gene flow drops to zero shortly after a subdivision event, standard measures of gene flow still (erroneously) indicate substantial genetic exchange, because they only measure the degree of allelic overlap. Practically, very recent splits can therefore only be inferred by testing the null hypothesis that the two considered subpopulations are random samples from a single well-mixed gene pool (e.g., see Hendry *et al.* 2000; Schliewen *et al.* 2001).
- In the context of phylogeny reconstruction, it has to be kept in mind that the splitting of alleles and lineages can predate the splitting of the corresponding populations. Hence, phylogenetic trees obtained from the analysis of sequence data are not necessarily in accordance with the true history of species splitting. A reliable trace of true phylogenies based on sequence data is only possible after lineage sorting is complete.

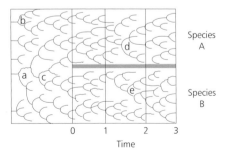

continued

Box 15.3 *continued*

The preceding figure (redrawn after Avise 1994) schematically shows the process of mitochondrial lineage sorting. At time 0, a population subdivision occurs such that gene flow between the ensuing species is interrupted (indicated by the gray bar). If these species are sampled at time 1, species A would be found to be of polyphyletic origin, since it harbors allelic lineages derived from node *a* as well as lineages derived from node *b*. If a sample is taken at at time 2, species A would be found to be monophyletic, because all lineages are derived from node *d*. Species B, on the other hand, at that time still harbors two separate lineages from node *a*, which predate both the species split at time 0 and the last common node with species A at *c*. Phylogenetic reconstruction would then suggest that species B is diphyletic (composed of two lineages). Monophyly for both species would only be inferred at time 3: at that time only the alleles derived from nodes *d* and *e* are left, which originated after the splitting event.

---

sexually driven models of sympatric speciation, subsequent allopatric segregation seems to be a logical consequence.

In models of sympatric adaptive speciation, however, ecological diversification occurs concomitantly with the emergence of assortativeness, which thus directly avoids the problem of ecologically equivalent species not being capable of long-term coexistence in the same area. A more complex scenario is therefore required to link this model with processes of spatial segregation. Several factors could at least promote such processes (Box 15.1). One is the simple fact that an individual looking for a matching mate has a higher chance of finding one if it is surrounded by individuals of its own type. Moreover, the recognition mechanisms involved in assortative mating may also be important for social interactions in general, which could further enhance adhesion properties (Kappeler *et al.* 2002). Finally, territories will show slight differences in their ecological qualities, which could favor their being inhabited by one of the newly formed species rather than by the other (Day 2000).

Although spatial segregation cannot be claimed to be an inevitable consequence of adaptive-speciation processes, such segregation seems at least to be a likely outcome. Therefore, some consequences expected from this can be considered:

- First, a relative cohesion of subpopulations and a displacement of the sister population from which it has separated should be seen.
- Accordingly, the most closely related subpopulations are expected initially to neighbor each other in sharp abutment without much overlap.
- Furthermore, such cohesion is expected to be maintained, even when climatic changes force populations away from their original habitat. Such situations occur typically during ice-age cycles. It is often assumed that populations that are driven into refugia will start to hybridize there. However, if they have already evolved mechanisms to attain an active cohesion, then it is expected that they could maintain their identity even in refugia, and in contact with closely related subpopulations.

The best test for the third prediction would be to analyze populations that are molecularly very closely related and that have been exposed to strong ecological disturbances. If these have maintained a populational cohesion in spite of the disturbance, this is a strong argument that an active isolation mechanism must exist to keep them apart. This active isolation mechanism can best be explained to have emerged under sympatry, since differential recognition systems can only be built up under conditions of contact.

To better understand the forces that act during the splitting phase of a sympatric population, it would evidently be most interesting to study phase 1 populations under natural conditions, as this would provide the opportunity to observe directly the dynamics of a splitting process. However, neutral markers would not be of much help in these situations, since significant differentiation cannot yet be expected (but see below for an exception). Only genes or genomic regions that are involved directly in causing the differential adaptation would show major changes. However, so far we know almost nothing about the genes that cause such adaptations, but it seems clear that they will become a major focus for research in the future (Tautz and Schmid 1998).

Adaptive speciation may therefore currently best be studied in phase 2 populations, in which molecular markers can indicate successful splitting, while the prevailing ecological conditions may be the same as at the time at which the splitting began. Measurement of the differential selection coefficients that caused the splits may thus still be possible, and at least some components of the model predictions may be verified. In particular, it should be possible in phase 2 populations to decide whether a sexually or ecologically driven model is more likely to apply, since only the latter predicts differential ecological adaptations in this phase.

To study allopatric scenarios, the focus would naturally be on phase 3 or phase 4 populations, since a sufficient amount of mutations would need to have accumulated to explain the separation. In fact, the vast majority of studies on speciation mechanisms have so far been done on phase 3 or even phase 4 situations, partly because sensitive markers that identify phase 2 populations have only become available in the past decade. To identify the factors that led to the initial split will evidently become more complicated the longer the period of time since the split occurred. Still, the allopatric paradigm remains prevalent and the fact that the majority of very closely related species or subspecies are geographically separated appears to support this notion. However, if the possibility that geographic subdivision patterns could be a secondary consequence of sympatric splitting is considered, this dynamic component of the process could easily be missed by focusing only on phase 3 or phase 4 populations.

15.4   Natural Populations

In the following, examples of natural populations that may be assigned to one of the four phases described above are discussed. Special attention is given to examples that appear specifically to support the adaptive-speciation scenario, as well as the secondary consequences of assortative mating discussed above.

Phase 1 populations

In phase 1, morphological differentiation should be found in the virtual absence of genetic differentiation. Such situations would most likely occur where new habitats have been colonized very recently, such as newly formed lakes or islands. In our times, this would often involve human-made habitats that were invaded or stocked some decades ago. Hendry *et al.* (2000) have studied such a case, in which salmon were introduced into Lake Washington in the 1940s. They found that, today, the salmon populations are clearly split into two different morphs that exploit different ecological niches. Intriguingly, using microsatellite markers, already a significant reduction of gene flow is found between these morphs, although only 13 generations have passed since the introduction. Although this is a telling example, it is probably on the extreme side. The introduced salmon originated from lakes that harbored similarly differentiated forms, although the forms were artificially interbred before release, which should have resulted in a complete mixing of the alleles. The fact that significant genetic differentiation of neutral markers can be found after such a short time may arise from highly unequal effective population sizes of the two forms (Hendry, personal communication).

Another potential phase 1 situation is the case of the fruit fly *Rhagoletis* described in detail in Chapter 11. Here one finds two distinct races that have specialized on hawthorn or apple trees. The splitting started only about 150 years ago, with the introduction of the apple trees into the area. Analysis of allozymes shows distinct allelic differences at some loci, although these might in part have been caused by differential selection. Still, mark–recapture experiments suggest the possibility of substantial gene flow between the races (Feder 1998), which would need to be counteracted by strong selection to maintain the racial distinction. Direct tests of selection effects show that this is indeed the case (Filchack *et al.* 2000).

The occurrence of different morphotypes in populations is a well-known phenomenon, and may arise from phenotypic plasticity in response to different ecological conditions, like different resource utilization (Meyer 1987) or environmental conditions (Chapter 16). This may be a reflection of competing life-history strategies within a population (Stearns 1992) in which different morphs exploit different parts of the resource spectrum of a given species. However, it would seem possible that such cases could alternatively be phase 1 situations, in which case whether there is assortative mating with respect to the different morphotypes would have to be investigated, as well as the initial differentiation of the gene pools.

Phase 2 populations

In phase 2, morphological differentiation, assortative mating in sympatry, and low, but significant, genetic differentiation are expected. Finding populations in such a situation should be the best indication for adaptive speciation, since this is much more difficult to interpret in any other model. Again, such situations are most likely to be found after recent colonization events, although these could be several hundred or thousand years ago, but well within the time frame of changes caused by modern civilization or by recolonizations after the last ice age. This suggests that the most interesting cases for the study of speciation might not be in remote tropical areas, but more or less on the doorstep of most researchers in this field (i.e., in Europe, as well as in North America). There are, of course, many well-studied populations in these areas, although the focus so far has almost always been on deeper splits (see the discussion of phase 3 populations below), but it should be relatively easy to enhance these studies in the future to find evidence for phase 2 situations.

Only a few explicit studies deal with phase 2 populations as described here. Two of these, namely the differentiation of salmonids in Icelandic lakes and the limnetic and benthic forms of sticklebacks in Canadian lakes, are discussed in detail in Chapters 10 and 9, respectively. Both studies deal with postglacial recolonization situations. In the case of the sticklebacks, it appears that independent colonization could explain the two forms. They would thus not have evolved in the respective lakes, although the maintenance of their differentiated status would be caused by ecological specialization and assortative mating. However, more detailed molecular-marker analysis is necessary to trace the origin and history of the different forms in the different lakes.

Cichlids in Cameroon. Another case that has been intensively studied, both for its morphological and ecological aspects, as well as on the basis of molecular markers, are cichlids in a small lake in Cameroon (Schliewen *et al.* 2001). This lake is only around 10 000 years old and very small (less than 0.5 km^2). Yet, it harbors at least five well-differentiated morphs that exploit different parts of the ecological resource spectrum. Schliewen *et al.* (2001) concentrated their studies on a particular pair of morphs that must undoubtedly have arisen within the lake and could not have been the product of independent colonization from the nearby river system. The ecological differentiation, however, is comparable to the stickleback case, with one form that specializes on the open water habitat, while the other mainly exploits the inshore resources. Still, as these fish are substrate breeders, both have to choose their mates and breed in the inshore area, next to each other. They show strong assortative mating, which can be studied easily in this case because the morphs differ significantly in body size. Interestingly, although some mixed mating pairs were found, microsatellite analysis shows that there is statistically significant genetic differentiation at neutral markers between the forms. Thus, effective gene flow is practically absent, which suggests that the offspring of the mixed pairs must have a strong ecological disadvantage, in line with the adaptive speciation model (Dieckmann and Doebeli 1999; Section 5.2). Thus, by all standard criteria of speciation research, these two morphs should be called good biological species: they show morphological differentiation, different ecological requirements, breed sympatrically, and there is effectively no gene flow between them. Still, they are extremely closely related (both morphologically and molecularly), and without the molecular analysis they would be called a perfect case of two life-history morphs.

Phase 3 populations

In phase 3, good morphological differentiation, assortative mating, and distinctive molecular differentiation should be found. Sympatry is not necessarily expected at this stage, at least if spatial sorting caused by assortative mating, as discussed above, has happened. In fact, spatial separation is found in the majority of phylogeographic studies available to date. It is also the situation in which parapatric and allopatric models can be applied equally well. Still, some patterns are more easily interpretable in a sympatric framework. For example, when the most closely related subpopulations are found in direct abutment, without any externally imposed geographic separation, or in cases of highly mobile animals that nonetheless maintain distinct populations in relatively close proximity. Particularly interesting are those situations in which continual genetic cohesion of subpopulations occurs, even under conditions of dramatic habitat shifts, as highlighted above. Two examples are discussed here in more detail. One concerns terrestrial salamanders and their postglacial colonization patterns in Europe. The other concerns the patterns of differentiation in brown bears on Hokkaido Island. Both of these examples testify to the strength of genetic cohesion of groups, even under conditions of drastic geographic displacements.

Terrestrial salamanders in Europe. Salamanders have long been a focus of interest for naturalists and biologists. Many subspecies morphotypes and color variants have been described, both for the genus *Salamandra* in Europe (Plate 8) and for the ecologically equivalent genus *Ambystoma* in North America. The great variety of forms means that both genera were called "polytypic species". However, re-examination of morphological characters, as well as molecular analyses, showed that *Salamandra* consists of at least six well-separated species, which must have split millions of years ago (Steinfartz *et al.* 2000). We focused our attention on the most widespread species in Europe, *S. salamandra* (Steinfartz *et al.* 2000). This species has been further subdivided on morphological grounds into at least 14 subspecies, most of which can also be differentiated at the molecular level. All of these live allopatrically, which would normally suggest that they are an excellent example of allopatric or parapatric differentiation, with or without concomitant speciation. Moreover, since salamanders are slow-moving animals and show strong site philopatry during their lifetime, a parapatric scenario would seem particularly likely at first sight.

However, the phylogeographic analysis tells a very different story. Using extensive mitochondrial D-loop sequencing, three separable clades can be shown within *S. salamandra*, which must have differentiated shortly before, or during, the Pleistocene glaciation cycles. The current geographic distribution of these clades shows that they must have had very different histories (Figure 15.2). Animals of clade A are only found in Southern Spain, those of clade B are found in Northern Spain and Southern Italy, while animals of clade C occur from Spain to Greece across all of Middle Europe. The disjunct distribution of clade B animals suggests that these, too, must have occupied at least most of Middle Europe in a previous interglacial

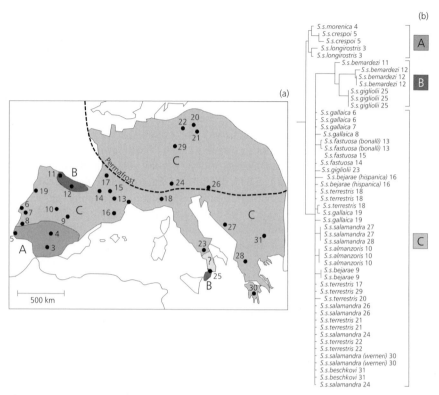

Figure 15.2 *S. salamandra* clades: (a) distribution and (b) phylogeny. The sampling sites are numbered and relate to the numbering in the phylogeny. The distribution ranges are primarily based on assessments of phenotypes and isoenzyme analysis. The phylogeny is based on mitochondrial D-loop sequences and is rooted with more basal *Salamandra* lineages (see Steinfartz *et al.* 2000 for the complete phylogeny).

period. However, they were later displaced by clade C animals, which have successfully recolonized Middle Europe in the previous and the current interglacial period. Intriguingly, at least in Northern Spain, clade B and clade C animals live next to each other under similar ecological conditions and without visible barriers, but separated by sharp borders (Alcobendas *et al.* 1996).

The distribution pattern of these clades suggests two conclusions. First, although individual salamanders are slow, they must be efficient colonizers and can spread into large territories, presumably by larval dispersal via streams and rivers. Thus, there is no *a priori* reason why clade B and C animals should not be able to invade each other's territory. Second, in spite of the ecological turmoil caused by the glaciation cycles, the clades have not become mixed in the refugia areas, which indicates that a strong isolation mechanism must exist between them to prevent such mixing.

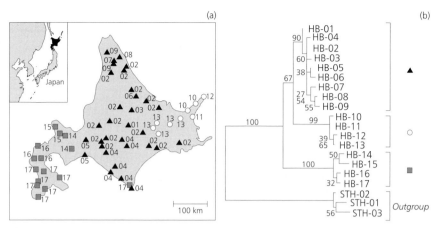

Figure 15.3 Mitochondrial lineages of bears on Hokkaido Island: (a) distribution and (b) phylogeny. Sampling sites are numbered and relate to the numbers in the phylogeny of the mitochondrial haplotypes. *Source*: Matsuhashi *et al.* (1999).

Brown bears on Hokkaido. Bears occur all across the northern hemisphere. They clearly constitute one of the most mobile terrestrial animal groups, with large home ranges, and are capable of long-range dispersal. Individuals can travel hundreds of kilometers easily within a short time. Still, they have split into several species and subspecies. Furthermore, mitochondrial analyses suggest further splits into distinct molecular lineages, which occur in defined, usually nonoverlapping, territories (Wooding and Ward 1997; Taberlet *et al.* 1998; Shields *et al.* 2000). An extreme case in this respect is Hokkaido Island, Japan (Matsuhashi *et al.* 1999). During the most recent interglacial period, this island was connected by a small land bridge to the mainland, and it is assumed that bears colonized Hokkaido during this time. Morphological analysis of skulls had already shown that at least two distinct morphotypes occur on the island. Sequencing of the mitochondrial control region has shown that there are three very distinct lineages, each restricted to a particular area of the island (Figure 15.3). These lineages are relatively old; thus, it is clear that they must have split originally on the continent. Still, that they have maintained their distinctness, despite the comparatively crowded condition on the island, suggests that strong isolation mechanisms are at work. Again, this evidence is only circumstantial, but even so, how is it possible in a parapatric or allopatric scenario that multiple differences build up on the continent, and yet several of the resultant lineages invade the island via a small land bridge and do not show signs of mixing. An explanation that invokes the active evolution of separation mechanisms during the splitting process seems a much more suitable explanation of such patterns than one that relies on the random occurrence of mutations that lead to prezygotic isolation, as would have to be the case for allopatric scenarios.

Still, at least for some populations it seems clear that mitochondrial patterns do not accurately reflect population subdivision, which indicates that caution is necessary when analyses are restricted to mitochondrial markers only (Paetkau *et al.*

1998; Waits *et al.* 2000). Nuclear markers have not yet been analyzed for the Hokkaido bears, but, given that the mitochondrial lineages correlate with morphological differences, it does not seem that much nuclear gene flow has occurred.

Phase 4 populations

It is expected that this phase will be characterized by very distinct molecular differences and often also by distinct morphology. Phase 4 is probably the stage at which a general agreement can be reached as to the species status of the considered types. However, it should be clear from the discussion above that to trace the mechanisms that lead to the initial split is most complicated in phase 4. If phase 4 populations live in allopatry for extended times, it is even possible that they have lost their ability to discriminate each other actively, even if such behavior initially evolved under sympatric conditions. When they come into secondary contact, they will form hybrids easily, although most likely these will be less fit than their parents. It is possible that many of the well-studied hybrid zones between closely related species are of this nature. Although those species that show such hybrid zones often appear to be closely related, molecular analysis shows in many cases that they split a few million years previously (i.e., within the time frame expected for phase 4 populations). Thus, the isolation factors measured across these hybrid zones would arise from the accumulation of random changes in the genome. The factors that play a role in the adaptive speciation scenario, however, would not allow for a hybrid zone to be formed in the first place. Rather, reinforcement of the isolation characters is expected, with a clear abutment of the borders between populations.

15.5 Concluding Comments

The goal of this chapter is to use the adaptive-speciation framework to interpret known phylogeographic data sets. Such data sets are traditionally interpreted within parapatric or allopatric frameworks. However, it appears that these data sets are often not only compatible with sympatric splitting scenarios, but also may even show distinct patterns that are interpreted more easily in an adaptive-speciation framework. The most crucial distinction between the models is clearly the existence of assortative mating invoked in one case, but not in the other.

For scenarios in which assortative mating is not considered, it should be assumed that closely related populations merge their gene pools upon secondary contact, because there are no mating barriers. In cases where mating barriers have evolved, these should be expected to arise from a few genes or alleles, which would have to be maintained by strong selection in the respective populations. Neutral genes or alleles, however, would still be exchanged in such secondary-contact populations.

In sympatric scenarios, adaptation toward differential resource utilization is assumed, accompanied by differential assortative mating. Therefore, individuals from different subpopulations are disposed to avoid each other, and gene flow is significantly reduced by this mechanism alone. Of course, F1 hybrids still occur, in particular in promiscuous mating systems, but there is a strong ecological selection against these, because they cannot compete effectively with either of the parent populations; further, backcrosses become very unlikely. Thus, both of these mechanisms also restrict gene flow for neutral genes and alleles.

A crucial assumption put forward in this chapter is that mutual geographic displacement of sister populations can be a secondary consequence of assortative mating that has evolved under sympatric conditions. This assumption remains very speculative at present, but can serve to explain many phylogeographic patterns that are otherwise difficult to reconcile with passive geographic subdivision effects.

Most of the currently available data sets have not employed sufficient depth in marker analysis to trace the discriminating distinctions between sympatric and allopatric models. Moreover, even in well-studied situations, it remains difficult to distinguish between ongoing gene flow and insufficient lineage sorting through recent ancestry. Still, the technical means exist to carry out such thorough analyses and to test different scenarios specifically. In the future it should also be possible to make more refined predictions about the patterns of subdivision expected under each model and for each organism. This will provide new and deep insights into speciation processes that go much beyond the usual passive scenarios currently so often assumed.

16

Evolutionary Diversification of Caribbean *Anolis* Lizards

16.1 Introduction

Jonathan B. Losos and Roger S. Thorpe

The diversification of the lizard genus *Anolis* on Caribbean islands surely represents one of the best-studied cases of adaptive radiation in evolutionary biology. Over the course of the past four decades, researchers have studied almost every aspect of anole evolutionary ecology. These include systematics; community, physiological, and behavioral ecology; functional morphology; ethology; and demography. Studies have been conducted in the laboratory and in the field, and have included basic natural history, geographic and temporal comparisons of populations, and a wide variety of experimental approaches to the study of phenotypic plasticity, ethology, ecology, and evolution [recent reviews include Losos (1994) and Roughgarden (1995)]. The result is an unusually broad and detailed understanding of the factors that promote and sustain evolutionary diversification and species coexistence.

Speciation and adaptation in anoles

Two conclusions from the current body of work are obvious. First, the genus *Anolis* has experienced extensive speciation. With more than 400 described species, and more being described every year, *Anolis* is the largest amniote genus, exceeded among tetrapods only by the potentially para- or polyphyletic frog genus *Eleutherodactylus*. The nearly 150 Caribbean species are descendants from as few as two initial colonizing species from the mainland (Jackman *et al.* 1997). Hence, the diversity of Caribbean species results almost entirely from speciation, rather than from repeated colonization.

Second, adaptive diversification has been rampant. Within assemblages of anoles, species are clearly specialized to occupy different niches. Physiological and functional studies have revealed evidence for adaptation to particular microclimatic and structural habitats. Moreover, intraspecific comparisons indicate that populations adapt to their particular environmental conditions (reviewed in Malhotra and Thorpe 2000).

Given the extensive adaptation and speciation exhibited by Caribbean anoles, one might wonder whether the two processes are linked. Can adaptive speciation – the topic of this volume and the theory that as a population diverges under the pressure of disruptive selection, speciation ultimately ensues – explain the evolutionary radiation of anoles? Alternatively, are adaptation and speciation in anoles largely unrelated, perhaps influencing each other to some extent (speciation making adaptation more likely, or vice versa), but not necessarily linked?

The anole radiations

Anoles are small, insectivorous, and primarily arboreal lizards found throughout the Caribbean, Central America, northern South America, and the southeastern United States. Two morphological features characterize anoles. First is the possession of expanded subdigital toe pads. These toe pads are composed of laterally expanded scales, termed lamellae. Each lamella is covered with millions of microscopic hair-like structures, called setae. These setae allow anoles to adhere to smooth surfaces by the forces that form between electrons on the setae and electrons on the surface (Irschick *et al.* 1996 and references therein; see also Autumn *et al.* 2000). Very similar structures have evolved independently in two other groups of lizards, the geckos and prasinohaemid skinks (Williams and Peterson 1982).

The second morphological structure that characterizes anoles is the presence of a gular throat fan, termed a dewlap. This structure, possessed by males of almost all species and females of many, is deployed in a variety of different contexts, including encounters with territorial rivals, potential mates, and predators. The color and pattern of the dewlap, as well as the specific pattern of head movements, is species specific and is used in species recognition (Rand and Williams 1970; Jenssen 1978; Losos 1985).

The closest relative of *Anolis* is the genus *Polychrus*, which contains five species in Central and South America (Frost and Etheridge 1989). Based on estimates from immunological studies, *Anolis* evolved at least 40 million years ago (Shochat and Dessauer 1981). Although the fossil record is sparse, several amber specimens from the Dominican Republic date to the Oligocene or Miocene (de Queiroz *et al.* 1998). Phylogenetic studies (reviewed in Jackman *et al.* 1997) indicate that *Anolis* originated in Central or South America and invaded the Caribbean twice. One lineage gave rise to the *roquet* group, which occupies the southern Lesser Antilles, whereas the second lineage gave rise to all other Caribbean anoles. In turn, the extensive radiation of the beta, or *Norops*, group on the mainland, composed of more than 150 species, appears to descend from a single colonist from the Caribbean back to the mainland.

As many as 11 species of anoles can occur sympatrically in the Caribbean; assemblages nearly as large are known from the mainland. Comparisons of sympatric species indicate that species are almost invariably differentiated from sympatric congeners in some aspect of habitat use and either morphology or physiology, or both. On Caribbean islands, interspecific competition appears to be a potent force in the regulation of assemblage structure (reviewed in Losos 1994); on the mainland, both competition and predation may be important (Andrews 1976, 1979; Guyer 1988).

The following two sections focus in turn on anole radiation in the Lesser Antilles (Section 16.2) and in the Greater Antilles (Section 16.3).

16.2 Adaptation and Speciation in Lesser Antillean Anoles

Roger S. Thorpe, Anita Malhotra, Andrew Stenson, and James T. Reardon

The Lesser Antilles archipelago is composed of an older outer arc and a younger inner arc, with some islands (e.g., Martinique) being composed of elements of both (Figure 16.1). Younger islands tend to have a high elevation (e.g., Dominica), that in turn results in pronounced ecological (both biotic and physical) altitudinal and longitudinal zonation. These islands are occupied by two nonoverlapping series of anoles. The northern islands, down to and including Dominica, are occupied by anoles from the *bimaculatus* series and the southern islands, up to and including Martinique, are occupied by anoles from the *roquet* series (Figure 16.1; Underwood 1959). No major island is unoccupied and, naturally, an island has either one or two species. The far northern and southern islands tend to have two species, even if the islands are small and ecologically homogeneous, while central, large, ecologically heterogeneous islands have only a single species (Figure 16.1). As these islands may have been colonized from the northern and southern extremities (Gorman and Atkins 1969), the number of species that occupy an island may be related to the temporal opportunity for two successful colonizations rather than to ecological complexity of the habitat. In any event, these communities are characterized by less congeneric competition than the Greater Antillean anole communities, in which many species may live in sympatry (Section 16.3).

While the number of sympatric species within a community is low, there are numerous allopatric species in this system. Here we attempt to elucidate the relative importance of adaptation in this system, the frequency and pattern of speciation, and the evidence for adaptive speciation.

Evidence for adaptation by natural selection

Evidence for the relative importance of adaptation in this system comes from three classes of study:

- Correlations between phenotype and habitat over space;
- Common-garden experiments;
- Natural selection experiments.

Correlational evidence. Geographic variations in the phenotype of anoles from mountainous, single-species islands (Montserrat, Basse Terre, Dominica, Martinique, and St Lucia) have been investigated (Malhotra and Thorpe 1991a, 1997a, 1997b; Thorpe and Malhotra 1996). These islands tend to have rather similar ecological zonation with windswept littoral woodland on at least part of the Atlantic coast, xeric woodland on the Caribbean coast, and rainforest (giving way to elfin woodland) at higher elevations. The anoles tend to adapt to these

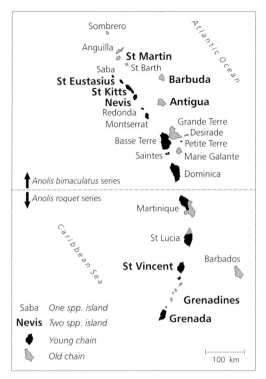

Figure 16.1 Distribution of the anole series in the Lesser Antilles, showing sympatry and simplified island age. The older chain arose primarily in the mid Eocene to mid Oligocene [about 45–25 million years before present (MYBP)] with some activity in the late Oligocene to late Miocene (ca. 35–5 MYBP). The younger chain showed some activity in the Miocene, but arose primarily in the Pliocene (ca. 5 to 2 MYBP).

habitats (Figure 16.2) in their scalation, body dimensions, and color pattern (including human-invisible ultraviolet markings), which results in conspecific intergrading ecotypes (i.e., ecological races of the same species adapted to different, spatially segregated habitats, as distinct from ecomorph species – see Section 16.3; Malhotra and Thorpe 1993b). The extent of this intraspecific geographic variation may be greater than the difference between allopatric species in some features. For example, there is a greater difference in body size between Caribbean lowland males and montane males within both Basse Terre and Dominica than there is between these island species (allowing for ecotype). Nevertheless, mitochondrial DNA (mtDNA) phylogenies for Dominica (Malhotra and Thorpe 2000) and Martinique (Thorpe and Stenson 2003) indicate that these ecotypes are not distinct lineages, although distinct mitochondrial lineages exist on these islands and partly correlate with color patterns.

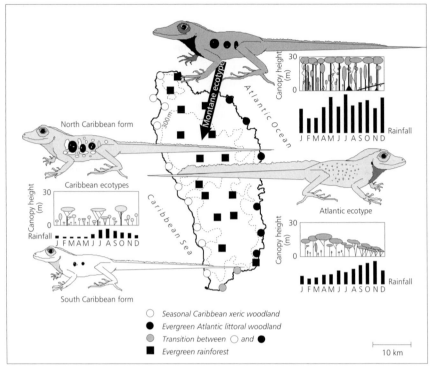

Figure 16.2 Ecotypes of *A. oculatus* on Dominica. Open circles indicate seasonal Caribbean xeric woodland, filled circles indicate evergreen Atlantic littoral woodland, gray circles indicate a transition between these former vegetation zones, and squares indicate evergreen rainforest. The 300 m contour is indicated by the dashed curve. For the main vegetation zones, the canopy height (in meters) is illustrated diagrammatically, and relative monthly rainfall is shown by a histogram (the maximum monthly rainfall depicted is about 850 mm). Some aspects of the size, shape, and color pattern are indicated for the ecotypes associated with these vegetation zones. The southern and northern Caribbean forms live in a broadly similar habitat, but have some differences in appearance, perhaps for historical reasons (see Malhotra and Thorpe 2000).

The pattern of geographic variation may be tested against several putative causes, taking into account phylogenetic relationships (when available) and geographic proximity, using partial matrix correspondence (Mantel) tests (Box 16.1). When this is done, the hypothesis that the geographic variation in the phenotype is determined by adaptation to the physical and biotic conditions in these habitats cannot be rejected; in these adaptations the scalation tends to be adapted to the rainfall and/or moisture levels, the body size and shape tend to be adapted to altitude, and the color pattern tends to be adapted to vegetation (Table 16.1). It appears that natural selection has resulted in a close association between phenotype and habitat with the same trends being paralleled on each independent island.

Box 16.1 Testing causal hypotheses for patterns of geographic variation

Matrix correspondence tests can be used to compare an observed pattern of variation (represented as a matrix of dissimilarities among localities) with a pattern derived from a causal hypothesis (also expressed as a dissimilarity matrix). If the measure of correspondence (e.g., regression or correlation) is not greater than, say, 95% of the correspondence values obtained when one of the matrices is randomized (e.g., 10 000 times), the null hypotheses of no correspondence may be accepted and the causal hypothesis rejected (Manly 1986a; Thorpe 1991). One complication is that alternative hypotheses may predict similar (correlated) patterns. With multiple hypotheses, this problem may be overcome by testing the hypotheses simultaneously using partial correlations (Smouse *et al.* 1986), or partial regressions (with the observed pattern as the dependent variable), as the measure of correspondence (Manly 1986b; Thorpe and Baez 1993; Thorpe and Malhotra 1996; Thorpe *et al.* 1996; Malhotra and Thorpe 2000).

The correspondence between patterns of geographic variation in morphology with alternative patterns of climate and vegetational zonation may be tested to investigate the role of natural selection (Malhotra and Thorpe 1991a, 1997a, 1997b, 2000; Thorpe and Baez 1993; Thorpe and Malhotra 1996), and a molecular phylogeny (the matrix could represent the patristic distances or clade membership) may also be tested against patterns derived from alternative vicariance and/or phylogeographic hypotheses (Thorpe *et al.* 1995, 1996; Daltry *et al.* 1996; Thorpe 1996; Malhotra and Thorpe 2000). Moreover, factors such as spatial proximity and phylogenetic relatedness could also be included as independent variables to allow for them when assessing alternative causal hypotheses, such as the adaptation of morphology to ecological conditions (Thorpe *et al.* 1996; Malhotra and Thorpe 2000).

Table 16.1 Partial matrix correspondence (Mantel) tests show statistically significant association between multivariate-generalized character systems in males and aspects of the environment (Box 16.1). Note the multiple parallels between independent mountainous island systems, in which scalation tends to adapt to rainfall and/or humidity, body dimensions tend to adapt to altitude, and color pattern tends to adapt to vegetation type. In both Dominica and Martinique, phylogeny is allowed for.

	Scalation	Body dimensions	Color pattern
Dominica	Rainfall, vegetation	Altitude, vegetation	Vegetation
Montserrat	Humidity	Altitude	Vegetation
Basse Terre	Rainfall		Vegetation
Martinique	Rainfall	Altitude	
St Lucia	Temperature		Vegetation

Common-garden experiments. However unlikely, it could be that the phenotype tracks habitat type through space entirely because of phenotypic plasticity, rather than natural selection (Section 16.3). To determine if the geographic pattern has a genetic component a common-garden experiment can be run in which individuals

---

Box 16.2 Common-garden experiments

In studies of geographic variation, knowing the percent heritability of a phenotypic character is of little value, as it is a within-population measure and does not predict whether the geographic pattern is genetically determined (unless heritability is absolutely zero or 100%, which it never is). To test if the pattern of geographic variation in morphology of the Dominican anole results entirely from phenotypic plasticity, rather than from natural selection acting on characters whose variation is influenced by genotype, Reardon and Thorpe (unpublished) ran a common-garden experiment. Pregnant females were sampled from 12 localities across the island and the resultant offspring were reared in a "common garden" situated in transitional woodland between northern Caribbean xeric woodland and montane forest. Insufficient offspring of both sexes were reared from three of the localities, but scalation and body dimension characters were recorded for the young adult offspring of the remaining nine localities. Their morphological characteristics were compared [using analysis of variance (ANOVA) and/or analysis of covariance (ANCOVA)]:

▪ Among localities, within offspring;
▪ Between offspring and adults previously sampled from the locality of origin.

The multivariate divergence (Mahalanobis D_M) between common-garden-reared offspring from a given locality and a sample that occured naturally at that locality was compared. Finally, a canonical analysis scatter diagram that summarized the affinities of the geographic localities based on offspring was similar to a scatter diagram for specimens from the original localities. These all reject the hypothesis that the geographic variation arises from phenotypic plasticity (see text).

---

from different localities are hatched and raised under the same natural conditions in the field (Garland and Adolph 1991). However, apart from the husbandry being excessively time consuming, these experiments can be problematic, because the more critically important natural selection is, the more likely it is that mortality will be high in individuals from habitats very different from the habitat at the site of the "common garden". For example, an anole adapted by natural selection for a damp cool rainforest may fail to flourish and so die in a common garden sited on a hot dry coast. This results in low sample sizes for offspring for some localities, a reduced number of localities, and a reduced character suite when some characters (e.g., sexual characteristics of mature males) take a long time to develop.

In spite of these difficulties, a successful common-garden experiment was run in Dominica on *A. oculatus*, using body dimensions and scalation (Box 16.2), and established that:

▪ No significant morphological differences occurred between experimentally reared offspring and adults from the original source locality.
▪ Significant morphological differences occurred between offspring from different localities.

Box 16.3 Natural selection experiments

The enclosures were constructed as in Malhotra and Thorpe (1991b, 1993a) and assiduously emptied of anoles. Specimens collected from control and test localities were anesthetized prior to individual labeling (toe clipping) and a multivariate profile was recorded from the scalation, body dimensions, and color pattern (the last via standardized high-quality Kodachrome macrophotographs). They were then released into the enclosures at the original biomass levels. At regular intervals, a census of the enclosures was taken to record those still alive.

Selection intensity was estimated by the multivariate distance (Mahalanobis D_M) between surviving and nonsurviving groups within an enclosure (i.e., of the same ecotype). Selection coefficients were estimated by partial regression coefficients, in which the morphological characters were the independent variables and fitness (number of census time periods survived) was the dependent variable (Brodie *et al.* 1995; Thorpe and Malhotra 1996).

■ The extent of morphological divergence between experimental offspring and anoles from the same locality did not depend on the extent of difference between the habitat at the common-garden site and the locality of origin.

■ A multivariate scatter diagram that summarized geographic variation in the morphology of offspring showed the same pattern as one based on anoles from the original localities.

This evidence strongly supports the proposal that the pattern of geographic variation is largely the product of natural selection working on characters that are, at least partially, genetically determined, rather than the pattern being a product of phenotypic plasticity.

Natural selection experiments. Components of natural selection may be measured directly in field experiments (Endler 1986; Brodie *et al.* 1995). To this end, four large-scale natural enclosures were established on the west coast of Dominica to investigate natural selection in *A. oculatus* in a series of chronologically and synchronously replicated translocation experiments.

Following the methodology outlined in Box 16.3, an initial experiment was set up to test for a significant selection intensity in montane, littoral woodland, south Caribbean, and north Caribbean ecotypes, with the last acting as the experimental control. After 60 days, a multiple analysis of variance (MANOVA) indicated significant selection within ecotypes, and it was shown that, although the controls had no significant selection acting on them, the translocated montane animals were subject to a significant selection intensity (Malhotra and Thorpe 1991b; Thorpe and Malhotra 1992, 1996). Moreover, the intensity of the selection correlated significantly to the extent of ecological difference between the locality of origin and the enclosure site. After more time had passed, significant selection (as shown by selection intensity) was shown to be acting on the littoral woodland ecotype.

This was supported by a partial reciprocal experiment with enclosures in the littoral woodland, which showed a significant selection intensity for the north xeric woodland ecotype, but no significant selection for the littoral woodland control (Thorpe and Malhotra 1996).

To investigate this further, critical aspects of the experiment were replicated chronologically by a pair of experiments with their own (internal) synchronous replication. This pair of experiments concentrated on montane translocates and north Caribbean controls, with each type in two enclosures. One experiment was run in the dry season and another in the wet season. A complete-block design MANOVA with interaction showed that, for both males and females, there was a significant selection effect (a significant difference between survivors and nonsurvivors) and that the intensity varied among ecotypes (a significant interaction between survivor–nonsurvivor and ecotype). For both the dry season and the wet season experiments, and for both males and females, the accumulated selection intensity tended to be high and significant for the translocated montanes and low and insignificant for xeric woodland controls. The targets of selection were also investigated by partial correlation (Box 16.3). Several characters were the targets of selection in the montane group and those characters with significant selection coefficients were shown to distributed among all three character types (scalation, color pattern, and body size and shape).

Finally, an experiment was run on 12 samples (replicated in the four enclosures) from a lowland-to-montane ecological gradient (using just females to maximize sample size). This investigated the relationship between the intensity of selection and the intensity of the difference between the ecological conditions at the site of the enclosures and the site of origin. This study showed a strong correlation between the two, irrespective of the time at which selection intensity was measured. This supports the contention that the lizards are adapting, by natural selection, to the spatially various ecological conditions.

This series of replicated experiments consistently indicates that natural selection is acting on A. oculatus and that the geographic variation in this anole is likely to result from adaptation to the geographic variation in physical and biotic conditions within Dominica.

The correlations between habitat and morphology (irrespective of phylogenetic history), the observed parallels in these correlations on comparable islands, the common-garden experiments, and the series of natural selection experiments, taken together, provide overwhelming evidence that adaptation is a very influential and important evolutionary factor in this system.

Evidence for speciation

In a sympatry-dominated system, such as the Greater Antillean anoles, species recognition is quite straightforward, but it is arbitrary when forms are allopatric, and interbreeding is untestable. Consequently, it has to be acknowledged that in an allopatry-dominated system, such as the Lesser Antillean anoles, to assess the extent of speciation will have an arbitrary element. This is illustrated by a series of

---

Box 16.4 Assessing gene-flow patterns using microsatellites

A series of five polymorphic microsatellites (Stenson *et al.* 2000) were screened for up to 50 specimens from each of 33 localities (as in Figure 16.2) across Dominica (Malhotra and Thorpe 2000). The gene flow between pairs of adjacent sites (adjusted for differential distance among sites by regression) was plotted on a minimum-contact, minimum-distance connectivity network that joined adjacent sites (Stenson *et al.* 2002). This is visualized by contouring these values to give a nonprobabilistic representation of the geographic pattern of reduction in gene flow (Figure 16.4). This shows that along the Caribbean coast a sharp reduction in gene flow is found between northern and southern types.

---

taxonomic changes. The species status of the two northern groups (one comprises *A. wattsi, pogus, foressti,* and *schwartzi,* and the other comprises *A. bimaculatus, gingivinus,* and *leachi*), as well as the southern group comprising *A. griseus* and *richardi,* have all changed, with more allopatric forms recognized as separate species (Schwartz and Henderson 1991; Malhotra and Thorpe 1999).

Patterns of between-island speciation. Molecular phylogenetic analysis based on mtDNA and supported by nuclear microsatellites established the relationships within the *A. bimaculatus* series and were used to hypothesize the colonization sequence and spaciotemporal pattern of speciation, following the procedures in Thorpe *et al.* (1994). Early on in the history of the group this shows a step-wise pattern of colonization of adjacent islands (or island banks) in a general north-to-south, old-to-young progression (Figure 16.3). The one interesting exception to this is the very recent colonization of Saba from southwest Basse Terre in a long-range, southeast-to-northwest track shown by many hurricanes.

Similar molecular phylogenies are available for the southern *A. roquet* series (Giannasi *et al.* 2000) and are compatible with the concept of speciation by early founder colonization between major islands and island banks. There is therefore a substantial amount of between-island speciation in this model.

Within-island speciation. In Dominica, the geographic variation in the morphology of *A. oculatus* generally shows all the indications that would result from adaptation by natural selection, such as incongruence in the patterns of geographic variation of different characters, gradual character change correlated to changes in environmental variables, and little association with the molecular phylogenetic history (Malhotra and Thorpe 2000). However, one exception to this is the change between the north and south Caribbean coastal forms (Malhotra and Thorpe 1994). Here the changes are very sharp and congruent, between both various types of morphological character and mtDNA lineage boundaries. Moreover, the sharp transition is not closely associated with sharp environmental changes, or extrinsic barriers to gene flow. Is this an example of partial reproductive isolation within an island? To test the extent of reproductive isolation, gene flow among populations

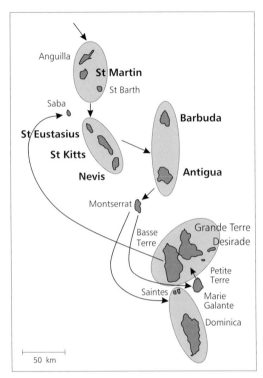

Figure 16.3 Hypothesized colonization pathways of the large-bodied anoles from the *A. bimaculatus* series. Ovals enclose island banks (or lineages within which colonization direction is equivocal) and arrows indicate the main colonization events associated with speciation. For anoles of the *bimaculatus* series, around 1200 base pairs from cytochrome b and cytochrome oxidase I were sequenced and a gene tree reconstructed using maximum parsimony and maximum likelihood. Several key aspects of the mitochondrial tree were corroborated by nuclear genes (microsatellites). The small-bodied *A. wattsi*, *pogus*, *foressti*, and *schwartzi* group was a separate lineage to the large-bodied *A. bimaculatus*, *gingivinus*, and *leachi* group. Taking the phylogeny of this latter group, it can be converted into a colonization sequence following the branch-length procedure of Thorpe *et al.* (1994), whereby a node is allocated a geographic locality based on the shortest patristic distance to the terminal node (geographic population). Nodes are then joined in phylogenetic sequence. This procedure is sensitive to the branch lengths of the tree and so only colonization steps common to both maximum parsimony and maximum likelihood are shown here. This procedure suggests a geographic sequential colonization in this anole series, with the speciation events associated with relatively early colonization of major islands or island banks, while later colonizations of islands within the same bank (e.g., within the St Martin, St Kitts, and Antigua banks) or colonization of adjacent islets (e.g., Petite Terre) are not associated with speciation. The important exception to this is the late, long-range colonization of Saba (from southwest Basse Terre) and speciation of *A. sabanus* after the colonization of much of the remaining Guadeloupe archipelago (Desirade, Petite Terre, Grande Terre, and Basse Terre) without speciation. The Saintes and Dominica have separate species, but the colonization sequence is equivocal, as is the colonization of Redonda.

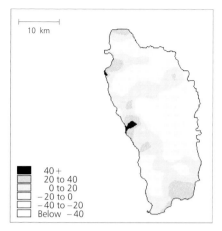

10 km

40 +
20 to 40
0 to 20
−20 to 0
−40 to −20
Below −40

Figure 16.4 Geographic variation in gene flow between adjacent localities in Dominica based on *A. oculatus* microsatellites. Residual values from the regression of linearized F_{ST} (between adjacent sites) against $\log(\Delta z)$, where Δz is the geographic distance between adjacent sites, were multiplied by 1000. The darker the area, the lower the gene flow. Based on 33 localities, as in Figure 16.2.

was measured using microsatellites and contoured across space (Box 16.4; Stenson *et al.* 2002). This shows, that while there is a very sharp reduction in gene flow at this point and some reproductive isolation between these forms along the coast, genes are exchanged via the inland populations at much the same rate as between contiguous population across the rest of the island (Figure 16.4). Consequently, this example does not provide convincing evidence of complete within-island speciation. One possible scenario is that volcanic activity has eradicated populations from some areas temporarily and that this coastal transition zone marks a point of secondary contact (Malhotra and Thorpe 2000). Factors may yet operate that will eventually result in speciation (e.g., there may be assortative mating at this coastal transition, and this warrants further investigation).

In Martinique, the molecular (mtDNA) phylogeny of *Anolis roquet* indicates that major phylogenetic divisions (about 11% uncorrected base pair differences) occur within the current island (Thorpe and Stenson 2003). Matrix correspondence tests and geological information indicate that this pattern most likely arises from several peripheral precursor islands (southern, southwestern, and northwestern) that formed Martinique when a central precursor island emerged and joined with each of the peripheral precursors (Thorpe and Stenson 2003). The phylogeographic pattern suggests secondary contact of the populations on these precursor islands, where at least the mtDNA has diverged in allopatry. This has resulted in three major secondary contact zones within the island marked by the distribution of the distinct mtDNA lineages. Moreover, there is distinct habitat zonation on the island, and the multivariate morphology (Giannasi 1997), color pattern, and hue relate largely to this zonation (Thorpe and Stenson 2003). Some of these forms are so divergent that there is an indication of multivariate bimodality at the point where at least one pair of parapatric forms meet (Thorpe and Stenson 2003), and a significant reduction in microsatellite gene flow between several pairs of parapatric forms (Ogden and Thorpe 2002). Moreover, the Martinique form is parapatric, as another full species, *A. extremus* from Barbados, is nested inside the Martinique complex (Thorpe and Stenson 2003). Hence, there is evidence of incipient speciation and some parapatric forms may warrant recognition as full species. However, given the occurrence of several potentially important factors (ecological zonation and past vicariance), and the difference between a significant reduction in gene flow and complete speciation, more research needs to be undertaken before it can be suggested that full parapatric speciation has taken place *in situ* within this island.

The evidence is overwhelming for numerous speciation events generally associated with early colonization. Anoles appear to speciate readily in geographic isolation. While there is clear evidence of strong morphological and molecular divergence within Lesser Antillean islands and some evidence of parapatric forms being incipent species, there is no adequate evidence for complete speciation *in situ* within islands at this stage. These studies on the differentiation between incipient speciation and speciation within islands are ongoing.

Has adaptive speciation occurred?

Clearly, both widespread adaptation and speciation occur within this Lesser Antillean anole system, but does adaptive speciation? This is not an easy question to answer, as it can be difficult to know the rate of some factors, irrespective of which model of speciation is suspected. For example, interpretation of the molecular phylogeny can suggest the timing and geographic origin of the Saba species, and its morphological differentiation can be measured, but what the relative roles of chance and adaptation were in this speciation event is not known. This is because the process is unique, as are so many of the facets of speciation events.

An area in which repetition may allow the cause to be tested is in the evolution of body size. It has long been known that on islands with only one lizard the body size tends to be intermediate, whereas on islands with two species, one species tends to be large and the other small (Schoener 1969, 1970; Plate 9). This may result from size assortment (i.e., the tendency for only species of dissimilar sizes to colonize and coexist on the same island because of competitive exclusion or interbreeding), or character displacement in which, after colonization of an island by precursors of similar size, they evolve into sympatric species of different sizes by character displacement to minimize the competition for resources (Losos 1990a; Giannasi *et al.* 2000). Ancestral character reconstruction methods (Maddison and Maddison 1992) allied to molecular phylogenies allow these two hypotheses to be tested in the southern *A. roquet* series. These tests show that an element of character displacement is likely, with size assortment playing a role (Giannasi *et al.* 2000).

To what extent does this interisland study of size offer evidence for the role of adaptation in speciation? In the north, the larger species are of the *bimaculatus* group and the smaller belong to the *wattsi* group. These are distinct lineages (Stenson 2000), so within a given island the two species will not be sister taxa. Hence there is no question of a common ancestor split by adaptive speciation. In the south, the phylogenetic relationships of the *roquet* group are resolved only partially (see Thorpe and Stenson 2003 and references therein), so it is not always possible to know if a pair of species on an island (Plate 9) are sister taxa that could have speciated *in situ* by adaptive speciation. Work on the phylogeny and colonization sequence of this group is ongoing.

Within islands there is evidence of parapatric forms with restricted gene flow among them on both Dominica (Stenson *et al.* 2002) and Martinique (Ogden and Thorpe 2002). The latter case was interpreted as evidence in favor of the process of ecological speciation, as the forms are defined by patterns of current habitat rather than past allopatry. Adaptive speciation may occur on an ecological gradient (Doebeli and Dieckmann 2003; Chapter 7), but as pointed out in Chapter 1, such a parapatric situation may, or may not, involve adaptive speciation. Consequently, whether the Martinique (Ogden and Thorpe 2002), and other (Thorpe and Richard 2001) lizard cases represent examples of adaptive speciation (Tautz 2003) remains to be investigated fully.

16.3   Adaptation and Speciation in Greater Antillean Anoles

Jonathan B. Losos

This section first reviews the evidence that both speciation and adaptation played important roles in anole diversification, focusing primarily on the anoles of the Greater Antilles (Cuba, Hispaniola, Jamaica, and Puerto Rico). It then addresses the extent to which the two processes are intimately linked. The theory of adaptive speciation presents one mechanism by which the two processes might occur in an integrated fashion, but there are other possibilities. In part this requires a discussion of what constitutes a species of *Anolis*, so that it is possible to recognize when two lineages have diverged to the level of different species. Finally, this section addresses some exciting new developments that affect our understanding of the processes important in anole adaptive radiation.

Evidence for speciation

The extent of speciation in Caribbean anoles can be viewed at several levels. Most generally, the number of Caribbean anole species indicates the high rates of speciation that have occurred. *Anolis* is the most species-rich genus of vertebrate in the Caribbean and phylogenetic studies indicate that Caribbean anoles are the descendants of only two colonizations from mainland taxa (Jackman *et al.* 1999). Hence, the Caribbean anole radiation has resulted almost entirely from *in situ* speciation, rather than from repeated colonization. Examination of anole phylogeny (Figure 16.5), however, can tell us considerably more about patterns of speciation.

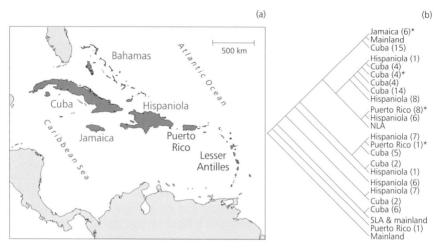

Figure 16.5 (a) Map of the Caribbean. (b) Phylogeny of *Anolis* based on mtDNA (Jackman *et al.* 1999). Relatively few interisland movements are required to account for the more than 100 species of anoles on the Greater Antilles. Numbers in parentheses indicate the number of species within each terminal clade. Monophyly of these terminal clades is based on traditional morphological data. Molecular phylogenetic analyses using DNA indicated that the placement of species into species groups based on morphology is nearly always correct (e.g., Jackman *et al.* 1999; Schneider *et al.* 2001; Jackman *et al.* 2002). Species from small islands in the Caribbean (e.g., *A. acutus* on St Croix, *A. conspersus* on Grand Cayman) occur within terminal clades (designated with an asterisk) and are not indicated. In all cases, these species are apparently derived from clades on one of the Greater Antilles [e.g., *A. conspersus* from Jamaica, *A. acutus* from Puerto Rico (Jackman *et al.* 2002)].

Many islands in the Lesser Antilles and elsewhere in the Caribbean, such as the Bahamas and the Virgin Islands, contain more than one anole species. However, these co-occurring species are almost invariably distantly related, which suggests that the ancestors of these species independently colonized the island, and are not the result of a speciation event that occurred on that island. By contrast, in the Greater Antilles, the situation is quite different. Many monophyletic clades are found only on a single island, which indicates that substantial speciation has occurred within an island. On Puerto Rico, for example, eight of the ten species are members of the *cristatellus* group, whereas in Jamaica, six of the seven species belong to the *grahami* group. Similarly, the *alutaceus* group of grass anoles, found only on Cuba, contains 14 species. In each case, the most parsimonious explanation is that the ancestor of the group occurred on the island in question and then, by a series of divergent speciation events, the single ancestral species gave rise to numerous descendants. Indeed, anole phylogeny suggests that as few as 11 interisland colonization events are necessary to explain the distribution of the 110 or more Greater Antillean species (Jackman *et al.* 1997); thus, all remaining species are the result of within-island diversification (Losos and Schluter 2000).

Evidence for adaptation

Several lines of evidence indicate the great extent of anole adaptive diversity. First, convergent habitat specialization is widespread in the Greater Antilles. Williams (1972, 1983) noted, for anole assemblages within the Greater Antilles, that species exhibit specializations to using different parts of the structural habitat. Thus, for example, twig species are streamlined with extremely short limbs and tails and trunk–ground species (which perch low on the trunk and frequently descend to the ground) are stocky with extremely long hind limbs and poorly developed toe pads. The similarities even extend to patterns of sexual size dimorphism: trunk–ground anoles, for example, are always highly dimorphic, whereas twig anoles exhibit little dimorphism (Butler *et al.* 2000). In all, six types of habitat specialists have been recognized.

Such a pattern of assemblage similarity could result either because each specialist evolved once only and made its way to each of the islands, or because highly similar assemblages evolved independently on each island. Phylogenetic studies make clear that the latter explanation is correct (Losos *et al.* 1998). Habitat specialists on different islands are almost never closely related; hence, these similar assemblages are the result of convergent evolution.

Patterns of divergent evolution also provide evidence of adaptation. For example, correlations exist between habitat use and morphology within both Lesser (Section 16.2) and Greater Antillean anoles (Lister 1976; Losos *et al.* 1994). Among populations of *A. sagrei*, for instance, a correlation exists between mean perch height and mean number of toe-pad lamellae (Lister 1976). Similar examples are presented from Lesser Antillean taxa (Section 16.2)

Convergent evolution and correlations between morphology and environment have long been taken as evidence of adaptation (reviewed in Harvey and Pagel 1991; Losos and Miles 1994). Nonetheless, to understand fully the adaptive basis of character evolution, we need to understand why particular traits evolved in particular environments. Doing so requires an understanding of the functional consequences of character variation, as well as detailed information of what organisms actually do in nature (Lauder 1981; Arnold 1983; Greene 1986; Wainwright 1988; Losos 1990b; Arnold 1994).

In the case of anoles, laboratory functional studies have revealed how variations in limb length and toe-pad dimensions affect the ability to run, jump, and cling (Irschick *et al.* 1996; Irschick and Losos 1999, and references therein). As might be expected, species with longer hind limbs have greater sprinting and jumping abilities, whereas toe-pad dimensions are related to clinging abilities. Moreover, functional capabilities are not independent of the environment. Long-legged lizards, for example, can run faster than short-legged species on broad surfaces, but not on narrow ones (Losos and Sinervo 1989; Losos and Irschick 1996).

In turn, field studies have revealed some of the contexts in which maximal performance is important. Most species run at maximal speed to escape predators, but not to capture prey (Irschick and Losos 1998). Moreover, species that run quickly tend primarily to use habitats in which they can run at top speed (i.e., broad sur-faces), whereas slow species, which rely more on crypsis and slow locomotion, are less constrained in their habitat use (Irschick and Losos 1999).

In a similar vein, physiological studies have established how species adapted to different thermal environments by altering their physiological capabilities. Thus, species that use cooler, forest habitats thermoregulate at lower temperatures, whereas species in hot, open habitats maintain higher temperatures. Further, re-search on Caribbean and mainland species has established that peak functional ca-pabilities correspond to preferred temperatures (e.g., Hertz 1979, 1992; Hertz *et al.* 1979; Huey 1983; Van Berkum 1986). Thus, although many questions remain unanswered, we do have a solid understanding of the functional consequences of observed variation in traits. This understanding supports the conclusion, based on convergence and correlation, that the extensive variation in traits – such as limb length, toe-pad dimensions, and thermal physiology – observed among anole species represents adaptation to living in different parts of the environment.

Has adaptive speciation occurred?

The observations that sympatric anoles almost always differ in habitat use, with attendant morphological or physiological adaptations, and that most speciation has occurred within islands in the Greater Antilles suggest that adaptive speciation has played an important role in anole adaptive radiation (Dieckmann and Doebeli 1999). Consider Jamaica, for example. Five species are widespread throughout the island, and often occur sympatrically. All five descended from a common ancestor on Jamaica, and each has adapted to its own specific niche. One ready explanation is that this diversity arose via adaptive speciation, as disruptive selection led first to polymorphism within a single ancestral species and then, as reproductive isolation evolved, to speciation and the production of an assemblage of species adapted to different parts of the environment. Indeed, given that the species now occur island wide, that it is not clear that the island was ever divided into separate parts by climate change or sea-level rise, and that there are few peripherally isolated offshore islands (Lazell 1996), a scenario of allopatric speciation is in some ways distinctly less parsimonious. Although opportunities for allopatric speciation are greater on the other Greater Antillean islands, which are larger and more varied topographically, the argument for adaptive, sympatric speciation still remains.

However, before the process of speciation can be discussed meaningfully, an understanding of what constitutes a species is necessary. Thus, before the evidence relative to modes of speciation in anoles is examined, first a digression is needed to discuss what defines an anole species and how one recognizes whether different taxa belong to the same species.

Concepts of anole species and speciation.   Recent years have seen much criticism of the biological species concept and a proliferation of other ideas about what a species is [e.g., Howard and Berlocher (1998), and references therein]. Nonethe-less, the vast majority of sympatric anoles coexist without interbreeding. Indeed, despite 40 years of extensive fieldwork in the Caribbean, hybridization has only

Table 16.2 Reported cases of hybridization in Caribbean *Anolis*.

Species	Island	Notes	References
A. aeneus × *A. trinitatis*	Trinidad	Infertile; species introduced to island, not naturally sympatric	Gorman and Atkins 1968; Gorman *et al.* 1971
A. allisoni × *A. porcatus*	Cuba	Hybridization suggested by morphological intermediacy	Ruibal and Williams 1961
A. brevirostris × *A. distichus*	Hispaniola	Abnormal meiosis, thus probably infertile; dewlaps very similar; hybridization possibly frequent	Webster 1977
A. caudalis × *A. websteri*	Hispaniola	Will intermate, but no documented hybrids; dewlaps different color	Jenssen 1996
A. chlorocyanus × *A. coelestinus*	Hispaniola	Hybridization suggested by the presence of morphologically intermediate forms and interspecific courtship in staged trials	Garcia *et al.* 1994
A. grahami × *A. lineatopus*	Jamaica	One individual; probably infertile because of chromosomal irregularities	Jenssen 1977

been suggested between six pairs of Caribbean species, most at a single locality only (Table 16.2). Further, the evidence in several of these cases is far from compelling. Thus, as a general rule, Caribbean anoles are characterized by premating reproductive isolation.

Fortunately, we have a good idea of what factors promote reproductive isolation in anoles. Anoles are visually oriented animals and communicate by visual displays that involve vertical movements of the head, termed head bobs, and by extending their colorful dewlaps. Several lines of evidence indicate that anoles are able to distinguish conspecifics from nonconspecifics by the form of their bobbing display and by the appearance of the dewlap. With regard to the head-bobbing patterns, each species has its own stereotyped pattern of movements, which differ, among species, in rhythm and amplitude (Jenssen 1978). A video playback experiment with one Central American species indicated that female *A. nebulosus* can discriminate males that perform their normal, unaltered displays from males whose displays have been experimentally altered (Jenssen 1970; see also Macedonia *et al.* 1993).

With regard to dewlaps, it is notable that sympatric species always differ in the size, color, or patterning of their dewlaps (Plate 10; Rand and Williams 1970; Williams and Rand 1977; Losos and Chu 1998). In itself, this decidedly nonrandom pattern suggests that the dewlap is used as a species-recognition signal. Moreover, an experimental study with a pair of sympatric species indicates that anoles use the color of the dewlap as a cue to determine how to respond to another male (Losos 1985; see also Macedonia and Stamps 1994).

This understanding of reproductive isolating mechanisms [or specific mate-recognition systems (Paterson 1982)] is important in two regards. First, it provides a mechanistic understanding of what goes on during speciation; before two species can coexist in sympatry, they apparently must evolve differences in these systems. Second, it permits an objective means of evaluating whether allopatric populations have differentiated to the extent that they would interact as distinct species were they to become sympatric: if the populations differ in dewlap color or head-bobbing pattern, they constitute distinct species. Of course, the converse is not always true. It is possible that allopatric populations may evolve postzygotic reproductive isolation and thus be isolated reproductively, even in the absence of premating isolating mechanisms. Thus, this method for determining specific status may, in some cases, fail to recognize populations that already constitute different biological species, a situation that may exist in some Lesser Antillean taxa (Section 16.2).

Evidence for adaptive speciation.   Given this understanding of what constitutes speciation in anoles, under what conditions has speciation occurred? The adaptive speciation hypothesis suggests that speciation was sympatric and that speciation and adaptation are related intimately, as discussed elsewhere in this volume (Chapter 1). Despite years of work on the evolutionary ecology of Caribbean anoles, little direct evidence exists to determine the geographic context in which speciation has occurred. However, data do exist from which the hypothesis that speciation and adaptation are related can be evaluated. Although far from conclusive, these data indicate that no necessary link exists between these two processes.

This argument has two lines of evidence. First, many examples indicate that adaptation can occur in the absence of speciation in Caribbean anoles. As discussed above, comparisons indicate that populations of the same species adapt to different environments in both the Lesser and Greater Antilles. Although detailed phylogeographic studies have been conducted for only a few of these species, such studies confirm that populations of several of these species are part of a single species, rather than distinct but unrecognized species (see Section 16.2). Thus, adaptive evolution can occur in the absence of speciation.

Conversely, speciation can occur, but it produces species that are adaptively undifferentiated. Evidence for this claim comes from complexes of allopatrically distributed species on the islands of Cuba and Hispaniola. On Cuba, for example, the monophyletic *alutaceus* group of grass anoles contains 14 species. Two of these species are found island wide, but the other 12 have narrow distributions, mostly centered on different mountain ranges. For the most part, these species are little differentiated, both ecologically and morphologically. However, differences in their dewlaps suggest that the species are isolated reproductively and, thus, valid species. Similar complexes of mostly allopatric species are seen in the *sagrei* group in Cuba (14 species) and the *cybotes* group in Hispaniola (eight species), as well as in a number of smaller complexes. This phenomenon reveals both the efficacy of allopatric speciation and the possibility that speciation can occur without substantial adaptive divergence (it could, of course, be suggested that these populations initially arose in sympatry and only later attained allopatry, but given the small and disjunct ranges – often centered on mountain ranges – of many of these species, such an explanation seems decidedly unlikely).

Adaptation can occur without speciation, and speciation can occur without adaptation. Does this indicate that adaptive, sympatric speciation does not occur? Of course not. No one would claim that adaptive speciation must account for all of the speciation and adaptation evident in the anole radiation. Moreover, although adaptive differentiation does occur intraspecifically, the amount of divergence is relatively limited and does not approach the extent of differentiation between different habitat specialists, such as the difference between twig and trunk–ground specialists. All that can be said is that no evidence exists to support the occurrence of adaptive speciation, and alternative mechanisms have been demonstrated that can produce adaptation and speciation in other ways.

An alternative hypothesis is that species arose in allopatry and perhaps differentiated adaptively to some extent. Then, when they came back into sympatry, interspecific interactions – most likely competition – lead to habitat partitioning and subsequent adaptation and specialization to different habitats. Such a scenario of character displacement is a standard view of the manner in which adaptive radiation proceeds (Grant 1986; Schluter 2000). Much experimental and observational data indicate that anole species do alter their habitat use in the presence of congeners and some data suggest that this leads to morphological change (reviewed in Losos 1994). However, the extent of habitat shifts and morphological change are minor relative to the differences between different habitat specialists. Thus, further investigation is required of both adaptive speciation and character displacement hypotheses for the adaptive radiation of anoles.

An alternative link between adaptation and speciation.   Adaptation and speciation may be linked directly in a manner different from that envisioned in the adaptive speciation hypothesis. As discussed above, speciation in anoles results when changes occur in the species-recognition systems, the dewlap and head-bobbing patterns. These systems may evolve, in turn, when populations adapt to new habitats. The reason is that these signals, being visual, must be seen to be effective. Consequently, if a population for some reason occupies a new habitat in which the visual environment is different, natural selection may favor evolutionary change in the signal to maximize the effectiveness of communication, both intra- and interspecifically [this is the sensory drive hypothesis of Endler (1992); see Fleishman (1992, 2000)].

For example, the most visible color for a dewlap varies depending on the environment. In closed forests, light levels are reduced and the light that does penetrate is mostly in the yellow and green parts of the spectrum. As a result, white or yellow dewlaps are optimal because they reflect the little light that is available. By contrast, in open areas there are few limits in the light availability and the most effective dewlaps are nonreflective and dark, producing a darkness contrast to the bright background (Fleishman 1992, 2000). Indeed, in a survey of Caribbean anole dewlap colors, Fleishman (1992, 2000) found that most closed-forest anole species have yellow or white dewlaps, whereas the majority of open-habitat species have orange, red, black, or blue dewlaps.

Thus, if an open-habitat species for some reason moved into a closed habitat [or perhaps the open area itself is transformed and the populations stay put, akin to Vanzolini and Williams' (1981) vanishing refugium hypothesis], selection would favor change in dewlap color to maximize the ability to communicate intraspecifically. An indirect by-product of this adaptation, however, may be that the population becomes reproductively isolated from other populations remaining in the original habitat. Thus, adaptation to a new visual environment may lead to speciation. This scenario could operate whether the populations are initially sympatric or allopatric; it could easily result as allopatric populations differentiate in different habitats, but it also might be the result of adaptive speciation in sympatry.

Similar arguments may apply to head-bobbing patterns. To be detected, the movements of a displaying lizard must differ from movements of the background vegetation. In habitats in which there is not much background vegetation, or in which the vegetation tends not to move much, small amplitude displays may be conspicuous. By contrast, when background vegetation moves greatly, more pronounced displays may be needed to catch the attention of other lizards (Fleishman 1992). As with dewlap color, adaptive change in head-bobbing patterns thus could produce speciation.

In addition, new habitats may select for other features as well as those used in communication, because the structural and thermal environments may differ also. Consequently, adaptations in a variety of different traits may all be linked to speciation because of the relationship between communication, light environments, and specific mate-recognition systems. This hypothesis is not easy to test, but important advances in our understanding of anole visual ecology and physiology are being made by Fleishman and colleagues (Fleishman *et al.* 1997; Persons *et al.* 1999; Leal and Fleishman 2002).

Phenotypic plasticity and anole adaptive radiation. Recent work has suggested an intriguing new hypothesis: phenotypic plasticity might play an important role in anole adaptation and adaptive radiation. The background for this work was a pair of studies on populations of *A. sagrei* in the Bahamas. A comparison of natural populations revealed a correlation between the mean diameter of perches used and mean relative limb length (Losos *et al.* 1994). Further, examination of populations experimentally established on tiny Bahamian cays 10–15 years previously (Schoener and Schoener 1983) revealed a similar trend (Losos *et al.* 1997). These correlations paralleled – at a lesser scale – the trend seen among anole habitat-specialists, exemplified by the extremely long-legged trunk–ground anoles that inhabit tree trunks and the short-legged anoles that inhabit twigs. Thus, one interpretation is that microevolutionary change over relatively short periods produced the same pattern as evident over macroevolutionary time; by extrapolation, the latter was simply an extension of the former over longer periods of time.

However, an alternative hypothesis is that limb length is a phenotypically plastic trait. Perhaps young *A. sagrei* that grow up using narrower supports develop shorter limbs than individuals that grow up using broader supports. Surprisingly, a laboratory study found just that: hind-limb length is a phenotypically plastic trait affected by perch diameter (Losos *et al.* 2000).

This study has two implications. More narrowly, it suggests that the differences observed between both experimental and natural populations in the Bahamas may be the result of plasticity, rather than genetic differentiation. Studies in which individuals from different islands are raised in the same environment are needed to test this hypothesis more directly.

More generally, these findings suggest the intriguing possibility that plasticity may play an important role in adaptive evolution. By permitting lizards to occupy a new habitat in which they otherwise might not be able to survive, plasticity may allow anoles to occupy new habitats. Once in these habitats, lizard behavior might change and, as new mutations arise, these could be selected for and thus greatly accentuate the initial, relatively minor, changes in limb length. In this way, plasticity might represent the first stage in major adaptive shifts. If nothing else, it is striking that phenotypic plasticity produces the same morphology–environment correlation as observed among habitat specialists [note that the differences in limb length among habitat specialists are vastly greater than those produced in the plasticity experiment and surely represent genetic differences (Losos *et al.* 2000)].

The hypothesis that plasticity might be important in adaptive differentiation was originally put forth 50 years ago by Schmalhausen (1949), Waddington (1953a, 1953b), and others. Long ignored, the idea has recently been revived (West-Eberhard 1989; Schlichting and Pigliucci 1998). Anoles may represent a good system in which to further explore these ideas.

16.4 Concluding Comments

Roger S. Thorpe and Jonathan B. Losos

The previous two sections on one of the most speciose genera of amniote vertebrates emphasize the Lesser Antilles (Section 16.2) and the Greater Antilles (Section 16.3) systems. These two systems have many basic differences. The former is dominated by solitary anole species (or at most two natural species in sympatry); no, or relatively low, congeneric competition; numerous colonization events between island banks associated with "speciation"; and somewhat arbitrary allopatric species. The latter is dominated by multiple-species sympatry, competition, few colonization events, speciation within islands, and a relatively high degree of confidence in what constitutes a species.

Despite these basic differences, the conclusions suggested by the two sections are very similar. In both the Lesser and Greater Antillean models, there is substantial speciation and substantial evidence of adaptation. In the former, evidence of adaptation comes from intraspecific, within-island adaptation to different habitat types (ecotypes) supported by correlational evidence, parallels, translocation experiments, and common-garden experiments. In the latter it comes from convergent habitat specialization among species (ecomorphs) supported by functional and physiological studies. Even though both speciation and adaptation are substantial in both systems, currently neither system shows hard evidence of adaptive speciation.

Perhaps we have not been able to reveal any clear-cut cases of adaptive speciation because it does not occur in this group. However, it may occur and our failure to reveal it may be because either the criteria used are too demanding or the appropriate studies have not been carried out. If the definition of adaptive speciation in Chapter 1 is used, an important subset of cases involve speciation in sympatry. In this subset of cases, anoles are unlikely models for adaptive speciation as there are no well-supported cases of sympatric speciation in squamates with normal sexual reproduction and without chromosomal changes. However, if adaptive speciation could occur where adaptation plays a key role in the speciation of populations in ecological (parapatric) contact, as indicated in Chapter 7, then adaptive speciation may play a role in the speciation of anoles and other squamates.

For the Greater Antilles, Losos (Section 16.3) suggests the possibility that adaptation to new habitats could trigger speciation indirectly by leading to changes in the behavioral or morphological facets of sexual signals, but this remains to be demonstrated. So in the Greater Antilles, perhaps, parapatric adaptive speciation may have played a part that future studies will reveal. However, the dewlap, which is thought to play a key role in sexual signaling, may show relatively little difference among habitat types within Lesser Antillean islands (e.g., *A. oculatus* in Malhotra 1992), an observation that does not provide support for the above proposal. Yet, more recent studies using spectrometric analysis of dewlap hue, in the *roquet* group, show distinct variation in relation to habitat type (Thorpe 2002; Thorpe and Stenson 2003).

With regard to the Lesser Antilles model, there is evidence (Ogden and Thorpe 2002) of a reduction in gene flow between parapatric habitat forms (incipient speciation), but this is not complete speciation, and even if these parapatric forms have become partially isolated *in situ*, this is not necessarily adaptive speciation (Chapter 1). With other contact zones within Martinique, which may warrant full species recognition, further work is required to exclude confidently a role for divergence in allopatry.

Acknowledgments   Roger S. Thorpe thanks the Natural Environment Research Council, the Linnean Society, and the Leverhulme Trust for support.

17

Adaptive Radiation of African Montane Plants

Eric B. Knox

17.1 Introduction

The mountains of central and eastern Africa constitute a model system for studying the ecological context and consequences of speciation and adaptation in plants.
Unlike most mountains elsewhere in the world, these form "islands in the sky" because their isolated, predominantly volcanic peaks rise far above the surrounding
plains and plateaus. These mountains are sufficiently tall to support an altitudinally stratified set of vegetation zones (Hedberg 1951; Figure 17.1), they are
sufficiently numerous to provide repeats of this "natural experiment" (Hedberg
1957; Mabberley 1973), and their age and arrangement around the Lake Victoria
basin permit the effects of time and position to be disentangled (Knox and Palmer
1998). Distances between mountains range from 50 to 1000 km, and the equatorial
location minimizes the effect of climatic variation, with the vegetation displaced
less than 1000 m downward during glacial maxima (Hamilton 1982). This model
system can be conceptualized as an array of insular habitats amenable to biogeographic analysis (Knox 1999). For groups of plants such as the giant senecios
(*Dendrosenecio*) and giant lobelias (*Lobelia*) that occupy cells of this array, the inherent bifurcating process of lineage separation during adaptive radiation necessitates some repetition of the evolutionary process as repeated altitudinal speciation
(into the set of habitats present on each mountain), repeated geographic speciation
(after colonization of a mountain), or a combination of both (Knox and Palmer
1998; Figure 17.2). This repetition of process within each group has generated a
morphological "mosaic of variation" (Mabberley 1973) through the combined effects of ancestry and adaptation, with occasional instances of reticulate evolution.

The stunning examples of parallel or convergent morphological evolution
among the giant senecios or giant lobelias are overshadowed by the repeated convergent evolution between these two groups (Plate 11). This convergence provides
a priori evidence that the unusual features of these plants are adaptations to proxi-
mal environmental factors. The alpine zone was characterized by Hedberg (1964)
as having "summer every day, winter every night" because the strong diurnal temperature fluctuation vastly exceeds the annual variation in average daily temperature. This unusual climate has important consequences for plant water-relations,
photosynthesis, and growth and development. In the alpine zone the most extreme
examples of the giant-rosette growth form are found, with adaptations that include:

Figure 17.1 The tall African mountains support an altitudinally stratified set of vegetation zones. Temperature and precipitation generate the broad patterns of zonation, but the topography of each mountain influences the fine patterns, particularly with respect to water and cold-air drainage.

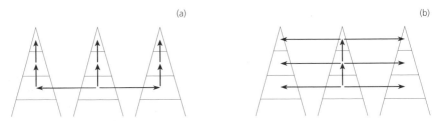

Figure 17.2 The mountains can be modeled as a two-dimensional array of insular habitats. Adaptive radiation throughout the cells of the array necessitates some repetition of evolutionary process, as (a) repeated altitudinal speciation, (b) repeated geographic speciation, or a combination of both.

- Massive leaf rosettes in which leaf development occurs in a large "apical bud" (Figure 17.3a);
- Internal water storage in the pith of the stem (Figure 17.3b);
- Insulation of the stem by marcescent foliage (Figure 17.4a);
- Secretion and impoundment of ice-nucleating polysaccharide fluids (Figure 17.4b);
- Nyctinastic leaf movement (Figure 17.5).

At successively lower elevations the diurnal temperature fluctuations are less extreme, the average daily temperature steadily increases, and the growth form and ecology of the giant senecios and giant lobelias reflect the increased influence of biotic factors (such as competition for light) over abiotic factors (such as nightly frost).

Two approaches are used to study adaptive speciation in plants. Part B discusses the ecological mechanisms responsible for reproductive isolation and speciation. In this chapter two phylogenetic case studies designed to assess common patterns of speciation and adaptation are presented. There are two reasons for studying adaptive speciation in a broader context of adaptive radiation:

- The underlying concept of adaptive radiation is that diversification was rapidly driven by natural selection.

(a)

(b)

Figure 17.3 (a) Leaf rosette of *L. telekii* at 3800 m on Mt Kenya. Leaves develop almost to maturity in the "apical bud" at the center. (b) Longitudinal section through an individual of *D. adnivalis* subsp. *adnivalis* var. *adnivalis* at 3975 m on the Ruwenzori. The primary stem has a wide pith that provides internal water storage.

■ Numerous species constitute a second dimension of "repeats" in a natural experiment, which can be assessed for common features indicative of general phenomena.

The comparison of the giant senecios and giant lobelias adds a third dimension to the repeated structure, and thereby extends the generality of the conclusions. The co-occurrence of representatives from each group in the various habitats (Plate 11) permits comparative studies of the functional significance of the putative adaptations, which eventually will bridge the two approaches.

In cladogenic speciation, one species enters the process and two species emerge, but how is adaptive speciation different from speciation in general? In allopatric speciation, initially the gene pool is subdivided geographically. If no divergence occurs, this subdivision simply creates a disjunction in the distribution of a single species. If natural selection results in the acquisition of adaptations in one or both of these initially isolated populations, a weak form of adaptive speciation has occurred (ecological speciation in an allopatric setting; see Chapter 1), but natural selection was not responsible for the initial separation of the gene pool. In these cases, it is necessary to demonstrate that the distinguishing features became prevalent because of selection for a particular function, and that they are not simply a consequence of genetic drift (strict allopatric speciation; see Chapter 1). A

(a)   (b)

Figure 17.4 (a) *D. kilimanjari* subsp. *cottonii* at 3775 m on Mt Kilimanjaro. The marcescent foliage creates a boundary layer of still air that insulates the stem from the extreme fluctuations of environmental temperature. The old, dead leaves at the base have been retained for more than 100 years, but some at the front have been lost, and the bark exposed. The vegetative and reproductive history is preserved in the architecture of each plant. This individual flowered the first time when approximately 2 m tall, produced two branches that grew about 1 m before reproducing again, with three terminal branches produced on the left fork and two terminal branches on the right. (b) *D. brassiciformis* at 3875 m on the Aberdares. The leaf rosette secretes and impounds an ice-nucleating polysaccharide fluid that buffers the temperatures experienced by the apical meristem and developing leaves.

strong form of adaptive speciation occurs when separation of the original gene pool is driven by natural selection. These cases must be sought in situations conducive for sympatric or parapatric speciation. Of particular interest are situations in which evolutionary innovation at the periphery of the species' range allows speciation into an adjacent habitat, but in which the initial isolation associated with peripatric speciation is not a prerequisite. The features that become prevalent because of natural selection for a particular function during speciation should show a goodness-of-fit with the ecological factors that distinguish the niche of a species from that of its close relative. Conversely, these ecological factors should be the selective forces responsible for the separation of the ancestral gene pool.

This chapter starts with an introduction to the African mountains and the unusual features of the giant senecios and giant lobelias. The giant senecios only occur on mountains over 3300 m within 4°C of the equator. The giant lobelias have a broader distribution, but this discussion focuses on one group that diversified on

(a)   (b)

Figure 17.5 Nyctinasty (a) slows radiative heat-loss at night, but (b) the leaf rosette reopens during the first hour after sunrise.

the same mountains as the giant senecios. The inferred biogeographic history of each group is summarized in brief narratives that precede more detailed discussion of adaptive speciation. The current taxonomy of each group (Knox 1993) includes some terminal taxa that originally were described as distinct species, but are now treated as subspecies to reflect the degree of morphological differentiation and/or the apparent degree of reproductive closure. Arguments could be made for reinstating most of these subspecies at a specific rank, but to the extent that evolution is an ongoing process and speciation is not instantaneous, the ranking of these terminal taxa is somewhat arbitrary. In all cases, the biogeographic separation of the terminal taxa is clear, and they are thought to be distinct evolutionary lineages. The discussion includes speciation and subspeciation in the analysis, with appropriate wording to reflect the current taxonomy.

17.2   Vegetation Zones on African Mountains

The African mountains support a series of vegetation zones. Ascending the mountains, one initially encounters the evergreen montane forest, characterized by relatively warm temperatures and abundant precipitation. Average temperature steadily decreases with increasing elevation, and mist forest grows where the clouds generally form. The frequency and severity of frost delimits a subalpine zone characterized by the presence of *Erica*, the giant heath. This ericaceous zone is present as scrubby moorland on the drier mountains, where fire is common, but on the wetter mountains and in protected sites the *Erica* functions as a frost-tolerant extension of the mist forest. The alpine zone extends to the upper limit of life (Figure 17.6). The average temperature is low, but the diurnal amplitude is large. Superimposed on this elevational zonation are the topographic effects of water and cold-air drainage. The mid-elevation meadows and the valley bottoms in the alpine zone are wetter and colder than the surrounding slopes.

Figure 17.6 Plants in the alpine zone must tolerate hard, nightly frosts to take advantage of the favorable daytime conditions. The drainage patterns of water and cold air in the Telekii Valley of Mt Kenya create an ecological abutment between the alpine vegetation on the slopes (with large populations of *D. keniodendron* and *L. telekii*) and the wet alpine vegetation in the valley (with large populations of *D. keniensis* and *L. gregoriana* subsp. *gregoriana*).

The island-like distribution of each vegetation zone across this montane system creates a stratified set of habitat archipelagoes. For ease of presentation, this three-dimensional system can be conceptualized as a two-dimensional array of insular habitats by treating geographic distance between mountains and adjacency of vegetation types as relevant axes (see Figure 17.2). Each adaptive radiation is analyzed as a space-filling process of successive colonization of the cells in the array, with the phylogenetic relationships, the proximity of mountains, and the adjacency of habitats used as constraints to seek the most parsimonious biogeographic reconstruction (Knox 1999). These biogeographically explicit models specify the location of inferred ancestral lineages, and can be used to assess the pattern of speciation.

Every species has limits to its biogeographic range. These limits result from the complex interactions of the intrinsic factors that determine a species' niche with the extrinsic factors that determine the distribution of suitable habitat. In the center of its range, each species obviously has an "ecological strategy" that works in that habitat, but at the periphery of its range, this ecological strategy becomes ill-suited to the local circumstances. Seed dispersal is passive, and plants are constantly "testing" their range limits. Evolutionary innovation along the periphery could enlarge the niche space, and thereby expand the range of suitable habitat. However, if organismal integration constrains the ecological strategy, evolutionary innovation may modify the ancestral strategy to produce a new strategy more suited to local circumstances. Adjustment of these constraints may allow the incipient species to flourish at the site of innovation and to expand its biogeographic range to the new limits as determined by its modified ecological strategy. The success of incipient speciation along the periphery of the range depends, in part, on the size of the reproductive neighborhood and the strength of genetic integration within the original species relative to the selective advantage of the innovation. If the innovation provides access to a habitat that was previously unavailable because of the constraints of the ecological strategy, then speciation enlarges the collective range of these diversifying lineages. In contrast to a process that partitions the ancestral range through more specialized resource utilization, an enlargement of the range involves utilization of a spatially distinct, alternative resource. In this situation, reproductive isolation might be attained easily, particularly if the constraints of organismal integration involve trade-offs that preclude occupation of the enlarged range with a single ecological strategy.

17.3 The Giant Senecios and Giant Lobelias

The giant senecios and giant lobelias share several broad similarities:

- Both groups have evolved from herbaceous ancestors, but in neither case have the giants evolved from low-elevation herbaceous ancestors in Africa (Knox *et al.* 1993; Knox and Palmer 1995a).
- Diversification of both groups involved no change in chromosome numbers (Knox and Kowal 1993).
- Practically no distributional overlap occurs between species within each group, although ecological abutments are common (see Figure 17.6). In contrast, pairs of giant senecio and giant lobelia have broadly overlapping distributions on each mountain, which suggests that they partition the habitats in similar ways.
- Both groups share a suite of unusual morphological features (Box 17.1) that show strong elevational patterns. Species at lower elevations have slender primary stems with small, lax leaf rosettes and relatively small inflorescences. Species at higher elevations have much wider primary stems, the leaves are more numerous and densely packed in the leaf rosette because of limited internode elongation, and the inflorescences are much larger. Within each species (or subspecies), these morphological features are relatively invariant and the patterns do not arise from ecophenotypic plasticity (as evidenced by growth under common greenhouse conditions). Additional morphological patterns are present within each group. The giant senecios at lower elevations have erect flower heads with relatively few disk florets and prominent ray florets. At higher elevations, the flower heads are pendulous with more numerous disk florets and the ray florets are greatly reduced or absent. The giant lobelias have flowers that are green or white at lower elevations, red or purple at middle elevations, and blue at the higher elevations. Unfortunately, little is known of the comparative pollination biology of these plants.

Box 17.1 Morphological adaptations of giant senecios and giant lobelias

Giant-rosette growth form. A giant leaf rosette is formed when numerous, large leaves are produced on a stem with limited internode elongation (Figure 17.7a). Developing leaves enlarge almost to the mature size within a large "apical bud" (Figure 17.3a). The large leaf area enables high levels of carbon assimilation, and the limited number of vegetative meristems results in the "cabbage-tree" appearance (Figure 17.7b).

Pith volume. The wide primary stem consists mostly of pith, which is an economic construction technique that only slightly diminishes certain biomechanical properties. At high elevation, the pith forms an internal water reservoir that permits photosynthesis during the early morning when low temperatures make soil water unavailable (Figure 17.3b). In vegetation with a closed canopy at lower elevation, the economic stem architecture facilitates competition for light.

Marcescent foliage. Retention of the old, dead leaves involves the maintenance of physical connections when the protective layer forms, which separates the living stem from the senescing leaf, followed by a process of controlled decomposition to yield the mummified marcescent foliage (Figure 17.7a). The resultant boundary layer of still air keeps the stem temperature virtually constant. The densely packed leaf bases also insulate the living stem tissue from the heat of fires that readily consume the mummified laminas.

Nyctinasty. The leaf rosettes of some species close during late afternoon and rapidly reopen just after sunrise (Figure 17.5). The closure slows radiative heat-loss, and thereby extends the period of favorable temperature for growth and development.

Ice-nucleating polysaccharides. In the high-elevation species, the mesophyll cells are coated with modified pectins that are effective ice nucleators. At night, these ice-nucleating polysaccharides induce the formation of small ice crystals in the intercellular space without causing frost damage. The protoplast plasmolyzes, but does not freeze. When the leaf warms, the ice melts, the water is reabsorbed, and physiological activity resumes. The developing leaves of the "apical bud" are also coated with a polysaccharide slime that may lubricate and provide frost protection for these delicate tissues. Rain washes this slime from the surface of the "apical bud", and this mixture can accumulate in the leaf rosette. However, some species secrete a polysaccharide-containing fluid in the leaf rosette and impound it in the boat-shaped leaf bases that have ciliate margins (Figure 17.4b). The thermal mass of this fluid buffers the temperature fluctuations experienced by the apical meristem and developing leaves, and the ice-nucleating compounds stabilize freezing at the highest possible subzero temperature. The volume of the fluid is sufficient enough for it not to freeze entirely during the night.

The alpine climate of "summer every day, winter every night" means that plants must endure the harsh conditions of the night to take advantage of the favorable daytime conditions. The giant-rosette growth form is a successful strategy in this habitat. The thermal mass buffers the temperature fluctuations, and thereby limits the harmful extremes and lengthens the daily period conducive to physiological

(a)

(b)

Figure 17.7 (a) The densely packed leaf rosette of *D. kilimanjari* subsp. *cottonii* at 3700 m on Mt Kilimanjaro (with leaves at the front cut to expose the arrangement). At the end of their photosynthetic life, senescing leaves undergo a process of controlled decomposition to form the mummified marcescent foliage. (b) An old individual of *D. elgonensis* subsp. *barbatipes* at 3975 m on Mt Elgon has undergone five cycles of reproduction and branching. These plants lack the marcescent foliage commonly found at high elevation, but rapidly produce a thick, furrowed bark.

activity. The internal water reservoir decouples the plants from the temporal availability of soil water, and the marcescent foliage keeps this water in an accessible state. The large leaf area provides the capacity for high levels of carbon assimilation, and this tight feedback loop generates the resources needed to construct the giant-rosette growth form. Nyctinasty (see Box 17.1) lengthens the period when newly acquired carbon can be reinvested in vegetative growth. The ice-nucleating polysaccharides prevent frost damage, and the impounded fluid buffers temperature fluctuations, which may allow the plant to optimize physiological activity within this narrower range of temperature it experiences.

17.4 Phylogenetic Patterns and Biogeographic Interpretation

The giant senecios have been surveyed for chloroplast DNA (cpDNA) restriction-site variation (Knox and Palmer 1995b) and nuclear ribosomal DNA (nrDNA) sequence variation (Knox and Panero, unpublished). A parallel survey of cpDNA restriction-site variation has been completed for the giant lobelias (Knox and Palmer 1998), but only preliminary nrDNA sequences are available for this group.

Interested readers should consult the original literature for the methods and a discussion of the potential limitations of cpDNA data for phylogenetic reconstruction (Knox and Palmer 1998). A closed, analytical method for biogeographic reconstruction of adaptive radiation in insular habitats (Knox 1999) formalizes the "reasoned argumentation" used in the previous work. This method integrates biogeographic information up to the appropriate level of biological organization and overcomes limitations of previous rule-based methods and character-state optimization that uses atomized data. In the discussion below, remember the following:

■ Empirical evidence is derived from *sampled* individuals, growing at particular sites, which are identified as members of a given species based on their morphological attributes.
■ A cladogram provides a phylogenetic *estimate* of the pattern of descent.
■ The existence and nature of *inferred* ancestral lineages depends on the historical accuracy of the phylogenetic estimate of the pattern of descent and the methods used to reconstruct the pattern of modification within this phylogenetic context.

In this brief synopsis of the origin and radiation of each group on the tall equatorial mountains, the appropriately cautionary language of sampling, estimation, and inference has been omitted. The resultant narratives provide a concise overview of the evolutionary history of each group, but the most parsimonious account is, by definition, the simplest interpretation, and future evidence may demonstrate instances in which the history was more complex.

The giant senecios

The giant senecios initially colonized Mt Kilimanjaro at high elevation within the past one million years to form *D. kilimanjari* (Figure 17.8). Downward altitudinal speciation resulted in the formation of *D. johnstonii* on Mt Kilimanjaro, and dispersal to Mt Meru established *D. meruensis*. Dispersal from Mt Kilimanjaro to the Aberdares established *D. battiscombei*, and speciation into the wet alpine habitat on the Aberdares resulted in the formation of *D. brassiciformis*. Dispersal from the Aberdares to Mt Kenya established a second insular population of *D. battiscombei*. Altitudinal speciation on Mt Kenya resulted in the formation of *D. keniodendron* and the "dwarf" *D. keniensis*. Dispersal from Mt Kenya back to the Aberdares established a second insular population of *D. keniodendron*. Dispersal from the Aberdares to the Cherangani Hills established *D. cheranganiensis* subsp. *cheranganiensis*, and altitudinal (sub)speciation into the wet alpine habitat resulted in formation of the "dwarf" subsp. *dalei*. Dispersal from the Aberdares to Mt Elgon established *D. elgonensis*, but prior to the altitudinal (sub)speciation of subsp. *barbatipes* from subsp. *elgonensis*, dispersal from Mt Elgon to the Virunga established *D. erici-rosenii*, and dispersal from Mt Elgon to the Ruwenzori established *D. adnivalis*. Dispersal from the Virunga to Mt Kahuzi established a second insular population of *D. erici-rosenii*, altitudinal (sub)speciation on the Virunga resulted in the formation of subsp. *alticola* from subsp. *erici-rosenii*, and dispersal from the Virunga to the Ruwenzori established a third insular population of

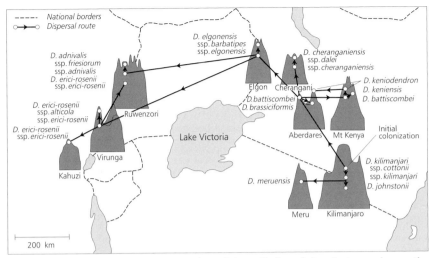

Figure 17.8 Biogeographic reconstruction of the radiation of the giant senecios on the mountains of central and eastern Africa. *Source*: Knox and Panero (unpublished).

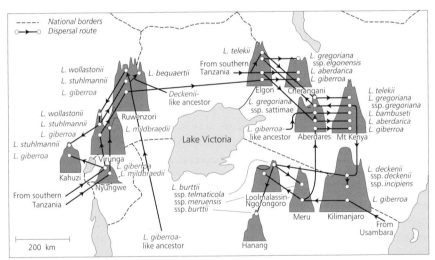

Figure 17.9 Biogeographic reconstruction of the radiation of the giant lobelias on the mountains of central and eastern Africa. The relationship of *L. gregoriana*, *L. deckenii*, and *L. burtii* to *L. aberdarica*, *L. bambuseti*, and *L. telekii* is not well resolved and is not indicated. *Source*: Knox and Palmer (1998).

subsp. *erici-rosenii*. Upward altitudinal (sub)speciation on the Ruwenzori resulted in the formation of subsp. *friesiorum* from subsp. *adnivalis*.

The giant lobelias

The giant lobelias initially colonized ancient upland features in Tanzania during the Miocene. *L. thuliniana* is a relictual species, endemic to the Iringa and Dabaga Highlands of central Tanzania, and is the closest relative of the species found on the tall mountains. From this *L. thuliniana*-like ancestral species, two early lineages emerged. One lineage gave rise to the widespread *L. giberroa*, and altitudinal speciation from an *L. giberroa*-like ancestral species resulted in the formation of the Western Rift endemics *L. wollastonii* and *L. stuhlmannii* (Figure 17.9). *L. bequaertii* had a hybrid origin on the Ruwenzori that involved an *L. wollastonii*-like ancestral species and an early colonist from the Eastern Rift *L. deckenii* group (discussed below). Dispersal from the Ruwenzori to the Virunga established a second insular population of *L. wollastonii*. Downward altitudinal speciation on the Virunga resulted in the formation of *L. stuhlmannii*, which subsequently dispersed to Mt Kahuzi and back to the Ruwenzori. Altitudinal speciation of the *L. thuliniana*-like ancestral species in southern Tanzania resulted in the formation of an *L. mildbraedii*-like ancestral species that dispersed to:

- Mt Elgon, where it established an *L. aberdarica*-like ancestral species that subsequently gave rise to the Eastern Rift endemics;
- Ethiopia, where it diversified into two endemic species (*L. acrochilus* and *L. rhynchopetalum*, not discussed in this treatment); and then
- The Western Rift mountains, which established an 800 km disjunction in the contemporary distribution of *L. mildbraedii*.

Altitudinal speciation from the *L. aberdarica*-like ancestral species on Mt Elgon resulted in the formation of *L. telekii*, which subsequently dispersed to the Aberdares and Mt Kenya. At about the same time, dispersal and a shift into the wet alpine habitat established the ancestral species of the traditionally recognized *L. deckenii* group, which rapidly dispersed to the remaining peaks of the Eastern Rift mountains, to establish the insular species and subspecies. Somewhere in this bout of "island hopping", one lineage dispersed to the Ruwenzori and contributed to the hybrid origin of *L. bequaertii*. Dispersal of *L. gregoriana* from Mt Kenya to the Aberdares established subsp. *sattimae*. Dispersal from the Aberdares to Mt Elgon established subsp. *elgonensis*, with a second insular population established by dispersal to the Cherangani Hills. Dispersal to Mt Kilimanjaro established *L. deckenii* subsp. *deckenii*, where downward altitudinal (sub)speciation resulted in the formation of subsp. *incipiens*. *L. burttii* originated on Loolmalassin, and dispersal to Mt Hanang and Mt Meru established subsp. *burttii* and subsp. *meruensis*, respectively. Dispersal of *L. aberdarica* established the insular populations on the Kenyan mountains. *L. bambuseti* had a hybrid origin (which involved *L. aberdarica* and an *L. giberroa*-like ancestral species) on the Aberdares, and subsequent dispersal to Mt Kenya established a second insular population.

17.5 Adaptive Speciation

The biogeographically explicit reconstruction of each adaptive radiation allows the pattern of speciation (and subspeciation) to be characterized. This characterization obviously depends on the historical accuracy of each reconstruction, but some aspects do not depend on the details of the narratives presented above. For example, one dispersal event is required to colonize each mountain in the system. The calculated minimum number of events that could account for the contemporary distribution of each group can be compared with the minimum number of inferred events based on the phylogenetic estimate for each radiation.

Dispersal and diversification

The phylogenetic estimate for the giant senecios is consistent with the hypothesis that each mountain was colonized once, with two exceptions. One additional dispersal event is required to account for the Aberdare population of *D. keniodendron* and a second additional dispersal event to account for the Ruwenzori population of *D. erici-rosenii*. In both cases, no morphological divergence occurred after the disjunct populations were established. In contrast, morphological divergence did occur after the initial colonization of each mountain, with one exception, namely, the colonization of Mt Kenya by *D. battiscombei*. Seven speciation events therefore occurred in allopatry. This morphological differentiation may be adaptive, but nonadaptive consequences of founder effect cannot be dismissed without investigating the functional significance of the traits in question. The remaining nine speciation (or subspeciation) events occurred on a given mountain as a lineage diversified to occupy different habitats. In the four cases of fully fledged speciation, one involved downward movement into the montane forest (*D. johnstonii*), two involved movement into the wet alpine habitat (*D. brassiciformis* and *D. keniensis*), and one involved upward movement into the upper alpine habitat (*D. keniodendron*). In the five cases of subspeciation, one involved downward movement into the mist forest (*D. kilimanjari* subsp. *kilimanjari*), one involved movement into the wet alpine habitat (*D. cheranganiensis* subsp. *dalei*), and three involved upward movement into the alpine habitat (*D. elgonensis* subsp. *barbatipes*, *D. erici-rosenii* subsp. *alticola*, and *D. adnivalis* subsp. *friesiorum*). In summary, seven of the 16 terminal taxa were derived in allopatry, nine were derived through habitat diversification, and in four instances colonization established an insular population with no morphological divergence.

The phylogenetic estimate for the giant lobelias indicates that each mountain was colonized more than once, except for Mt Hanang, which only has one recorded species. It requires 34 colonization events to account for the 37 insular populations on the 12 mountains under consideration. The inferred origins of *L. giberroa* and *L. mildbraedii* in southern Tanzania leave 14 species and heteronymic subspecies that originated on the tall mountains, nine of which evolved in allopatry (*L. wollastonii*, *L. aberdarica*, and seven of the eight Eastern Rift terminal taxa in the *L. deckenii* group). The origins of *L. wollastonii* and the ancestor of the *L. deckenii* group each involved the colonization of a new mountain and a shift in habitat from that of the inferred ancestral species. *L. bequaertii* originated from hybridization between one resident and one colonizing species, and it still occupies the wet alpine habitat of the inferred colonist from the Eastern Rift. *L. bambuseti* also originated from hybridization, and its occupation of the mist forest involved a habitat shift predicated on the amalgamation of traits from parental species that grew in wet meadows and montane forest, respectively. The three remaining speciation (or subspeciation) events involved one case of upward movement into the alpine zone (*L. telekii*), and two cases of downward movement into the mist forest (*L. stuhlmannii* and *L. deckenii* subsp. *incipiens*). In summary, nine of the 14 terminal taxa were derived in allopatry, three were derived through habitat diversification, two had hybrid origins, and in 23 instances colonization established an insular population with no morphological divergence.

Convergent evolution

The giant-rosette growth form is a plesiomorphic feature of the giant senecios that evolved during the origin of *Dendrosenecio* at high elevation on Mt Kilimanjaro. During the radiation of this group, the most extreme development occurred in *D. keniodendron*. The convergent evolution of the giant-rosette growth form occurred in the "dwarfed" species that occupy the wet alpine habitat (*D. keniensis* and *D. cheranganiensis* subsp. *dalei*) and the species with slender primary stems that occupy the mist forest and montane forest. In contrast, the common ancestral species of the giant lobelias grew at relatively low elevation, and convergent evolution of the compact leaf rosette occurred in *L. wollastonii*, *L. telekii*, and the *L. deckenii* group as they reached the alpine zone independently.

A wide primary stem is necessary to support a massive leaf rosette, but the "dwarfed" species of giant senecio and the relatively small alpine species of giant lobelia on the Eastern Rift mountains maintain a large pith volume for water control, rather than mechanical support. Relatively slender stems are a plesiomorphic feature of the giant lobelia, but convergent evolution of more slender stems has occurred in all cases of downward speciation (or subspeciation) as an adaptation for rapid vertical growth.

Marcescent foliage is also a plesiomorphic feature of the giant senecios that has been lost in *D. johnstonii*. The leaves of *D. meruensis* and *D. erici-rosenii* are not deciduous, but the dead laminas readily decay to leave the leaf bases only. The alpine *D. elgonensis* subsp. *barbatipes* does not retain its dead leaves, but develops a thick, furrowed bark that provides insulation and fire protection (Figure 17.7b). Convergent evolution of marcescent foliage has occurred in *L. wollastonii* and the *L. deckenii* group. The leaves of *L. telekii* are not deciduous, but reproduction is initiated when the leaf rosette is still close to the soil surface.

Strong nyctinasty has evolved independently in *D. keniensis*, *L. telekii*, and the *L. deckenii* group. It has not been reported for other species, but systematic observations are still required.

Small quantities of fluid that contains ice-nucleating polysaccharides accumulate in the leaf-bases of *D. keniensis* and *D. cheranganiensis* subsp. *dalei*. *D. brassiciformis* secretes and impounds several liters of fluid in its modified leaf rosette, but this impoundment capacity is lost during reproduction when the apical meristem produces the inflorescence. Lateral branches are initiated at the base of the inflorescence, but these die before functional impoundment can be re-established. As a consequence, these plants are monocarpic. The subspecies of *L. gregoriana* also secrete and impound large quantities of fluid in the leaf rosette, and these subspecies grow adjacent to the giant senecios with this capacity. *L. telekii* secretes and impounds a similar fluid inside its hollow inflorescence, filling this closed cylinder about half way. When the fluid freezes, the phase change from liquid to solid releases the latent heat of fusion, which stabilizes the entire inflorescence at about 0°C. This species has unusually small flowers nestled within a mantle of long pubescent bracts. The resultant boundary layer creates one of the few cases in which plant trichomes function like mammalian hair, because the core temperature is warmer than the air throughout the night.

17.6 Concluding Comments

The giant senecios originated relatively recently at high elevation, they colonized other mountains at high elevation, and most colonization events led to allopatric speciation. The repeated evolutionary process occurred as species diversified to occupy the various habitats present on each mountain. The giant lobelias are an older group that initially colonized ancient upland sites in Tanzania, and diversified on the tall mountains after they arose. Allopatric speciation, speciation into adjacent habitats, and hybrid speciation contributed to their diversification. The giant lobelias colonized almost every mountain more than once, and in many cases no morphological or ecological divergence has occurred in these isolated populations.

On most of these mountains, the giant lobelias are inferred to have colonized sites before the radiation of the giant senecios, but Mt Meru and most of the Virunga volcanoes formed after both groups were present in the region. Colonization of each mountain presumably occurred via rare instances of long-distance seed dispersal. The giant senecios have seeds with a pappus of hairs, and the giant lobelias have very small seeds that shake out of the dry capsules when the wind blows. In both cases, the median distance of seed dispersal involves meters, not kilometers. The rare instances of long-distance dispersal may have been achieved when unusually large storms passed through the region, or when seeds became attached to migratory birds. Colonization of the mountains has obviously occurred, but not with sufficient frequency to maintain the disjunct populations as metapopulations that evolve cohesively. Although the probability of intermountain dispersal may have remained constant over time, the initial colonization of a site may limit the potential for establishment by seeds that arrive subsequently.

The lack of divergence in disjunct populations on different mountains raises interesting questions about the role of genetic integration within a species. Morphological and ecological constancy in disjunct populations could result if colonization occurred recently, and sufficient time has not elapsed for divergence to occur. However, if the amount of molecular variation provides some indication of the timing of colonization, then the apparent stasis cannot be attributed solely to the recency of events. The disjunct populations of *L. giberroa* have as much molecular variation as the species and subspecies that comprise the *L. deckenii* group (Knox and Palmer 1998). This comparison might suggest that conditions in the montane forest are more similar across mountains, and a common regime of stabilizing selection might account for the lack of divergence in *L. giberroa*. At higher elevations, the "summit effect" becomes more pronounced, and the greater variation in abiotic factors among mountains, with corresponding influences on biotic factors, might account for the divergence in the *L. deckenii* group through local adaptation. Alternatively, there may be differences in the capacity for change. If the constancy of disjunct populations does not arise from genetic integration within a species or the recency of colonization, then to what extent does genetic integration account for the morphological and ecological constancy within a species on a given mountain?

The model of punctuated equilibrium (Eldredge and Gould 1972; Gould and Eldredge 1977, 1993) predicts that such patterns of stasis are common, and change, when it does occur, proceeds rapidly in small, isolated populations. Colonization of a mountain would provide the small, isolated population at the periphery of a species' geographic range in which founder effect and adaptation might rapidly result in allopatric speciation, but how much genetic isolation at the periphery of a species' ecological range is required for speciation? The most dramatic evolutionary changes occurred during diversification into the various habitats present on a mountain, and the limits on ecological range should be the type of situation in which evolutionary innovation is highly favored by natural selection. Innovation that provides access to a new resource should concomitantly foster the assortative mating needed to drive adaptive speciation. This is particularly true in parapatric situations in which an incipient species can physically move into an adjacent habitat, but sympatric speciation involves a similar enlargement of collective niche space, even though individuals may share a broadly construed habitat.

The comparative patterns of convergent evolution in the giant senecios and giant lobelias indicate that both groups responded in similar ways to the prevailing environmental factors that structure their habitats (Plate 11). This occurred despite marked differences in the time frames and events that characterize their respective evolutionary histories. The broad patterns of range overlap between pairs of ecologically similar species from each group suggest that similar ecological strategies entail similar constraints. Not enough is known about the physiological performance of each species over its ecological range, the specific factors that limit each species' ecological range, or the trade-offs in form and function that may have been affected during speciation. The results to date provide the phylogenetic relationships needed to pose specific questions about individual speciation events and to analyze comparative data appropriately. Future work should provide more detailed insight into the speciation processes that have occurred in these fascinating groups of plants.

18

Diversity and Speciation of Semionotid Fishes in Mesozoic Rift Lakes

Amy R. McCune

18.1 Introduction

Prolific speciation of fishes within lakes has long fascinated evolutionary biologists (e.g., Brooks 1950; Martens 1997), in part because of the difficulty in identifying extrinsic barriers to gene exchange that might permit allopatric speciation of vagile organisms such as fishes in lakes. The cichlid fishes of Lakes Victoria and Malawi (Fryer and Iles 1972; Greenwood 1981b) are the most celebrated examples of intralacustrine speciation, but its generality has been established by many other examples of endemic fishes, both extant (Kornfield and Carpenter 1984; Parenti 1984; Parker and Kornfield 1995; Strecker *et al.* 1996; Seegers *et al.* 1999) and fossil (McCune *et al.* 1984). Whether such intralacustrine speciation could be sympatric has engendered considerable debate (Worthington 1954; Kosswig 1963; Ribbink 1994; Turner 1994), despite previous theoretical arguments that the evolution of reproductive isolation is extremely unlikely without geographic separation (Tregenza and Butlin 1999). Empirical evidence that speciation in some of these fishes occurred in sympatry has been growing, however (Meyer *et al.* 1990; Schliewen *et al.* 1994; Johnson *et al.* 1996a; McCune and Lovejoy 1998), and recent theoretical results suggest that sympatric speciation is more plausible than previously believed (e.g., Dieckmann and Doebeli 1999; Kondrashov and Kondrashov 1999; Chapter 5). According to these latter theoretical studies, nonrandom mating and ecological interactions, like intraspecific competition for resources, can initiate sympatric speciation.

In this chapter, intralacustrine radiation of semionotid fishes during the Mesozoic is considered through a paleobiological perspective. While it is impossible to assess nonrandom mating in the fossil record, paleobiological studies do have the potential to illuminate pertinent aspects of ecology over a temporal scale impossible to study in extant organisms. First, some background on semionotid fishes and the ecological setting for their lives and preservation as fossils is given in Sections 18.2 and 18.3. In Section 18.4, parallel radiations of *Semionotus* in distinct lakes are described. These lacustrine radiations do, indeed, suggest an important role for ecology. Through detailed study of the history of a single lake (Section 18.5), it is shown that both intraspecific variation that subsequently became incorporated into new species and the first appearances of new species are concentrated in early lake history. The evidence, based on an exceptional lacustrine fossil record, that speciation by *Semionotus* was extremely rapid – as rapid as

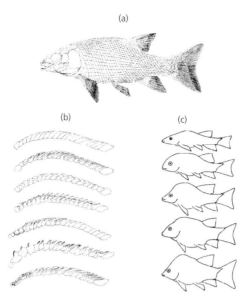

Figure 18.1 Morphology and variability of *Semionotus*. (a) Reconstruction of *Semionotus*. Note particularly the dorsal ridge scales (DRSs), the row of modified scales along the dorsal midline between the nape and the origin of the dorsal fin. (b) Variation in morphology of DRSs in Newark *Semionotus*. From top to bottom, these DRS types are simple, modified simple, short spined, thin spined, globular, robust, and concave. (c) Representative array of body shapes in Newark *Semionotus*. Note that fish with very slender bodies are found as well as deep-bodied ones, and some have humps or steep-sloping foreheads. *Sources*: (a) Olsen and McCune (1991); (b) and (c) McCune (1987a, 1990, 1996).

that experienced by extant cichlid fishes – is reviewed (Section 18.6). Finally, the macroecological view that the evolution of lacustrine fishes can be entrained by the formation of lakes, with each new lake providing new ecological opportunities under changed selective regimes that may lead to speciation, is discussed.

18.2 Semionotid Fishes

Semionotids are basal rayfinned fishes (Actinopterygii), most closely related among living fishes to the heavily armored gars (*Lepisosteus*) of North America. Like gars, semionotids have asymmetric tails and are encased by heavy, ganoid scales (Figure 18.1a; Plate 12). In contrast to the elongated jaws filled with sharp teeth that characterize predatory gars, however, semionotids have small jaws and undifferentiated peg-like teeth. The family name Semionotidae, which literally means signalback, refers to the modified scales with spines directed posteriorly, which rim the dorsal midline between the nape and the dorsal fin. Seven different types of dorsal ridge scales (DRSs) characterize species or groups of species (McCune 1987a; Figure 18.1b). Different DRS types are distinguished by the size of the spines, the size and shape of the scale base, and whether the scales are concave or convex when viewed from a dorsal perspective. The scales within the dorsal ridge series vary characteristically from anterior to posterior, with the anterior scales being most differentiated.

Semionotids have been found throughout the Mesozoic, in both freshwater and marine deposits, and on all continents except Antarctica. Although semionotids were cosmopolitan in their day, it is only in a series of Mesozoic rift-lake deposits in eastern North America that *Semionotus* is diverse. This pattern parallels that of cichlids and a number of other species flocks in which, despite a widespread geographic distribution for the family, diverse complexes of species occur in only one or a few lakes.

18.3   Newark Lake Paleolimnology

Semionotus dominated a number of large rift-valley lakes in eastern North America during the Late Triassic and Early Jurassic. At this time, the North American and African plates were adjacent and the tectonic activity that eventually opened the Atlantic Ocean between them created great rift valleys near the North American continental margin. The deep basins of the rift valleys were sometimes filled with water, to form great rift lakes similar to Lakes Malawi and Tanganyika of today's African rift. The sediments laid down in these ancient rift-valley lakes are now preserved as part of the Newark Supergroup, which comprises eight major and several minor basins exposed in eastern North America (Figure 18.2). Continental positions during the Late Triassic and Early Jurassic placed these North American rift lakes in tropical and subtropical latitudes.

In each of the Newark Supergroup basins, exposed sediments show a strikingly cyclic pattern, in which finely layered, dark sediments deposited in deep lakes are sandwiched between lighter, less structured sediments deposited in swamps or terrestrial habitats. These sedimentary cycles, named Van Houten cycles to honor the geologist who first described them in the 1960s, provide a remarkable ecological and temporal framework in which to study the evolution of the fishes they preserved.

Van Houten cycles have been studied extensively by Paul Olsen (Olsen *et al.* 1978; Olsen 1980, 1986; Olsen *et al.* 1996; Olsen and Kent 1999), who described three main divisions of a cycle, which reflect the formation, maximum lake stand, and subsequent evaporation of a large deep lake (Figure 18.3). At the bottom of a cycle, poorly layered sediments of division 1 reflect the expansion of marshy ponds and swamps. These sediments grade into division 2, which consists of very finely layered sediments deposited in deep perennially stratified lakes. Division 2 sediments preserve complete, fully articulated fishes and annual layers of sediment because they were preserved in sediments too deep to sustain physical disturbance from waves, and in anoxic bottom waters that did not mix with the upper waters of the lake (Plate 12). As a result, no scavengers and decomposers disturbed carcasses that fell to the lake floor and no benthic organisms disturbed the fine layers of sediments laid down annually. These deep-lake sediments comprise fine-grained, organic-rich deposits marked annually by a precipitate of sediment rich in calcium

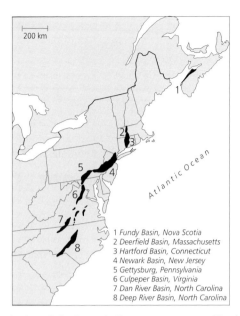

1 *Fundy Basin, Nova Scotia*
2 *Deerfield Basin, Massachusetts*
3 *Hartford Basin, Connecticut*
4 *Newark Basin, New Jersey*
5 *Gettysburg, Pennsylvania*
6 *Culpeper Basin, Virginia*
7 *Dan River Basin, North Carolina*
8 *Deep River Basin, North Carolina*

Figure 18.2 Major basins of the Newark Supergroup, eastern North America. Newark Basin, New Jersey (4) is the site of the focal lake deposit (cycle P4) discussed in the text. *Sources*: McCune (1996); see also Olsen (1986) and Olsen *et al.* (1996).

carbonate. Each pair of layers (dark organic layer plus white calcium carbonate layer), termed a varve, was deposited during a period of one year, thus providing an annual chronology for the fishes they preserve (Plate 12; Olsen 1980, 1986).

Above division 2 of a Van Houten cycle, the sediments grade into poorly laminated sediments of division 3 deposited as the lake level declined, mixing of lake waters resumed, the lake became shallower, and finally disappeared. In these uppermost sediments, there are preserved fossil mudcracks, salt casts left from evaporation, roots of plants preserved *in situ*, fossil soil profiles, and dinosaur footprints, all evidence that the lake dried up completely.

The cyclic nature of Newark lacustrine sediments has been documented in a number of different basins of the Newark Supergroup, but studied most in the Newark Basin of New Jersey (Olsen 1986, 1988; Olsen and Kent 1999). In any substantial exposure (quarries, road cuts, railroad cuts), multiple sedimentary cycles – lake histories stacked one on top of another – are visually striking. The lacustrine history of the Newark Basin includes more than 60 Van Houten cycles of distinct lake histories.

Why would such a regular pattern of sedimentation be found in multiple basins of the Newark Supergroup? It is well established that the cyclic formation and evaporation of lakes in the Newark Basin was caused by periodic climatic change, in turn driven by the precession cycle of the equinox (Olsen 1986; Olsen and Kent 1996, 1999). Over a period of approximately 22 000 years, the orientation of the

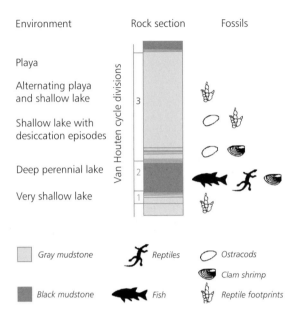

Figure 18.3 Generalized Van Houten cycle. Rock section shows characteristic progression of sedimentary structure from poorly bedded sediments, which grades into microlaminated and then back to poorly bedded ones. Whole fishes are found in the microlaminated sediments of division 2, which sometimes also yield articulated reptiles and even insects (Olsen *et al.* 1978). Sediments of division 2 were deposited in deep perennial lakes. Divisions 1 and 3 represent the early expansion and later contraction–evaporation phases, respectively, of the lakes. *Sources*: McCune (1996); see also Olsen (1986) and Olsen *et al.* (1996).

earth's axis of rotation varies, and so produces regular, predictable effects on climate by changing the seasonal distribution of sunlight received at different latitudes. The important point for the study of semionotid evolution is that the duration of each episode of lake formation and evaporation was about 22 000 years, making it possible to sample successive radiations from successive lakes that occupy the same basin at different times. Within a lake's history, it is also possible to follow changes in species' composition and morphology using the yearly chronology of varved sediments.

18.4   *Semionotus* Radiations in the Newark Supergroup

More than 50 species of *Semionotus* have been described from nine different basins of the Newark Supergroup. *Semionotus* exhibits the greatest diversity in Early Jurassic lake cycles, in which there are few or no nonsemionotid fishes, in contrast to Late Triassic lake deposits that contain one semionotid and a variety of paleonisciform and coelacanth fishes (McCune *et al.* 1984). The most diverse semionotid fauna in the Newark Supergroup is found in sedimentary cycle "P4" of the Towaco Formation (Plate 12; McCune 1987a, 1987b). In this single lake deposit, 21 described species of *Semionotus* exhibit all seven described types of

---

Box 18.1 Shape diversity in fishes

Body shapes of fishes are exceedingly diverse, and range from tubular to fusiform to very deep bodied. Some are highly intricate (e.g., sea horses), or have huge heads (e.g., goosefish and anglerfishes). However, families of fishes can usually be characterized by shape because species within families are quite similar in body form (Matthews 1998). For example, sunfishes (Centrarchidae) are deep bodied and laterally compressed. Pikes and pickerels (Esocidae) are elongate tubular with spatulate snouts (Keast and Webb 1966). Of course, there are exceptions to this generalization: the Cichlidae and Characidae are families that include species with a great diversity of body shape.

   Given this diversity of form in fishes, it is natural to wonder whether generalizations can be made about the ecological roles of species with similar shapes. In freshwater ecosystems, attempts to depict "typical" arrays of body form, (e.g., Keast and Webb 1966; Lowe-McConnell 1987) show forms that range from tubular species to fusiform, to deep bodied, and to species flattened dorsally or those with dorsal humps. Convincing interpretations of the ecological significance of these different forms are yet to be made, except at the most general level. There are clearly hydrodynamic consequences of gross differences in shapes, say between eel-shaped and fusiform species (for an introduction, see Helfman *et al.* 1997) and most investigators would agree that deep-bodied fishes are generally found in still water and more fusiform fishes tend to be pelagic or to live in currents. Unfortunately, the suggested associations between body shape and particular ecological roles rely on perceived correlations of particular taxa to their habitats or lifestyles [for a recent review, see Matthews (1998)] and are not usually generalizable to other taxa in different ecosystems. However, when the array of shapes found in a temperate lake (Keast and Webb 1966), tropical freshwaters, or one of the Newark Supergroup fossil lakes are compared, the feeling that different species may be playing some of the same ecological roles in different bodies of freshwater cannot be avoided.

DRSs and a broad array of body shapes (Figure 18.1b,c). While the function of DRSs remains elusive (could they have a hydrodynamic function or be involved in mate recognition?), it is clear that in living fishes body shape is an ecologically important trait (Matthews 1998). In general, fusiform (torpedo-shaped) fishes tend to be pelagic or live in currents, while deep-bodied fishes live in still water such as lakes or pools within rivers (Keast and Webb 1966). More detailed associations of shape and ecology have been made for living fishes (e.g., Keast and Webb 1966; Winemiller 1992), but it has not been possible to generalize these associations to other taxa that live in other places, and, of course, there is no way to test these functional associations for extinct taxa. However, it is probably safe to conclude that for semionotids, like all other fishes, body shape is ecologically important.

   The diversity of body form exhibited in the single genus *Semionotus* is greater than that found in most families of living fishes. In lake P4, *Semionotus* species ranged from slender fusiform to deep bodied, some with dorsal humps, some with steep foreheads, and some with large heads (Figure 18.1c; McCune *et al.* 1984;

McCune 1987a). In most families of fishes, the overall shape of the fish is quite characteristic (Box 18.1), but the range of shapes in *Semionotus* approaches that of even remarkably diverse families like characids and cichlids. While it is difficult to specify precisely the lifestyle represented by a particular body shape, the diversity of *Semionotus* suggests that these species played a broad range of ecological roles in lake P4.

At least some of the several deep-bodied *Semionotus* evolved independently in different Newark lakes. In cycle P4, colonist species tend to be fusiform and the most deep-bodied fishes are endemics (McCune 1996). This is consistent with the expectation that riverine colonists would be fusiform. Endemic distributions of several deep-bodied species of *Semionotus* having different types of DRS suggests repeated evolution of deep-bodied lacustrine forms in several distinct Newark lakes. For example, in older (Feltville Formation) and younger (Boonton Formation) lake deposits in the Newark Basin, there are deep-bodied species of *Semionotus*, which are distinct from each other and from P4 *Semionotus*, as evidenced by their different DRS morphologies (McCune *et al.* 1984; McCune 1987b).

18.5   Ecological Context of Evolutionary Novelty and Speciation

Studies of speciation are often hampered because the most relevant time scale is generally greater than the time available for ecological studies and less than the temporal resolution of most fossil sequences – at best, usually tens or hundreds of thousands of years. Varved lake sediments that retain annual layers, as in the Newark, are an important exception, and they make it possible to examine the historical pattern of species distributions and morphology of *Semionotus* with virtually unrivaled temporal resolution (McCune 1990).

Evolutionary novelty in *Semionotus*

As described above, the DRS series is a striking feature of *Semionotus*, and while the morphology of these scales varies anteriorly to posteriorly, the pattern of variation is stereotypic for a species or group of species (Figure 18.1b). In about 5.5% of individuals examined, however, anomalous individual scales of one type are mixed into a series of another type (Figure 18.4; McCune 1990). These anomalies are sometimes atavistic or a doubling of scales, but the most interesting cases are novel anomalies that foreshadow future DRS types. For example, the first known occurrences of concave DRS are the simultaneous occurrence of an individual with a single concave scale and the first individual of a species that has a complete concave DRS series. Both of these individuals are from lake cycle P4 in which, 90 years later, a second species with concave DRS scales appears, followed by, approximately 44 000 years later (two cycles) in the same basin (Boonton Formation), a radiation of nine species, all of which are characterized by concave DRS (Figure 18.5; McCune 1990).

DRS anomalies are not evenly distributed through the history of lake cycle P4. A plot of the microstratigraphic distribution of anomalies shows that their frequency is high during the earliest quarter of deep-lake history, and subsequently

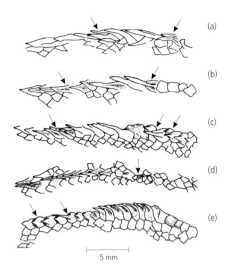

5 mm

Figure 18.4 DRS anomalies. Camera-lucida drawings of the DRSs of five individuals. Arrowheads indicate anomalies, classified as follows: (a) scale doubling, two concave scales in simple scale series; (b) mix of concave and thin-spined scales; (c) mix of concave, thin-spined, and short-spined scales in simple series; (d) short-spined scales in simple series; (e) short-spined and thin-spined scales in globular series. For a key to the individual specimens, see McCune (1990). *Source*: McCune (1990).

declines during the maximum lake stand (Figure 18.6; McCune 1990). It also appears that increases in species richness coincide with times when high proportions of individuals that sport anomalies were recovered. The higher frequency of DRS anomalies during the early maximum lake stand is arguably a case of classic ecological release, in which selection on DRSs has been relaxed (McCune 1990). Of course, this does not mean that selection was relaxed on all traits, it simply means that the dorsal ridge scales were not under strong selection during this period, as evidenced by the higher frequency of DRS anomalies. Overall, however, selection may have been quite strong during the initial colonization of the expanding lake. Models of the process of ecological release suggest that selection gradients are higher immediately after colonization than later, when niche space is more fully subdivided (Box 18.2). Hypothetically, the genotypes that produce the most offspring in source rivers would be quite different to the genotypes that produce the most offspring in the expanding lake. Thus, in the new lake environment other sets of traits may contribute to large differences in fitness, while the DRS do not. As discussed in Box 18.2, a high dimensional trait space allows a great deal of room for play in directions orthogonal to the direction of strongest selection, at least during situations of ecological release. This "adaptive dynamicist" view is consistent with the empirical data for the history of the *Semionotus* radiation: the new and different environment of an expanding lake relative to source rivers may have produced very strong selection on some traits, and at the same time allowed

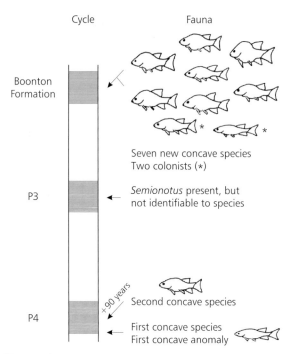

Figure 18.5 History of concave scale anomalies and species with concave DRSs in the Newark Basin. The first anomalous concave scale from the Newark Supergroup occurred early in the history of cycle P4, at the same microstratigraphic level as the first individual *Semionotus* that showed the stereotypic concave DRS. In microstrata of sediments about 90 years younger, there is an individual of a second species with concave DRS. In the Boonton Formation lake deposit (two cycles above, or about 44 000 years younger), nine species with concave DRS occur. Two of these species are presumed colonists, indicated by an asterisk. In an intermediate age deposit (cycle P3), specimens of *Semionotus* have been found, but they are not sufficiently well preserved to identify DRS type or species.

a great deal of play in other traits (i.e., relaxed selection on DRSs). What is perhaps most interesting about this case, however, is that variation not under strong selection (i.e., DRS anomalies) is variation that later became incorporated into the *S. elegans* clade of nine species that lived in Boonton Lake some 44 000 years later (Figure 18.5).

Timing of speciation in *Semionotus* relative to lake formation

First appearances of endemic species in cycle P4 occur mainly during the first quarter of the deep-lake history (Figure 18.6). While no comparable record is available for living species of fishes, inferences from DNA sequence data and geological data suggest a similar pattern for some cichlids. In Lake Tanganyika, for example, the monophyletic Ectodini and Tropheini are older than many other

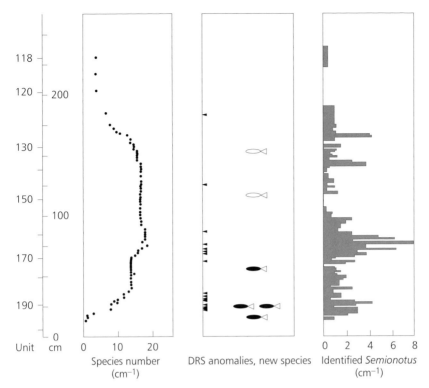

Figure 18.6 Microstratigraphic distribution of species number and DRS anomalies through 3 m of sediment from division 2, cycle P4. The oldest sediments are on the bottom and younger sediments are toward the top. "Units" are microstratigraphic units of varying thickness (generally 1–2 cm). Each centimeter of sediment corresponds to about nine years of time (McCune 1990). In the right panel the plot of identified *Semionotus* shows the number of individuals, given in terms of number of individuals per centimeter of sediment, for which the DRSs are visible. The left panel shows the number of species ranges that pass through a given microstratigraphic unit. The middle panel indicates with arrows the layers in which individuals with DRS anomalies were found. Individuals that sport anomalies are concentrated in older sediments ($\chi^2 = 7.9$, one degree of freedom, $P < 0.005$). The first appearances of new species are indicated in the same panel by fish symbols. Fish outlines mark the first appearances in which there is only one specimen (and thus the species range must be considered unknown); filled outlines are species known from multiple individuals. *Source*: McCune (1990).

extant species flocks, but approximately equal branch lengths for many species within these clades suggest that speciation was concentrated during the early history of the clade (Sturmbauer *et al.* 1994). Both empirical results again appear to be consistent with theory, which predicts relatively rapid speciation after ecological release, with successively longer times for speciation in successive events (Box 18.2).

Box 18.2 Predictions from the changing geometry of fitness landscapes
Johan A.J. Metz and Amy R. McCune

As discussed in Chapter 4, adaptive dynamics theory has as its starting point the idea that a species in fitness terms always "tries to walk uphill", but at the same time it treads the fitness landscape flat as it goes (Marrow *et al.* 1996). For to persist on an evolutionary time scale, a phenotype may, on the population dynamic time scale, on average neither grow nor decline in numbers, which is equivalent to it having a fitness of zero. That it can stay so, seemingly poised on the brink between eternal growth and extinction, is because of the so-called environmental feedback loop: populations change the environment of their individuals in a dynamic fashion so that the community necessarily ends up at an equilibrium point or, more generally, at a so-called attractor; in the environment generated by that attractor the fitness of all types still present is zero. The fitness landscape, therefore, is anchored to zero at those points at which there are species. Between anchor points, valleys and summits occur. Substitution steps are always in directions that have an uphill component at that particular moment in evolutionary time, since only phenotypes with positive fitness can invade.

The picture sketched above can be made rigorous only for populations that reproduce asexually. However, the considerations expounded below are only of a global geometric nature, relating to the steepness of the ecologically determined fitness gradients, and therefore may be expected *grosso modo* to hold good for both asexual and sexual populations (Box 4.8 and Chapter 5).

The theory of adaptive dynamics presented in Chapters 4 and 5 largely considers one-dimensional trait spaces. This suffices for demonstration purposes, and makes the theory easier to comprehend. However, to make predictions about the fossil record higher-dimensional trait spaces must be considered. In these trait spaces all the phenomena of one-dimensional trait spaces occur in roughly the same manner, and some of them in combination, plus a few more.

To see what insights adaptive dynamics has to offer for the issues of this chapter we discuss the changes in the fitness landscape that can be expected to occur in the course of the evolution of a newly formed lake. The initial colonization of the lake may be supposed to occur with one or a few species. Later on, the lake fauna diversifies through speciation and, possibly, the immigration of additional species. The more species that are around, the larger is the number of points at which the fitness surface is anchored to zero, the more the fitness landscape is dominated by ridges, and the lower are its summits. As a consequence, the landscape initially has a few very high summits and correspondingly steep slopes. Later on the summits become lower, but the landscape becomes more and more crinkled.

Evolution halts once the landscape and the traits of the species create a situation in which each species sits on its own summit. All the summits then have a height (fitness) of exactly zero and are all surrounded by a below-zero region. (The population dynamics guarantees that the latter is the case. For constant-fitness landscapes, not molded by population dynamics, it is highly unlikely that even two summits would have exactly the same height.) After this point, only environmental changes imposed from the outside make the system move again, by forcing a change in the topography of the fitness landscape.

continued

Box 18.2 *continued*

Tight versus relaxed selection. Natural selection is, on average, stronger under conditions of ecological release because there are more opportunities for large fitness differences. For one-dimensional trait-space models, such as the examples from Chapters 4 and 5, the selection gradients initially are much stronger than later on, when niche space is more finely partitioned. In the direction of the selection gradient (the direction of steepest increase in fitness), the same holds for higher-dimensional trait spaces. However, the picture changes if one concentrates on the directions orthogonal to the most strongly selected one. The higher the dimension, the more room there is for those other directions. Under conditions of ecological release fitness changes only slightly in these directions. Therefore, variation in these directions hitchhikes with the trait under strong selection. Later, when the niche space is more finely partitioned, the curvature of the boundaries of the sets of potential invaders begins to play a role (see figure). Initially, when there is little curvature, only the upward component of a mutational step contributes to its success, as the sideways components are effectively neutral. Later, when the curvature of the boundary of the set of potential invaders increases, even small sideways swerves become detrimental. It is as if, at first, a zigzag path up a steep slope can be followed, while in the final phase a narrow ridge has to be followed to the top. Along that ridge the uphill slope is not steep, but zigzag movement is effectively prohibited.

Generic local shapes of the adaptive landscape at different stages of the history of a lake. The grey horizontal plane intersects the fitness surface along the thin curve. (a) Early stage of colonization (ecological release), characterized by strong directional selection (tapping into unexploited ecological opportunities) and weak transversal selection (relaxed selection in empty habitat). (b) Late stage of colonization (ecological fine-tuning), characterized by weak directional selection (honing of the design) and strong transversal selection (because of the nearly filled niche space).

Changes in speciation rates. Adaptive dynamics models tend to show rather fast speciation immediately after ecological release. The next speciations take longer, subsequent ones longer again, and so on. This change in speciation behavior can be understood as follows. Initially the fitness landscape has steep slopes up to the valley bottoms, including any branching points. Later on, as the landscape becomes further divided through the presence of more anchoring points, overall the slopes become less steep. Hence evolution proceeds more slowly to the branching points, and branching itself is slower because the valleys are less steep. [For more species than one, branching points are points in the combined trait spaces of all the species together such that the fitness landscape engendered by that species combination places some species at the bottom of a valley (making them speciation prone), and the remaining species on summits. This scenario is not as odd as it may seem, for the theory of adaptive dynamics shows that the population dynamics more or less forces the existence and, often, also the attractivity of such configurations.]

18.6   Time Required for Speciation

How long it takes for reproductive isolation (often inferred from phenotypic differentiation) to evolve is highly variable (Futuyma 1998) and difficult to measure. Cichlid fishes from the African great lakes have long been considered to lie at the rapid end of the continuum in animals (e.g., Futuyma 1979), and recent estimates of the age of Lake Victoria suggest that more than 300 species of cichlids evolved in only 12 400 years (Johnson *et al.* 1996a). Given that the geological dating of lakes is generally more precise than for many other geological events (McCune 1997), the evolution of endemic fishes in lakes can be used to estimate how long speciation takes. This has been possible for *Semionotus*, and the data can be compared to those for living species flocks of fishes, reviewed below.

Colonization, endemism, and time for speciation in P4 *Semionotus*

To the extent that some of the 21 species of *Semionotus* found in lake deposit P4 were endemic (and assuming that endemics evolved within lake P4), they must have arisen in less than the duration of one episode of lake formation–evaporation. A broad-scale biogeographic and stratigraphic survey of >2000 individuals of *Semionotus* revealed that, of the 21 species found in P4, eight species were colonists and six were endemic; data are equivocal for the remaining seven species because it is difficult to identify incomplete specimens to the species level (McCune 1996). The time of origin for these six *Semionotus* (and possibly as many as 12 if equivocal species are included) can be further refined, based on microstratigraphic distribution, to the first 5000–8000 years of lake history. This timing is remarkably similar to that for the origin of five cichlids derived from Lake Victoria cichlids after the separation of the satellite Lake Nabugabo from Lake Victoria about 5000 years ago (Fryer and Iles 1972).

Times to speciation in *Semionotus* and other species flocks

In the case of *Semionotus* and the Nabugabo cichlids referred to above, it is easy to regard the net rate of diversification (speciation less extinction) in the two groups as quite similar, because both the number of taxa and the time frame are comparable. To make broader comparisons among clades, an heuristic empirical estimate of time for speciation (TFS) can reveal variation, among taxa and under different environmental circumstances, in the tempo of speciation (Box 18.3). TFS can also provide a temporal framework within which speciation can be related to the historical and geological events involved in speciation. Minimally, to estimate TFS, data on age of the clade and number of species are required. However, estimates of TFS will differ with the branching topology assumed. (Box 18.3; McCune 1997). Thus, while estimates of TFS for particular clades would be much improved with information about phylogenetic topology and branch lengths of the tree, in the absence of these data for most groups, an heuristic estimate of TFS has been computed, assuming a fully dichotomous, symmetrical phylogenetic topology and a constant average rate of divergence (Box 18.3).

Box 18.3 An empirical estimate of time for speciation (TFS)

The length of time it takes for species-level divergence to arise, that is, the time for speciation (TFS), surely varies among organisms and the circumstances of speciation. Empirical estimates of TFS can reveal to what extent speciation has proceeded at different tempos in different taxa or under different circumstances, and such estimates can be useful for relating speciation to historical or geological events. Ideally, the average TFS would be computed for a particular clade, given a completely resolved phylogeny with known branch lengths. However, species-level phylogenies that are completely sampled, or even nearly so, are rare (McCune and Lovejoy 1998), especially for diverse clades like species flocks. Rather, often only reasonable knowledge of the number of species in a clade and an estimate of the age of a clade, derived from DNA sequence data or from fossil data, are available. Additionally, for endemic fishes in lakes, the age of lake origin can serve as a maximum estimate of clade age (McCune 1997), just as the age of island origin can be used as a maximum estimate of clade age for island endemics. Given clade age and the number of species in that clade, it is possible to obtain a very rough empirical estimate of TFS, simply as

$$\text{TFS}_{\text{linear}} = T/(N-1) \tag{a}$$

where T is the age of the clade and N is the number of taxa. This simplistic computation of TFS assumes that speciation occurred repeatedly in one lineage (yielding a pattern of asymmetric branching or an unbalanced tree) and this pattern of branching can be visualized as a comb-shaped phylogeny (left panel below).

Alternatively, one could assume that, within a clade, TFS was approximately constant and that all lineages speciated equally. In this special case, the branching pattern would be fully dichotomous and symmetrical (a balanced tree), as described by

$$\text{TFS}_{\text{log}} = (T \ln 2)/\ln N \tag{b}$$

continued

142 857 years   333 333 years

Differences in estimated TFS for different phylogenetic topologies. For a hypothetical clade of eight species arising in 1 million years, the average TFS differs according to the phylogenetic topology assumed. Divergences are numbered, with simultaneous sets of divergences indicated by dashed lines. The average estimated TFS for a clade is given below each cladogram. (a) Comb-shaped phylogeny assumed by computation of $\text{TFS}_{\text{linear}}$. (b) Fully dichotomous, symmetrical phylogeny assumed by computation of TFS_{log}. For this hypothetical case, estimates of TFS with other phylogenetic branching patterns are given in McCune (1997).

Box 18.3 *continued*

(right panel on preceding page; for derivation, see McCune 1997). In reality, most phylogenies have topologies somewhere between these two extremes, with published phylogenies tending to be more imbalanced than expected under an equal-rate, random-speciation model (Guyer and Slowinski 1991; Heard 1996). While both estimates of TFS are potentially useful as comparative metrics, the most realistic estimate of average TFS for a given clade is obtained if the phylogenetic topology and its branch lengths is known. For the sake of comparisons between clades, and in the absence of well-resolved species-level phylogenies for the groups considered, the logarithmic estimate of TFS [Equation (b)] is used here to compute heuristic empirical estimates of TFS. This estimate gives the more conservative (slower) estimate of TFS. Note also that both estimates of TFS [Equations (a) and (b)] reflect net speciation: neither estimate incorporates extinction and thus TFS may be overestimated (with speciation appearing to take longer) if extinct species have been excluded.

Comparisons of TFS among species flocks of fishes show that many of these, not just cichlids, speciate rapidly (Table 18.1; McCune 1997). For most species flocks, estimates of TFS fall within 1500–300 000 years, with the lower and upper bounds being cichlids from Lake Victoria and Barombi Mbo, respectively. Values for other species flocks, like pupfishes from Lake Chichancanab and Newark semionotids, fall between with values of 5000–8000 years.

All values of TFS for these lacustrine fishes (except for the highly variable outlying estimates for Tanganyikan cichlids) are substantially lower than the TFS of 0.6–2.4 million years for radiations of arthropods and birds on oceanic islands (Table 18.1). While the difference in TFS between lacustrine species flocks of fishes and island radiations could be an artifact of dating or time scale, it may also reflect the mode of speciation (whether allopatric or sympatric). According to population genetic theory (Bush 1993; Kondrashov *et al.* 1998), if sympatric speciation occurs, it will involve selection that acts directly on mating systems and thus will happen much more quickly than allopatric divergence that produces reproductive isolation as an incidental by-product of drift or local adaptation (see Chapter 1 for a discussion of the role of ecological divergence in selection that acts on mating systems). Interestingly, DNA sequence divergence is markedly greater for allopatric sister species of fishes than it is for sympatric sister species, which suggests that the difference between TFS for island animals (0.6–1.4 million years) and lacustrine species flocks of fishes (0.015–0.3 million years) may, in part, reflect a difference in mode of speciation (Table 18.1; McCune and Lovejoy 1998).

If this difference in TFS between allopatric sister species and sympatric sister species is real, it has an interesting implication for the problem of cichlid speciation. In the past, it has been unclear whether speciation by cichlids in Lakes Malawi and Victoria could have been sympatric (Ribbink 1994). Some investigators have favored a microallopatric model of populations isolated by habitat

Table 18.1 Time for speciation (TFS): animals on islands compared to fishes within lakes.

Taxon	Lake or island	Endemics (number of species)	Age of clade (million years)	TFS_{log} (years/ speciation)
Crickets	Hawaiian Islands	250	5.6	700 000
Drosophilids	Hawaiian Islands	860	5.6	570 000
Passerines	Hawaiian Islands	59	5.6	950 000
Finches	Galapagos Islands	14	5.0–9.0	1.3–2.4 10^6
Cichlids	Lake Victoria	300	0.012	1500
Cichlids	Lake Nabugabo	5	0.004	1700
Cichlids	Lake Malawi	400	0.7–2.1	81 000–240 000
Cichlids	Lake Tanganyika	30–171	9–12	1.2–2.4 10^6
Cichlids	Barombi Mbo	11	1.0–1.1	288 000
Cyprinodontids	Lake Titicaca	~22	0.020–0.15	4500–34 000
Cyprinodontids	Lago Chichancanab	5	0.008	3400
Semionotids	Lake P4, Newark Basin	6–12	0.005–0.008	5000–8000

Source: McCune (1997).

patches and/or fluctuating lake levels. However, "ordinary" allopatric speciation, in which reproductive isolation is a by-product of separation by an extrinsic barrier, could not alone explain speciation in cichlids from Lakes Malawi and Victoria, because it appears that there simply has not been enough time. This is absolutely clear in the case of Lake Victoria, the age of which (12 400 years) is far shorter than the minimum estimated time necessary for a single allopatric speciation event (600 000 years), let alone the 8–9 sets of simultaneous dichotomous divergences that would yield 300 species. Even for the much older Lake Malawi, with an estimated age of 1–2 million years, it would seem that reproductive isolation as a by-product of an extrinsic barrier is not sufficient to explain the current diversity of cichlid fishes. Using the most conservative estimate of 2.0 million years for the age of Lake Malawi, for 392 endemic species (and some estimates for number of cichlids in Malawi are as high as 1000) to diversify in that time (three episodes of about 0.6 million years), eight species would have to diverge simultaneously, and each of these would have to produce seven daughter species, which in turn would also have to produce seven daughter species (Figure 18.7). Furthermore, for allopatric divergence to be completed, isolated habitat patches or lake-level fluctuations would have to persist for 600 000 years each time to enable each set of simultaneous divergences. Indeed, there have been substantial lake-level fluctuations in Lakes Malawi and Tanganyika, but the period of fluctuation is about 100 000 years (Lezzar *et al.* 1996; Cohen *et al.* 1997; Soreghan *et al.* 1999), far less than the required persistence time of 0.6 million years for each episode of allopatric divergence.

Interesting as the empirical data on TFS are for allopatrically and sympatrically distributed fishes, there remains a theoretical possibility that rapid microallopatric or parapatric speciation might have produced sympatrically distributed

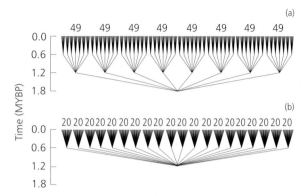

Figure 18.7 Could allopatric speciation account for speciation of Malawi cichlids? Depicted here are possible phylogenetic topologies for clades implied by a TFS$_{allopatric}$ of 600 000 years (McCune and Lovejoy 1998) and an age of 1.2 or 1.8 million years for Lake Malawi. (a) Three episodes of simultaneous speciation result in eight species after 600 000 years, each of which split into an additional 49 species after an additional 1.2 million years to produce a total of 392 species in 1.8 million years. (b) Two episodes of simultaneous speciation result in 20 species after 600 000 years, each of which split into an additional 20 species after another 600 000 years. For these scenarios to be viable, either 7–8 extrinsic barriers must be maintained for 600 000 years three times, or 20 extrinsic barriers must be maintained for 600 000 years for two episodes of speciation. Fluctuations in lake level have often been proposed as the source of extrinsic barriers, but the periodicity of these fluctuations is only about 100 000 years, apparently not nearly enough time to allow allopatric speciation unassisted by other factors (e.g., sexual selection, secondary reinforcement, nonrandom mating).

species flocks. Simulations suggest that rapid parapatric or allopatric speciation is possible, in principle, if many loci affect reproductive isolation and the rate of mutation is not too small relative to the rate of migration (Gavrilets 2000a; Gavrilets *et al.* 2000a; Chapter 6). Furthermore, apparently even relatively weak selection for local adaptation could dramatically shorten the "waiting time" to speciation (Gavrilets 2000a). However, if endemic species flocks of fishes are the result of rapid parapatric speciation, what accounts for the differences in TFS between allopatrically and sympatrically distributed fishes? Furthermore, with the Gavrilets model, new species would be expected to arise at a constant rate through time, a prediction that potentially could be tested for living species flocks with DNA sequence data, assuming clocklike sequence evolution. For fossil semionotids from lake P4, this prediction is not borne out in that endemic semionotids all make their first appearances during the first quarter of lake history (Section 18.5).

Another possibility suggested by theory is that cases of rapid speciation in endemic species flocks reflect the relatively recent colonization of recently formed habitats. Adaptive dynamicists argue that speciation will be most rapid immediately after colonization, during a period of ecological release, because of strong selection associated with colonization of a new habitat (Box 18.2). This scenario is consistent with the high-resolution semionotid data reviewed in Section 18.5, in that all endemic semionotids first appear within the first quarter of lake history. Again, for living taxa the prediction of decelerating speciation potentially could be tested using molecular data, assuming clocklike sequence evolution. With this theoretical result and the empirical data in mind, the nature of lakes appears to be a key factor in promoting rapid ecological speciation. Lakes are more ephemeral than most ocean islands and continental ecosystems (McCune *et al.* 1984; McCune 1997), and it may be that these geologically young lakes (even extant "ancient lakes" are young compared to most oceanic islands) simply provide ecological opportunity, allow ecological release, and thus engender subsequent rapid speciation.

18.7 Concluding Comments

The theme of this volume might be characterized simply as to explore the role of ecology in speciation. In the realm of fishes that live in modern lakes, natural selection can drive the adaptive divergence of different forms of sticklebacks (Chapter 9) and charr (Chapter 10) within the lakes. After initiation of speciation by sexual selection, disruptive selection on resource utilization may be important to complete speciation in cichlids (Chapter 8). Common to all of these cases is that diversification has followed colonization of a new lake by ancestral forms. From a paleobiological perspective that encompasses a greater temporal scope, the pattern is even more dramatic. As seen in the fossil history of Newark *Semionotus*, evolutionary novelties and species do not arise throughout the lake's history, but in the early expansion phase up to the earliest phase of maximum lake stand, as described in Section 18.5. Formation of new species happens rapidly (Section 18.6) and multiple times in distinct lakes (Section 18.4).

The geologically ephemeral nature of lakes, even of the deep rift-valley Newark lakes in which *Semionotus* diversified, suggests that the formation and colonization of new habitats is key in providing the ecological opportunity for subsequent diversification. This pattern is consistent with theory that suggests the selection gradients are strongest after colonization of a new habitat, and that speciation may occur rapidly under these conditions. Climatic cycles that caused the repeated formation and evaporation of Newark lakes apparently entrained the evolution of *Semionotus* by repeatedly providing ecological opportunity. This resulted in rapid speciation and caused the subsequent demise of many of these new species as their habitat declined and disappeared, which in turn paved the way for new lakes and new species.

Acknowledgments   With thanks to S. Ellner, N. Hairston, S. Heard, N.R. Lovejoy, P.F. Olsen, and D.W. Winkler for enlightening discussion and/or comments on the manuscript. Partial support was provided by the National Science Foundation, DEB9981445.

19

Epilogue

Ulf Dieckmann, Diethard Tautz, Michael Doebeli,
and Johan A.J. Metz

When Terry Erwin from the Smithsonian National Museum of Natural History examined the diversity of beetles that lived on a single species of tropical trees, he found 682 different beetle species, 163 of which he classified as specialist species that lived exclusively on the particular tree species used in his study. Since there are around 50 000 tropical trees species, Erwin extrapolated that there must be on the order of 7 million specialist beetle species (Erwin 1982). Using similar extrapolations, Erwin (1982) also estimated the total number of tropical arthropod species as about 30 000 000. While these estimates may be too high (Schilthuizen 2000; Ødegaard 2000; Novotny *et al.* 2002), they are mind-boggling nevertheless and serve as an illustration of the incredible amount of species diversity that exists on our planet: estimates for the total number of extant species of plants and animals range from 10 million to 100 million (May 1990; Schilthuizen 2000). It is also estimated that the number of extant species represents only about 1% of the total number of species that ever existed during the history of life on earth. Together with the common phylogenetic ancestry usually inferred for the tree of life for higher organisms, this implies that speciation must have been truly rampant during the creation and evolution of our biosphere.

Indeed, there was ample space and time for the evolutionary generation of diversity. After all, the past 3.5 billion years, during which life presumably evolved from some self-replicating molecules, has seen large-scale geographic changes, including the rise, shift, and disappearance of continents and oceans. For example, the fauna and flora of Australia is in many ways very different from that of any other place in the world, presumably because of the long-lasting geographic isolation of this continent from other landmasses.

19.1 The Allopatric Dogma

By scaling down to smaller geographic areas, one arrives at the allopatric speciation model. According to this perspective, speciation occurs when subpopulations of a single ancestral population become geographically isolated and embark on different evolutionary trajectories. During separate long-term evolution in geographic isolation, reproductive isolation evolves as a by-product of divergence in other aspects of an organism's phenotype or genotype. When, at some later point in time, secondary contact occurs between the diverging species, so that their ranges overlap again, hybrid inferiority may reinforce reproductive isolation, but the primary cause of speciation in this scenario remains long-term geographic isolation.

Prominent evolutionary biologists, such as Theodosius Dobzhansky and Ernst Mayr, strongly promoted this type of speciation as the main mode of evolutionary diversification, to the point of ridiculing alternative scenarios, as evidenced by Dobzhansky's remark that sympatric speciation, that is, speciation unfolding in the absence of geographic isolation, "is like the measles; everyone gets it and we all get over it" (Bush 1998). Indeed, it is now the widely accepted common wisdom that most evolutionary diversification occurred according to the allopatric speciation scenario.

However, even though allopatric speciation, with its apparent simplicity, is an intuitively appealing idea, a number of problems reveal themselves upon closer inspection. For one thing, allopatric speciation is not really simple conceptually, because the mechanisms that underlie the evolution of reproductive isolation as a by-product of divergence in other traits are understood only poorly, both empirically and theoretically. In particular, some closely related species (e.g., North American and European buffaloes) have evolved in allopatry for millions of years, but readily interbreed when brought into contact. Moreover, other species, such as oak, have long evolved in sympatric ranges and clearly maintained their phenotypic and genetic identity, yet they can interbreed easily. It seems, then, that by-product reproductive isolation is a far more complicated concept than implied by allopatric speciation theories. It is also commonly acknowledged that – except when driven by sexual arms races (Schilthuizen 2001) – allopatric speciation is a very slow process, because it involves neither inherent selection for differentiation, nor selection for isolating mating mechanisms. This, combined with need to postulate billions of geographic events to create the isolation between subpopulations of ancestral species, casts serious doubts on the ubiquity of allopatric speciation.

19.2 Adaptive Speciation

Perhaps the most convincing counterarguments against the supreme rule of allopatric mechanisms come from theoretical developments which show that speciation in sympatry, that is, under conditions of ecological contact, is an entirely plausible evolutionary scenario. Models of adaptive speciation show that many types of frequency-dependent biological interactions can readily cause the dynamic emergence of disruptive selection in an evolving population, and disruptive selection can in turn readily induce adaptations that result in diminished gene flow between sympatrically diverging subpopulations. Part A of this book is devoted to explaining the theory of adaptive speciation in some detail. The theory is developed within the framework of adaptive dynamics, and is based on the phenomenon of evolutionary branching. The basic notions used in this theoretical framework for the study of evolutionary dynamics in phenotype space are explained in Chapter 4, which focuses on asexual populations. Combining this with population genetic modeling leads to a fully fledged theory of adaptive speciation in sexual populations, both under completely sympatric conditions (Chapter 5) and in geographically structured populations with spatially localized gene flow (Chapter 7). The models presented in Part A of this book show that adaptive speciation under sympatric conditions can no longer be dismissed on theoretical grounds.

The theory of adaptive speciation advanced here arises from a confluence of earlier developments. In particular, the notion that frequency-dependent selection can induce sympatric speciation has been highlighted already in previous models, based on the assumption that reproductive isolation can be pleiotropically induced by adaptation to different local habitats. These Levene-type models are reviewed in Chapter 3, and many of them are, in turn, based on one of the early paradigms of sympatric speciation: host shifts and host-race formation in insects (Bush 1969; Feder 1998). The theory of adaptive speciation extends these early models by showing that disruptive selection does not occur only under rather special assumptions about pleiotropic fitness interactions in two different habitats. Rather, the dynamic emergence of disruptive selection during the course of gradual evolution is a robust consequence of frequency-dependent interactions of many different kinds, including all the basic types of ecological interactions, as well as interactions that lead to sexual selection and sexual conflict (Chapter 5). That frequency-dependent ecological interactions can often lead to the dynamic emergence of fitness minima has been foreshadowed in earlier work (Rosenzweig 1978; Eshel 1983; Taylor 1989; Christiansen 1991; Brown and Pavlovic 1992; Abrams *et al.* 1993a), but until recently neither its ubiquity nor its significance for the theory of speciation had been appreciated fully.

This may, in part, arise from the focus on studying the mean and variance of quantitative genetic traits (Lande 1979b), an approach that makes it difficult to model evolutionary processes through which the trait's frequency distribution can become bimodal. Also, when approaching the problem from the population genetics' aspect, the simplifying assumptions turn out to be restrictive. Study of the evolution of isolating mating mechanisms in Levene-type models with deterministic dynamics that involved at most a few loci had initially cast doubt on the feasibility of sympatric speciation (Felsenstein 1981; Seger 1985a), and thus reinvigorated the case for allopatric speciation. In rather stark contrast, individual-based stochastic models of adaptive dynamics that incorporate multilocus genetics reveal that the evolution of various types of isolating mating mechanisms occurs generically and with relative ease once disruptive selection has emerged dynamically in a sexual population.

We thus see the theory of adaptive speciation as an extension of this earlier work, based on a less restrictive genetic modeling of reproductive processes, as well as on advances in our understanding of the ecological causes of evolutionary diversification that result from an integrated approach to phenotypic evolution under frequency-dependent selection. These developments have led to a unifying framework for theoretical investigations of adaptive diversification, which may challenge the perception of allopatric speciation as the only viable scenario.

19.3   Diversity of Speciation Processes

Freeing research on species formation from the straitjacket imposed by considering such processes allopatric unless unequivocally proved otherwise refocuses our attention onto the great richness and exciting complexity of speciation processes. To appreciate this richness it must be recognized that speciation processes – far from being the single events they present themselves as to the evolutionary taxonomist – have a temporal and spatial extension that allows them to involve multiple phases in time and/or multiple domains in space. In particular, species may originate against a background of allopatric, parapatric, and sympatric distribution patterns, and the results of such a pattern-oriented classification may differ when applied to successive stages of the speciation process. The same holds for the process-oriented distinction between adaptive and nonadaptive mechanisms of speciation: here, too, the classification may differ between stages, or possibly even between domains, of unfolding speciation dynamics. In general, the perception of speciation as a potentially multilayered process offers a healthy antidote against becoming caught up in semantic controversies about rigid dichotomies.

We are thus compelled to be more pragmatic about the classification of speciation processes, because it is unlikely that the complexity of these processes can be captured in a single binary distinction. The long-standing debate as to the prevailing mode of speciation focused primarily on spatial patterns of population distributions and championed nonadaptive mechanisms of speciation (genetic drift, or separate local adaptation to disconnected habitats with pleiotropic implications for reproductive isolation). In this traditional view, reproductive isolation emerges as a by-product of other factors and is not by itself adaptive. By contrast, many chapters in this book illustrate how reproductive isolation can be selected for directly through natural and sexual selection. This alternative view incorporates processes of reinforcement as special cases that can occur after a primary allopatric phase has already created a situation of partial reproductive isolation, which is brought to completion through selection against hybrids and for assortative mating. Discussions in this volume highlight that a primary allopatric phase is entirely dispensable: reproductive isolation can also be selected for in continual sympatry.

This leads to the conclusion that we can characterize speciation processes adequately only by utilizing various dichotomies complementarily: allopatric versus sympatric, nonadaptive versus adaptive, speciation driven by natural versus sexual selection, speciation with and without ecological character displacement, etc. It is also clear that, sometimes, these distinctions can be attributed meaningfully to individual process phases only and not necessarily to the speciation process as a whole. At first sight, this may seem like a plea for unwieldy conceptual intricacy. Instead, we suggest that this recognition provides a fascinating opportunity to achieve a greater (and more encompassing) unity in our description of speciation processes. Aspects of spatial structure and pattern formation, of ecological character displacement and limiting similarity, and of reproductive isolation and assortative mating are often dynamically and inextricably linked when speciation processes run their course. Based on this recognition, we propose to unfold the resultant continuum of possible speciation scenarios along three fundamental axes: spatial differentiation, ecological differentiation, and mating differentiation. As explained in more detail in Box 19.1, this allows us to look at alternative evolutionary pathways of species formation within a common conceptual framework. All traditionally acknowledged speciation mechanisms, as well as those described in this book, are accommodated readily in this broadened classification scheme.

19.4 Empirical Studies of Speciation

The natural diversity of speciation patterns and processes is captured in the empirical chapters of Parts B and C. These deal with examples from very different systems and perspectives, and reflect the broadness of the speciation scenarios encountered in nature. None of these studies were conducted with the theory of adaptive speciation in mind, but they can be seen as starting points to disentangle the diversity of processes into the basic components depicted in Box 19.1.

The chapters in Part B deal with cases of recent or ongoing genetic divergence under conditions of contact, which is where one can hope to find the best evidence for the adaptive speciation scenario. However, each of the chapters deals with settings in which alternative or additional components must be considered. For example, although the three chapters on fish systems deal with very recent splitting events, they come to different conclusions with respect to the primacy of the process that causes speciation. While sexual selection is suggested as the primary driving force for the haplochromine cichlids in African lakes (Chapter 8), the plasticity of the pharyngeal jaw apparatus is also implicated as an additional factor that facilitates quick adaptation to new ecological niches. The stickleback study (Chapter 9) is an excellent example of the interplay between primary allopatric divergence and secondary contact that leads to niche partitioning. And in the arctic charr (Chapter 10), the complexity of the available niches after the postglacial reinvasion of volcanic lakes provides a level of resource polymorphism that is expected to promote the fast generation of newly adapted morphs. Adaptive speciation mechanisms are likely to play a role in all three of these cases, but the actual natural settings will inevitably always be more complex than the idealized world of theoretical abstractions.

The chapters in Part C look at the natural patterns of speciation, including those from phylogeography and paleontology. To infer past processes from extant pattern is an old exercise in speciation research, but to use the knowledge of processes to interpret patterns may be equally enlightening. For example, the paleontological analysis of the repeated radiations of semionotid fish in mesozoic rift lakes (Chapter 18) shows that most evolutionary novelties arose very quickly after the colonization of the lake. This is difficult to explain with a conventional allopatric model, but it is perfectly in line with predictions from the theory of adaptive speciation. Niche partitioning under conditions of disruptive selection is expected to be most efficient during an initial colonization phase, when ecological opportunities abound and evolutionary pathways are less constrained.

Box 19.1 A process-based classification of speciation routes

Realizing that the dichotomies traditionally applied to describe speciation processes are too coarse, here we propose an extended classification scheme. It encompasses that speciation, in general, can be driven by or lead to differentiation between the incipient species in terms of their spatial distributions, their ecological role, and their mating and interbreeding options.

The graphs below therefore simultaneously utilize continuous axes for spatial differentiation (front to back), ecological differentiation (left to right), and mating differentiation (bottom to top). At the onset of speciation, populations are undifferentiated, which corresponds to a starting point at the origin (i.e., in the lower left front corner) of each panel (open circles).

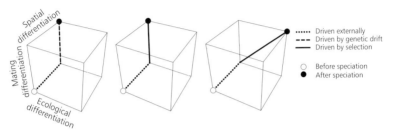

In classic allopatric speciation scenarios, external causes first result in geographic isolation between two incipient species, and thus introduce a high degree of spatial differentiation (dotted lines). After that, either genetic drift (dashed line in left panel above; Sections 6.2 and 6.6, and Box 6.5) or sexual selection and/or conflict (continuous line in middle panel above; Section 6.6, and Box 6.5) can increase mating differentiation. Alternatively, local adaptation with pleiotropic effects on mating can increase ecological and mating differentiation concomitantly (continuous line in right panel above; Sections 7.2, 8.4, 10.4, 11.6, and 13.2 to 13.3, and Box 13.1). In all three cases, the incipient species become reproductively isolated at the end of these speciation trajectories (filled circles).

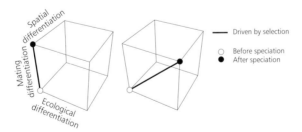

Sympatric speciation scenarios, in contrast, do not require that external causes, as a first step, lead to geographic isolation. For populations that lack any spatial structure, two scenarios have been suggested: either evolution driven by sexual selection and/or conflict induces reproductive isolation in the absence of concomitant ecological differentiation (left panel above; Sections 3.3, 5.4, and 8.2 to 8.4) or such

continued

Box 19.1 *continued*

ecological differentiation is accompanied by the evolution of assortative mating (right panel above; Sections 3.3, 5.2, and 5.3). While the first of these cases draws its motivation from the explosive radiation of cichlid color morphs in African lakes, it is doubtful that morphs differentiated only with respect to their mating characteristics, and not ecologically or spatially, can coexist on an ecological time scale: a large proportion of the resultant species are thus likely to be ephemeral only.

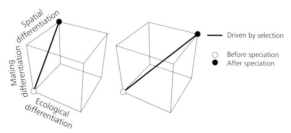

   Introducing a spatial extension, and thus the opportunity for populations to differentiate spatially, extends models of sympatric speciation. If the resultant models are classified according to where the speciation mechanism operates, they can still be called sympatric, whereas judged from the spatial pattern that the speciation process generates they might be termed parapatric. This ambiguity underscores that features like allopatric, parapatric, and sympatric can be attributed meaningfully only to particular stages of speciation processes. Again, we have to consider two cases: either evolution driven by sexual selection and/or conflict induces reproductive isolation and spatial differentiation by giving rise to mating domains (left panel above; Section 15.3; Boxes 7.5 and 15.1), or ecological differentiation is accompanied by the evolution of assortative mating and the emergence of spatial differentiation (right panel above). The latter type of speciation process can occur at least in two guises: first, in the course of host-race formation (Sections 3.4, 11.5 to 11.6, 12.2 to 12.4, and Box 11.1) and, second, through local adaptation and speciation along environmental gradients (Sections 7.3 and 7.4).

   The speciation processes considered so far are all examples of one-phase processes (provided we do not count the imposition of geographic isolation in allopatric scenarios as a separate phase). Given these process "atoms", as the next step we can classify those slightly more complex speciation processes in which two phases are involved. This is accomplished easily, as the examples below illustrate.

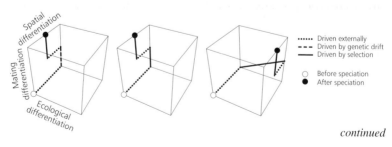

continued

Box 19.1 *continued*

The three panels above show alternative scenarios in which allopatric speciation is brought to completion by reinforcement (Sections 7.2, 9.4, and 11.6). In the wake of geographic isolation (dotted long lines), the incipient species develop partial reproductive isolation, through genetic drift (left panel above), through sexual selection and/or conflict (middle panel above), or through local ecological adaptation (right panel above). This first phase is followed by the establishment of secondary contact (dotted short lines) and subsequent reinforcement (upper continuous lines).

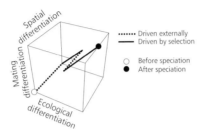

The panel above shows another two-phase speciation process. This time, evolution during a first phase after geographic isolation results in partial ecological differentiation and partial mating differentiation (lower continuous line). In a second phase, contact between the incipient species is reestablished, and further ecological and mating differentiation ensues (upper continuous line); the second phase may also involve an increase in spatial differentiation. A process of this type is favored currently to explain the sympatric occurrence of limnetic and benthic forms of sticklebacks in some Canadian lakes (Sections 9.3 to 9.4, and Boxes 9.1, 9.2, 9.4, and 9.5).

It is evident that the classification scheme proposed here can accommodate even more complex types of speciation processes that involve, for example, three consecutive phases until speciation is completed. Speciation in asexual organisms is another special case: since no mating differentiation evolves, speciation trajectories are restricted to the bottom plane in the graphs above.

Phylogeography based on the analysis of DNA sequences from spatially distributed populations is another source of patterns that can be used to evaluate alternative models of speciation. These patterns enable us to estimate the time scale of divergence events, as well as vicariance patterns and population histories. Such studies often show that closely related species or populations do not occupy the same spatial area (i.e., do not occur in sympatry). This is interpreted habitually as the strongest evidence for the ubiquity of allopatric speciation. However, distinct spatial segregation is often maintained in spite of species ranges that shift in response to environmental fluctuations, such as ice-age cycles (Chapter 15). In such situations segregation must be maintained actively and cannot be considered as a simple by-product of previous allopatry. Such active separation mechanisms are more likely to evolve during an initial sympatric phase under the adaptive speciation scenario, in which assortative mating builds up and results in discrimination between the differently adapted forms. Chapter 7 explains how such sympatric processes of local adaptive diversification can result in patterns of spatial segregation of newly forming species.

The empirical studies in this book were chosen as examples in which adaptive speciation might play an important role, but none of them provides unequivocal proof for the mechanism. Of all the possible empirical approaches, the experiments that involve microorganisms (Chapter 14) will probably enable the most careful control of the conditions required for adaptive speciation. However, we think that adaptive speciation will also be a satisfying explanation for the results of many other studies, even when alternatives cannot be ruled out completely. Future investigations will benefit highly from an intensified interplay between theoretical and empirical work, in particular because this will diminish the risk of biased data acquisition and interpretation that results from self-imposed conceptual restrictions.

19.5   Continuous Splitting and Radiations

The adaptive speciation process has an inherent tendency to lead to the continuous splitting of populations. The reason is that, when a population has reached a new adaptive peak, it may well again come into a situation in which intraspecific competition causes disruptive selection. This can lead to a further split, and also the resultant new populations may go through a new cycle. This tendency for continuous splitting will be limited only by the ecological opportunities initially available or becoming available through changes in the biotic environment caused by the diversification process itself. However, once such saturation is attained, many of the then extant lineages are likely to be highly specialized. Since these lineages will be sensitive to even mild environmental perturbations, they can be lost again easily. On this basis, we should expect a pattern of lineage splitting as depicted in Figure 19.1: while a large diversity of lineages can exist at any given time, only a few of these will survive to form deeper splits.

At first sight, it might appear that such a continuous-splitting process is not compatible with the phylogenetic patterns found in nature. At least in the official taxonomy, species are well defined and clearly delimited. But for almost any species it is possible to identify subspecies, or races, that differ in certain characters. Some species appear so variable that they are officially even called "polytypic". We suggest that this inherent variability can be explained by the continuous-splitting scenario outlined here.

If an environment offers completely new ecological opportunities, the expectation is that newly split lineages are less likely to be lost. Going beyond the small changes that occur under relatively stable conditions, these initial lineages could, rather, be the basis for further specific adaptations. Such cases would become manifest as radiations in the evolutionary history and in the fossil record. Paleontological patterns have been described as punctuated equilibria with periods of

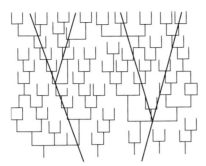

Figure 19.1 The continuous-splitting scenario. If populations have a continuous tendency to split, closely related sister groups are present at any time. However, because of environmental fluctuations, many of these splits are either lost or merge again. Hence, only a relatively few deeper splits persist in the long run. For example, at the top of the tree, we would count 18 genetically different populations, subspecies, or species (thin lines), but only four taxonomically distinguished species that have descended from deeper splits (thick lines).

fast radiation and generation of morphological diversity, followed by relative stasis with little morphological change. If one envisages evolution and speciation as a continuous process of divergence, this would seem like a paradox. However, this is the expected pattern under the adaptive speciation scenario: when new ecological opportunities open up – for example, when a species colonizes a new area (see Boxes 4.8 and 18.2) – it quickly adapts to the new condition and starts cycles of adaptive speciation that lead to a fast succession of splits until all the available ecological opportunities are taken. Since each cycle of splitting may take less than a thousand generations, such radiation will seem very sudden in the fossil record. As long as there is no further ecological change, new adaptations cannot happen and relative morphological stasis is expected. Moreover, as discussed in Box 4.8, adaptively driven punctuation events also can be triggered by quantitatively minor changes in the geometry of the fitness landscape, which occur in the wake of slow changes in the overall environment. Thus, there is no need to invoke a special macro-evolutionary mechanism, or divergence in peripheral isolates, to explain the pattern of punctuated equilibria encountered in the fossil record.

19.6 Future Directions

It is evident that new, dedicated studies are required to better explore the applicability of the adaptive speciation process to a wide range of natural systems. The preceding chapters also highlight a suite of unanswered questions that need to be tackled. Advances are needed with regard to both theoretical and empirical approaches.

Speciation models have come a long way over the past five decades – yet major challenges still remain. The ultimate goal of theoretical endeavors is to catalog the relationship between potential mechanisms and potential phenomena. In the

Box 19.2 Future theoretical research

In this book we survey a number of processes that can lead to adaptive speciation, and discuss the patterns such processes are expected to generate. We clearly are in the initial phases of this endeavor, and most of the models put forward still require further investigation. Below we outline some of the most interesting options for further theoretical research.

Many questions below extend to sympatric, parapatric, and allopatric speciation and address the following two fundamental issues. First, what genetic, ecological, mating, and spatial structures are particularly conducive to speciation that progresses along certain routes? Second, what are the phylogenetic and biogeographic consequences of such speciation? Answers to these questions will help us evaluate the probabilities of past speciation routes from present observations.

Speciation genetics. Responses of alternative genetic architectures to the selection pressures encountered at evolutionary branching points should be analyzed. Forays in this direction have been made by Kawecki (Chapter 3), and by Geritz and Kisdi (2000), who considered adaptive dynamics in allele space in Levene-type models (see Box 4.7). Corresponding multilocus models, comparable to those considered by Dieckmann and Doebeli (Chapter 5), should also be investigated. In this context departures from additive genetics have to be explored and their implications for the potential and pace of adaptive speciation analyzed. In addition, we need better insight into the potential of sex-ratio distorters and selfish elements to promote or prevent adaptive speciation (Chapter 12).

Evolving mate choice. Models in which mate choice is based on separate loci for mating signals, mating preferences, and the strength of these preferences need to be integrated with ecological models of adaptive speciation. Also, mate finding and mate recognition could be modeled separately from mate choice. In the real world, the space of phenotypes potentially involved in such processes is often very highly dimensional. The implications of such high dimensionality for the speed at which fitness minima can be escaped need to be analyzed (e.g., Van Doorn *et al.* 2001). Also, general models for speciation based on sexual selection alone have to be developed and studied in greater detail (Van Doorn and Weissing 2001); such models should start to incorporate the frequency dependence that often arises from mate choice. Sexual arms races are one potential mechanism that enables fast allopatric speciation [Schilthuizen 2001; see also Gavrilets (2000b) and Section 5.4]: more mechanistic models that underpin this idea are needed, as well as parapatric and sympatric variants of these models to assess the amount of gene flow under which pairs of arms races can still diverge.

Speciation, learning, and plasticity. Models that combine genetic evolution with learning of the ecological role, or with plasticity in an ecologically relevant trait, should be developed and studied. In this context, links with optimal foraging theory should be explored. The same applies to models that combine ecological branching with learned mate selection (e.g., through song learning in birds, or parental imprinting). In general, under what conditions does plasticity or learning help speciation to take off, by allowing species to enter a wider variety of niches, and when does it actually preclude speciation by broadening resource specialization (Chapter 12)? The answer necessarily depends on details of the fitness landscape: plasticity can make a bumpy landscape smoother and so more conducive to evolutionary change (as species do not become stuck on any little hill), but plasticity can also just supply the phenotypic variation otherwise provided by genetic diversification (Chapter 10).

continued

Box 19.2 *continued*

Patterns, biogeographic and other. Predictions are needed about the spatial patterns of genetic variation of and linkage disequilibria between ecological and mating characters expected from adaptive speciation. These patterns are likely to depend on the speciation route followed by the process (Chapter 7; Box 19.1); therefore, mechanisms need to be grouped according to the patterns they engender. Predictions that can distinguish ongoing sympatric speciation from secondary contact after incomplete allopatric speciation would be particularly valuable. Also, conditions for the evolution of spatial mosaics through adaptive speciation have to be better understood; these should refer to the ecological mechanism that causes divergence, the movement pattern of individuals, and the underlying environmental heterogeneity. At the temporal end, in allopatric speciation as envisaged in Chapter 6, the average number of speciation events per time unit will probably decrease only slowly over evolutionary time, while the arguments put forward in Box 18.2 suggest that speciation rates decrease very rapidly in sympatric adaptive speciation scenarios. In general, certain phylogenetic features may indicate certain speciation processes – exactly how remains to be determined.

Speciation time scales. We need better insight into the time scales of speciation processes as they unfold along different routes. How robust are the salient numerical results, that is, to what extent do they depend on particular model assumptions? For example, how predictions from the models discussed in Chapter 6 depend on the specific assumptions made about the dependence of genetic incompatibility on genetic distance should be resolved. When considering alternative mechanisms – like dominance, sexual dimorphisms, or step-like phenotypic plasticity (Van Dooren 1999; Matessi *et al.* 2001) – that could accomplish splitting a population into a number of discrete morphs, we have to assess how these mechanisms fare when raced against each other in more elaborate ecological models.

Long-term phylogenetic implications. It will also be interesting to understand how patterns caused by the various speciation routes considered in this book can themselves act as historical constraints on later developments. This can involve ecological processes (speciation events change the community and may thus prepare the way for further speciation events, either in the descendant species or in other parts of the community), spatial patterns (particular distributions may be more conducive to further speciation than others), and genetic architectures affected by earlier speciation (some such architectures may allow escapes from fitness minima more readily than others, as, for example, when earlier mechanisms for assortative mating help new sorts of assortativeness to start).

More complex speciation models. Integration of the various speciation mechanisms into more general models is required. In particular, models for the generation of ecological sister species through sexual selection, like that discussed in Chapter 8, should be combined with models for adaptive speciation caused by ecological factors, to better understand how the processes may interact (Van Doorn and Weissing 2001). Also, with regard to the underlying ecology it is of great importance to assess the robustness of predictions within more extended model families. An interesting development in this direction is that for very large classes of ecological models the local form of invasion fitness is the same (up to second-order terms) as that of a Lotka–Volterra system (Durinx *et al.*, unpublished). This suggests that for particular theoretical problems we can restrict our attention to such approximations of more complex ecologies.

Box 19.3 Future empirical research

The chapters in this book make clear that, to gain access to more conclusive evidence for particular speciation scenarios, much more attention has to be paid to the earliest phases of the splitting process, which are confined to a few hundred generations after the conditions for splitting have been established. Identifying such situations will be a challenge, but the continuous-splitting scenario (Figure 19.1) suggests that they are not rare. Studying newly colonized habitats will probably provide the best entry point to such situations in nature. Furthermore, because the splitting process is expected to be relatively quick, it will be possible to set up laboratory experiments with rapidly reproducing species that may allow the full speciation process to be followed. Several aspects should be at the top of the agenda for such future empirical research.

Assortative mating. For sexually reproducing species, the emergence of assortative mating is a prerequisite for the splitting process. Although assortativeness is a well-known phenomenon in nature, it must now be studied in much more detail and with new conceptual approaches. The simplest assumption would be that signals for assortative mating are coupled pleiotropically to the ecological trait under selection. However, this is not a strict requirement since Dieckmann and Doebeli (1999) showed that such coupling between the ecological trait and specific mating signals can also evolve from scratch. A further alternative is that an assortative mating system is already present, because it evolved in the previous round of speciation. Only minor modifications may be necessary to achieve a further differentiation. It should thus be rewarding to identify situations in which successive speciation events have taken place within a short time and compare the signaling and recognition mechanisms in each of the sister groups. Ethologists know that animals are generally rather choosy when it comes to mating, for good reasons. The choice of the right partner has a direct fitness consequence because it determines the genetic quality of the offspring. The optimal partner for an organism is one that is most compatible with its own genotype, in the sense that the joint offspring are able to compete effectively for the ecological niche that was also used by the parents. This leads to "genotypic assortativeness", which needs to be coupled to specific signals. Thus, genetic assortativeness can be considered as something that is already built into the system of a sexually reproducing species, and that does not have to evolve newly during a speciation process. In general, a better understanding of the genetics of recognition and signaling is required to investigate ongoing processes of adaptive speciation.

Genetic incompatibilities. Although the initial splitting may only involve prezygotic mechanisms and the sorting of pre-existing alleles, it is clear that postzygotic genetic incompatibility evolves at some point. There is increasing evidence that such postzygotic effects are not simply a consequence of the random accumulation of differences, but can be driven by specific genes. One of these, the Odysseus locus in *Drosophila*, is particularly well studied and there is evidence for continuous strong positive selection at this locus, which suggests an active role in the separation process (Ting *et al.* 1998). In future studies, it will be very important to trace the role of such "speciation genes" during the separation process. Can they act as selfish elements that drive separation on its own, such as *Wolbachia* bacteria in insects? Are they part of the assortative mating process by affecting signaling or recognition? Or are they only recruited during a later phase of the separation process?

continued

Box 19.3 *continued*

Spatial context. One of the largest challenges for the future is to understand the connection between the splitting processes under conditions of ecological contact and subsequent spatial separation. Clearly, most closely related species do not occur in sympatry and there must therefore be a mechanism that leads to these spatial splits. Modeling efforts in this direction are already well underway (Chapter 7; Doebeli and Dieckmann 2003; M. Rost and M. Lässig, personal communication). For empirical studies, it is particularly important to apply refined molecular markers in the phylogeographic reconstruction of existing patterns of species and population distribution. This will enable situations to be identified at different stages of the separation, or secondary contact process, which can then be studied in detail. In particular, a number of interesting differences in the gene flow patterns for mitochondrial and nuclear markers have now been documented (Shaw 2002; Ogden and Thorpe 2002). These findings might provide vital clues about the role of male and female migration in establishing or obliterating spatial patterns of genetic differentiation.

Experimental systems. To study the full process of separation in the laboratory, one will have to resort to organisms with short generation times. The first choice would be viruses, bacteria, or unicellular eukaryotes, such as yeast or algae. They potentially allow each generation to be retained by freezing samples, to carry out replicate experiments under controlled conditions, and possibly to study genetic changes at the genome level. Chapter 14 provides salient examples of how such studies can be used to investigate adaptive speciation and radiation. However, while experimental systems of microorganisms have many advantages for the study of ecological mechanisms of diversification, they have some obvious disadvantages when it comes to the study of assortative mating mechanisms – if only because many microorganisms have irregular and often quite complicated modes of sexual reproduction. It should therefore be attractive to further develop short-lived, sexually reproducing organisms of higher taxa into laboratory models for speciation experiments. *Drosophila* has already been used often for speciation studies, but not in the context of explicit scenarios for adaptive speciation processes. With the availability of the *Drosophila* genome sequence, this system might offer an attractive alternative to studies that focus on unicellular organisms.

---

best of all worlds we would thus be able to write down and analyze a single, all encompassing family of eco-evolutionary speciation models. This family's parameter space would be parceled up into different regions that correspond to different speciation routes and mechanisms. In practice, however, speciation processes at present appear far too complex and diverse for such an exercise to be feasible yet. At best we can analyze small subsets of the larger family in which we stress some mechanisms at the cost of neglecting others. In a next stage, we may tentatively combine two, or sometimes even more, mechanisms to explore their interactions. Box 19.2 gives an overview of the present and future challenges offered by such a research program.

Future empirical studies will benefit from the analysis of carefully controlled laboratory populations, as well as from the identification of natural situations of ongoing splitting. We need to better understand the specific biological mechanisms that underlie evolutionary branching and the evolution of assortative mating. Although assortative mating and specific mate choice are well known as such, their genetic basis and evolutionary origins now have to be explored in the context of speciation. As discussed in Chapter 1, there are many possibilities as to how assortativeness can come about, so the empirical studies should pay particular attention to these mechanisms. The implied challenges for further empirical research are summarized in Box 19.3.

A key issue for both theory and empirical studies is the inclusion of spatial context. After all, parapatric or allopatric patterns of species distribution are prevalent in nature and need to be explained. The model discussed in Chapter 7 shows that gradients of environmental resource distribution lead to spatial splits under the adaptive speciation scenario (Doebeli and Dieckmann 2003). This line of investigation bears further exploration, and there is a need for empirical studies that analyze the ecological settings prone to such spatial splitting.

The study of speciation mechanisms can be freed finally from the conceptual chains that external causes always have to be invoked as the driving forces of speciation processes. Instead, frequency-dependent selection and evolutionary branching emerge as plausible mechanisms of lineage splits that are adaptive and immanently arise in speciating populations. When the conditions for adaptive speciation are met, the splitting of a population becomes an inescapable consequence of its interacting and reproducing constituents. Speciation is thus a law of nature, rather than an accident.

References

Page numbers of reference citations in this volume are given in square brackets.

Abe Y (1991). Host race formation in the gall wasp *Andricus mukaigawae*. *Entomologia Experimentalis et Applicata* **58**:15–20 [*236, 241*]

Abrams PA (1983). The theory of limiting similarity. *Annual Review of Ecology and Systematics* **4**:359–376 [*311*]

Abrams PA (1986). Character displacement and niche shift analyzed using consumer–resource models of competition. *Theoretical Population Biology* **29**:107–160 [*45*]

Abrams PA (2001a). Modelling the adaptive dynamics of traits involved in inter- and intraspecific interactions: An assessment of three methods. *Ecology Letters* **4**:166–176 [*72*]

Abrams PA (2001b). Adaptive dynamics: Neither F nor G. *Evolutionary Ecology Research* **3**:369–373 [*72*]

Abrams PA & Matsuda H (1996). Fitness minimization and dynamic instability as a consequence of predator–prey coevolution. *Evolutionary Ecology* **10**:167–186 [*98*]

Abrams PA, Matsuda H & Harada Y (1993a). Evolutionarily unstable fitness maxima and stable fitness minima of continuous traits. *Evolutionary Ecology* **7**:465–487 [*42, 60, 70, 75, 107, 143, 382*]

Abrams PA, Harada Y & Matsuda H (1993b). On the relationship between quantitative genetic and ESS models. *Evolution* **47**:982–985 [*60, 74*]

Adams C, Woltering C & Alexander G (2003). Epigenetic regulation of trophic anatomy through feeding behaviour in Arctic charr, *Salvelinus alpinus*. *Biological Journal of the Linnean Society* **78**:43–49 [*220, 225*]

Agrawal AA, Vala F & Sabelis MW (2002). Induction of preference and performance after acclimation to novel hosts in a phytophagous spider mite: Adaptive plasticity. *The American Naturalist* **159**:553–565 [*255*]

Akimoto S (1990). Local adaptation and host race formation of a gall-forming aphid in relation to environmental heterogeneity. *Oecologia* **83**:162–170 [*242*]

Alarcón R & Campbell DR (2000). Absence of conspecific pollen advantage in the dynamics of an *Ipomopsis* (Polemoniaceae) hybrid zone. *American Journal of Botany* **87**:819–824 [*265*]

Alcobendas M, Dopazo H & Alberch A (1996). Geographic variation in allozymes of populations of *Salamandra salamandra* (Amphibia: Urodela) exhibiting distinct reproductive modes. *Journal of Evolutionary Biology* **9**:83–102 [*318*]

Alekseyev SS (1995). Formation of morphological difference between the sharp-snouted and blunt-snouted lenok (genus: Brachymystax, Slamonidae) during ontogeny and the role of heterochrony in their divergence. *Journal of Ichthyology* **35**:1–21 [*218*]

Andersson M (1994). *Sexual Selection*. Princeton, NJ, USA: Princeton University Press [*44, 98*]

Andersson S & Widén B (1993). Pollinator-mediated selection on floral traits in a synthetic population of *Senecio integrifolius* (Asteraceae). *Oikos* **66**:72–79 [*266*]

Andow DA & Imura O (1994). Specialization of phytophagous arthropod communities on introduced plants. *Ecology* **75**:296–300 [*252*]

Andrews RM (1976). Growth rate in island and mainland anoline lizards. *Copeia* **1976**:477–482 [*323*]

Andrews RM (1979). Evolution of life histories: A comparison of *Anolis* lizards from matched island and mainland habitats. *Breviora* **454**:1–51 [*323*]

Antonovics J (1968). Evolution in closely adjacent plant populations VI. Manifold effects of gene flow. *Heredity* **23**:507–524 [*272*]

Antonovics J (1971). The effects of a heterogeneous environment on the genetics of natural populations. *American Scientist* **59**:593–599 [*277*]

Armbruster WS (1990). Estimating and testing the shapes of adaptive surfaces: The morphology and pollination of *Dalechampia* blossoms. *The American Naturalist* **135**:14–31 [*277*]

Armbruster WS (1996). Evolution of floral morphology and function: An integrative approach to adaptation, constraint, and compromise in *Dalechampia* (Euphorbiaceae). In *Floral Biology: Studies on Floral Evolution in Animal-Pollinated Plants*, eds. Lloyd DG & Barrett SCH, pp. 241–272. London, UK: Chapman & Hall [*276*]

Armstrong RA & McGehee R (1980). Competitive exclusion. *The American Naturalist* **115**:151–170 [*306*]

Arnegard ME, Markert JA, Danley PD & Kocher TD (1999). Phylogeny of a rapidly evolving clade: The cichlid fishes of Lake Malawi, East Africa. *Proceedings of the National Academy of Sciences of the USA* **96**:5107–6110 [*174, 185*]

Arnold SJ (1983). Morphology, performance, and fitness. *American Zoologist* **23**:1347–1361 [*337*]

Arnold EN (1994). Investigating the origins of performance advantage: Adaptation, exaptation and lineage effects. In *Phylogenetics and Ecology*, eds. Eggleton P & Vane-Wright R, pp. 123–168. London, UK: Academic Press [*337*]

Arnold ML, Hamrick JL & Bennett BD (1990). Allozyme variation in Louisiana irises: A test for introgression and hybrid speciation. *Heredity* **65**:279–306 [*265*]

Arnold ML, Hamrick JL & Bennett BD (1993). Interspecific pollen competition and reproductive isolation in *Iris. Journal of Heredity* **84**:13–16 [*275*]

Arnqvist G (1998). Comparative evidence for the evolution of genitalia by sexual selection. *Nature* **393**:784–786 [*125*]

Arnquist G & Rowe L (2002). Antagonistic coevolution between the sexes in a group of insects. *Nature* **415**:787–789 [*102*]

Askew RR (1968). Considerations on speciation in Chalcidoidea (Hymenoptera). *Evolution* **22**:642–645 [*240*]

Askew RR (1971). *Parasitic Insects*. New York, NY, USA: American Elsevier [*240*]

Asquith A (1993). Patterns of speciation in the genus *Lopidea* (Heteroptera: Miridae Orthotylinae). *Systematic Entomology* **18**:169–

180 [*239*]

Asquith A (1995). Evolution of *Sarona* (Heteroptera, Miridae): Speciation on geographic and ecological islands. In *Hawaiian Biogeography: Evolution on a Hot Spot Archipelago*, eds. Wagner WL & Funk VA, pp. 90–120. Smithsonian Series in Comparative Evolutionary Biology, Washington, DC, USA: Smithsonian Institution Press [*239*]

Autumn K, Liang YA, Hsieh T, Fisher RN, Zesch W, Chan WP, Kenny TW, Fearing R & Full RJ (2000). Adhesive force of a single gecko foot-hair. *Nature* **405**:681–685 [*323*]

Avise JC (1994). *Molecular Markers, Natural History and Evolution*. London, UK: Chapman & Hall [*298, 305, 313*]

Avise JC (2000a). *Phylogeography*. Cambridge, MA, USA: Harvard University Press [*115*]

Avise JC (2000b). Cladists in wonderland. *Evolution* **54**:1828–1832 [*18, 24*]

Avise JC, Arnold J, Ball RM, Bermingham E, Lamb T, Neigel JE, Reeb CA & Saunders NC (1987). Intraspecific phylogeography: The mitochondrial DNA bridge between population genetics and systematics. *Annual Review of Ecology and Systematics* **18**:489–522 [*305*]

Babajide A, Hofacker IL, Sippl MJ & Stadler PF (1997). Neutral networks in protein space: A computational study based on knowledge-based potentials of mean force. *Folding and Design* **2**:261–269 [*128*]

Baker HG (1963). Evolutionary mechanisms in pollination biology. *Science* **139**:877–883 [*264*]

Baldwin JM (1896). A new factor in evolution. *The American Naturalist* **30**:441–451 [*250*]

Balkau B & Feldman MW (1973). Selection for migration modification. *Genetics* **74**:171–174 [*71*]

Balon EK (1985). *Early Life Histories of Fishes: New Developmental, Ecological and Evolutionary Perspectives*. Volume 5 of Developments in Environmental Biology of Fishes. The Hague, Netherlands: Dr W Junk Publishers [*213*]

Barigozzi C, ed. (1982). *Mechanisms of Speciation*. New York, NY, USA: Alan Liss [*23*]

Barker SC (1994). Phylogeny and classification, origins, and evolution of host associations of lice. *International Journal of Para-*

sitology **24**:1285–1291 [*231*]

Barnard CJ (1998). Sexual selection. In *The Encyclopedia of Ecology and Environmental Management*, ed. Calow P, pp. 681–682. Oxford, UK: Blackwell Science [*98*]

Barraclough TG & Nee S (2001). Phylogenetics and speciation. *Trends in Ecology and Evolution* **16**:391–399 [*110*]

Barraclough TG & Vogler AP (2000). Detecting the geographical pattern of speciation from species-level phylogenies. *The American Naturalist* **155**:419–434 [*140, 235*]

Barton NH (1979). Gene flow past a cline. *Heredity* **43**:333–339 [*115*]

Barton NH (1989a). Founder effect speciation. In *Speciation and Its Consequences*, eds. Otte D & Endler JA, pp. 229–256. Sunderland, MA, USA: Sinauer Associates Inc. [*117, 119, 135*]

Barton NH (1989b). The divergence of a polygenic system subject to stabilizing selection, mutation and drift. *Genetical Research* **54**:59–77 [*129*]

Barton NH (1999). Clines in polygenic traits. *Genetical Research* **74**:223–236 [*141*]

Barton NH & Bengtsson BO (1986). The barrier to genetic exchange between hybridizing populations. *Heredity* **56**:357–376 [*115, 125*]

Barton NH & Charlesworth B (1984). Genetic revolutions, founder effects, and speciation. *Annual Review of Ecology and Systematics* **15**:133–164 [*118–119, 121, 135*]

Barton NH & Gale KS (1993). Genetic analysis of hybrid zones. In *Hybrid Zones and the Evolutionary Process*, ed. Harrison RG, pp. 13–45. New York, NY, USA: Oxford University Press [*115*]

Barton NH & Hewitt GM (1989). Adaptation, speciation, and hybrid zones. *Nature* **341**:497–503 [*141, 165*]

Bateson W (1909). Heredity and variation in modern lights. In *Darwin and Modern Science*, ed. Seward AC, pp. 85–101. Cambridge, UK: Cambridge University Press [*123*]

Baur A, Chalwatzis N & Buschinger A (1995). Mitochondrial DNA sequences reveal close relationships between social parasitic ants and their host species. *Current Genetics* **28**:242–247 [*241*]

Becerra J (1997). Insects on plants: Macroevolutionary chemical trends in host use. *Sci-*

ence **276**:253–256 [*236, 240*]

Bengtsson BO (1979). Theoretical models of speciation. *Zoologica Scripta* **8**:303–304 [*222*]

Bengtsson BO (1985). The flow of genes through a genetic barrier. In *Evolution: Essays in Honor of John Maynard Smith*, eds. Greenwood PJ, Harvey PH & Slatkin M, pp. 31–42. Cambridge, UK: Cambridge University Press [*115, 125*]

Bengtsson BO & Christiansen FB (1983). A two-locus mutation selection model and some of its evolutionary implications. *Theoretical Population Biology* **24**:59–77 [*125*]

Bennington CC & McGraw JB (1995). Natural selection and ecotypic differentiation in *Impatiens pallida*. *Ecological Monographs* **65**:303–324 [*272*]

Berenbaum MR & Zangerl AR (1998). Chemical phenotype matching between a plant and its insect herbivore. *Proceedings of the National Academy of Sciences of the USA* **95**:13743–13748 [*252*]

Berlocher SH (1998). Can sympatric speciation via host or habitat shift be proven from phylogenetic and biogeographic evidence? In *Endless Forms: Species and Speciation*, eds. Howard DJ & Berlocher SH, pp. 99–113. New York, NY, USA: Oxford University Press [*110*]

Berlocher SH (2000). Radiation and divergence in the *Rhagoletis pomonella* species group: Inferences from allozymes. *Evolution* **54**:543–553 [*241*]

Berlocher SH & Feder JL (2002). Sympatric speciation in phytophagous insects: Moving beyond controversy? *Annual Review of Entomology* **47**:773–815 [*91, 236–237*]

Bernatchez L & Dodson JJ (1990). Allopatric origin of sympatric population of lake whitefish (*Coregonus clupeaformis*) as revealed by mitochondrial-DNA restriction analysis. *Evolution* **44**:1263–1271 [*219*]

Bernays EA (1995). Effects of experience on host-plant selection. In *Chemical Ecology of Insects 2*, eds. Cardé RT & Bell WJ, pp. 47–64. London, UK: Chapman & Hall [*254*]

Berry PE & Calvo RN (1989). Wind pollination, self-incompatibility, and altitudinal shifts in pollination systems in the high Andean genus *Espeletia* (Asteraceae). *American Journal of Botany* **76**:1602–1614 [*272*]

Bess HA (1974). *Hedylepta blackburni* (Butler), a perennial pest of coconut on wind-swept sites in Hawaii. *Proceedings of the Hawaiian Entomological Society* **21**:343–353   [*239*]

Beuning KRM, Kelts K & Ito E (1997). Paleo-hydrology of Lake Victoria, East Africa, inferred from $^{18}O/^{16}O$ ratios in sediment cellulose. *Geology* **25**:1083–1086   [*173–174*]

Bever JD (1999). Dynamics within mutualism and the maintenance of diversity: inference from a model of interguild frequency dependence. *Ecology Letters* **2**:52–61   [*94*]

Blair WF (1955). Mating call and stage of speciation in the *Microhyla olivacea* – *M. carolinensis* complex. *Evolution* **9**:469–480   [*193, 203*]

Boag PT & Grant PR (1981). Intense natural selection in a population of Darwin's finches (*Geospizinae*) in the Galapagos. *Science* **214**:82–85   [*194*]

Bodaly RA, Clayton JW, Lindsey CC & Vuorinen J (1992). Evolution of the lake whitefish (*Coregonus clupeaformis*) in North America during the Pleistocene: Genetic differentiation between sympatric populations. *Canadian Journal of Fisheries and Aquatic Sciences* **49**:769–779   [*219, 221*]

Boots M & Haraguchi Y (1999). The evolution of costly resistance in host–parasite systems. *The American Naturalist* **153**:359–370   [*73*]

Bordenstein S, O'Hara FP & Werren JH (2001). *Wolbachia*-induced incompatibility precedes other hybrid incompatibilities in *Nasonia*. *Nature* **409**:707–710   [*260*]

Borland M (1986). *Size-assortative Mating in Threespine Sticklebacks from Two Sites on the Salmon River, British Columbia*. MSc Thesis. Vancouver, BC, Canada: University of British Columbia   [*209*]

Boughman JW (2001). Divergent sexual selection enhances reproductive isolation in sticklebacks. *Nature* **411**:944–948   [*98*]

Boughman JW (2002). How sensory drive can promote speciation. *Trends in Ecology and Evolution* **17**:571–577   [*9, 98, 193*]

Bourke P, Magnan P & Rodriguez MA (1997). Individual variations in habitat use and morphology in brook charr. *Journal of Fish Biology* **51**:783–794   [*216*]

Bouton N, Witte F, van Alphen JJM, Schenk A & Seehausen O (1999). Local adaptations in populations of rock-dwelling haplochromines (Pisces: Cichlidae) from south-

ern Lake Victoria. *Proceedings of the Royal Society of London B* **266**:355–360   [*174, 190*]

Bowler PJ (1983). *The Eclipse of Darwinism: Anti-Darwinian Evolution Theories in the Decades Around 1900*. Baltimore, MD, USA: Johns Hopkins University Press   [*17*]

Bradshaw AD (1960). Population differentiation in *Agrostis tenuis* Sibth. III. Populations in varied environments. *New Phytologist* **59**:92–103   [*272*]

Bradshaw AD (1972). Some of the evolutionary consequences of being a plant. *Evolutionary Biology* **15**:25–47   [*277*]

Bradshaw HD Jr, Wilbert SM, Otto KG & Schemske DW (1995). Genetic mapping of floral traits associated with reproductive isolation in monkeyflowers (*Mimulus*). *Nature* **376**:762–765   [*265, 276*]

Breeuwer JAJ (1997). *Wolbachia* and cytoplasmic incompatibility in the spider mites *Tetranychus urticae* and *T. turkestani*. *Heredity* **79**:41–47   [*253*]

Breeuwer JAJ & Werren JH (1990). Microorganisms associated with chromosome destruction and reproductive isolation between two insect species. *Nature* **346**:558–560   [*260*]

Brenner T (1980). The arctic charr, *Salvelinus alpinus*, in the prealpine Attersee, Austria. In *Charrs: Salmonid Fishes of the Genus Salvelinus*, ed. Balon EK, pp. 765–772. The Hague, Netherlands: Dr W Junk Publishers   [*213*]

Bridle JR & Jiggins CD (2000). Adaptive dynamics: Is speciation too easy? *Trends in Ecology and Evolution* **15**:225–226   [*76*]

Bridle JR & Ritchie MG (2001). Assortative mating and the genic view of speciation. *Journal of Evolutionary Biology* **14**:878–879   [*311*]

Briese DT, Espiau C & Pouchot-Lermans A (1996). Micro-evolution in the weevil genus *Larinus*: The formation of host biotypes and speciation. *Molecular Ecology* **5**:531–545   [*233*]

Briggs JC (1999). Coincident biogeographic patterns: Indo-West Pacific Ocean. *Evolution* **53**:326–335   [*116*]

Brodie ED III, Moore AJ & Janzen FJ (1995). Visualising and quantifying natural selection. *Trends in Ecology and Evolution* **10**:313–318   [*329*]

Brody AK (1997). Effects of pollinators, herbivores, and seed predators on flowering phenology. *Ecology* **78**:1624–1631 [*273–274*]

Brooks JL (1950). Speciation in ancient lakes. *Quarterly Review of Biology* **25**:30–60, 131–176 [*362*]

Brooks DR & McLennan DA (1991). *Phylogeny, Ecology, and Behavior: A Research Program in Comparative Biology*. Chicago, IL, USA: University of Chicago Press [*112–114*]

Brown JL Jr (1957). Centrifugal speciation. *Quarterly Review of Biology* **32**:247–277 [*116, 136*]

Brown JS & Pavlovic NB (1992). Evolution in heterogeneous environments: Effects of migration on habitat specialization. *Evolutionary Ecology Research* **6**:360–382 [*72, 75, 107, 382*]

Brown JS & Vincent TL (1987). Predator–prey coevolution as an evolutionary game. In *Applications of Control Theory in Ecology*, ed. Cohen Y, pp. 83–101. Lecture Notes in Biomathematics 73. Berlin, Germany: Springer-Verlag [*95*]

Brown JS & Vincent TL (1992). Organization of predator–prey communities as an evolutionary game. *Evolution* **46**:1269–1283 [*75, 95*]

Brown WL & Wilson EO (1956). Character displacement. *Systematic Zoology* **5**:49–64 [*203*]

Brown GE, Browm JA & Srivastava RK (1992). The effect of stocking density on the behavior of arctic charr *Salvelinus alpinus* L. *Journal of Fish Biology* **41**:995–963 [*217*]

Brown JM, Abrahamson WG & Way PA (1996). Mitochondrial DNA phylogeography of host races of the goldenrod ball gallmaker, *Eurosta solidaginis* (Diptera: Tephritidae). *Evolution* **50**:777–786 [*242*]

Brown JM, Leebens-Mack JH, Thompson JN, Pellmyr O & Harrison RG (1997). Phylogeography and host association in a pollinating seed parasite *Greya politella* (Lepidoptera: Prodoxidae). *Molecular Ecology* **6**:215–224 [*235, 239*]

Bullini L (1994). Origin and evolution of animal hybrid species. *Trends in Ecology and Evolution* **9**:422–426 [*230–231*]

Bulmer M (1989). Structural instability of models of sexual selection. *Theoretical Population Biology* **35**:195–206 [*99*]

Burbrink FT, Lawson R & Slowinski JB (2000). Mitochondrial DNA phylogeography of the polytypic North American Rat Snake (*Elaphe obsoleta*): A critique of the subspecies concept. *Evolution* **54**:2107–2118 [*26–28*]

Burger W (1995). Montane species-limits in Costa Rica and evidence for local speciation on attitudinal gradients. In *Biodiversity and Conservation of Neotropical Montane Forests*, ed. Churchill SP, pp. 127–133. New York, NY, USA: The New York Botanical Garden [*115*]

Bush GL (1969). Sympatric host race formation and speciation in frugivorous flies of the genus *Rhagoletis* (Diptera: Tephritidae). *Evolution* **23**:237–251 [*49, 237, 382*]

Bush GL (1975). Modes of animal speciation. *Annual Review of Ecology and Systematics* **6**:339–364 [*79*]

Bush GL (1992). Host race formation and sympatric speciation in *Rhagoletis* fruit flies (Diptera: Tephritidae). *Psyche* **99**:335–357 [*236*]

Bush GL (1993). A reaffirmation of Santa Rosalia, or Why are there so many kinds of *small* animals? In *Evolutionary Patterns and Processes*, eds. Lees DR & Edwards D, pp. 229–249. Linnean Society Symposium Series No. 14. London, UK: Academic Press [*229, 376*]

Bush GL (1994). Sympatric speciation in animals: New wine in old bottles. *Trends in Ecology and Evolution* **9**:285–288 [*49–50, 53, 211, 237*]

Bush GL (1998). The conceptual radicalization of an evolutionary biologist. In *Endless Forms: Species and Speciation*, eds. Howard D & Berlocher S, pp. 425–438. Oxford, UK: Oxford University Press [*76, 381*]

Bush GL & Smith JJ (1997). The sympatric origin of phytophagous insects. In *Vertical Food Web Interactions: Evolutionary Patterns and Driving Forces*, eds. Dettner K & Völkl W, pp. 3–19. Heidelberg, Germany: Springer-Verlag [*230, 241*]

Bush GL & Smith JJ (1998). The genetics and ecology of sympatric speciation: A case study. *Researches on Population Ecology* **40**:175–187 [*232, 235, 241*]

Bush GL, Feder JL, Berlocher SH, McPheron BA, Smith DC & Chilcote CA (1989). Sympatric origins of *R. pomonella*. *Nature*

339:346 [*237*]

Butler MA, Schoener TW & Losos JB (2000). The relationship between sexual size dimorphism and habitat use in Greater Antillean *Anolis* lizards. *Evolution* **54**:259–272 [*337*]

Butlin RK (1987). Speciation by reinforcement. *Trends in Ecology and Evolution* **2**:8–13 [*206*]

Butlin RK (1989). Reinforcement of premating isolation. In *Speciation and Its Consequences*, eds. Otte D & Endler JA, pp. 158–179. Sunderland, MA, USA: Sinauer Associates Inc. [*206, 208, 246*]

Butlin RK (1990). Swallowtail performance. *Nature* **344**:716 [*230*]

Butlin RK (1996). Co-ordination of the sexual signaling system and the genetic basis of differentiation between populations in the brown planthopper, *Nilaparvata lugens*. *Heredity* **77**:369–377 [*238, 247*]

Butlin RK (1998). What do hybrid zones in general, and the *Chorthippus parallelus* zone in particular, tell us about speciation? In *Endless Forms: Species and Speciation*, eds. Howard DJ & Berlocher SH, pp. 367–378. Oxford, UK: Oxford University Press [*247*]

Butlin RK & Tregenza T (1997). Is speciation no accident? *Nature* **387**:551–553 [*246*]

Cabot EL, Davis AW, Johnson NA & Wu C-I (1994). Genetics of reproductive isolation in the *Drosophila simulans* clade: Complex epistasis underlying hybrid male sterility. *Genetics* **137**:175–189 [*124–125*]

Callaini G, Dallai R & Riparbelli MG (1997). *Wolbachia*-induced delay of paternal chromatin condensation does not prevent maternal chromosomes from entering anaphase in incompatible crosses of *Drosophila simulans*. *Journal of Cell Science* **110**:271–280 [*260*]

Caillaud MC & Via S (2000). Specialized feeding behavior influences both ecological specialization and assortative mating in sympatric host races of pea aphids. *The American Naturalist* **156**:606–621 [*49, 254, 257*]

Campbell DR (1989). Measurements of selection in a hermaphroditic plant: Variation in male and female pollination success. *Evolution* **43**:318–334 [*266, 269, 277*]

Campbell DR (1991a). Comparing pollen dispersal and gene flow in a natural population. *Evolution* **45**:1965–1968 [*273*]

Campbell DR (1991b). Effects of floral traits on sequential components of fitness in *Ipomopsis aggregata*. *The American Naturalist* **137**:713–737 [*274*]

Campbell DR (1996). Evolution of floral traits in a hermaphroditic plant: Field measurements of heritabilities and genetic correlations. *Evolution* **50**:1442–1453 [*269, 277*]

Campbell DR & Waser NM (1989). Variation in pollen flow within and among populations of *Ipomopsis aggregata*. *Evolution* **43**:1444–1455 [*273*]

Campbell DR & Waser NM (2001). Genotype-by-environment interaction and the fitness of plant hybrids in the wild. *Evolution* **55**:669–676 [*275*]

Campbell DR, Waser NM, Price MV, Lynch EA & Mitchell RJ (1991). A mechanistic analysis of phenotypic selection: Pollen export and flower corolla width in *Ipomopsis aggregata*. *Evolution* **45**:1458–1467 [*269*]

Campbell DR, Waser NM & Price MV (1996). Mechanisms of hummingbird-mediated selection for flower width in *Ipomopsis aggregata*. *Ecology* **77**:1463–1472 [*269*]

Campbell DR, Waser NM & Meléndez-Ackerman EJ (1997). Analyzing pollinator-mediated selection in a plant hybrid zone: Hummingbird visitation patterns on three spatial scales. *The American Naturalist* **149**:295–315 [*268–269, 275*]

Campbell DR, Waser NM & Wolf PG (1998). Pollen transfer by natural hybrids and parental species in an *Ipomopsis* hybrid zone. *Evolution* **52**:1602–1611 [*268, 275*]

Campbell DR, Waser NM & Pederson GT (2002). Predicting patterns of mating and rates of hybridization from pollinator behavior. *The American Naturalist* **159**:438–450 [*268, 275*]

Carey K (1983). Breeding system, genetic variability, and response to selection in *Plectritis* (Valerianaceae). *Evolution* **37**:947–956 [*272*]

Carney SE, Cruzan MB & Arnold ML (1994). Reproductive interactions between hybridizing irises: Analyses of pollen tube growth and fertilization success. *American Journal of Botany* **81**:1169–1175 [*275*]

Carroll RL (2000). Towards a new evolutionary synthesis. *Trends in Ecology and Evolution* **15**:27–32 [*137*]

Carroll SP & Boyd C (1992). Host race radiation in the soapberry bug: Natural history with the history. *Evolution* **46**:1052–1069 [242]

Carroll SP, Dingle H & Klassen SP (1997). Genetic differentiation of fitness-associated traits among rapidly evolving populations of the soapberry bug. *Evolution* **51**:1182–1188 [242]

Case TJ & Taper ML (2000). Interspecific competition, environmental gradients, gene flow, and the coevolution of species' borders. *The American Naturalist* **155**:583–605 [158, 160–161]

Carson HL (1968). The population flush and its genetic consequences. In *Population Biology and Evolution*, ed. Lewontin RC, pp. 123–137. Syracuse, NY, USA: Syracuse University Press [119]

Caswell H (1989). *Matrix Population Models: Construction, Analysis, and Interpretation*. Sunderland, MA, USA: Sinauer Associates Inc. [56]

Champagnat N, Ferrière R & Arous BG (2001). The canonical equation of adaptive dynamics: A mathematical view. *Selection* **2**:73–83 [67]

Chapman RF (1982). Chemoreception: The significance of receptor number. *Advances in Insect Physiology* **16**:247–356 [252]

Charlesworth B (1990). Speciation. In *Paleobiology: A Synthesis*, eds. Briggs DEG & Crowther PR, pp. 100–106. Oxford, UK: Blackwell Scientific Publications [124]

Cheptou PO & Mathias A (2001). Can varying inbreeding depression select for intermediary selfing rate? *The American Naturalist* **157**:361–373 [73]

Chesson P (1994). Multispecies competition in variable environments. *Theoretical Population Biology* **45**:227–276 [54]

Chirov PA & Ozerova RA (1997). Taksonomiia golovnoi i platianoi vshei cheloveka [The taxonomy of human head and body lice]. *Meditsinskaia Parazitologiia I Parazitarnye Bolezni (Moskva)* **April–June**:38–40 [231]

Chittka L, Thomson JD & Waser NM (1999). Flower constancy, insect psychology, and plant evolution. *Naturwissenschaften* **86**:361–377 [276]

Chown SL & Smith VR (1993). Climate change and the short-term impact of feral house mice at the sub-Antarctic Prince Edwards Islands. *Oecologia* **96**:508–516 [71]

Christiansen FB (1975). Hard and soft selection in a subdivided population. *The American Naturalist* **109**:11–16 [39, 41]

Christiansen FB (1985). Selection and population regulation with habitat variation. *The American Naturalist* **126**:418–429 [42]

Christiansen FB (1991). On conditions for evolutionary stability for a continuously varying character. *The American Naturalist* **138**:37–50 [55, 70, 75, 107, 143, 382]

Christiansen FB & Feldman MW (1975). Subdivided populations: A review of the one- and two-locus deterministic theory. *Theoretical Population Biology* **7**:13–38 [78]

Christiansen FB & Fenchel TM (1977). *Theories of Populations in Biological Communities*. Heidelberg, Germany: Springer-Verlag [57]

Christiansen FB & Loeschcke V (1980). Evolution and intraspecific exploitative competition. I. One locus theory for small additive gene effects. *Theoretical Population Biology* **18**:297–313 [54, 57, 67, 75]

Christie P & Macnair MR (1984). Complementary lethal factors in two North American populations of the yellow monkey flower. *Journal of Heredity* **75**:510–511 [194]

Claessen D & Dieckmann U (2002). Ontogenetic niche shifts and evolutionary branching in size-structured populations. *Evolutionary Ecology Research* **4**:189–217 [225–226]

Claridge MF (1985). Acoustic signals in the Homoptera. *Annual Review of Entomology* **30**:297–317 [238]

Clarke BC (1975). The contribution of ecological genetics to evolutionary theory: Detecting the direct effects of natural selection on particularly polymorphic loci. *Genetics* **79**:101–113 [199]

Clausen J, Keck DD & Hiesey WM (1940). *Experimental Studies of the Nature of Species. I. Effect of Varied Environments on Western North American Plants*. Carnegie Institution of Washington Publication No. 520. Washington, DC, USA: Carnegie Institution of Washington [275]

Clayton DH, Price RD & Page RDM (1996). Revision of *Dennyus (Collodennyus)* lice (Phthiraptera: Menoponidae) from swiftlets, with descriptions of new taxa and a comparison of host–parasite relationships. *Systematic Entomology* **21**:179–204   [*231*]

Cohen D & Levin SA (1991). Dispersal in patchy environments: The effects of temporal and spatial structure. *Theoretical Population Biology* **39**:63–99   [*75*]

Cohen AS, Soreghan MJ & Scholtz CA (1993). Estimating the age of formation of lakes: An example from Lake Tanganyika, East African Rift System. *Geology* **21**:511–514   [*173*]

Cohen AS, Lezzar KE, Tiercelin JJ & Soreghan M (1997). New palaeogeographic and lake level reconstructions of Lake Tanganyika: Implications for tectonic, climatic and biologic evolution in a rift lake. *Basin Research* **9**:107–132   [*377*]

Cohen Y, Vincent TL & Brown JS (1999). A G-function approach to fitness minima, fitness maxima, evolutionarily stable strategies and adaptive landscapes. *Evolutionary Ecology Research* **1**:923–942   [*72*]

Condon MA & Steck GJ (1997). Evolution of host use in fruit flies of the genus *Blepharoneura* (Diptera: Tephritidae): Cryptic species on sexually dimorphic host plants. *Biological Journal of the Linnean Society* **60**:443–466   [*231, 241*]

Conner JK, Rush S, Kercher S & Jennetten P (1996). Measurements of natural selection on floral traits in wild radish (*Raphanus raphanistrum*). II. Selection through lifetime male and total fitness. *Evolution* **50**:1137–1146   [*277*]

Conrad M (1990). The geometry of evolution. *Biosystems* **24**:61–81   [*106*]

Conway-Morris S (1998). *The Crucible of Creation*. Oxford, UK: Oxford University Press [*286*]

Cooper VS & Lenski RE (2000). The population genetics of ecological specialization in evolving *E. coli* populations. *Nature* **407**:736–739   [*280*]

Coyne JA (1992). The genetics of speciation. *Nature* **355**:511–515   [*24, 192, 259–260*]

Coyne JA (1994). Ernst Mayr and the origin of species. *Evolution* **48**:19–30   [*24*]

Coyne JA & Orr HA (1989). Patterns of speciation in *Drosophila*. *Evolution* **43**:362–381 [*71, 203*]

Coyne JA & Orr HA (1997). "Patterns of speciation in *Drosophila*," revisited. *Evolution* **51**:295–303   [*71, 203, 277*]

Coyne JA & Orr HA (1998). The evolutionary genetics of speciation. *Philosophical Transactions of the Royal Society of London B* **353**:287–305   [*37, 124–125*]

Coyne JA, Barton NH & Turelli M (1997). A critique of Sewall Wright's shifting balance theory of evolution. *Evolution* **51**:643–671 [*119*]

Craig TP, Itami JK, Abrahamson WG & Horner JD (1993). Behavioral evidence for host-race formation in *Eurosta solidaginis*. *Evolution* **4**:1696–1710   [*237*]

Craig TP, Horner JD & Itami JK (1997). Hybridization studies on the host races of *Eurosta solidaginis*: Implications for sympatric speciation. *Evolution* **51**:1552–1560   [*49, 242*]

Crandell PA & Gall GAE (1993). The genetics of age and weight at sexual maturity based on individually tagged rainbow trout *Oncorhynchus mykiss*. *Aquaculture* **117**:95–105   [*217*]

Crepet WL (1984). Advanced (constant) insect pollination mechanisms: Pattern of evolution and implications vis-à-vis angiosperm diversity. *Annals of the Missouri Botanical Garden* **71**:607–630   [*264*]

Crosby JL (1970). The evolution of genetic discontinuity: Computer models of the selection of barriers to interbreeding between species. *Heredity* **25**:253–297   [*265*]

Crow JF & Kimura M (1970). *An Introduction to Population Genetics Theory*. New York, NY, USA: Harper & Row   [*48, 58*]

Crozier RH (1985). Adaptive consequences of male haploidy. In *Spider Mites: Their Biology, Natural Enemies and Control*, eds. Helle W & Sabelis MW, pp. 201–222. Amsterdam, Netherlands: Elsevier Science Publishers   [*253*]

Crozier WW & Ferguson A (1986). Electrophoretic examination of the population structure of brown trout, *Salmo trutta* L., from the Lough Neagh catchment, Northern Ireland. *Journal of Fish Biology* **28**:459–477 [*213*]

Daltry J, Wüster W & Thorpe RS (1996). Diet and snake venom evolution. *Nature* **379**:537–540   [*327*]

Danley PD, Markert JM, Arnegard ME & Kocher TD (2000). Divergence with gene flow in the rock-dwelling cichlids of Lake Malawi. *Evolution* **54**:1725–1737   [*174*]

Danzmann RG, Ferguson MM, Skúlason S, Snorrason SS & Noakes DLG (1991). Mitochondrial DNA diversity among four sympatric morphs of Arctic charr, *Salvelinus alpinus*, from Thingvallavatn, Iceland. *Journal of Fish Biology* **39**:649–659   [*221*]

Darlington CD & Mather K (1949). *The Elements of Genetics*. London, UK: Allen and Unwin   [*265*]

Darwin C (1859). *On the Origin of Species by Means of Natural Selection or the Preservation of Favoured Races in the Struggle for Life*. Reprinted 1964. Cambridge, MA, USA: Harvard University Press   [*1–3, 16, 18–20, 98, 115*]

Darwin C (1871). *The Descent of Man, and Selection in Relation to Sex*. London, UK: John Murray   [*98*]

Darwin C (1876). *The Effects of Cross and Self Fertilisation in the Vegetable Kingdom*. London, UK: John Murray   [*265*]

Davies MS & Snaydon RW (1976). Rapid population differentiation. III. Measures of selection pressures. *Heredity* **36**:59–66   [*272, 277*]

Day T (2000). Competition and the effect of spatial resource heterogeneity on evolutionary diversification. *The American Naturalist* **155**:790–803   [*57, 67, 73, 140, 313*]

Day T, Pritchard J & Schluter D (1994). A comparison of two sticklebacks. *Evolution* **48**:1723–1734   [*218–219*]

De Boer R (1980). Genetic affinities between spider mite *Tetranychus urticae* (Acari: Tetranychidae) populations in a non-agricultural area. *Entomologia Experimentalis et Applicata* **28**:22–28   [*253*]

De Boer R (1985). Reproductive barriers. In *Spider Mites: Their Biology, Natural Enemies and Control*, eds. Helle W & Sabelis MW, pp. 193–199. Amsterdam, Netherlands: Elsevier Science Publishers   [*253*]

De Cara MAR & Dieckmann U. Speciation by pattern formation revisited. Unpublished   [*166*]

De Jong TJ & Geritz SAH (2001). The role of geitonogamy in the gradual evolution towards dioecy in cosexual plants. *Selection* **2**:133–146   [*73*]

De Meeûs T, Michalakis Y, Renaud F & Olivieri I (1993). Polymorphism in heterogeneous environments, evolution of habitat selection and sympatric speciation: Hard and soft selection models. *Evolutionary Ecology Research* **7**:175–198   [*50, 244*]

Dempster ER (1955). Maintenance of genetic heterogeneity. *Cold Spring Harbor Symposia on Quantitative Biology* **20**:25–32   [*41*]

Dennert E (1903). *Vom Sterbelager des Darwinismus*. Stuttgart, Germany: Kielmann   [*17*]

Denno RF, McClure MS & Ott JR (1995). Interspecific interactions in phytophagous insects: Competition reexamined and resurrected. *Annual Review of Entomology* **40**:297–331   [*244*]

De Queiroz K (1998). The general lineage concept of species, species criteria, and the process of speciation: A conceptual unification and terminological recommendations. In *Endless Forms: Species and Speciation*, eds. Howard DJ & Berlocher SH, pp. 57–75. Oxford, UK: Oxford University Press   [*11*]

De Queiroz K, Chu LR & Losos JB (1998). A second *Anolis* lizard in Dominican amber and the systematics and ecological morphology of Dominican amber anoles. *American Museum Novitates* **3249**:1–23   [*323*]

Dercole F, Ferrière R & Rinaldi S (2002). Ecological bistability and evolutionary reversals under asymmetrical competition. *Evolution* **56**:1081–1090   [*74, 98*]

Deutsch JC (1997). Colour diversification in Malawi cichlids: Evidence for adaptation, reinforcement or sexual selection? *Biological Journal of the Linnean Society* **62**:1–14   [*176*]

Devlin B & Ellstrand NC (1990). The development and application of a refined method for estimating gene flow from angiosperm paternity analysis. *Evolution* **44**:248–259   [*273*]

Dickinson H & Antonovics J (1973). Theoretical considerations of sympatric divergence. *The American Naturalist* **107**:256–274 [*82*]

Dieckmann U & Doebeli M (1999). On the origin of species by sympatric speciation. *Nature* **400**:354–357   [*42–43, 46–47, 57, 70–71, 75, 78, 80, 83, 86, 89–93, 110, 138, 155, 222, 225, 246, 249, 259, 265–266, 270, 305, 308, 311, 316, 338, 362, 392*]

Dieckmann U & Ferrière R (2004). Adaptive dynamics and evolving biodiversity. In *Evolutionary Conservation Biology*, eds. Ferrière R, Dieckmann U & Couvet D, pp. 188–

230. Cambridge, UK: Cambridge University Press [*98*]

Dieckmann U & Law R (1996). The dynamical theory of coevolution: A derivation from stochastic ecological processes. *Journal of Mathematical Biology* **34**:579–612 [*55, 58, 60, 65, 67, 144–145, 222*]

Dieckmann U, Marrow P & Law R (1995). Evolutionary cycling in predator–prey interactions: Population dynamics and the Red Queen. *Journal of Theoretical Biology* **176**:91–102 [*74, 95*]

Dieckmann U, Law R & Metz JAJ (2000). *The Geometry of Ecological Interactions: Simplifying Spatial Complexity.* Cambridge, UK: Cambridge University Press [*155*]

Diehl SR & Bush GL (1984). An evolutionary and applied perspective of insect biotypes. *Annual Review of Entomology* **29**:471–504 [*237, 252–253*]

Diehl SR & Bush GL (1989). The role of habitat preference in adaptation and speciation. In *Speciation and Its Consequences*, eds. Otte D & Endler JA, pp. 345–365. Sunderland, MA, USA: Sinauer Associates Inc. [*49–50, 53, 71, 211, 259*]

Dill LM (1983). Adaptive flexibility in the foraging behavior of fishes. *Canadian Journal of Fisheries and Aquatic Sciences* **40**:398–408 [*220*]

Dobler S & Farrell BD (1999). Host use evolution in *Chrysochus* milkweed beetles: Evidence from behaviour, population genetics and phylogeny. *Molecular Ecology* **8**:1297–1307 [*235, 240*]

Dobler S, Mardulyn P, Pasteels JM & Rowell-Rahier M (1996). Host-plant switches and the evolution of chemical defense and life history in the leaf beetle genus *Oreina*. *Evolution* **50**:2373–2386 [*240*]

Dobzhansky TG (1937). *Genetics and the Origin of Species.* New York, NY, USA: Columbia University Press [*113, 123–125*]

Dobzhansky TG (1940). Speciation as a stage in evolutionary divergence. *The American Naturalist* **74**:312–321 [*193, 203*]

Dobzhansky TG (1951). *Genetics and the Origin of Species*, Third Edition. New York, NY, USA: Columbia University Press [*109, 192*]

Dodd DMB (1989). Reproductive isolation as a consequence of adaptive divergence in *Drosophila pseudoobscura*. *Evolution*

43:1308–1311 [*194, 196*]

Doebeli M (1996a). A quantitative genetic competition model for sympatric speciation. *Journal of Evolutionary Biology* **9**:893–909 [*46, 73, 75, 230, 305, 308*]

Doebeli M (1996b). An explicit genetic model for ecological character displacement. *Ecology* **77**:510–520 [*57, 75*]

Doebeli M (1997). Genetic variation and the persistence of predator–prey interactions in the Nicholson–Bailey model. *Journal of Theoretical Biology* **188**:109–120 [*96*]

Doebeli M & Dieckmann U (2000). Evolutionary branching and sympatric speciation caused by different types of ecological interactions. *The American Naturalist* **156**:S77–S101 [*57, 73, 94–98, 143, 271*]

Doebeli M & Dieckmann U (2003). Speciation along environmental gradients. *Nature* **421**:259–263 [*110, 140, 166, 335, 393–394*]

Doebeli M & Ruxton GD (1997). Evolution of dispersal rates in metapopulation models: Branching and cyclic dynamics in phenotype space. *Evolution* **51**:1730–1741 [*73*]

Doebley J (1993). Genetics, development and plant evolution. *Current Opinion in Genetics and Development* **3**:865–872 [*276*]

Dominey W (1984). Effects of sexual selection and life history evolution on speciation: Species flocks in African cichlids and Hawaiian *Drosophila*. In *Evolution of Fish Species Flocks*, eds. Echelle AA & Kornfield I, pp. 231–250. Orono, ME, USA: University of Maine Press [*176, 214*]

Doucette LI (2001). *Variability in Resource Based Behaviour of the Threespine Stickleback* (Gasterosteus aculeatus) *in Iceland*. MSc Thesis. Reykjavik, Iceland: University of Iceland [*212, 216, 219*]

Douglas AE (1997). Provenance, experience and plant utilization by the polyphagous aphid, *Aphis fabae*. *Entomologia Experimentalis et Applicata* **83**:161–170 [*254*]

Drossel B & McKane A (1999). Ecological character displacement in quantitative genetic models. *Journal of Theoretical Biology* **196**:363–376 [*57*]

Drossel B & McKane A (2000). Competitive speciation in quantitative genetic models, *Journal of Theoretical Biology* **204**:467–478 [*70–71, 92*]

Duffy JE (1996). Resource-associated population subdivision in a symbiotic coral-reef shrimp. *Evolution* **50**:360–373 [*49*]

Dukas R & Bernays EA (2000). Learning improves growth rate in grasshoppers. *Proceedings of the National Academy of Sciences of the USA* **97**:2637–2640 [*255*]

Durinx M, Metz JAJ & Meszéna G. Adaptive dynamics for physiologically structured population models. Unpublished [*391*]

Dykhuizen DE (1990). Experimental studies of natural selection in bacteria. *Annual Review of Ecology and Systematics* **21**:373–398 [*278*]

Dykhuizen DE & Davies M (1980). An experimental model: Bacterial specialization and generalists competing in chemostats. *Ecology* **61**:1213–1227 [*289*]

Echelle AA & Kornfield I, eds. (1984). *Evolution of Fish Species Flocks*. Orono, ME, USA: University of Maine Press [*210*]

Edelstein-Keshet L (1988). *Mathematical Models in Biology*. New York, NY, USA: Random House [*292*]

Edmands S (1999). Heterosis and outbreeding depression in interpopulation crosses spanning a wide range of divergence. *Evolution* **53**:1757–1768 [*277*]

Edwards SV & Beerli P (2000). Perspective: Gene divergence, population divergence, and the variance in coalescence time in phylogeographic studies. *Evolution* **54**:1839–1854 [*312*]

Egas M & Sabelis MW (2001). Adaptive learning of host preference in a herbivorous arthropod. *Ecology Letters* **4**:190–195 [*255, 262*]

Egas M, Dieckmann U & Sabelis MW. Evolution restricts the coexistence of specialists and generalists – the role of trade-off structure. *The American Naturalist*. In press [*249–250, 253, 256*]

Egas M, Norde DJ & Sabelis MW (2003). Adaptive learning in arthropods: Spider mites learn to distinguish food quality. *Experimental and Applied Acarology* **30**:233–247 [*255*]

Egas M, Sabelis MW & Dieckmann U. Evolution of specialization and ecological character displacement of herbivores along a gradient of plant quality. Unpublished [*249–251, 253, 256*]

Ehrendorfer F (1953). Ökologisch-geographische Mikro-Differenzierung einer Pop-

ulation von *Galium pumilum* Murr. s. str. (Eco-geographical microdifferentiation of a population of *Galium pumilum* Murr. s. str.). *Österreichische Botanische Zeitschrift* **100**:616–638 [*274*]

Eiríksson GM (1999). *Heterochrony in bone development and growth in two morphs of arctic charr,* Salvelinus alpinus *L., from Thingvallavatn, Iceland*. MSc Thesis. Reykjavik, Iceland: University of Iceland [*218*]

Eiríksson GM, Skúlason S & Snorrason SS (1999). Heterochrony in skeletal development and body size in progeny of two morphs of Arctic charr from Thingvallavatn, Iceland. *Journal of Fish Biology* **55**:175–185 [*217–218*]

Eiríksson GM, Snorrason SS & Skúlason S. Heterochrony in bone formation: A developmental basis of morph formation in arctic charr, *Salvelinus alpinus*. *Biological Journal of the Linnean Society*. In press [*218*]

Eldredge N (1971). The allopatric model and phylogeny in Paleozoic invertebrates. *Evolution* **25**:156–167 [*127*]

Eldredge N (1985). *Unfinished Synthesis: Biological Hierarchies and Modern Evolutionary Thought*. New York, NY, USA: Oxford University Press [*137*]

Eldredge N (2002). The sloshing bucket: How the physical realm controls evolution. In *Towards a Comprehensive Dynamics of Evolution: Exploring the Interplay of Selection, Neutrality, Accident, and Function*, eds. Crutchfield J & Schuster P, pp. 3–32. New York, NY, USA: Oxford University Press [*138*]

Eldredge N & Gould SJ (1972). Punctuated equilibria: An alternative to phyletic gradualism. In *Models in Paleobiology*, ed. Schopf TJ, pp. 82–115. San Francisco, CA, USA: Freeman & Cooper [*18, 127, 360*]

Elena SF, Cooper VS & Lenski RE (1996). Punctuated evolution caused by selection of rare beneficial mutations. *Science* **272**:1802–1804 [*280*]

Elwood R, Gibson J & Neil S (1987). The amorous *Gammarus*: Size-assortative mating in *Gammarus pulex*. *Animal Behavior* **35**:1–6 [*46*]

Emelianov I, Mallet J & Baltensweiler W (1995). Genetic differentiation in *Zeiraphera diniana* (Lepidoptera: Tortricidae, the larch budmoth): Polymorphism, host races or sibling species? *Heredity* **75**:416–424 [*239*]

Emms SK & Arnold ML (1997). The effect of habitat on parental and hybrid fitness: Transplant experiments with Louisiana irises. *Evolution* **51**:1112–1119   [*275*]

Endler JA (1977). *Geographic Variation, Speciation and Clines*. Princeton, NJ, USA: Princeton University Press   [*114–115, 142, 165, 167*]

Endler JA (1986). *Natural Selection in the Wild*. Monographs in Population Biology 21. Princeton, NJ, USA: Princeton University Press   [*18, 199, 278, 298, 329*]

Endler JA (1989). Conceptual and other problems in speciation. In *Speciation and Its Consequences*, eds. Otte D & Endler JA, pp. 625–648. Sunderland, MA, USA: Sinauer Associates Inc.   [*99*]

Endler JA (1992). Signals, signal conditions, and the direction of evolution. *The American Naturalist* **139**:S125–S153   [*98, 341*]

Eriksson O & Bremer B (1992). Pollination systems, dispersal modes, life forms, and diversification rates in angiosperm families. *Evolution* **46**:258–266   [*264*]

Erwin TL (1982). Tropical forests: Their richness in Coleoptera and other arthropod species. *Coleopterist Bulletin* **36**:74–75 [*229, 380*]

Eshel I (1983). Evolutionary and continuous stability. *Journal of Theoretical Biology* **103**:99–111   [*55, 62–63, 107, 143, 382*]

Eshel I & Motro U (1981). Kin selection and strong stability of mutual help. *Theoretical Population Biology* **19**:420–433   [*107*]

Eshel I, Motro U & Sansone E (1997). Continuous stability and evolutionary convergence. *Journal of Theoretical Biology* **185**:333–343 [*72*]

Faegri K & van der Pijl L (1979). *The Principles of Pollination Ecology*, Third Edition. Oxford, UK: Pergamon Press   [*265*]

Farrell BD (1998). "Inordinate fondness" explained: Why are there so many beetles? *Science* **281**:555–559   [*240*]

Feder JL (1998). The apple maggot fly, *Rhagoletis pomonella*: Flies in the face of conventional wisdom about speciation? In *Endless Forms: Species and Speciation*, eds. Howard DJ & Berlocher SH, pp. 130–144. Oxford, UK: Oxford University Press   [*71, 236–237, 245, 247, 249, 254, 256, 315, 382*]

Feder JL & Filchack K (1999). It's about time: The evidence for host plant-mediated selection in the apple maggot fly, *Rhagoletis pomonella*, and its implications for fitness trade-offs in phytophagous insects. *Entomologia Experimentalis et Applicata* **91**:211–225   [*237, 244–245*]

Feder JL, Chilcote CA & Bush GL (1988). Genetic differentiation between sympatric host races of *Rhagoletis pomonella*. *Nature* **336**:61–64   [*79, 237, 242*]

Feder JL, Opp SB, Wlazlo B, Reynolds K, Go W & Spisak S (1994). Host fidelity is an effective pre-mating barrier between sympatric races of the apple maggot fly, *Rhagoletis pomonella*. *Proceedings of the National Academy of Sciences of the USA* **91**:7990–7994   [*236, 247*]

Feder JL, Reynolds K, Go W & Wang EC (1995). Intra- and interspecific competition and host race formation in the apple maggot fly, *Rhagoletis pomonella* (Diptera: Tephritidae). *Oecologia* **101**:416–425   [*245*]

Feder JL, Roethele JB, Wlazlo B & Berlocher SH (1997a). Selective maintenance of allozyme differences among sympatric host races of the apple maggot fly. *Proceedings of the National Academy of Sciences of the USA* **94**:11417–11421   [*237*]

Feder JL, Stolz U, Lewis KM, Perry WM, Roethele JB & Rogers A (1997b). The effects of winter length on the genetics of apple and hawthorn races of *Rhagoletis pomonella* (Diptera: Tephritidae). *Evolution* **51**:1862–1876   [*237*]

Felsenstein J (1976). The theoretical population genetics of variable selection and migration. *Annual Review of Genetics* **10**:253–280   [*39, 47*]

Felsenstein J (1981). Skepticism towards Santa Rosalia, or why are there so few kinds of animals? *Evolution* **35**:124–238   [*37, 42–43, 46–47, 49, 52, 71, 73, 76–81, 84–85, 89, 91–92, 108, 221, 259, 265–266, 382*]

Ferguson A & Hynes RA (1995). Molecular approaches to the study of genetic variation in Salmonid fishes. *Nordic Journal of Freshwater Research* **71**:23–32   [*213, 219*]

Ferguson J & Taggart JB (1991). Genetic differentiation among the sympatric brown trout (*Salmo trutta*) populations of Lough Melvin, Ireland. *Biological Journal of the Linnean Society* **43**:221–237   [*219, 221*]

Ferrière R (2000). Adaptive responses to environmental threats: evolutionary suicide, insurance, and rescue. *Options* Spring 2000, pp. 12–16. Laxenburg, Austria: International Institute for Applied Systems Analysis [*98*]

Filchack KE, Feder JL, Roethele JB & Stolz U (1999). A field test for host-plant dependent selection on larvae of the apple maggot fly, *Rhagoletis pomonella*. *Evolution* **53**:187–200 [*237*]

Filchack KE, Roethele JB & Feder JL (2000). Natural selection and sympatric divergence in the apple maggot *Rhagoletis pomonella*. *Nature* **407**:739–742 [*254, 256, 315*]

Fineblum WL & Rausher M (1997). Do floral pigmentation genes also influence resistance to enemies? The W locus in *Ipomoea purpurea*. *Ecology* **78**:1646–1654 [*274, 276*]

Fisher RA (1915). The evolution of sexual preference. *Eugenics Review* **7**:184–192 [*98*]

Fisher RA (1930). *The Genetical Theory of Natural Selection*. Oxford, UK: Oxford University Press [*98, 192*]

Fitch WM & Margoliash E (1967). The construction of phylogenetic trees: A generally applicable method utilizing estimates of the mutation distance obtained from cytochrome c sequences. *Science* **155**:279–284 [*298*]

Fleishman LJ (1992). The influence of sensory system and the environment on motion patterns in the visual displays of anoline lizards and other vertebrates. *The American Naturalist* **139**:S36–S61 [*341–342*]

Fleishman LJ (2000). Signal function, signal efficiency and the evolution of anoline lizard dewlap color. In *Animal Signals: Signalling and Signal Design in Animal Communication*, eds. Espmark Y, Amundsen T & Rosenqvist G, pp. 209–236. Trondheim, Norway: Tapir Academic Press [*341–342*]

Fleishman LJ, Bowman M, Saunders D, Miller WE, Rury MJ & Loew ER (1997). The visual ecology of Puerto Rican anoline lizards: Habitat light and spectral sensitivity. *Journal of Comparative Physiology A* **181**:446–460 [*342*]

Fontana W & Schuster P (1987). A computer model of evolutionary optimization. *Biophysical Chemistry* **26**:123–147 [*128*]

Foote CJ, Wood CC & Withler RE (1989). Biochemical genetic comparison of sockeye and kokanee, the anadromous and nonanadromous forms of *Oncorhynchus nerka*. *Canadian Journal of Fisheries and Aquatic Sciences* **46**:149–158 [*219, 221*]

Forseth T, Ugedal O & Jonsson B (1994). The energy budget, niche shift, reproduction and growth in a population of Arctic charr, *Salvelinus alpinus*. *Journal of Animal Ecology* **63**:116–126 [*220*]

Foster S (1994). Evolution of the reproductive behavior of the threespine stickleback. In *The Evolutionary Biology of the Threespine Stickleback*, eds. Bell M & Foster S, pp. 381–398. Oxford, UK: Oxford University Press [*208*]

Frey JK (1993). Modes of peripheral isolate formation and speciation. *Systematic Biology* **42**:373–381 [*116*]

Frias D & Atria J (1998). Chromosomal variation, macroevolution and possible parapatric speciation in *Mepraia spinolai* (Porter) (Hemiptera: Reduviidae). *Genetics and Molecular Evolution* **21**:179–184 [*115*]

Friesen VL & Anderson DJ (1997). Phylogeny and evolution of the Sulidae (Aves: Pelecaniformes): A test of alternative modes of speciation. *Molecular Phylogenetics and Evolution* **7**:252–260 [*115*]

Frost DR & Etheridge R (1989). A phylogenetic analysis and taxonomy of iguanian lizards (Reptilia: Squamata). *Miscellaneous Publications, University of Kansas Museum of Natural History* **81**:1–65 [*323*]

Fry J (1992). On the maintenance of genetic variation by disruptive selection among hosts in a phytophagous mite. *Evolution* **46**:279–283 [*253–254*]

Fryer G (1997). Biological implications of a suggested late-Pleistocene desiccation of Lake Victoria. *Hydrobiologia* **354**:177–182 [*173*]

Fryer G & Iles TD (1972). *The Cichlid Fishes of the Great Lakes of Africa: Their Biology and Evolution*. London, UK: Oliver and Boyd [*173–174, 176, 181, 187, 191, 362, 374*]

Fulton M & Hodges SA (1999). Floral isolation between *Aquilegia formosa* and *A. pubescens*. *Proceedings of the Royal Society of London B* **266**:2247–2252 [*267*]

Funk DJ (1998). Isolating a role for natural selection in speciation: Host adaptation and sexual isolation in *Neochlamisus bebbianae* leaf beetles. *Evolution* **52**:1744–1759 [*202, 233*]

Futuyma DJ (1979). *Evolutionary Biology.* Sunderland, MA, USA: Sinauer Associates Inc. [*23, 374*]

Futuyma DJ (1998). *Evolutionary Biology,* Third Edition. Sunderland, MA, USA: Sinauer Associates Inc. [*112–114, 192, 374*]

Futuyma DJ & Gould F (1979). Associations of plants and insects in a deciduous forest. *Ecological Monographs* **49**:33–50 [*252*]

Futuyma DJ & Mayer GC (1980). Non-allopatric speciation in animals. *Systematic Zoology* **29**:254–271 [*36–38, 107, 233*]

Futuyma DJ & Shapiro LH (1995). Hybrid zones. *Evolution* **49**:222–226 [*277*]

Futuyma DJ, Keese MC & Funk DJ (1995). Genetic constraints on macroevolution: The evolution of host affiliation in the leaf beetle genus *Ophraella*. *Evolution* **49**:797–809 [*233, 243*]

Galen C (1983). The effects of nectar thieving ants on seed set in floral scent morphs of *Polemonium viscosum*. *Oikos* **41**:245–249 [*273*]

Galen C (1996). Rates of floral evolution: Adaptation to bumblebee pollination in an alpine wildflower, *Polemonium viscosum*. *Evolution* **50**:120–125 [*266, 273–274*]

Galen C & Cuba J (2001). Down the tubes: Pollinators, predators, and the evolution of flower shape in the alpine skypilot, *Polemonium viscosum*. *Evolution* **55**:1963–1971 [*273–274*]

Galen C & Stanton ML (1991). Consequences of emergence phenology for reproductive success in *Ranunculus adoneus* (Ranunculaceae). *American Journal of Botany* **78**:978–988 [*272–273*]

Galen C, Sherry RA & Carroll AB (1999). Are flowers physiological sinks or faucets? Costs and correlates of water use by flowers of *Polemonium viscosum*. *Oecologia* **118**:461–470 [*273–274*]

Galis F (1992). A model for biting in the pharyngeal jaws of a cichlid fish: *Haplochromus piceatus*. *Journal of Theoretical Biology* **155**:343–368 [*187–188, 190*]

Galis F (1993a). Interactions between the pharyngeal jaw apparatus, feeding behaviour and ontogeny in the cichlid fish *Haplochromus piceatus*: A study of morphological constraints in evolutionary ecology. *Jour-*

nal of Experimental Zoology **267**:137–154 [*187*]

Galis F (1993b). Morphological constraints on behaviour through ontogeny: The importance of developmental constraints. *Marine Behaviour and Physiology* **23**:119–135 [*187*]

Galis F & Drucker EG (1996). Pharyngeal biting mechanisms in centrarchids and cichlids: Insights into a key evolutionary innovation. *Journal of Evolutionary Biology* **9**:641–670 [*187–188, 190*]

Galis F & Metz JAJ (1998). Why are there so many cichlid species? *Trends in Ecology and Evolution* **13**:1–2 [*72, 188, 190*]

Gallez GP & Gottlieb LD (1982). Genetic evidence for the hybrid origin of the diploid plant *Stephanomeria diegensis*. *Evolution* **36**:1158–1167 [*265*]

Garcia R, Queral A, Powell R, Parmerlee JS Jr, Smith DD & Lathrop A (1994). Evidence of hybridization among green anoles (Lacertilia: Polychrotidae) from Hispaniola. *Caribbean Journal of Science* **30**:279–281 [*339*]

Garland T Jr & Adolph SC (1991). Physiological differentiation of vertebrate populations. *Annual Review of Ecology and Systematics* **22**:193–228 [*328*]

Gause GJ (1934). *The Struggle for Existence.* Baltimore, MD, USA: Williams & Wilkins [*306*]

Gavrilets S (1996). On phase three of the shifting-balance theory. *Evolution* **50**:1034–1041 [*119*]

Gavrilets S (1997). Evolution and speciation on holey adaptive landscapes. *Trends in Ecology and Evolution* **12**:307–312 [*106, 115, 125, 128*]

Gavrilets S (1999). A dynamical theory of speciation on holey adaptive landscapes. *The American Naturalist* **154**:1–22 [*134, 142, 165*]

Gavrilets S (2000a). Waiting time to parapatric speciation. *Proceedings of the Royal Society of London B* **267**:2483–2492 [*102–104, 125, 127, 134, 138, 378*]

Gavrilets S (2000b). Rapid evolution of reproductive isolation driven by sexual conflict. *Nature* **403**:886–889 [*136, 265, 277, 390*]

Gavrilets S (2002). Evolution and speciation in a hyperspace: The roles of neutrality, selection, mutation and random drift. In *Towards a Comprehensive Dynamics of Evolution: Exploring the Interplay of Selection, Neutrality, Accident, and Function*. eds. Crutchfield J & Schuster P, pp. 135–162. New York, NY, USA: Oxford University Press [*124, 128*]

Gavrilets S & Boake CRB (1998). On the evolution of premating isolation after a founder event. *The American Naturalist* **152**:706–716 [*135*]

Gavrilets S & Cruzan MB (1998). Neutral gene flow across single locus clines. *Evolution* **52**:1277–1284 [*115*]

Gavrilets S & Gravner J (1997). Percolation on the fitness hypercube and the evolution of reproductive isolation. *Journal of Theoretical Biology* **184**:51–64 [*129, 133*]

Gavrilets S & Hastings A (1996). Founder effect speciation: A theoretical reassessment. *The American Naturalist* **147**:466–491 [*135*]

Gavrilets S & Waxman D (2002). Sympatric speciation by sexual conflict. *Proceedings of the National Academy of Sciences of the USA* **99**:10533–10538 [*45, 102–107*]

Gavrilets S, Li H & Vose MD (1998). Rapid parapatric speciation on holey adaptive landscapes. *Proceedings of the Royal Society of London B* **265**:1483–1489 [*116, 134–136*]

Gavrilets S, Li H & Vose MD (2000a). Patterns of parapatric speciation. *Evolution* **54**:1126–1134 [*116, 134–136, 378*]

Gavrilets S, Acton R & Gravner J (2000b). Dynamics of speciation and diversification in a metapopulation. *Evolution* **54**:1493–1501 [*135*]

Geritz SAH & Kisdi É (2000). Adaptive dynamics in diploid sexual populations and the evolution of reproductive isolation. *Proceedings of the Royal Society of London B* **267**:1671–1678 [*67, 69–71, 85, 390*]

Geritz SAH, Metz JAJ, Kisdi É & Meszéna G (1997). Dynamics of adaptation and evolutionary branching. *Physical Review Letters* **78**:2024–2027 [*55, 72, 107, 222*]

Geritz SAH, Kisdi É, Meszéna G & JAJ Metz (1998). Evolutionarily singular strategies and the adaptive growth and branching of the evolutionary tree. *Evolutionary Ecology Research* **12**:35–57 [*6, 55–56, 59, 62–63, 65, 67–68, 72, 107, 143, 222*]

Geritz SAH, Van der Meijden E & Metz JAJ (1999). Evolutionary dynamics of seed size and seedling competitive ability. *Theoretical Population Biology* **55**:324–343 [*55, 67, 73–74*]

Giannasi N (1997). *Morphological, Molecular and Behavioural Evolution of the* Anolis roquet *group*. PhD thesis. Bangor, UK: University of Wales [*333*]

Giannasi N, Thorpe RS & Malhotra A (2000). A phylogenetic analysis of body size evolution in the *Anolis roquet* group (Sauria: Iguanidae): Character displacement or size assortment? *Molecular Ecology* **9**:193–202 [*331, 334*]

Gibbons JRH (1979). A model for sympatric speciation in *Megarhyssa* (Hymenoptera: Ichneumonidae): Competitive speciation. *The American Naturalist* **114**:719–741 [*222, 230, 241*]

Gillespie JH (1991). *The Causes of Molecular Evolution*. Oxford, UK: Oxford University Press [*118*]

Gillespie JH (1999). The role of population size in molecular evolution. *Theoretical Population Biology* **55**:145–156 [*23*]

Gillespie JH (2000a). Genetic drift in an infinite population: The pseudohitchhiking model. *Genetics* **155**:909–919 [*23*]

Gillespie JH (2000b). The neutral theory in an infinite population. *Gene* **155**:11–18 [*23*]

Giordano R, Jackson JJ & Robertson HM (1997). The role of *Wolbachia* bacteria in reproductive incompatibilities and hybrid zones of *Diabrotica* beetles and *Gryllus* crickets. *Proceedings of the National Academy of Sciences of the USA* **94**:11439–11444 [*262*]

Gíslason D (1998). *Genetic and Morphological Variation in Polymorphic Arctic Charr, Salvelinus alpinus, from Icelandic Lakes*. MSc Thesis. Guelph, Ontario, Canada: University of Guelph [*223*]

Gíslason D, Ferguson MM, Skúlason S & Snorrason SS (1999). Rapid and coupled phenotypic and genetic divergence in Icelandic Arctic charr (*Salvelinus alpinus*). *Canadian Journal of Fisheries and Aquatic Sciences* **56**:2229–2234 [*219, 221–222*]

Gjedrem T (1983). Genetic variation in quantitative traits and selective breeding in fish and shellfish. *Aquaculture* **86**:51–72 [*217*]

Godfray HCJ (1994). *Parasitoids: Behavioral and Evolutionary Ecology*, Princeton, NJ, USA: Princeton University Press [*231, 240*]

Godfray HCJ & Waage JK (1988). Learning in parasitic wasps. *Nature* **331**:211 [*240*]

Goel NS & Richter-Dyn N (1974). *Stochastic Models in Biology*. New York, NY, USA: Academic Press [*58*]

Goldschmidt T (1991). Egg mimics in haplochromine cichlids (Pisces, Perciformes) from Lake Victoria. *Ethology* **88**:177–190 [*176*]

Goldschmidt T (1996). *Darwin's Dream Pond: Drama in Lake Victoria*. Cambridge, MA, USA: MIT Press [*25*]

Goldschmidt T & de Visser J (1990). On the possible role of egg mimics in speciation. *Acta Biotheoretica* **38**:125–134 [*176*]

Gorman GC & Atkins L (1968). Natural hybridization between two sibling species of *Anolis* lizards: Chromosome cytology. *Science* **159**:1358–1360 [*339*]

Gorman GC & Atkins L (1969). The zoogeography of the Lesser Antilles *Anolis* lizards: An analysis based on chromosomes and lactic dehydrogenases. *Bulletin of the Museum of Comparative Zoology* **138**:53–80 [*324*]

Gorman GC, Licht P, Dessauer HC & Boos JO (1971). Reproductive failure among the hybridizing *Anolis* lizards of Trinidad. *Systematic Zoology* **20**:1–12 [*339*]

Gotoh T, Bruin J, Sabelis MW & Menken SBJ (1993). Host race formation in *Tetranychus urticae*: Genetic differentiation, host plant preference and mate choice in a tomato and a cucumber strain. *Entomologia Experimentalis et Applicata* **68**:171–178 [*249, 253–255, 257*]

Gottlieb LD (1984). Genetics and morphological evolution in plants. *The American Naturalist* **123**:681–709 [*265, 276*]

Gould SJ (1977). *Ontogeny and Phylogeny*. London, UK: Cambridge and Belknap Press [*190*]

Gould SJ (1989). *Wonderful Life: The Burgess Shale and the Nature of History*. New York, NY, USA: WW Norton & Company [*286–287*]

Gould SJ & Eldredge N (1977). Punctuated equilibria: The tempo and mode of evolution reconsidered. *Paleobiology* **3**:115–151 [*360*]

Gould SJ & Eldredge N (1993). Punctuated equilibrium comes of age. *Nature* **366**:223–227 [*360*]

Gould SJ & Lewontin RC (1979). The spandrels of San Marco and the Panglossian paradigm: A critique of the adaptationist programme. *Proceedings of the Royal Society of London B* **205**:581–598 [*286*]

Gradshteyn IS & Ryzhik IM (1994). *Tables of Integrals, Series and Products*. San Diego, CA, USA: Academic Press [*120*]

Grant V (1949). Pollination systems as isolating mechanisms in angiosperms. *Evolution* **3**:82–97 [*265, 267*]

Grant V (1981). *Plant Speciation*, Second Edition. New York, NY, USA: Columbia University Press [*265, 276*]

Grant PR (1986). *Ecology and Evolution of Darwin's Finches*. Princeton, NJ, USA: Princeton University Press [*341*]

Grant PR (1999). *Ecology and Evolution of Darwin's Finches*, New Edition. Princeton, NJ, USA: Princeton University Press [*125*]

Grant V & Grant A (1960). Genetic and taxonomic studies in *Gilia*. XI. Fertility relationships of the diploid cobwebby gilias. *Aliso* **4**:435–481 [*273*]

Grant V & Grant KA (1965). *Flower Pollination in the* Phlox *Family*. New York, NY, USA: Columbia University Press [*265, 267*]

Grant PR & Grant BR (1997). Genetics and the origin of bird species. *Proceedings of the National Academy of Sciences of the USA* **94**:7768–7775 [*125*]

Grant PR, Grant BR & Petren K (2000). The allopatric phase of speciation: The sharp-beaked ground finch (*Geospiza difficilis*) on the Galapagos islands. *Biological Journal of the Linnean Society* **69**:287–317 [*71*]

Gratton C & Welter SC (1998). Oviposition preference and larval performance of *Liriomyza helianthi* (Diptera: Agromyzidae) on normal and novel host plants. *Environmental Entomology* **27**:926–935 [*245–246*]

Greene HW (1986). Diet and arboreality in the emerald monitor, *Varanus prasinus*, with comments on the study of adaptation. *Fieldiana Zoology, New Series* **31**:1–12 [*337*]

Greenwood PH (1965). The cichlid fishes of Lake Nabugabo, Uganda. *Bulletin of the British Museum of Natural History (Zoology)* **12**:315–359 [*187*]

Greenwood PH (1974). The cichlid fishes of Lake Victoria, East Africa: The biology and evolution of a species flock. *Bulletin of the British Museum of National History (Zool-*

ogy) **6**:S1–S134   [*187, 190*]

Greenwood PH (1981a). Species flocks and explosive speciation. In *Chance, Change and Challenge: The Evolving Biosphere*, eds. Greenwood PH & Florey PL, pp. 61–74. London, UK: Cambridge University Press and British Museum (Natural History) [*190*]

Greenwood PH (1981b). *The Haplochromine Fishes of the East African Lakes*. Munich, Germany: Kraus International Publications [*362*]

Greenwood PH (1994). The species flock of cichlid fishes in Lake Victoria – and those of other African Great Lakes. *Archiv für Hydrobiologie – Advances in Limnology* **44**:347–354   [*190*]

Griffiths D (1994). The size structure of lacustrine arctic charr (Pisces: Salmonidae) populations. *Biological Journal of the Linnean Society* **43**:221–237   [*216, 220*]

Grimaldi DA (1999). The co-radiations of pollinating insects and angiosperms in the Cretaceous. *Annals of the Missouri Botanical Garden* **86**:373–406   [*264*]

Grossniklaus U, Vielle-Calzada JP, Hoeppner MA & Gagliano WB (1998). Maternal control of embryogenesis by MEDEA, a polycomb group gene in *Arabidopsis*. *Science* **280**:446–450   [*277*]

Grover JP (1997). *Resource Competition*. London, UK: Chapman & Hall   [*306*]

Grüner W, Giegerich R, Strothmann D, Reidys C, Weber J, Hofacker IL, Stadler PF & Schuster P (1996a). Analysis of RNA sequence structure maps by exhaustive enumeration: Neutral networks. *Monatshefte für Chemie* **127**:355–374   [*128*]

Grüner W, Giegerich R, Strothmann D, Reidys C, Weber J, Hofacker IL, Stadler PF & Schuster P (1996b). Structure of neutral networks and shape space covering. *Monatshefte für Chemie* **127**:375–389   [*128*]

Guldemond JA & Dixon AFG (1994). Specificity and daily cycle of release of sex pheromones in aphids: A case of reinforcement? *Biological Journal of the Linnean Society* **52**:287–303   [*236*]

Guldemond JA & Mackenzie A (1994). Sympatric speciation in aphids. I. Host race formation by escape from gene flow. In *Individuals, Populations and Patterns in Ecology*, eds. Leather SR, Walters KFA, Mills NJ

& Watt AD, pp. 367–378. Andover, Hampshire, UK: Intercept Ltd.   [*242–243, 247*]

Guyer C (1988). Food supplementation in a tropical mainland anole, *Norops humilis*: Effects on individuals. *Ecology* **69**:362–369 [*323*]

Guyer C & Slowinski JB (1991). Comparisons of observed phylogenetic topologies with null expectations among three monophyletic lineages. *Evolution* **45**:340–350   [*376*]

Gyllenberg M, Parvinen K & Dieckmann U (2002). Evolutionary suicide and evolution of dispersal in structured metapopulations. *Journal of Mathematical Biology* **45**:79–105 [*98*]

Hafner MS, Sudman PD, Villablanca FX, Spradling TA, Demastes JW & Nadler SA (1994). Disparate rates of molecular evolution in cospeciating hosts and parasites. *Science* **265**:1087–1090   [*231*]

Haig D & Westoby M (1989). Parent-specific gene expression and the triploid endosperm. *The American Naturalist* **134**:147–155 [*265, 277*]

Hamilton AC (1982). *Environmental History of East Africa*, New York, NY, USA: Academic Press   [*345*]

Hamilton WD & Zuk M (1982). Heritable true fitness and bright birds: A role for parasites? *Science* **218**:384–387   [*99, 257*]

Harari AR, Handler AM & Landolt PJ (1999). Size-assortative mating, male choice and female choice in the curculionid beetle *Diaprepes abbreviatus*. *Animal Behavior* **58**:1191–1200   [*46*]

Hardin G (1960). The competitive exclusion principle. *Science* **131**:1292–1298   [*306*]

Harper JL (1977). *Population Biology of Plants*. London, UK: Academic Press [*277*]

Harrison RG (1990). Hybrid zones: Windows on evolutionary process. *Oxford Surveys in Evolutionary Biology* **7**:69–128   [*114*]

Harrison RG (1991). Molecular changes at speciation. *Annual Review of Ecology and Systematics* **22**:181–308   [*276*]

Harrison RG (1993). *Hybrid Zones and the Evolutionary Process*. New York, NY, USA: Oxford University Press   [*275*]

Harrison RG (1998). Linking evolutionary pattern and process: The relevance of species concepts for the study of speciation. In *Endless Forms: Species and Speciation*, eds. Howard DJ & Berlocher SH, pp. 19–31. Oxford, UK: Oxford University Press [*24*]

Harrison RD & Vawter AT (1977). Allozyme differentiation between pheromone strains of the European corn borer, *Ostrinia nubilalis*. *Annals of the Entomological Society of America* **70**:717–720 [*239*]

Hartley SE (1992). Mitochondrial DNA analysis of Scottish populations of Arctic charr, *Salvelinus alpinus* (L.). *Journal of Fish Biology* **4**:219–224 [*221*]

Hartley SE, McGowan C, Greer RB & Walker AF (1992). The genetics of sympatric Arctic charr, *Salvelinus alpinus* (L.) population from Loch Rannoch, Scotland. *Journal of Fish Biology* **41**:1021–1031 [*221*]

Hatfield T (1997). Genetic divergence in adaptive characters between sympatric species of stickleback. *The American Naturalist* **149**:1009–1029 [*196*]

Hatfield T & Schluter D (1999). Ecological speciation in sticklebacks: Environment-dependent hybrid fitness. *Evolution* **53**:866–873 [*196, 208–209, 264, 275*]

Hartl DL & Clark AG (1997). *Principles of Population Genetics*. Sunderland, MA, USA: Sinauer Associates Inc. [*38*]

Harvey PH & Pagel MD (1991). *The Comparative Method in Evolutionary Biology*. Oxford, UK: Oxford University Press [*337*]

Heard S (1996). Patterns in phylogenetic tree balance with variable and evolving speciation rates. *Evolution* **50**:2141–2148 [*376*]

Heath BD, Butcher RDJ, Whitfield WGF & Hubbard SF (1999). Horizontal transfer of *Wolbachia* between phylogenetically distant insect species by a naturally occurring mechanism. *Current Biology* **9**:313–316 [*262*]

Hedberg O (1951). Vegetation belts of the East African mountains. *Svensk Botanisk Tidskrift* **45**:140–202 [*345*]

Hedberg O (1957). Afroalpine vascular plants: A taxonomic revision. *Symbolae Botanicae Upsaliensis* **15**:1–411 [*345*]

Hedberg O (1964). Features of afroalpine plant ecology. *Acta Phytogeographica Suecica* **49**:1–144 [*345*]

Hedrick PW (1986). Genetic polymorphism in heterogeneous environments: A decade later. *Annual Review of Ecology and Systematics* **17**:535–566 [*39, 291*]

Hedrick PW, Ginevan ME & Ewing EP (1976). Genetic polymorphism in heterogeneous environments. *Annual Review of Ecology and Systematics* **7**:1–32 [*39*]

Heed WB & Kircher HW (1965). Unique sterol in the ecology and nutrition of *Drosophila pachea*. *Science* **149**:758–761 [*232*]

Heino M, Metz JAJ & Kaitala V (1997). Evolution of mixed maturation strategies in semelparous life-histories: The crucial role of dimensionality of feedback environment. *Philosophical Transactions of the Royal Society of London B* **352**:1647–1655 [*226*]

Helfman GS, Collette BB & Facey DE (1997). *The Diversity of Fishes*. Malden, MA, USA: Blackwell Science [*367*]

Helle W (1962). Genetics of resistance to organophosphorus compounds and its relation to diapause in *Tetranychus urticae*. *Tijdschrift voor Plantenziekten* **68**:155–195 [*253*]

Helle W & Overmeer WPJ (1973). Variability in Tetranychid mites. *Annual Review of Entomology* **18**:97–120 [*253*]

Helle W & Pieterse AH (1965). Genetic affinities between adjacent populations of spider mites (*Tetranychus urticae* Koch). *Entomologia Experimentalis et Applicata* **8**:305–308 [*253*]

Helle W & Sabelis MW (1985). *Spider Mites: Their Biology, Natural Enemies and Control*. Amsterdam, Netherlands: Elsevier Science Publishers [*253*]

Helling RB, Vargas CN & Adams J (1987). Evolution of *Escherichia coli* during growth in a constant environment. *Genetics* **116**:349–358 [*290–291*]

Hendry AP (2001). Adaptive divergence and the evolution of reproductive isolation in the wild: An empirical demonstration using introduced sockeye salmon. *Genetica* **112–113**:515–534 [*214, 222*]

Hendry AP & Kinnison MT (1999). The pace of modern life: Measuring rates of contemporary microevolution. *Evolution* **53**:1637–1653 [*73*]

Hendry AP, Wenburg JK, Bentzen P, Volk EC & Quinn TP (2000). Rapid evolution of reproductive isolation in the wild: Evidence from introduced salmon. *Science* **290**:516–518 [*222, 225, 312, 315*]

Herre EA, Machado CA, Bermingham E, Nason JD, Windsor DM, McCafferty SS, Van Houten W & Bachmann K (1996). Molecular phylogenies of figs and their pollinator wasps. *Journal of Biogeography* **23**:521–530 [*271*]

Herrera CM (1993). Selection on floral morphology and environmental determinants of fecundity in a hawk moth-pollinated violet. *Ecological Monographs* **63**:251–276 [*276*]

Herrera CM (1996). Floral traits and plant adaptation to insect pollinators: A devil's advocate approach. In *Floral Biology: Studies on Floral Evolution in Animal-Pollinated Plants*, eds. Lloyd DG & Barrett SCH, pp. 65–87. London, UK: Chapman & Hall [*267, 271*]

Hert E (1989). The function of egg-spots in an African mouth-brooding cichlid fish. *Animal Behaviour* **37**:229–237 [*176*]

Hertz PE (1979). Sensitivity to high temperatures in three West Indian grass anoles (Sauria: Iguanidae), with a review of heat sensitivity in the genus *Anolis*. *Journal of Comparative Biochemistry and Physiology* **62A**:217–222 [*338*]

Hertz PE (1992). Temperature regulation in Puerto Rican *Anolis* lizards: A field test using null hypotheses. *Ecology* **73**:1405–1417 [*338*]

Hertz PE, Arce-Hernandez A, Ramirez-Vazquez J, Tirado-Rivera W & Vazquez-Vives L (1979). Geographical variation of heat sensitivity and water loss rates in the tropical lizard, *Anolis gundlachi*. *Comparative Biochemistry and Physiology* **62A**:947–953 [*338*]

Hickey DA & McNeilly T (1975). Competition between metal tolerant and normal plant populations: A field experiment on normal soil. *Evolution* **29**:458–464 [*272*]

Hiesey WM, Nobs MA & Bjoerkman O (1971). Experimental studies on the nature of species V. Biosystematics, genetics and physiological ecology of the Erythranthe section of *Mimulus*. Publication 628. Washington, DC, USA: Carnegie Institution of Washington [*274*]

Higashi M, Takimoto G & Yamamura N (1999). Sympatric speciation by sexual selection. *Nature* **402**:523–526 [*44–45, 72, 100–101, 305, 311*]

Higgie M, Chenoweth S & Blows MW (2000). Natural selection and the reinforcement of mate recognition. *Science* **290**:519–521 [*203*]

Hindar K (1994). Alternative life histories and genetic conservation. In *Conservation Genetics*, eds. Loeschcke V, Tomink J & Jain SK, pp. 323–336. Basel, Switzerland: Birkhäuser Verlag [*219, 221*]

Hindar K & Jonsson B (1982). Habitat and food segregation of dwarf and normal arctic charr (*Salvelinus alpinus*) from Vangsvatnet Lake, western Norway. *Canadian Journal of Fisheries and Aquatic Sciences* **39**:1030–1045 [*211, 216, 218*]

Hindar K & Jonsson B (1993). Ecological polymorphism in Arctic charr. *Biological Journal of the Linnean Society* **48**:63–74 [*217, 221*]

Hindar K, Ryman N & Stahl G (1986). Genetic differentiation among local populations and morphotypes of Arctic charr, *Salvelinus alpinus*. *Biological Journal of the Linnean Society* **27**:269–285 [*219, 221*]

Hinton GE & Nowlan SJ (1987). How learning can guide evolution. *Complex Systems* **1**:495–502 [*250*]

Hoekstra RF, Bijlsma R & Dolman J (1985). Polymorphism from environmental heterogeneity: Models are only robust if the heterozygote is close in fitness to the favoured homozygote in each environment. *Genetical Research* **45**:299–314 [*68–69*]

Hofbauer J & Sigmund K (1990). Adaptive dynamics and evolutionary stability. *Applied Mathematics Letters* **3**:75–79 [*54*]

Holt RD (1984). Spatial heterogeneity, indirect interactions, and the coexistence of prey species. *The American Naturalist* **124**:377–406 [*39*]

Holzberg S (1978). A field and laboratory study of the behavior and ecology of *Pseudotropheus zebra* (Boulenger), an endemic cichlid of Lake Malawi (Pisces; Cichlidae). *Zeitschrift für Zoologische Systematik und Evolutionsforschung* **16**:171–187 [*181*]

Hoogerhoud RJC & Barel CDN (1978). Integrated morphological adaptations in piscivorous and mollusc crushing *Haplochromus* species. In *Proceedings of the Zodiac Symposium on Adaptation*, ed. Osse JWM, pp. 52–56. Wageningen, Netherlands: Pudoc [*187*]

Hopkins AD (1917). A discussion of C.C. Hewitt's paper on "Insect Behavior". *Journal of Economical Entomology* **10**:92–93 [*257*]

Hori M (1993). Frequency-dependent natural selection in the handedness of scale-eating cichlid fish. *Science* **260**:216–219 [*212, 217, 221*]

Howard DJ (1993). Reinforcement: Origin, dynamics, and fate of an evolutionary hypothesis. In *Hybrid Zones and the Evolutionary Process*, ed. Harrison RG, pp. 46–69. New York, NY, USA: Oxford University Press [*203, 208*]

Howard DJ (1999). Conspecific sperm and pollen precedence and speciation. *Annual Review of Ecology and Systematics* **30**:109–132 [*125*]

Howard DJ & Berlocher SH, eds. (1998). *Endless Forms: Species and Speciation*. Oxford, UK: Oxford University Press [*23–24, 338*]

Huey RB (1983). Natural variation in body temperature and physiological performance in a lizard (*Anolis cristatellus*). In *Advances in Herpetology and Evolutionary Biology: Essays in Honor of Ernest E. Williams*, eds. Rhodin AGJ & Miyata K, pp. 484–490. Cambridge, MA, USA: Museum of Comparative Zoology, Harvard University [*338*]

Huey RB & Bennett AF (1987). Phylogenetic studies of coadaptation: Preferred temperatures versus optimal performance temperatures of lizards. *Evolution* **41**:1098–1115 [*298*]

Hughes KW & Vickery RK Jr (1974). Patterns of heterosis and crossing barriers from increasing genetic distance between populations of the *Mimulus luteus* complex. *Journal of Genetics* **61**:235–245 [*273*]

Huigens ME, Luck RF, Klaassen RHG, Maas MFPM, Timmermans MJTN & Stouthamer R (2000). Infectious parthenogenesis. *Nature* **405**:178–179 [*262*]

Huisman J & Weissing FJ (1999). Biodiversity of plankton by species oscillations and chaos. *Nature* **402**:407–410 [*306*]

Hurst GDD & Schilthuizen M (1998). Selfish genetic elements and speciation. *Heredity* **80**:2–8 [*252, 259, 261*]

Husband BC & Schemske DW (2000). Ecological mechanisms of reproductive isolation between diploid and tetraploid *Chamerion angustifolium*. *Journal of Ecology* **88**:689–701 [*44*]

Hutchinson GE (1961). The paradox of the plankton. *The American Naturalist* **95**:137–145 [*306*]

Huynen MA (1996). Exploring phenotype space through neutral evolution. *Journal of*

Molecular Evolution **43**:165–169 [*128*]

Huynen MA, Stadler PF & Fontana W (1996). Smoothness within ruggedness: The role of neutrality in adaptation. *Proceedings of the National Academy of Sciences of the USA* **93**:397–401 [*128*]

Ingimarsson F & Snorrason SS. Variation in skull bones and dentition of four arctic charr morphs from Thingvallavatn, Iceland. Unpublished [*215*]

Ippolito A & Holtsford T (1999). Selection of *Nicotiana* (Solanaceae) floral morphology in the wild. *Abstracts of the XVI International Botanical Congress*, p. 442. St Louis, MO, USA: America's Center [*268*]

Irschick DJ & Losos JB (1998). A comparative analysis of the ecological significance of maximal locomotor performance in Caribbean *Anolis* lizards. *Evolution* **52**:219–226 [*337*]

Irschick DJ & Losos JB (1999). Do lizards only use habitats in which performance is maximal? The relationship between sprinting capabilities and structural habitat use in Caribbean anoles. *The American Naturalist* **154**:293–305 [*337–338*]

Irschick DJ, Austin CC, Petren K, Fisher RN, Losos JB & Ellers O (1996). A comparative analysis of clinging ability among pad-bearing lizards. *Biological Journal of the Linnean Society* **59**:21–35 [*323, 337*]

Irwin RE & Brody AK (1999). Nectar-robbing bumble bees reduce the fitness of *Ipomopsis aggregata* (Polemoniaceae). *Ecology* **80**:1703–1712 [*274*]

Irwin DE & Price T (1999). Sexual imprinting, learning and speciation. *Heredity* **82**:347–354 [*256*]

Itami JK, Craig TP & Horner JD (1998). Factors affecting gene flow between the host races of *Eurosta solidaginis*. In *Local Genetic Structure in Natural Insect Populations: Effects of Host Plant and Life History*, eds. Mopper S & Strauss S, pp. 375–407. London, UK: Chapman & Hall [*242*]

Iwasa I & Pomiankowski A (1995). Continual change in mate preferences. *Nature* **377**:420–422 [*136*]

Iwasa Y, Pomiankowski A & Nee S (1991). The evolution of costly mate preferences. II. The "handicap" principle. *Evolution* **45**:1431–1442 [*99*]

Jackman T, Losos JB, Larson A & de Queiroz

K (1997). Phylogenetic studies of convergent adaptive radiations in Caribbean *Anolis* lizards. In *Molecular Evolution and Adaptive Radiation*, eds. Givinish TJ & Sytsma KJ, pp. 535–557. Cambridge, UK: Cambridge University Press [*322–323, 336*]

Jackman TR, Larson A, de Queiroz K & Losos JB (1999). Phylogenetic relationships and the tempo of early diversification in *Anolis* lizards. *Systematic Biology* **48**:254–285 [*335–336*]

Jackman TR, Irschick DJ, de Queiroz K, Losos JB & Larson A (2002). Molecular phylogenetic perspective on evolution of lizards of the *Anolis grahami* series. *Journal of Experimental Zoology: Molecular and Developmental Evolution* **294**:1–16 [*336*]

Jacobs FJ, Metz JAJ, Geritz SAH & Meszéna G. Bifurcation analysis for adaptive dynamics based on Lotka–Volterra community dynamics. Unpublished [*73*]

Jaenike J (1983). Induction of host preference in *Drosophila melanogaster*. *Oecologia* **58**:320–325 [*257*]

Jaenike J (1990). Host specialization in phytophagous insects. *Annual Review of Ecology and Systematics* **21**:243–273 [*229*]

Jaenike J & Holt RD (1991). Genetic variation for habitat preference: Evidence and explanations. *The American Naturalist* **137**:67–90 [*52*]

Jaenike J & Papaj DR (1992). Behavioral plasticity and patterns of host use by insects. In *Insect Chemical Ecology*, eds. Roitberg BD & Isman MB, pp. 245–264. London, UK: Chapman & Hall [*250–251, 253*]

Jain SK & Bradshaw AD (1966). Evolutionary divergence among adjacent plant populations. I. The evidence and its theoretical analysis. *Heredity* **20**:407–441 [*272, 277*]

Jansen D (1988). Ecological characterization of a Costa Rican dry forest caterpillar fauna. *Biotropica* **20**:120–135 [*239*]

Jansen VAA & Mulder GSEE (1999). Evolving biodiversity. *Ecology Letters* **2**:379–386 [*73*]

Janz N & Nylin S (1998). Butterflies and plants: A phylogenetic study. *Evolution* **52**:486–502 [*239, 243*]

Jensen JS (1990). Plausibility and testability: Assessing the consequences of evolutionary innovation. In *Evolutionary Innovations*, ed. Nitecki MH, pp. 171–190. Chicago, IL,

USA: University of Chicago Press [*187*]

Jenssen TA (1970). Female response to filmed displays of *Anolis nebulosus* (Sauria: Iguanidae). *Animal Behaviour* **18**:640–647 [*339*]

Jenssen TA (1977). Morphological, behavioral and electrophoretic evidence of hybridization between the lizards, *Anolis grahami* and *Anolis lineatopus neckeri*, on Jamaica. *Copeia* **1977**:270–276 [*339*]

Jenssen TA (1978). Display diversity in anoline lizards and problems of interpretation. In *Behavior and Neurology of Lizards*, eds. Greenberg N & MacLean PD, pp. 269–285. Rockville, MD, USA: National Institute of Mental Health [*323, 339*]

Jenssen TA (1996). A test of assortative mating between sibling lizard species, *Anolis websteri* and *A. caudalis*, in Haiti. In *Contributions to West Indian Herpetology: A Tribute to Albert Schwartz*, eds. Powell R & Henderson RW, pp. 303–315. Ithaca, NY, USA: Society for the Study of Amphibians and Reptiles [*339*]

Jiggins CD & Mallet J (2000). Bimodal hybrid zones and speciation. *Trends in Ecology and Evolution* **15**:250–255 [*46*]

Jobling M, Jørgensen EH, Arnesen AM & Ringø E (1993). Feeding, growth and environmental requirements of arctic charr: A review of aquaculture potential. *Aquaculture International* **1**:20–46 [*217*]

Johnson NA & Porter AH (2000). Rapid speciation via parallel, directional selection on regulatory genetic pathways. *Journal of Theoretical Biology* **205**:527–542 [*125*]

Johnson CD & Siemens DH (1991). Expanded oviposition range by a seed beetle (Coleoptera: Bruchidae) in proximity to a normal host. *Environmental Entomology* **20**:1577–1582 [*246*]

Johnson TC, Scholtz CA, Talbot MR, Kelts K, Ricketts RD, Ngobi G, Beuning K, Ssemmanda I & McGill JW (1996a). Late Pleistocene desiccation of Lake Victoria and rapid evolution of cichlid fishes. *Science* **273**:1091–1093 [*173, 362, 374*]

Johnson PA, Hoppensteadt FC, Smith JJ & Bush GL (1996b). Conditions for sympatric speciation: A diploid model incorporating habitat fidelity and non-habitat assortative mating. *Evolutionary Ecology Research* **10**:187–205 [*43, 49, 53, 72, 82,*

174, 222, 230, 237, 244, 247, 305]
Johnson TC, Kelts K & Odada E (2000). The Holocene history of Lake Victoria. *Ambio* **29**:2–11 [*174*]

Johnston MO (1991). Natural selection on floral traits in two species of *Lobelia* with different pollinators. *Evolution* **45**:1468–1479 [*266, 277*]

Jones KN (1996). Pollinator behavior and postpollination reproductive success in alternative floral phenotypes of *Clarkia gracilis* (Onagraceae). *International Journal of Plant Science* **157**:733–738 [*274*]

Jones CD (1998). The genetic basis of *Drosophila sechellia*'s resistance to a host plant toxin. *Genetics* **149**:1899–1908 [*230*]

Jonsson B & Jonsson N (2001). Polymorphism and speciation in Arctic charr. *Journal of Fish Biology* **58**:605–638 [*46*]

Jonsson B & Skúlason S (2000). Polymorphic segregation in arctic charr *Salvelinus alpinus* from lake Vatnshlíðarvatn, Iceland. *Biological Journal of the Linnean Society* **69**:55–74 [*213, 215–216, 220*]

Jonsson B, Skúlason B, Snorrason SS, Sandlund OT, Malmquist HJ, Jónasson PM, Gydemo R & Linden T (1988). Life history variation of polymorphic Arctic charr (*Salvelinus alpinus*) in Thingvallavatn, Iceland. *Canadian Journal of Fisheries and Aquatic Sciences* **45**:1537–1547 [*216*]

Kacelnik A & Krebs JR (1985). Learning to exploit patchily distributed food. In *Behavioral Ecology: Ecological Consequences of Adaptive Behavior, 25th British Ecological Society Symposium*, eds. Sibly RM & Smith RH, pp. 189–205. Oxford, UK: Blackwell Scientific Publications [*256*]

Kambysellis MP, Ho KF, Craddock EM, Piano F, Parisi M & Cohen J (1995). Pattern of ecological shifts in the diversification of Hawaiian *Drosophila* inferred from a molecular phylogeny. *Current Biology* **5**:1129–1139 [*232*]

Kaneko K & Yomo T (2000). Sympatric speciation: Compliance with phenotype diversification from a single genotype. *Proceedings of the Royal Society of London B* **267**:2367–2373 [*125*]

Kaplan NL, Hudson RR & Iizuka M (1991). The coalescent process in models with selection, recombination and geographic subdivision. *Genetical Research* **57**:83–91 [*298*]

Kappeler PM, Wimmer B, Zinner D & Tautz D (2002). The hidden matrilineal structure of a solitary lemur: Implications for primate social evolution. *Proceedings of the Royal Society of London B* **269**:1755–1763 [*313*]

Kareiva P (1999). Coevolutionary arms races: Is victory possible? *Proceedings of the National Academy of Sciences of the USA* **96**:8–10 [*252*]

Karlin S & Campbell RB (1981). The existence of a protected polymorphism under conditions of soft as opposed to hard selection in a multideme population system. *The American Naturalist* **117**:262–275 [*39*]

Karron JD, Tucker R, Thumser NN & Reinartz JA (1995). Comparison of pollinator flight movements and gene dispersal patterns in *Mimulus ringens*. *Heredity* **75**:612–617 [*273*]

Kassen R, Schluter D & McPhail JD (1995). Evolutionary history of threespine sticklebacks (*Gasterosteus* spp.) in British Columbia: Insights from a physiological clock. *Canadian Journal of Zoology* **73**:2154–2158 [*198, 201*]

Katakura H (1997). Species of *Epilachna* ladybird beetles. *Zoological Science* **14**:869–881 [*240*]

Kauffman SA & Levin S (1987). Towards a general theory of adaptive walks on rugged landscapes. *Journal of Theoretical Biology* **128**:11–45 [*117*]

Kaufman L & Liem KF (1982). Fishes of the suborder labroidae (Pisces; Perciformes): Phylogeny, ecology, and evolutionary significance. *Breviora* **472**:1–19 [*187*]

Kawecki TJ (1994). Accumulation of deleterious mutations and the evolutionary cost of being a generalist. *The American Naturalist* **144**:833–838 [*50, 52*]

Kawecki TJ (1996). Sympatric speciation driven by beneficial mutations. *Proceedings of the Royal Society of London B* **263**:1515–1520 [*49–50, 52–53, 71, 82*]

Kawecki TJ (1997). Sympatric speciation via habitat specialization driven by deleterious mutations. *Evolution* **51**:1751–1763 [*49–50, 52, 71, 82, 219*]

Kawecki TJ (1998). Red Queen meets Santa Rosalia: Arms races and the evolution of host specialization in organisms with parasitic lifestyles. *The American Naturalist* **152**:635–651 [*49–50, 52, 229, 245*]

Keast A & Webb P (1966). Mouth and body form relative to feeding ecology in the fish fauna of a small lake, Lake Opinicon, Ontario. *Journal of the Fisheries Research Board of Canada* **23**:1845–1874   [*367*]

Khibnik AI & Kondrashov AS (1997). Three mechanisms of Red Queen dynamics. *Proceedings of the Royal Society of London B* **264**:1049–1056   [*74*]

Kiester AR, Lande R & Schemske DW (1984). Models of coevolution and speciation in plants and their pollinators. *The American Naturalist* **124**:220–243   [*95*]

Kilias G, Alahiotus SN & Pelecanos M (1980). A multifactorial genetic investigation of speciation theory using *Drosophila melanogaster*. *Evolution* **34**:730–737   [*194, 196*]

Kimura M (1983). *The Neutral Theory of Molecular Evolution*. New York, NY, USA: Cambridge University Press   [*18, 117, 125, 127, 133, 284, 286*]

Kimura M & Crow JF (1964). The theory of genetic loads. *Proceedings of the XI International Congress of Genetics* **3**:495–506   [*129*]

Kimura M & Weis GH (1964). The stepping stone model of population structure and the decrease of genetic correlation with distance. *Genetics* **49**:561–576   [*112*]

Kirkpatrick M (1982a). Quantum evolution and punctuated equilibrium in continuous genetic characters. *The American Naturalist* **119**:833–848   [*119*]

Kirkpatrick M (1982b). Sexual selection and the evolution of female choice. *Evolution* **36**:1–12   [*44, 99*]

Kirkpatrick M & Barton NH (1997). Evolution of a species' range. *The American Naturalist* **150**:1–23   [*141, 158, 160–161*]

Kirkpatrick M & Ryan MJ (1991). The paradox of the lek and the evolution of mating preferences. *Nature* **350**:33–38   [*99*]

Kirkpatrick M & Servedio MR (1999). The reinforcement of mating preferences on an island. *Genetics* **151**:865–884   [*142*]

Kisdi É (1999). Evolutionary branching under asymmetric competition. *Journal of Theoretical Biology* **197**:149–162   [*73, 94*]

Kisdi É (2001). Long-term adaptive diversity in Levene-type models. *Evolutionary Ecology Research* **3**:721–727   [*82*]

Kisdi É & Geritz SAH (1999). Adaptive dynamics in allele space: Evolution of genetic polymorphism by small mutations in a heterogeneous environment. *Evolution* **53**:993–1008   [*55, 67–69, 73, 85*]

Kisdi É & Meszéna G (1993). Density dependent life history evolution in fluctuating environments. In *Adaptation in a Stochastic Environment*, eds. Yoshimura J & Clark C, pp. 26–62, Lecture Notes in Biomathematics Vol. 98. Heidelberg, Germany: Springer-Verlag   [*54*]

Kisdi É, Jacobs FJA & Geritz SAH (2001). Red Queen evolution by cycles of evolutionary branching and extinction. *Selection* **2**:161–176   [*74*]

Klemetsen A, Amundsen PA, Knudsen R & Bjorn HA (1997). Profundal, winter-spawning morph of arctic charr *Salvelinus alpinus* (L.) in Fjellfrøsvatn, northern Norway. *Nordic Journal of Freshwater Research* **73**:13–73   [*213*]

Knight ME, Turner GF, Rico C, van Oppen MJH & Hewitt GM (1998). Microsatellite paternity analysis on captive Lake Malawi cichlids supports reproductive isolation by direct mate choice. *Molecular Ecology* **7**:1605–1610   [*176*]

Knowles LL, Futuyma DJ, Eanes WF & Rannala B (1999). Insight into speciation from historical demography in the phytophagous beetle genus *Ophraella*. *Evolution* **53**:1846–1856   [*233*]

Knox EB (1993). The species of giant senecio (*Compositae*) and giant lobelia (*Lobeliaceae*) in eastern Africa. *Contributions from the University of Michigan Herbarium* **19**:241–257   [*349*]

Knox EB (1999). Reconstruction of adaptive radiation in insular habitats using integrated biogeographic information. *Journal of Biogeography* **26**:983–991   [*345, 350, 354*]

Knox EB & Kowal RR (1993). Chromosome numbers of the East African giant senecios and giant lobelias and their evolutionary significance. *American Journal of Botany* **80**:847–853   [*351*]

Knox EB & Palmer JD (1995a). The origin of *Dendrosenecio* within the Senecioneae (Asteraceae) based on chloroplast DNA evidence. *American Journal of Botany* **82**:1567–1573   [*351*]

Knox EB & Palmer JD (1995b). Chloroplast DNA variation and the recent radiation of the giant senecios (Asteraceae) on the tall mountains of eastern Africa. *Proceedings of the National Academy of Sciences of the USA* **92**:10349–10353 [*353*]

Knox EB & Palmer JD (1998). Chloroplast DNA on the origin and radiation of the giant lobelias in eastern Africa. *Systematic Botany* **23**:109–149 [*345, 353–355, 360*]

Knox EB & Panero JL. Radiation of the giant senecios (Asteraceae) revisited: Comparison of phylogenetic estimates from chloroplast DNA and nuclear ribosomal DNA. Unpublished [*353, 355*]

Knox EB, Downie SR & Palmer JD (1993). Chloroplast genome rearrangements and the evolution of giant lobelias from herbaceous ancestors. *Molecular Biology and Evolution* **10**:414–430 [*351*]

Koella JC & Doebeli M (1999). Population dynamics and the evolution of virulence in epidemiological models with discrete host generations. *Journal of Theoretical Biology* **198**:461–475 [*73*]

Kölreuter JG (1763). *Vorläufige Nachricht von einigen das Geschlecht der Pflanzen betreffenden Versuchen und Beobachtungen: Fortsetzung* (Preliminary report of some experiments and observations concerning the sexuality of plants: Continuation). Leipzig, Germany: Gleditschische Handlung [*265*]

Kondrashov AS (1983a). Multilocus model of sympatric speciation I. One character. *Theoretical Population Biology* **24**:121–135 [*85*]

Kondrashov AS (1983b). Multilocus model of sympatric speciation II. Two characters. *Theoretical Population Biology* **24**:136–144 [*85*]

Kondrashov AS (1986). Multilocus model of sympatric speciation. III. Computer simulations. *Theoretical Population Biology* **29**:1–15 [*42–43, 46–47, 85*]

Kondrashov AS (2001). Speciation: Darwin revisited. *Trends in Ecology and Evolution* **16**:412 [*3*]

Kondrashov AS & Kondrashov F (1999). Interactions among quantitative traits in the course of sympatric speciation. *Nature* **400**:351–354 [*71, 73, 85, 110, 138, 246, 305, 362*]

Kondrashov AS & Mina MV (1986). Sympatric speciation: When is it possible? *Biological Journal of the Linnean Society* **27**:201–223 [*36, 42, 47, 52, 85*]

Kondrashov AS & Shpak M (1998). On the origin of speciation by means of assortative mating. *Proceedings of the Royal Society of London B* **265**:2273–2278 [*43, 85*]

Kondrashov AS, Yampolsky LY & Shabalina SA (1998). On the sympatric origin of species by means of natural selection. In *Endless Forms: Species and Speciation*, eds. Howard DJ & Berlocher SH, pp. 172–184. Oxford, UK: Oxford University Press [*222, 376*]

Kooijman SALM & Metz JAJ (1984). On the dynamics of chemically stressed populations: the deduction of population consequences from effects on individuals. *Ecotoxicology and Environmental Safety* **8**:254–274 [*226*]

Kornfield I (1991). Genetics. In *Cichlid Fishes: Behaviour, Ecology, and Evolution*, ed. Keenleyside MHA, pp. 103–128. London, UK: Chapman & Hall [*179*]

Kornfield I & Carpenter KE (1984). Cyprinids of Lake Lanao, Philippines: Taxonomic validity, evolutionary rates and speciation scenarios. In *Evolution of Fish Species Flocks*, eds. Echelle AA & Kornfield I, pp. 69–84. Orono, ME, USA: University of Maine Press [*362*]

Korona R, Nakatsu CH, Forney LJ & Lenski RE (1994). Evidence for multiple adaptive peaks from populations of bacteria evolving in a structure habitat. *Proceedings of the National Academy of Sciences of the USA* **91**:9037–9041 [*280*]

Kosswig C (1947). Selective mating as a factor for speciation in cichlid fish of East African lakes. *Nature* **159**:604–605 [*176*]

Kosswig C (1963). Ways of speciation in fishes. *Copeia* **1963**:238–244 [*362*]

Kottelat M (1997). European freshwater fishes. *Biologia* **52**(Suppl. 5):1–271 [*173*]

Kreslavskiy AG & Mikheyev AV (1994). Gene geography of racial differences in *Lochmaea capreae* L. (Coleoptera, Chrysomelidae), and the problem of sympatric speciation. *Entomological Review* **73**:85–92 [*240*]

Kristjánsson BK, Skúlason S & Noakes DLG (2002a). Morphological segregation of Icelandic threespine stickleback (*Gasterosteus aculeatus* L.). *Biological Journal of the Linnean Society* **76**:247–257 [*219*]

Kristjánsson BK, Skúlason S & Noakes DLG (2002b). Rapid divergence in a recently isolated population of threespine stickleback (*Gasterosteus aculeatus* L.). *Evolutionary Ecology Research* **4**:695–672   [*222*]

Kruckeberg A (1957). Variation in infertility of hybrids between isolated populations of the serpentine species *Streptanthus glandulosus* Hook. *Evolution* **11**:185–211   [*273*]

Krysan JL, McDonald IC & Tumlinson JH (1989). Phenogram based on allozymes and its relationship to classical biosystematics and pheromone structure among eleven diabroticites (Coleoptera: Chrysomelidae). *Annals of the Entomological Society of America* **82**:574–581   [*240*]

Kubitschek HE (1970). *Introduction to Research with Continuous Cultures*. Englewood Cliffs, NJ, USA: Prentice-Hall   [*280*]

Kuhner MK & Felsenstein J (1994). A simulation comparison of phylogeny algorithms under equal and unequal evolutionary rates. *Molecular Biology and Evolution* **11**:459–468   [*298*]

Lande R (1976). Natural selection and random genetic drift in phenotypic evolution. *Evolution* **30**:314–334   [*70, 135*]

Lande R (1979a). Effective deme size during long-term evolution estimated from rates of chromosomal rearrangement. *Evolution* **33**:234–251   [*118, 120*]

Lande R (1979b). Quantitative genetic analysis of multivariate evolution, applied to brain:body size allometry. *Evolution* **33**:402–416   [*65, 144, 160, 382*]

Lande R (1980). Sexual dimorphism, sexual selection, and adaptation in polygenic characters. *Evolution* **34**:292–305   [*99*]

Lande R (1981). Models of speciation by sexual selection on polygenic characters. *Proceedings of the National Academy of Sciences of the USA* **78**:3721–3725   [*99, 135–137*]

Lande R (1982). Rapid origin of sexual isolation and character divergence in a cline. *Evolution* **36**:213–223   [*36, 99, 136, 166, 306*]

Lande R (1985). Expected time for random genetic drift of a population between stable phenotypic states. *Proceedings of the National Academy of Sciences of the USA* **82**:7641–7645   [*118, 121*]

Lande R & Kirkpatrick M (1988). Ecological speciation by sexual selection. *Journal of Theoretical Biology* **133**:85–98   [*99, 186*]

Lande R, Seehausen O & van Alphen JJM

(2001) Mechanisms of rapid sympatric speciation by sex reversal and sexual selection in cichlid fish. *Genetica* **112–113**:435–443   [*181, 184, 214*]

Langeland A & L'Abee-Lund JH (1998). An experimental test of the genetic component of the ontogenetic habitat shift in Arctic charr (*Salvelinus alpinus*). *Ecology of Freshwater Fish* **7**:200–207   [*226*]

Langlet O (1971). Two hundred years genecology. *Taxon* **20**:653–721   [*272, 275*]

Larson GL (1976). Social behaviour and feeding ability of two phenotypes of *Gasterosteus aculeatus* in relation to their spatial and trophic segregation in a temperate lake. *Canadian Journal of Zoology* **54**:107–121   [*216*]

Lauder GV (1981). Form and function: Structural analysis in evolutionary morphology. *Paleobiology* **7**:430–442   [*337*]

Lauder GV (1983a). Functional and morphological basis of trophic specialization in sunfishes (Teleostei, Centrarchidae). *Journal of Morphology* **178**:1–21   [*187*]

Lauder GV (1983b). Neuromuscular patterns and the origin of trophic specialization in fishes. *Science* **219**:1235–1237   [*187*]

Law R & Dieckmann U (1998). Symbiosis through exploitation and the merger of lineages in evolution. *Proceedings of the Royal Society of London B* **265**:1245–1253   [*54*]

Law R & Dieckmann U (2000). Moment approximations of individual-based models. In *The Geometry of Ecological Interactions: Simplifying Spatial Complexity*, eds. Dieckmann U, Law R & Metz JAJ, pp. 252–269. Cambridge, UK: Cambridge University Press   [*306*]

Law R, Marrow P & Dieckmann U (1997). On evolution under asymmetric competition. *Evolutionary Ecology* **11**:485–501   [*75, 94*]

Law R, Bronstein JL & Ferrière R (2001). On mutualists and exploiters: Plant–insect coevolution in pollinating seed–parasite systems. *Journal of Theoretical Biology* **212**:373–389   [*73*]

Lazell J (1996). Careening Island and the Goat Islands: Evidence for the arid–insular invasion wave theory of dichopatric speciation in Jamaica. In *Contributions to West Indian Herpetology: A Tribute to Albert Schwartz*, eds. Powell R & Henderson RW, pp. 195–205. Ithaca, NY, USA: Society for the Study of Amphibians and Reptiles [*338*]

Leal M & Fleishman LJ (2002). Evidence for habitat partitioning based on adaptation to environmental light in a pair of sympatric lizard species. *Proceedings of the Royal Society of London B* **269**:351–359 [*342*]

Leimar O (2001). Evolutionary change and Darwinian demons. *Selection* **2**:65–72 [*67*]

Leimar O. Multidimensional convergence stability and the canonical adaptive dynamics. In *Elements of Adaptive Dynamics*, eds. Dieckmann U & Metz JAJ. Cambridge, UK: Cambridge University Press. In press [*67*]

Lenski RE (1995). Evolution, theory and experiments. In *Encyclopedia of Microbiology*, ed. Lederberg J, pp. 283–298. San Diego, CA, USA: Academic Press [*278*]

Lenski RE & Travisano M (1994). Dynamics of adaptation and diversification: A 10,000 generation experiment with bacterial populations. *Proceedings of the National Academy of Sciences of the USA* **91**:6808–6814 [*280–281, 287*]

Lenski RE, Rose MR, Simpson SC & Tadler SC (1991). Long-term experimental evolution in *Escherichia coli*. I. Adaptation and divergence during 2000 generations. *The American Naturalist* **138**:1315–1341 [*279–281*]

Lenski RE, Mongold JA, Sniegowski PD, Travisano M, Vasi F, Gerrish PJ & Schmidt TM (1998). Evolution of competitive fitness in experimental populations of *E. coli*: What makes one genotype a better competitor than another? *Antonie van Leeuwenhoek* **73**:35–47 [*280*]

Lesna I & Sabelis MW (1999). Diet-dependent female choice for males with "good genes" in a soil predatory mite. *Nature* **401**:581–584 [*258–259, 263*]

Levene H (1953). Genetic equilibrium when more than one ecological niche is available. *The American Naturalist* **87**:331–333 [*39, 54, 68–69, 73, 82*]

Levene H & Dobzhansky T (1959). Possible genetic differences between the head louse and the body louse (*Pediculus humanus* L.). *The American Naturalist* **63**:347–353 [*231*]

Levin DA (1972). The adaptedness of corolla color variants in experimental and natural populations of *Phlox drummondii*. *The American Naturalist* **106**:57–70 [*274*]

Levin DA (1993). Local speciation in plants: The rule not the exception. *Systematic Botany* **18**:197–208 [*265*]

Levin DA (2000). *The Origin, Expansion and Demise of Plant Species*. New York, NY, USA: Oxford University Press [*264–265, 273*]

Levin DA & Brack ET (1995). Natural selection against white petals in *Phlox*. *Evolution* **49**:1017–1022 [*274*]

Levin DA & Kerster HW (1967). Natural selection for reproductive isolation in *Phlox*. *Evolution* **21**:679–687 [*267*]

Levin DA & Kerster HW (1974). Gene flow in seed plants. *Evolutionary Biology* **7**:139–220 [*273*]

Levin DA & Schaal BA (1970). Corolla color as an inhibitor of interspecific hybridization in *Phlox*. *The American Naturalist* **104**:273–283 [*267*]

Levins R (1962). Theory of fitness in a heterogeneous environment. I. The fitness set and adaptive function. *The American Naturalist* **96**:361–373 [*249*]

Lewontin RC (1974). *The Genetic Basis of Evolutionary Change*. New York, NY, USA: Columbia University Press [*298*]

Lezzar KE, Tiercelin JJ, De Batist M, Cohen AS, Bandora T, Van Rensbergen P, Le Turdu C, Wafula M & Klerkx J (1996). New seismic stratigraphy and Late Tertiary history of the North Tanganyika Basin, East African Rift system, deduced from multifold reflection and high resolution seismic data and piston core evidence. *Basin Research* **8**:1–28 [*377*]

Li WH (1997). *Molecular Evolution*. Sunderland, MA, USA: Sinauer Associates Inc. [*118*]

Liem KF (1970). Comparative functional anatomy of the Nandidae (Pisces: Teleostei). *Fieldiana Zoology* **56**:1–166 [*187*]

Liem KF (1973). Evolutionary strategies and morphological innovations: Cichlid pharyngeal jaws. *Systematic Zoology* **22**:425–441 [*187–188*]

Liem KF (1978). Modulatory multiplicity in the functional repertoire of the feeding mechanisms in cichlid fishes. *Journal of Morphology* **158**:323–360 [*187*]

Liem KF (1979). Modulatory multiplicity in the feeding mechanism in cichlid fishes, as exemplified by the invertebrate pickers of Lake Tanganyika. *Journal of Zoology, Lon-*

don **189**:93–125 [*187*]

Liem KF & Greenwood PH (1981). A functional approach to the phylogeny of the pharyngognath teleosts. *American Zoology* **15**:83–101 [*187*]

Liem KF & Kaufman LS (1984). Intraspecific macroevolution: Functional biology of the polymorphic Cichlid species *Cichlasoma minckleyi*. In *Evolution of Fish Species Flocks*, eds. Echelle AA & Kornfield I, pp. 203–215. Orono, ME, USA: University of Maine Press [*217*]

Liem KF & Sanderson SL (1986). The pharyngeal jaw apparatus of labroid fishes: A functional morphological perspective. *Journal of Morphology* **187**:143–158 [*187*]

Liou LW & Price TD (1994). Speciation by reinforcement of premating isolation. *Evolution* **48**:1451–1459 [*142, 159*]

Lippitsch E (1993). A phyletic study on lacustrine haplochromine fishes (Perciformes, Cichlidae) of East Africa, based on scale and squamation characters. *Journal of Fish Biology* **51**:284–299 [*173*]

Lister BC (1976). The nature of niche expansion in West Indian *Anolis* lizards: II. Evolutionary components. *Evolution* **30**:677–692 [*337*]

Littlejohn MJ (1981). Reproductive isolation: A critical review. In *Evolution and Speciation: Essays in Honor of MJD White*, eds. Atchley W & Woodruff D, pp. 298–334. Cambridge, MA, USA: Cambridge University Press [*203*]

Losos JB (1985). An experimental demonstration of the species recognition role of *Anolis* dewlap color. *Copeia* **1985**:905–910 [*323, 340*]

Losos JB (1990a). Ecomorphology, performance capability, and scaling of West Indian *Anolis* lizards: An evolutionary analysis. *Ecological Monographs* **60**:369–388 [*334*]

Losos JB (1990b). A phylogenetic analysis of character displacement in Caribbean *Anolis* lizards. *Evolution* **44**:558–569 [*337*]

Losos JB (1994). Integrative approaches to evolutionary ecology: *Anolis* lizards as model systems. *Annual Review of Ecology and Systematics* **25**:467–493 [*322–323, 341*]

Losos JB & Chu LR (1998). Examination of factors potentially affecting dewlap size in Caribbean anoles. *Copeia* **1998**:430–438 [*339*]

Losos JB & Irschick DJ (1996). The effect of perch diameter on escape behaviour of *Anolis* lizards: Laboratory-based predictions and field tests. *Animal Behaviour* **51**:593–602 [*337*]

Losos JB & Miles DB (1994). Adaptation, constraint, and the comparative method: Phylogenetic issues and methods. In *Ecological Morphology: Integrative Organismal Biology*, eds. Wainwright PC & Reilly SM, pp. 60–98. Chicago, IL, USA: University of Chicago Press [*337*]

Losos JB & Schluter D (2000). Analysis of an evolutionary species–area relationship. *Nature* **408**:847–850 [*163–164, 336*]

Losos JB & Sinervo B (1989). The effect of morphology and perch diameter on sprint performance of *Anolis* lizards. *Journal of Experimental Biology* **145**:23–30 [*337*]

Losos JB, Irschick DJ & Schoener TW (1994). Adaptation and constraint in the evolution of specialization of Bahamian *Anolis* lizards. *Evolution* **48**:1786–1798 [*337, 342*]

Losos JB, Warheit KI & Schoener TW (1997). Adaptive differentiation following experimental island colonization in *Anolis* lizards. *Nature* **387**:70–73 [*342*]

Losos JB, Jackman TR, Larson A, de Queiroz K & Rodríguez-Schettino L (1998). Historical contingency and determinism in replicated adaptive radiations of island lizards. *Science* **279**:2115–2118 [*73, 337*]

Losos JB, Creer DA, Glossip D, Goellner R, Hampton A, Roberts G, Haskell N, Taylor P & Etling J (2000). Evolutionary implications of phenotypic plasticity in the hindlimb of the lizard *Anolis sagrei*. *Evolution* **54**:301–305 [*343*]

Lovett Doust L (1981). Population dynamics and local specialization in a clonal perennial (*Ranunculus repens*). II. The dynamics of leaves, and a reciprocal transplant replant experiment. *Journal of Ecology* **69**:757–768 [*272, 277*]

Lowe-McConnell RH (1987). *Ecological Studies in Tropical Fish Communities*. Cambridge, UK: Cambridge University Press [*173, 367*]

Lu G & Bernatchez L (1999). Correlated trophic specialization and genetic divergence in sympatric lake whitefish ecotypes

(*Coregonus clupeaformis*): Support for the ecological speciation hypothesis. *Evolution* **53**:1491–1505 [*219, 222–223*]

Ludwig D & Levin SA (1991). Evolutionary stability of plant communities and the maintenance of multiple dispersal types. *Theoretical Population Biology* **40**:285–307 [*75*]

Lundberg P, Ranta E, Kaitala V & Jonzén N (1999). Coexistence and resource competition. *Nature* **407**:694 [*306*]

Lynch JD (1989). The gauge of speciation: On the frequencies of modes of speciation. In *Speciation and Its Consequences*, eds. Otte D & Endler JA, pp. 527–553. Sunderland, MA, USA: Sinauer Associates Inc. [*116*]

Lynch M & Force AG (2000). The origin of interspecific genomic incompatibility via gene duplication. *The American Naturalist* **156**:590–605 [*125*]

Mabberley DJ (1973). Evolution in the giant groundsels. *Kew Bulletin* **28**:61–96 [*345*]

MacArthur RH (1969). Species packing, and what interspecies competition minimizes. *Proceedings of the National Academy of Sciences of the USA* **64**:1369–1371 [*270*]

MacArthur RH (1970). Species packing and competitive equilibrium for many species. *Theoretical Population Biology* **1**:1–11 [*306*]

MacArthur RH & Levins R (1967). The limiting similarity, convergence, and divergence of coexisting species. *The American Naturalist* **101**:377–385 [*94, 306*]

MacArthur RH & Wilson EO (1967). *Theory of Island Biogeography*. Princeton, NJ, USA: Princeton University Press [*163, 291*]

Macedonia JM & Stamps JA (1994). Species recognition in *Anolis grahami* (Sauria: Iguanidae): Evidence from responses to video playbacks of conspecific and heterospecific displays. *Ethology* **98**:246–264 [*340*]

Macedonia JM, Evans CS & Losos JB (1993). Male *Anolis* lizards discriminate video-recorded conspecific and heterospecific displays, performance, and fitness. *Animal Behaviour* **47**:1220–1223 [*339*]

Machado CA, Herre EA, McCafferty S & Bermingham E (1996). Molecular phylogenies of fig pollinating and non-pollinating wasps and the implications for the origin and evolution of fig–fig wasp mutualism. *Journal of Biogeography* **23**:531–542 [*231*]

Mackenzie A (1996). A trade-off for host plant utilization in the black bean aphid, *Aphis fabae*. *Evolution* **50**:155–162 [*236, 242, 245*]

Mackenzie A & Guldemond JA (1994). Sympatric speciation in aphids. II. Host race formation in the face of gene flow. In *Individuals, Populations and Patterns in Ecology*, eds. Leather SR, Walters KFA, Mills NJ & Watt AD, pp. 379–395. Andover, Hampshire, UK: Intercept Ltd. [*236*]

Macnair MR & Christie P (1983). Reproductive isolation as a pleiotropic effect of copper tolerance in *Mimulus guttatus*? *Heredity* **50**:295–302 [*194, 273*]

Macnair MR & Gardner M (1998). The evolution of edaphic endemics. In *Endless Forms: Species and Speciation*, eds. Howard DJ & Berlocher SH, pp. 157–171. Oxford, UK: Oxford University Press [*71, 115, 272*]

Maddison WP & Maddison DR (1992). *MacClade, Version 3 – Analysis of Phylogeny and Character Evolution*. Sunderland, MA, USA: Sinauer Associates Inc. [*334*]

Magnusson KP & Ferguson MM (1987). Genetic analysis for four sympatric morphs of the Arctic charr, *Salvelinus alpinus*, from Thingvallavatn, Iceland. *Environmental Biology and Fishery* **20**:67–73 [*221*]

Magowski W, Egas M, Bruin J & Sabelis MW (2003). Intraspecific variation in induction of feeding preference and performance in a herbivorous mite. *Experimental and Applied Acarology* **29**:13–25 [*255*]

Maire N, Ackermann M & Doebeli M (2001). Evolutionary branching and the evolution of anisogamy. *Selection* **2**:119–132 [*75*]

Malhotra A (1992). *What Causes Geographic Variation: A Case Study of* Anolis *oculatus*. PhD thesis. Aberdeen, UK: University of Aberdeen [*344*]

Malhotra A & Thorpe RS (1991a). Microgeographic variation in *Anolis oculatus* on the island of Dominica, West Indies. *Journal of Evolutionary Biology* **4**:321–335 [*324, 327*]

Malhotra A & Thorpe RS (1991b). Experimental detection of rapid evolutionary response in natural lizard populations. *Nature* **353**:347–348 [*329*]

Malhotra A & Thorpe RS (1993a). An experimental field study of an eurytopic anole, *Anolis oculatus*. *Journal of Zoology* **229**:163–170 [*329*]

Malhotra A & Thorpe RS (1993b). *Anolis oculatus. Catalogue of American Amphibians and Reptiles, Society for the Study of Amphibians and Reptiles* **540**:1–4 [*325*]

Malhotra A & Thorpe RS (1994). Parallels between island lizards suggests selection on mitochondrial DNA and morphology. *Proceedings of the Royal Society of London B* **257**:37–42 [*331*]

Malhotra A & Thorpe RS (1997a). Size and shape variation in a Lesser Antillean anole, *Anolis oculatus* (Sauria: Iguanidae) in relation to habitat. *Biological Journal of the Linnean Society* **60**:53–72 [*324, 327*]

Malhotra A & Thorpe RS (1997b). Microgeographic variation in scalation of *Anolis oculatus* (Dominica, West Indies): A multivariate analysis. *Herpetologica* **53**:49–62 [*324, 327*]

Malhotra A & Thorpe RS (1997c). Size and shape variation in a Lesser Antillean anole *Anolis oculatus* (Sauria: Iguanidae) in relation to habitat. *Biological Journal of the Linnean Society* **60**:53–72 [*164*]

Malhotra A & Thorpe RS (1999). *Reptiles and Amphibians of the Eastern Caribbean.* London, UK: MacMillan Press [*331*]

Malhotra A & Thorpe RS (2000). The dynamics of natural selection and vicariance in the Dominican anole: Patterns of within-island molecular and morphological divergence. *Evolution* **54**:245–258 [*322, 325–327, 331, 333*]

Mallet J (1995). A species definition for the modern synthesis. *Trends in Ecology and Evolution* **10**:294–299 [*10, 237*]

Mallet J (2001). The speciation revolution. *Journal of Evolutionary Biology* **14**:887–888 [*3*]

Mallet J, McMillan WO & Jiggins CD (1998). Mimicry and warning color at the boundary between races and species. In *Endless Forms: Species and Speciation*, eds. Howard DJ & Berlocher SH, pp. 390–403. Oxford, UK: Oxford University Press [*233*]

Malmquist HJ (1992). Phenotype-specific feeding behaviour of two arctic charr *Salvelinus alpinus* morphs. *Oecologia* **92**:354–361 [*216, 218*]

Malmquist HJ, Snorrason SS, Skúlason S, Jonsson B, Sandlund OT & Jónasson PM (1992). Diet differentiation in polymorphic Arctic charr in Thingvallavatn, Iceland.

Journal of Animal Ecology **61**:21–35 [*211, 218*]

Mani GS & Clarke BC (1990). Mutational order: A major stochastic process in evolution. *Proceedings of the Royal Society of London B* **240**:29–37 [*129*]

Manly BJF (1986a). *Multivariate Statistical Methods: A Primer.* London, UK: Chapman & Hall [*327*]

Manly BJF (1986b). Randomization and regression methods for testing associations with geographical, environmental and biological distances between populations. *Researches on Population Ecology* **28**:201–218 [*327*]

Margolies DC (1995). Evidence of selection on spider mite dispersal rates in relation to habitat persistence in agroecosystems. *Entomologia Experimentalis et Applicata* **76**:105–108 [*253*]

Markert JA, Danley DD & Argengard ME (2001). New markers for new species: Microsatellite loci and the East African cichlids. *Trends in Ecology and Evolution* **16**:100–107 [*25–26*]

Marrow P, Law R & Cannings C (1992). The coevolution of predator–prey interactions: ESSs and Red Queen dynamics. *Proceedings of the Royal Society of London B* **250**:133–141 [*54, 95*]

Marrow P, Dieckmann U & Law R (1996). Evolutionary dynamics of predator–prey systems: An ecological perspective. *Journal of Mathematical Biology* **34**:556–578 [*60, 65, 95, 372*]

Marsh AC, Ribbink AJ & Marsh BA (1981). Sibling species complexes in sympatric populations of Petrotilapia Trewavas (Cichlidae, Lake Malawi). *Zoological Journal of the Linnean Society* **71**:253–264 [*176*]

Martens K (1997). Speciation in ancient lakes. *Trends in Ecology and Evolution* **12**:177–181 [*362*]

Martienssen R (1996). Paramutation and gene silencing in plants. *Current Biology* **6**:810–813 [*277*]

Martin OY & Hosken DJ (2003). The evolution of reproductive isolation through sexual conflict. *Nature* **423**:979–982 [*102*]

Matessi C & Di Pasquale C (1996). Long term evolution of multi-locus traits. *Journal of Mathematical Biology* **34**:613–653 [*65, 67*]

Matessi C, Gimelfarb A & Gavrilets S (2001).

Long-term buildup of reproductive isolation promoted by disruptive selection: How far does it go? *Selection* **2**:41–64 [*70–71, 91, 391*]

Mather K (1947). Species crosses in *Antirrhinum* I. Genetic isolation of the species *majus, glutinosa* and *orontium*. *Heredity* **1**:175–187 [*265, 267*]

Mathias A & Kisdi É (2002). Adaptive diversification of germination strategies. *Proceedings of the Royal Society of London B* **269**:151–156 [*73*]

Mathias A & Kisdí É. Evolutionary branching and coexistence of germination strategies. In *Elements of Adaptive Dynamics*, eds. Dieckmann U & Metz JAJ. Cambridge, UK: Cambridge University Press. In press [*74*]

Mathias A, Kisdi É & Olivieri I (2001). Divergent evolution of dispersal in a heterogeneous landscape. *Evolution* **55**:246–259 [*73*]

Matsuda H (1985). Evolutionarily stable strategies for predator switching. *Journal of Theoretical Biology* **115**:351–366 [*58*]

Matsuda H & Abrams PA (1994a). Runaway evolution to self-extinction under asymmetrical competition. *Evolution* **48**:1764–1772 [*98*]

Matsuda H & Abrams PA (1994b). Timid consumers – self-extinction due to adaptive change in foraging and anti-predator effort. *Theoretical Population Biology* **45**:76–91 [*98*]

Matsuhashi T, Masuda R, Mano T & Yoshida MC (1999). Microevolution of the mitochondrial control region in the Japanese brown bear (*Ursus arctos*) population. *Molecular Ecology* **8**:676–684 [*319*]

Matthews WJ (1998). *Patterns in Freshwater Fish Ecology*. London, UK: Chapman & Hall [*367*]

May RM (1990). How many species? *Philosophical Transactions of the Royal Society of London B* **330**:293–304 [*380*]

Mayley G (1997). Guiding or hiding: Explorations into the effects of learning on the rate of evolution. In *Fourth European Conference on Artificial Life*, eds. Husbands P & Harvey I, pp. 156–173. Cambridge, MA, USA: MIT Press [*250*]

Maynard Smith J (1966). Sympatric speciation. *The American Naturalist* **100**:637–650 [*33, 42–43, 46, 50, 52, 68, 76–77, 81–82, 221, 230*]

Maynard Smith J (1982). *Evolution and the Theory of Games*. Cambridge, UK: Cambridge University Press [*55, 61–62*]

Maynard Smith J (1987). When learning guides evolution. *Nature* **329**:761–762 [*250*]

Maynard Smith J (1991). Theories of sexual selection. *Trends in Ecology and Evolution* **6**:146–151 [*98*]

Maynard Smith J & Brown RL (1986). Competition and body size. *Theoretical Population Biology* **30**:166–179 [*54*]

Maynard Smith J & Hoekstra R (1980). Polymorphism in a varied environment: How robust are the models? *Genetical Research* **35**:45–57 [*39–40*]

Maynard Smith J & Price GR (1973). The logic of animal conflict. *Nature* **246**:15–18 [*54*]

Mayr E (1942). *Systematics and the Origin of Species*. New York, NY, USA: Columbia University Press [*25, 112, 114, 116, 119, 136, 192*]

Mayr E (1947). Ecological factors in speciation. *Evolution* **1**:263–288 [*233*]

Mayr E (1959a). Where are we? *Cold Spring Harbor Symposia on Quantitative Biology* **24**:1–14 [*17*]

Mayr E (1959b). Isolation as an evolutionary factor. *Proceedings of the American Philosophical Society* **103**:221–230 [*22*]

Mayr E (1963). *Animal Species and Evolution*. Cambridge, MA, USA: Harvard University Press [*10, 18, 21–23, 37–38, 42, 76, 107, 112–114, 116, 119, 136, 222, 233*]

Mayr E, ed. (1964). *Charles Darwin: On the Origin of Species*. A facsimile of the First Edition, with an introduction by Ernst Mayr. Cambridge, MA, USA: Harvard University Press [*20*]

Mayr E (1982). *The Growth of Biological Thought: Diversity, Evolution, and Inheritance*. Cambridge, MA, USA: The Belknap Press of Harvard University Press [*1, 16*]

McCune AR (1987a). Toward the phylogeny of a fossil species flock: Semionotid fishes from a lake deposit in the Early Jurassic Towaco Formation, Newark Basin. *Bulletin of the Yale Peabody Museum of Natural History* **43**:1–108 [*363, 366, 368*]

McCune AR (1987b). Lakes as laboratories of evolution: Endemic fishes and environmental cyclicity. *Palaios* **2**:446–454 [*366, 368*]

McCune AR (1990). Evolutionary novelty and atavism in the *Semionotus* complex: Relaxed selection during colonization of an expanding lake. *Evolution* **44**:71–85 [*363, 368–369, 371*]

McCune AR (1996). Biogeographic and stratigraphic evidence for rapid speciation in semionotid fishes. *Paleobiology* **22**:34–48 [*363, 365–366, 368, 374*]

McCune AR (1997). How fast do fishes speciate? Molecular, geological, and phylogenetic evidence from adaptive radiations of fishes. In *Molecular Evolution and Adaptive Radiation*, eds. Givinish TJ & Sytsma KJ, pp. 585–610. Cambridge, UK: Cambridge University Press [*374–377, 379*]

McCune AR & Lovejoy NR (1998). The relative rate of sympatric and allopatric speciation in fishes: Tests using DNA sequence divergence between sister species and among clades. In *Endless Forms: Species and Speciation*, eds. Howard DJ & Berlocher SH, pp. 172–185. Oxford, UK: Oxford University Press [*73, 362, 375–376, 378*]

McCune AR, Thomson KS & Olsen PE (1984). Semionotid fishes from the Mesozoic Great Lakes of North America. In *Evolution of Fish Species Flocks*, eds. Echelle AA & Kornfield I, pp. 22–44. Orono, ME, USA: University of Maine Press [*362, 366–368, 379*]

McElroy DM & Kornfield I (1990). Sexual selection, reproductive behaviour, and speciation in the Mbuna species flock of Lake Malawi (Pisces: Cichlidae). *Environmental Biology of Fishes* **28**:273–284 [*176*]

McElroy DM, Kornfield I & Everett J (1991). Colouration in African cichlids: Diversity and constraints in Lake Malawi endemics. *Netherlands Journal of Zoology* **41**:250–268 [*176*]

McKaye KR (1991). Sexual selection and the evolution of the cichlid fishes of Lake Malawi, Africa. In *Cichlid Fishes: Behavior, Ecology, and Evolution*, ed. Keenleyside MHA, pp. 241–257. London, UK: Chapman & Hall [*176*]

McKaye KR & Gray WN (1984). Extrinsic barriers to gene flow in rock-dwelling cichlids of Lake Malawi: Macro-habitat heterogeneity and reef colonization. In *Evolution of Fish Species Flocks*, eds. Echelle AA & Kornfield I, pp. 169–183. Orono, ME, USA:

University of Maine Press [*174*]

McKaye KR, Kocher TD, Reinthal P, Harrison R & Kornfield I (1982). A sympatric sibling species complex of *Petrotilapia trewavas* (Cichlidae) from Lake Malawi analyzed by enzyme electrophoresis. *Zoological Journal of the Linnean Society* **76**:91–96 [*176*]

McKaye KR, Kocher TD, Reinthal P, Harrison R & Kornfield I (1984). Genetic evidence of allopatric and sympatric differentiation among color morphs of a Lake Malawi cichlid fish. *Evolution* **38**:215–219 [*176*]

McKinnon JS, Mori S & Schluter D. Natural selection and parallel speciation in freshwater anadromous sticklebacks. Unpublished [*202*]

McLaughlin RL (2001). Behavioural diversification in brook charr: Adaptive responses to local conditions. *Journal of Animal Ecology* **70**:325–337 [*216*]

McLaughlin RL, Ferguson MM & Noakes DLG (1999). Adaptive peaks and alternative foraging tactics in brook charr: Evidence of short-term divergent selection for sitting-and-waiting and actively searching. *Behavioural Ecology and Sociobiology* **45**:386–395 [*216, 220*]

McMillan WO, Jiggins CD & Mallet J (1997). What initiates speciation in passion-vine butterflies? *Proceedings of the National Academy of Sciences of the USA* **94**:8628–8633 [*233*]

McNeilly T & Antonovics J (1968). Evolution in closely adjacent plant populations IV. Barriers to gene flow. *Heredity* **23**:205–218 [*272, 274*]

McPeek MA & Wellborn GA (1998). Genetic variation and reproductive isolation among phenotypically divergent amphipod populations. *Limnology and Oceanography* **43**:1162–1169 [*202*]

McPhail JD (1984). Ecology and evolution of sympatric sticklebacks (*Gasterosteus*): Morphological and genetic evidence for a species pair in Enos Lake, British Columbia. *Canadian Journal of Zoology* **62**:1402–1408 [*196, 198, 214*]

McPhail JD (1992). Ecology and evolution of sympatric sticklebacks (*Gasterosteus*): Evidence for a species pair in Paxton Lake, Texada Island, British Columbia. *Canadian Journal of Zoology* **70**:361–369 [*196, 198*]

McPhail JD (1993). Ecology and evolution of sympatric sticklebacks (*Gasterosteus*): Origins of the species pairs. *Canadian Journal of Zoology* 71:515–523   [*196, 198*]

McPhail JD (1994). Speciation and the evolution of reproductive isolation in the sticklebacks (*Gasterosteus*) of south-western British Columbia. In *The Evolutionary Biology of the Threespine Stickleback*, eds. Bell M & Foster S, pp. 399–437. Oxford, UK: Oxford University Press   [*196, 218, 221*]

McPheron BA, Smith DC & Berlocher SH (1988). Genetic differences between host races of *Rhagoletis pomonella*. *Nature* 336:64–66   [*237*]

Mehrhoff LA & Turkington R (1996). Growth and survival of white clover (*Trifolium repens*) transplanted into patches of different grass species. *Canadian Journal of Botany* 74:1243–1247   [*272*]

Meléndez-Ackerman E & Campbell DR (1998). Adaptive significance of flower color and inter-trait correlations in an *Ipomopsis* hybrid zone. *Evolution* 52:1293–1303   [*269*]

Memmott J (1999). The structure of a plant–pollinator food web. *Ecology Letters* 2:276–280   [*267*]

Menken SBJ & Roessingh P (1998). Evolution of insect–plant associations: Sensory perception and receptor modifications direct food specialization and host shifts in phytophagous insects. In *Endless Forms: Species and Speciation*, eds. Howard DJ & Berlocher SH, pp. 145–156. Oxford, UK: Oxford University Press   [*231, 239, 254*]

Meszéna G & Christiansen FB. On the adaptive emergence of reproductive isolation. Unpublished   [*70–71*]

Meszéna G & Szathmáry E (2001). Adaptive dynamics of parabolic replicators. *Selection* 2:147–159   [*75*]

Meszéna G, Czibula I & Geritz SAH (1997). Adaptive dynamics in a 2-patch environment: A toy model for allopatric and parapatric speciation. *Journal of Biological Systems* 5:265–284   [*66–67, 73*]

Metcalfe NB (1993). Behavioral causes and consequences of life history variation in fish. *Marine Behaviour and Physiology* 23:205–217   [*220*]

Metz JAJ, Nisbet RM & Geritz SAH (1992). How should we define "fitness" for general ecological scenarios? *Trends in Ecology and Evolution* 7:198–202   [*56, 73–74, 107*]

Metz JAJ, Geritz SAH, Meszéna G, Jacobs FJA & van Heerwaarden JS (1996). Adaptive dynamics: A geometrical study of the consequences of nearly faithful reproduction. In *Stochastic and Spatial Structures of Dynamical Systems, Proceedings of the Royal Dutch Academy of Science (KNAW Verhandelingen)*, eds. van Strien SJ & Verduyn Lunel SM, pp. 183–231. Dordrecht, Netherlands: North Holland   [*6, 55–57, 59, 65, 67, 72–74, 107, 143, 222, 270*]

Meyer A (1987). Phenotypic plasticity and heterochrony in *Cichlasoma managuense* (Pisces, Cichlidae) and their implications for speciation in cichlid fishes. *Evolution* 41:1357–1369   [*217, 220, 225, 315*]

Meyer A (1990). Ecological and evolutionary consequences of the trophic polymorphism in *Cichlasoma citrinellum* (Pisces: Cichlidae). *Biological Journal of the Linnean Society* 39:279–299   [*217*]

Meyer A (1993). Phylogenetic relationships and evolutionary processes in East African cichlid fishes. *Trends in Ecology and Evolution* 8:279–284   [*210*]

Meyer A, Kocher TD, Basasibwaki P & Wilson AC (1990). Monophyletic origin of Lake Victoria cichlid fishes suggested by mitochondrial DNA sequences. *Nature* 347:550–553   [*173, 362*]

Mikheev VN, Adams CE, Huntinford EA & Thorpe JE (1996). Behavioural responses of benthic and pelagic arctic charr to substratum heterogeneity. *Journal of Fish Biology* 49:494–500   [*216–218*]

Mitchell PC (1911). Species. In *The Encyclopedia Britannica*, Eleventh Edition, Volume XXV, pp. 616–617. Chicago, IL, USA: Encyclopedia Britannica   [*19*]

Mitchell R, Bleakly D, Cabin R, Chan R, Enquist B, Evans A, Lowrey T, Marshall D, Reed S, Stevens C & Waser NM (1993). Species concepts. *Nature* 364:20   [*277*]

Mittelbach GG (1981). Foraging efficiency and body size: A study of optimal diet and habitat use by bluegills. *Ecology* 62:1370–1386   [*226*]

Mitter C, Farrell B & Futuyma DJ (1991). Phylogenetic studies of insect–plant interactions: Insights into the genesis of diversity. *Trends in Ecology and Evolution* 6:290–293

[*231, 233*]
Miyatake T & Shimizu T (1999). Genetic correlations between life-history and behavioral traits can cause reproductive isolation. *Evolution* **53**:201–208 [*237*]

Mizera F & Meszéna G (2003). Spatial niche packing, character displacement and adaptive speciation along an environmental gradient. *Evolutionary Ecology Research* **5**:363–382 [*67, 73, 156*]

Møller AP (1994). *Sexual Selection and the Barn Swallow*. Oxford, UK: Oxford University Press [*257*]

Mopper S (1996). Adaptive genetic structure in phytophagous insect populations. *Trends in Ecology and Evolution* **11**:235–238 [*237*]

Moran NA, Kaplan ME, Gelsey MJ, Murphy TG & Scholes EA (1999). Phylogenetics and evolution of the aphid genus *Uroleucon* based on mitochondrial and nuclear DNA sequences. *Systematic Entomology* **24**:85–93 [*238*]

Morgan CL (1896). On modification and variation. *Science* **4**:733–740 [*250*]

Morin PJ (1999). Productivity, intraguild predation, and population dynamics in experimental food webs. *Ecology* **80**:752–760 [*300*]

Morrell V (1999). Ecology returns to speciation studies. *Science* **284**:2106–2108 [*202*]

Muller HJ (1940). Bearing of the *Drosophila* work on systematics. In *The New Systematics*, ed. Huxley J, pp. 185–268. Oxford, UK: Oxford University Press [*123*]

Muller HJ (1942). Isolating mechanisms, evolution and temperature. *Biological Symposia* **6**:71–125 [*123, 125, 192*]

Nagel L & Schluter D (1998). Body size, natural selection, and speciation in sticklebacks. *Evolution* **52**:209–218 [*46, 71, 194, 196, 201, 206, 209*]

Nagelkerke LAJ & Sibbing FA (1996). Reproductive segregation among the large barbs (*Barbus intermedius* complex) of Lake Tana, Ethiopia: An example of intralacustrine speciation? *Journal of Fish Biology* **49**:1244–1266 [*191*]

Nagy ES (1997). Selection for native characters in hybrids between two locally adapted plant subspecies. *Evolution* **51**:1469–1480 [*275*]

Nault LR (1985). Evolutionary relationships between maize leafhoppers and their host plants. In *The Leafhoppers and Planthoppers*, eds. Nault LR & Rodriguez JG, pp. 309–330. New York, NY, USA: John Wiley & Sons [*238*]

Naumann ID, ed. (1991). *The Insects of Australia*, Volumes 1 and 2, Second Edition. Melbourne, Australia: Melbourne University Press [*229*]

Navajas M (1998). Host plant associations in the spider mite *Tetranychus urticae* (Acari: Tetranychidae): Insights from molecular phylogeography. *Experimental and Applied Acarology* **22**:201–214 [*253*]

Nei M (1976). Mathematical models of speciation and genetic distance. In *Population Genetics and Ecology*, eds. Karlin S & Nevo E, pp. 723–768. New York, NY, USA: Academic Press [*125*]

Nei M (1989). *Molecular Evolutionary Genetics*. New York, NY, USA: Columbia University Press [*18*]

Nei M, Maruyama T & Wu CI (1983). Models of evolution of reproductive isolation. *Genetics* **103**:557–579 [*129, 131–132, 134–135*]

Neigel JE & Avise JC (1986). Phylogenetic relationships of mitochondrial DNA under various demographic models of speciation. In *Evolutionary Processes and Theory*, eds. Nevo E & Karlin S, pp. 515–534. New York, NY, USA: Academic Press [*312*]

Newman MEJ & Engelhardt R (1998). Effects of selective neutrality on the evolution of molecular species. *Proceedings of the Royal Society of London B* **265**:1333–1338 [*129*]

Nielsen JK (1999). Specificity of a Y-linked gene in the flea beetle *Phyllotreta nemorum* for defenses in *Barbarea vulgaris*. *Entomologia Experimentalis et Applicata* **91**:359–368 [*245*]

Nilson J (1990). Heritability estimates of growth-related traits in Arctic charr (*Salvelinus alpinus*). *Aquaculture* **84**:211–217 [*217*]

Nishida M (1997). Phylogenetic relationships and evolution of Tanganyikan cichlids: A molecular perspective. In *Fish Communities in Lake Tanganyika*, eds. Kawabane H, Hori M & Makoto N, pp. 1–23. Kyoto, Japan: Kyoto University Press [*173*]

Nishida T, Pudjiastuti LE, Nakano S, Abbas I, Kahono S, Nakamura K & Katakura H (1997). The eggplant beetle on a leguminous weed: Host race formation in progress? *Tropics* **7**:115–121   [*242*]

Noakes DLG (1989). Early life history and behaviour of charrs. In *Biology of Charrs and Masu Salmon: Proceedings of the International Symposium on Charrs and Masu Salmon*, eds. Kawanabe H, Yamazaki F & Noakes DLG, pp. 173–186. Special Volume 1 of *Physiology and Ecology of Japan* [*220*]

Noakes DLG, Skúlason S & Snorrason SS (1989). Alternative life-history styles in salmonid fishes with emphasis on Arctic char (*Salvelinus alpinus*). In *Alternative Life-History Styles of Animals*, ed. Bruton MN, pp. 329–346. Dordrecht, Netherlands: Kluwer Academic Publishers   [*217*]

Noor MAF (1995). Speciation driven by natural selection in *Drosophila*. *Nature* **375**:674–675   [*71, 142, 203*]

Noor MA (1997). How often does sympatry affect sexual isolation in *Drosophila*? *The American Naturalist* **149**:1156–1163 [*203*]

Noor MAF (1999). Reinforcement and other consequences of sympatry. *Heredity* **83**:503–508   [*246*]

Nordeng H (1983). Solution to the "charr problem" based on arctic charr (*Salvelinus alpinus*) in Norway. *Canadian Journal of Fisheries and Aquatic Sciences* **40**:1372–1387 [*217, 221–223*]

Novotny V, Basset Y, Miller SE, Weiblen GD, Bremer B, Cizek L & Drozd P (2002). Low host specificity of herbivorous insects in a tropical forest. *Nature* **416**:841–844. [*380*]

Nowak M (1990). An evolutionarily stable strategy may be inaccessible. *Journal of Theoretical Biology* **142**:237–241   [*55*]

Ødegaard F (2000). How many species of arthropods? Erwin's estimate revised. *Biological Journal of the Linnean Society* **71**:583–597   [*380*]

O'Donald P (1962). The theory of sexual selection. *Heredity* **22**:499–518   [*99*]

O'Donald P (1967). A general model of sexual and natural selection. *Theoretical Population Biology* **12**:298–334   [*99*]

O'Donald P (1977). Theoretical aspects of sexual selection. *Theoretical Population Biology* **12**:298–334   [*99*]

O'Donald P (1980). *Genetic Models of Sexual Selection*. Cambridge, UK: Cambridge University Press   [*99*]

Ogden R & Thorpe RS (2002). Molecular evidence for ecological speciation in tropical habitats. *Proceedings of the National Academy of Sciences of the USA* **99**:13612–13615   [*166, 334–335, 344, 393*]

Ohta T (1992). The nearly neutral theory of molecular evolution. *Annual Review of Ecology and Systematics* **23**:263–286   [*118*]

Ohta T (1998). Evolution by nearly-neutral mutations. *Genetica* **102/103**:83–90   [*118*]

Ohta T & Kimura M (1973). A model of mutation appropriate to estimate the number of electrophoretically detectable alleles in a finite population. *Genetical Research* **22**:201–204   [*129*]

Okubo A (1980). *Diffusion and Ecological Problems: Mathematical Models*. Berlin, Germany: Springer-Verlag   [*306*]

Ollerton J (1996). Reconciling ecological processes with phylogenetic patterns: The apparent paradox of plant–pollinator systems. *Journal of Ecology* **84**:767–769   [*270*]

Olsen PE (1980). Fossil great lakes of the Newark Supergroup in New Jersey. In *Field Studies of New Jersey Geology and Guide to Field Trips*, ed. Manspeizer W, pp. 352–398. Newark, NJ, USA: Newark College of Arts and Sciences, Rutgers University   [*364–365*]

Olsen PE (1986). A 40-million year record of Early Mesozoic orbital climatic forcing. *Science* **234**:842–848   [*364–366*]

Olsen PE (1988). Continuity of strata in the Newark and Hartford Basins. *US Geological Survey Bulletin* **1776**:6–18   [*365*]

Olsen PE & Kent DV (1996). Milankovitch climate forcing in the tropics of Pangaea during the Late Triassic. *Palaeogeography, Palaeoclimatology, Palaeoecology* **122**:1–26 [*365*]

Olsen PE & Kent DV (1999). Long-period Milankovitch cycles from the Late Triassic and Early Jurassic of eastern North America and their implications for the calibration of the Early Mesozoic time-scale and the long-term behaviour of the planets. *Philosophical Transactions of the Royal Society of London B* **357**:1761–1786   [*364–365*]

Olsen PE & McCune AR (1991). Morphology of the *Semionotus elegans* group from the Early Jurassic part of the Newark Supergroup of eastern North America, with comments on the family Semionotidae (Pisces: Neopterygii). *Journal of Vertebrate Paleontology* **11**:269–292   [*363*]

Olsen PE, Remington CL, Cornet B & Thomson KS (1978). Cyclic change in Late Triassic lacustrine communities. *Science* **201**:729–733 [*364, 366*]

Olsen PE, Kent DV, Cornet B, Witte WK & Schlische RW (1996). High-resolution stratigraphy of the more than 5000 m Newark Rift Basin section (Early Mesozoic, eastern North America). *Geological Society of America Bulletin* **108**:40–77 [*364–366*]

Omholt SW, Plahte E, Oyehaug L & Xiang KF (2000). Gene regulatory networks generating the phenomena of additivity, dominance and epistasis. *Genetics* **155**:969–980 [*125*]

Orr HA (1995). The population genetics of speciation: The evolution of hybrid incompatibilities. *Genetics* **139**:1803–1813 [*124, 127, 132–133, 135*]

Orr HA & Orr LH (1996). Waiting for speciation: The effect of population subdivision on the waiting time to speciation. *Evolution* **50**:1742–1749 [*132*]

Orzack SH, Sohn JJ, Kallman KD, Levin SA & Johnston R (1980). Maintenance of the three sex chromosome polymorphism in the platyfish, *Xiphophorus maculatus*. *Evolution* **14**:663–772 [*181*]

Osborn HF (1896). Ontogenic and phylogenic variation. *Science* **4**:786–789 [*250*]

Otte D & Endler JA, eds. (1989). *Speciation and Its Consequences*. Sunderland, MA, USA: Sinauer Associates Inc. [*23*]

Pacala SW & Tilman D (1994). Limiting similarity in mechanistic and spatial models of plant competition in heterogeneous environments. *The American Naturalist* **143**:222–257 [*306*]

Paetkau D, Shields GF & Strobeck C (1998). Gene flow between insular, coastal and interior populations of brown bears in Alaska. *Molecular Ecology* **7**:1283–1292 [*319*]

Page RDM & Hafner MS (1996). Molecular phylogenies and host–parasite cospeciation: Gophers and lice as a model system. In *New Uses for New Phylogenies*, eds. Harvey PH, Brown AJL, Smith JM & Nee S, pp. 255–270. Oxford, UK: Oxford University Press [*231*]

Pál C & Miklós I (1999). Epigenetic inheritance, genetic assimilation and speciation. *Journal of Theoretical Biology* **200**:19–37 [*121*]

Palumbi SR (1992). Marine speciation on a small planet. *Trends in Ecology and Evolution* **7**:114–118 [*72*]

Palumbi SR (1998). Species formation and the evolution of gamete recognition loci. In *Endless Forms: Species and Speciation*, eds. Howard DJ & Berlocher SH, pp. 271–278. Oxford, UK: Oxford University Press [*125*]

Panhuis T, Butlin R, Zuk M & Tregenza T (2001). Sexual selection and speciation. *Trends in Ecology and Evolution* **16**:364–371 [*99, 103*]

Papaj DR (1993). Automatic behavior and the evolution of instinct: Lessons from learning in parasitoids. In *Insect Learning: Ecology and Evolutionary Perspectives*, eds. Papaj DR & Lewis AC, pp. 243–272. London, UK: Chapman & Hall [*250*]

Papaj DR & Prokopy RJ (1986). Phytochemical basis of learning in *Rhagoletis pomonella* and other herbivorous insects. *Journal of Chemical Ecology* **12**:1125–1143 [*254*]

Papaj DR & Prokopy RJ (1988). The effect of prior adult experience on components of habitat preference in the apple maggot fly (*Rhagoletis pomonella*). *Oecologia* **76**:538–543 [*254*]

Papaj DR & Prokopy RJ (1989). Ecological and evolutionary aspects of learning in phytophagous insects. *Annual Review of Entomology* **34**:315–350 [*254–256*]

Parenti LR (1984). Biogeography of the Andean killifish genus *Orestias* with comments on the species flock concept. In *Evolution of Fish Species Flocks*, eds. Echelle AA & Kornfield I, pp. 85–92. Orono, ME, USA: University of Maine Press [*362*]

Parker MA (1985). Local population differentiation for compatibility in an annual legume and its host-specific fungal pathogen. *Evolution* **39**:713–723 [*272*]

Parker A & Kornfield I (1995). A molecular perspective on evaluation and zoogeography of cyprinodontid killifishes (Teleostei, Atherinomorpha). *Copeia* **1995**:8–21 [*362*]

Parker GA & Partridge L (1998). Sexual conflict and speciation. *Philosophical Transactions of the Royal Society of London B* **353**:261–274 [*102*]

Parvinen K (1999). Evolution of migration in a metapopulation. *Bulletin of Mathematical Biology* **61**:531–550 [*73*]

Paterniani E (1969). Selection for reproductive

isolation between two populations of maize, *Zea mays* L. *Evolution* **23**:534–547   [*265*]

Paterson HEH (1982). Perspective on speciation by reinforcement. *South African Journal of Science* **78**:53–57   [*340*]

Paterson HEH (1985). The recognition concept of species. In *Species and Speciation*, ed. Vrba ES, pp. 21–29, Transvaal Museum Monograph No. 4. Pretoria, South Africa: Transvaal Museum   [*10, 125*]

Payne RJH & Krakauer DC (1997). Sexual selection, space, and speciation. *Evolution* **51**:1–9   [*45, 72, 99, 166, 306*]

Peichel CL, Nereng KS, Ohgi KA, Cole BLE, Colosimo PF, Buerkle CA, Schluter D & Kingsley DM (2001). The genetic architecture of divergence between threespine stickleback species. *Nature* **414**:901–905   [*218, 228*]

Persons MH, Fleishman LJ, Frye MA & Stimphil ME (1999). Sensory response patterns and the evolution of visual signal design in anoline lizards. *Journal of Comparative Physiology A* **184**:585–607   [*342*]

Persson L & Greenberg LA (1990). Optimal foraging and habitat shifts in perch, *Perca fluviatilis*, in a resource gradient. *Ecology* **71**:1699–1713   [*226*]

Persson L, Leonardsson K, de Roos AM, Gyllenberg M & Christensen B (1998). Ontogenetic scaling of foraging rates and the dynamics of a size-structured consumer-resource model. *Theoretical Population Biology* **54**:270–293   [*226*]

Pfennig DW (1992). Polyphenism in spadefoot toad tadpoles as a locally adjusted evolutionarily stable strategy. *Evolution* **46**:1408–1420   [*212*]

Phelan PL & Baker TC (1987). Evolution of male pheromones in moths: Reproductive isolation through sexual selection? *Science* **235**:205–207   [*203*]

Phillips PA & Barnes MM (1975). Host race formation among sympatric apple, walnut, and plum populations of the codling moth *Laspeyresia pomonella*. *Annals of the Entomological Society of America* **68**:1053–1060   [*239–240*]

Phillips PC & Johnson NA (1998). The population genetics of synthetic lethals. *Genetics* **150**:449–458   [*125*]

Pierce NE (1987). The evolution and biogeography of associations between lycaenid butterflies and ants. In *Oxford Surveys in Evolutionary Biology*, eds. Harvey PH & Partridge L, pp. 89–116. Oxford, UK: Oxford University Press   [*239*]

Pimm ST (1979). Sympatric speciation: A simulation model. *Biological Journal of the Linnean Society* **11**:131–139   [*39, 42, 54, 70, 222*]

Pomiankowski A (1987). The costs of choice in sexual selection. *Journal of Theoretical Biology* **128**:195–218   [*99*]

Pomiankowski A, Iwasa Y & Nee S (1991). The evolution of costly mate preferences. I. Fisher and biased mutation. *Evolution* **45**:1422–1430   [*99*]

Power G (2002). Charrs, glaciations and seasonal ice. *Environmental Biology of Fishes* **64**:17–35   [*219*]

Pratt GF, Wright DM & Pavulaan H (1994). The various taxa and hosts of the North American *Celastrina* (Lepidoptera: Lycaenidae). *Proceedings of the Entomological Society of Washington* **96**:566–578   [*239*]

Price TD, Grant PR, Gibbs HL & Boag PT (1984). Recurrent patterns of natural selection in a population of Darwin's finches. *Nature* **309**:787–789   [*194*]

Price T, Turelli M & Slatkin M (1993). Peak shift by correlated response to selection. *Evolution* **47**:280–290   [*121*]

Prokopy RJ, Averill AL, Cooley SS & Roitberg CA (1982). Associative learning in egglaying site selection by apple maggot flies. *Science* **218**:76–77   [*254*]

Prokopy RJ, Diehl SR & Cooley SS (1988). Behavioral evidence for host races in *Rhagoletis pomonella* flies. *Oecologia* **76**:138–147   [*254*]

Prokopy RJ, Bergweiler C, Galarza L & Schwerin J (1994). Prior experience affects the visual ability of *Rhagoletis pomonella* flies (Diptera: Tephritidae) to find host fruit. *Journal of Insect Behavior* **7**:663–677 [*254*]

Prout T (1968). Sufficient conditions for multiple niche polymorphism. *The American Naturalist* **102**:493–496   [*39–40*]

Provine WB (1986). *Sewall Wright and Evolutionary Biology*. Chicago, IL, USA: University of Chicago Press   [*117*]

Provine WB (2001). *Origins of Theoretical Population Genetics*. Second Edition with new afterword. Chicago, IL, USA: University of Chicago Press [22]

Prowell DP (1998). Sex linkage and speciation in Lepidoptera. In *Endless Forms: Species and Speciation*, eds. Howard DJ & Berlocher SH, pp. 309–319. Oxford, UK: Oxford University Press [239]

Pulliam HR (1981). Learning to forage optimally. In *Foraging Behavior: Ecological, Ethological and Psychological Approaches*, eds. Kamil AC & Sargent TD, pp. 379–388. New York, NY, USA: Garland STPM Press [249, 256]

Qvarnström A (2001). Context-dependent genetic benefits from mate choice. *Trends in Ecology and Evolution* **16**:5–7 [251, 259]

Rainey PB & Travisano M (1998). Adaptive radiation in a heterogeneous environment. *Nature* **394**:69–72 [291]

Ramadevan S & Deakin MAB (1990). The Gibbons speciation mechanism. *Journal of Theoretical Biology* **145**:447–456 [241]

Ramírez CC & Niemeyer HM (2000). The influence of previous experience and starvation on aphid feeding behavior. *Journal of Insect Behavior* **13**:699–709 [254]

Rand AS & Williams EE (1970). An estimation of redundancy and information content of anole dewlaps. *The American Naturalist* **104**:99–103 [323, 339]

Rand DA & Wilson HB (1993). Evolutionary catastrophes, punctuated equilibria and gradualism in ecosystem evolution. *Proceedings of the Royal Society of London B* **253**:137–141 [73–74]

Ratcliffe LM & Grant PR (1983). Species recognition in Darwin's finches (*Geospiza, Gould*). I. Discrimination by morphological cues. *Animal Behaviour* **31**:1139–1153 [194]

Rausher MD (2001). Co-evolution and plant resistance to natural enemies. *Nature* **411**:857–864 [252]

Rausher MD & Fry JD (1993). Effects of a locus affecting floral pigmentation in *Ipomoea purpurea* on female fitness components. *Genetics* **134**:1237–1247 [274]

Reardon JT & Thorpe RS. Natural selection and common garden experiments on *Anolis oculatus*. Unpublished [328]

Reidys CM (1997). Random induced subgraphs of generalized n-cubes. *Advances in Applied Mathematics* **19**:360–377 [129]

Reidys CM, Stadler PF & Schuster P (1997). Generic properties of combinatory maps: Neutral networks of RNA secondary structures. *Bulletin of Mathematical Biology* **59**:339–397 [129]

Reznick DN, Shaw FH, Rodd FH & Shaw RG (1997). Evaluation of the rate of evolution in natural populations of guppies (*Poecilia reticulata*). *Science* **275**:1934–1937 [214]

Ribbink AJ (1994). Alternative perspectives on some controversial aspects of cichlid fish speciation. *Archiv für Hydrobiologie – Advances in Limnology* **44**:101–125 [362, 376]

Ribbink AJ, Marsh BA, Marsh AC, Ribbink AC & Sharp BJ (1983). A preliminary survey of the cichlid fishes of rocky habitats in Lake Malawi. *South African Journal of Zoology* **18**:149–309 [174]

Rice WR (1984a). Disruptive selection on habitat preference and the evolution of reproductive isolation: A simulation study. *Evolution* **38**:1251–1260 [49–50]

Rice WR (1984b). Sex chromosomes and the evolution of sexual dimorphism. *Evolution* **38**:735–742 [211]

Rice WR (1987). Speciation via habitat specialization: The evolution of reproductive isolation as a correlated character. *Evolutionary Ecology* **1**:301–314 [43, 49–50, 211, 230, 259, 305]

Rice WR (1998). Intergenomic conflict, interlocus antagonistic coevolution, and the evolution of reproductive isolation. In *Endless Forms: Species and Speciation*, eds. Howard DJ & Berlocher SH, pp. 261–270. Oxford, UK: Oxford University Press [277]

Rice WR & Hostert EE (1993). Laboratory experiments on speciation: What have we learned in 40 years? *Evolution* **47**:1637–1653 [114, 211, 214, 221–222, 265]

Rice WR & Salt GW (1988). Speciation via disruptive selection on habitat preference: Experimental evidence. *The American Naturalist* **131**:911–917 [194]

Rice WR & Salt GW (1990). The evolution of reproductive isolation as a correlated character under sympatric conditions: Experimental evidence. *Evolution* **44**:1140–1152 [50, 194]

Ricklefs RE & Renner SS (2000). Evolutionary flexibility and flowering plant familial diversity: A comment on Dodd, Silvertown, and Chase. *Evolution* **54**:1061–1065 [264]

Ridgway MS & McPhail JD (1984). Ecology and evolution of sympatric sticklebacks (*Gasterosteus*): Mate choice and reproductive isolation in the Enos Lake species pair. *Canadian Journal of Zoology* **62**:1813–1818 [*196*]

Ridley M (1993). *Evolution*. Boston, MA, USA: Blackwell Scientific Publications [*70, 112, 114, 117*]

Ridley M (1996). *Evolution*, Second Edition. Cambridge, MA, USA: Blackwell Science Inc. [*285*]

Rieseberg LH (1995). The role of hybridization in evolution: Old wine in new skins. *American Journal of Botany* **82**:944–953 [*265*]

Rieseberg LH (1997). Hybrid origins of plant species. *Annual Review of Ecology and Systematics* **27**:359–389 [*265*]

Rieseberg LH & Brouillet L (1994). Are many plant species paraphyletic? *Taxon* **43**:21–32 [*265*]

Ripley SD & Beehler BM (1990). Patterns of speciation in Indian birds. *Journal of Biogeography* **17**:639–648 [*115*]

Robinson BW & Dukas R (1999). The influence of phenotypic modifications on evolution: The Baldwin effect and modern perspectives. *Oikos* **85**:582–589 [*250*]

Robinson BW & Schluter D (2000). Natural selection and the evolution of adaptive genetic variation in northern freshwater fishes. In *Adaptive Genetic Variation in the Wild*, eds. Mousseau TA, Sinervo B & Endler JA, pp. 65–94. New York, NY, USA: Oxford University Press [*211, 215, 219*]

Robinson BW & Wilson DS (1994). Character release and displacement in fishes, a neglected literature. *The American Naturalist* **144**:596–627 [*210, 212, 214*]

Robinson BW & Wilson DS (1996). Genetic variation and phenotypic plasticity in a trophically polymorphic population of pumpkinseed sunfish (*Lepomis gibbosus*). *Evolutionary Ecology* **10**:631–652 [*217*]

Robinson BW, Wilson DS, Margosian SS & Lotito PT (1993). Ecological and morphological differentiation of pumpkinseed sunfish in lakes without bluegill sunfish. *Evolutionary Ecology* **7**:451–464 [*211–212*]

Roderick GK & Gillespie RG (1998). Speciation and phylogeography of Hawaiian terrestrial arthropods. *Molecular Ecology* **7**:519–531 [*232*]

Roessingh P, Hora KH, Loon JJA & Menken SBJ (1999). Evolution of gustatory sensitivity in *Yponomeuta* caterpillars: Sensitivity to the stereoisomers dulcitol and sorbitol is localised in a single sensory cell. *Journal of Comparative Physiology A* **184**:119–126 [*246*]

Rogers SM, Campbell D, Baird SJE, Danzmann RG & Bernatchez L (2001). Combining the analysis of introgressive hybridization and linkage mapping to investigate the genetic architecture of population divergence in the lake whitefish *Coregonus clupeaformis*, Mitchill. *Genetica* **111**:25–114 [*228*]

Rokas A (2000). *Wolbachia* as a speciation agent. *Trends in Ecology and Evolution* **15**:44–45 [*262*]

Rolán-Alvarez E, Johannesson K & Erlandsson J (1997). The maintenance of a cline in the marine snail *Littorina saxatilis*: The role of home site advantage and hybrid fitness. *Evolution* **51**:1838–1847 [*115*]

Rolán-Alvarez E, Erlandsson J, Johannesson K & Cruz R (1999). Mechanisms of incomplete prezygotic reproductive isolation in an intertidal snail: Testing behavioural models in wild populations. *Journal of Evolutionary Biology* **12**:879–890 [*46*]

Ronsheim ML (1997). Distance-dependent performance of asexual progeny in *Allium vineale* (Liliaceae). *American Journal of Botany* **84**:1279–1284 [*272*]

Rosenzweig ML (1978). Competitive speciation. *Biological Journal of the Linnean Society* **10**:275–289 [*9, 39, 42, 53, 81, 83–85, 87, 107, 222, 382*]

Rosenzweig ML (1995). *Species Diversity in Space and Time*. Cambridge, UK: Cambridge University Press [*112, 114, 162*]

Rosenzweig RF, Sharp RR, Treves DS & Adams J (1994). Microbial evolution in a simple unstructured environment: Genetic differentiation in *Escherichia coli*. *Genetics* **137**:903–917 [*290–291*]

Ross HH (1958). Evidence suggesting a hybrid origin of certain leafhopper species. *Evolution* **12**:337–446 [*238*]

Roughgarden J (1995). Anolis *Lizards of the Caribbean: Ecology, Evolution, and Plate Tectonics*. New York, NY, USA: Oxford University Press [*164, 322*]

Ruibal R & Williams EE (1961). Two sympatric Cuban anoles of the *carolinensis* group. *Bulletin of the Museum of Comparative Zoology* **125**:183–208 [*339*]

Ruiz MAM, Eiríksson GM & Skúlason S. Behavior and growth of two sympatric morphs of arctic charr incubated at different temperatures. Unpublished [*218*]

Rundle HD & Schluter D (1998). Reinforcement of stickleback mate preferences: Sympatry breeds contempt. *Evolution* **52**:200–208 [*71, 203, 206–208, 214*]

Rundle HD, Nagel L, Boughman JW & Schluter D (2000). Natural selection and parallel speciation in sympatric sticklebacks. *Science* **287**:306–308 [*201–202*]

Rundle HD, Breden F, Griswold C, Mooers AØ, Vos RA & Whitton J (2001). Hybridization without guilt: Gene flow and the biological species concept. *Journal of Evolutionary Biology* **14**:868–869 [*192*]

Ryan MJ & Rand AS (1990). The sensory basis of sexual selection for complex calls in the túngara frog, *Physalaemus pustulosus* (sexual selection for sensory exploitation). *Evolution* **44**:305–314 [*98*]

Saemundsson K (1992). Geology of the Thingvallavatn area. *Oikos* **64**:40–68 [*220–221*]

Sætre GP, Moum T, Bures S, Král M, Adamjan M & Moreno J (1997). A sexually selected character displacement in flycatchers reinforces premating isolation. *Nature* **387**:589–592 [*71, 203*]

Saloniemi I (1993). A coevolutionary predator–prey model with quantitative characters. *The American Naturalist* **141**:880–896 [*95*]

Schaal BA, Hayworth DA, Olsen KM, Rouscher JT & Smith WA (1998). Phytogeographic studies in plants: problems and prospects. *Molecular Ecology* **7**:465–474 [*277*]

Schemske DW (2000). Understanding the origin of species. *Evolution* **54**:1069–1073 [*24*]

Schemske DW & Bradshaw HD (1999). Pollinator preference and the evolution of floral traits in monkeyflowers (*Mimulus*). *Proceedings of the National Academy of Sciences of the USA* **96**:11910–11915 [*268*]

Schilthuizen M (2000). Ecotone: Speciation-prone. *Trends in Ecology and Evolution* **15**:130–131 [*73, 380*]

Schilthuizen M (2001). *Frogs, Flies and Dan-* *delions: The Making of Species*. Oxford, UK: Oxford University Press [*174, 381, 390*]

Schlichting CD & Pigliucci M (1998). *Phenotypic Evolution: A Reaction Norm Perspective*. Sunderland, MA, USA: Sinauer Associates Inc. [*217, 343*]

Schliewen UK, Tautz D & Pääbo S (1994). Sympatric speciation suggested by monophyly of crater lake cichlids. *Nature* **368**:629–632 [*191, 210, 362*]

Schliewen U, Rassmann K, Markmann M, Markert J, Kocher T & Tautz D (2001). Genetic and ecological divergence of a monophyletic cichlid species pair under fully sympatric conditions in Lake Ejagham, Cameroon. *Molecular Ecology* **10**:1471–1488 [*91, 312, 316*]

Schluter D (1993). Adaptive radiation in sticklebacks: Size, shape, and habitat use efficiency. *Ecology* **74**:699–709 [*197, 209*]

Schluter D (1994). Experimental evidence that competition promotes divergence in adaptive radiation. *Science* **266**:798–800 [*196, 198, 206*]

Schluter D (1995). Adaptive radiation in sticklebacks: Trade-offs in feeding performance and growth. *Ecology* **76**:82–90 [*196, 209*]

Schluter D (1996a). Ecological causes of adaptive radiation. *The American Naturalist* **148**:S40–S64 [*73, 192, 206, 208, 234, 278*]

Schluter D (1996b). Ecological speciation in postglacial fishes. *Philosophical Transactions of the Royal Society of London B* **351**:807–814 [*210, 219*]

Schluter D (1998). Ecological causes of speciation. In *Endless Forms: Species and Speciation*, eds. Howard DJ & Berlocher SH, pp. 114–129. Oxford, UK: Oxford University Press [*210*]

Schluter D (2000). *The Ecology of Adaptive Radiation*. Oxford, UK: Oxford University Press [*18, 23–25, 192–193, 208, 249, 341*]

Schluter D (2001). Ecology and the origin of species. *Trends in Ecology and Evolution* **16**:372–380 [*192–193, 208, 210, 214, 221*]

Schluter D & McPhail JD (1992). Ecological character displacement and speciation in sticklebacks. *The American Naturalist* **140**:85–108 [*196–198, 206*]

Schluter D & McPhail JD (1993). Character displacement and replicate adaptive radiation. *Trends in Ecology and Evolution*

8:197–200 [*197, 212*]

Schluter D & Nagel L (1995). Natural selection and parallel speciation. *The American Naturalist* **146**:292–301 [*73, 199, 201*]

Schmalhausen II (1949). *Factors of Evolution.* Philadelphia, PA, USA: Blakiston [*343*]

Schneider CJ, Smith TB, Larison B & Moritz C (1999). A test of alternative models of diversification in tropical rainforests: Ecological gradients vs. rainforest refugia. *Proceedings of the National Academy of Sciences of the USA* **96**:13869–13873 [*73*]

Schneider CJ, Losos JB & de Queiroz K (2001). Evolutionary relationships of the *Anolis bimaculatus* group from the northern Lesser Antilles. *Journal of Herpetology* **35**:1–12 [*336*]

Schoener TW (1969). Size patterns in West Indian *Anolis* lizard: I. Size and species diversity. *Systematic Zoology* **18**:386–401 [*334*]

Schoener TW (1970). Size patterns in West Indian *Anolis* lizard: II. Correlations with the sizes of particular sympatric species – displacement and convergence. *The American Naturalist* **104**:155–174 [*334*]

Schoener TW & Schoener A (1983). The time to extinction of a colonizing propagule of lizards increases with island area. *Nature* **302**:332–334 [*342*]

Schoonhoven LM, Jermy T & Van Loon JJA (1998). *Insect–Plant Biology – From Physiology to Evolution.* London, UK: Chapman & Hall [*257*]

Schuster P (1996). How does complexity arise in evolution? *Complexity* **2**:22–30 [*106*]

Schuster P, Fontana W, Stadler PF & Hofacker IL (1994). From sequences to shapes and back: A case study in RNA secondary structures. *Proceedings of the Royal Society of London B* **255**:279–284 [*128*]

Schwartz A & Henderson RW (1991). *Amphibians and Reptiles of the West Indies: Descriptions, Distributions, and Natural History.* Gainsville, FL, USA: University of Florida Press [*331*]

Secord D & Kareiva P (1996). Perils and pitfalls in the host specificity paradigm. *BioScience* **46**:448–453 [*237*]

Seegers L, Sonnenberg R & Yamamoto R (1999). Molecular analysis of the Alcolapia flock from Lakes Natron and Magadi, Tanzania and Kenya (Teleostei: Cichlidae), and implications for their systematics and evolution. *Ichthyological Exploration in Freshwater* **10**:175–200 [*362*]

Seehausen O (1996). *Lake Victoria Rock Cichlids: Taxonomy, Ecology and Distribution.* Zevenhuizen, Netherlands: Verduijn Cichlids [*174*]

Seehausen O (1997). Distribution of, and reproductive isolation among, colour morphs of a rock-dwelling Lake Victoria cichlid (*Haplochromus nyererei*). *Ecology of Freshwater Fish* **6**:59–66 [*177*]

Seehausen O (2000). Explosive speciation rates and unusual species richness in haplochromine cichlid fishes: Effects of sexual selection. *Advances in Ecological Research* **31**:237–274 [*176, 178*]

Seehausen O & Bouton N (1997). Microdistribution and fluctuations in niche overlap in a rocky shore cichlid community in Lake Victoria. *Ecology of Freshwater Fish* **6**:161–173 [*177, 191*]

Seehausen O & van Alphen JJM (1998). The effect of male coloration on female mate choice in closely related Lake Victoria cichlids (*H. nyererei* complex). *Behavioural Ecology and Sociobiology* **42**:1–8 [*176–177*]

Seehausen O & van Alphen JJM (1999). Can sympatric speciation by disruptive sexual selection explain rapid evolution of cichlid diversity in Lake Victoria? *Ecology Letters* **2**:262–271 [*174, 186*]

Seehausen O & van Alphen JJM (2000). Reply: Inferring modes of speciation from distribution patterns. *Ecology Letters* **3**:169–171 [*174, 186*]

Seehausen O, van Alphen JJM & Witte F (1997). Cichlid fish diversity threatened by eutrophication that curbs sexual selection. *Science* **277**:1808–1811 [*72, 176–177, 185*]

Seehausen O, Witte F, van Alphen JJM & Bouton N (1998). Direct mate choice maintains diversity among sympatric cichlids in Lake Victoria. *Journal of Fish Biology* **53**(Suppl. A):37–55 [*174, 185, 214*]

Seehausen O, van Alphen JJM & Lande R (1999). Color polymorphism and sex-ratio distortion in a cichlid fish as a transient stage in sympatric speciation by sexual selection. *Ecology Letters* **2**:367–378 [*179, 183–184*]

Seger J (1985a). Intraspecific resource competition as a cause of sympatric speciation. In *Evolution: Essays in Honour of John May-*

nard Smith, eds. Greenwood PJ, Harvey PH & Slatkin M, pp. 43–53. Cambridge, UK: Cambridge University Press [*84–85, 109, 222, 230, 382*]

Seger J (1985b). Unifying genetic models for the evolution of female choice. *Evolution* **39**:1185–1193 [*99*]

Servedio MR (2000). Reinforcement and the genetics of nonrandom mating. *Evolution* **54**:21–29 [*159, 246*]

Servedio MR (2001). Beyond reinforcement: The evolution of premating isolation by direct selection on preferences and postmating, zygotic incompatibilities. *Evolution* **55**:1909–1920 [*208*]

Sezer M & Butlin RK (1998a). The genetic basis of host plant adaptation in the brown planthopper (*Nilaparvata lugens*). *Heredity* **80**:499–508 [*238, 246*]

Sezer M & Butlin RK (1998b). The genetic basis of oviposition preference differences between sympatric host races of the brown planthopper (*Nilaparvata lugens*). *Proceedings of the Royal Society of London B* **265**:2399–2405 [*238, 244, 246*]

Shaw KL (2002). Conflict between nuclear and mitochondrial DNA phylogenies of a recent species radiation: What mtDNA reveals and conceals about modes of speciation in Hawaiian crickets. *Proceedings of the National Academy of Sciences of the USA* **99**:16122–16127 [*393*]

Shaw D, Arnold M, Marchant A & Contreras N (1988). Chromosomal rearrangements, ribosomal genes and mitochondrial DNA: Contrasting patterns of introgression across a narrow hybrid zone. In *Kew Chromosome Conference III*, ed. Brandham PE, pp. 121–129. London, UK: HMSO [*241*]

Shaw PW, Turner GF, Idid MR, Robinson RL & Carvalho GR (2000). Genetic population structure indicates sympatric speciation of Lake Malawi pelagic cichlids. *Proceedings of the Royal Society of London B* **267**:2273–2280 [*174*]

Shields GF, Adams D, Garner G, Labelle M, Pietsch J, Ramsay M, Schwartz C, Titus K & Williamson S (2000). Phylogeography of mitochondrial DNA variation in brown bears and polar bears. *Molecular Phylogenetics and Evolution* **15**:319–326 [*319*]

Shochat D & Dessauer HC (1981). Comparative immunological study of albumins of *Anolis* lizards of the Caribbean islands. *Comparative Biochemistry and Physiology* **68A**:67–73 [*323*]

Shoemaker DD & Ross KG (1996). Effects of social organization on gene flow in the fire ant *Solenopsis invicta*. *Nature* **383**:613–616 [*241*]

Shoemaker DD, Katju V & Jaenike J (1999). *Wolbachia* and the evolution of reproductive isolation between *Drosophila recens* and *Drosophila subquinaria*. *Evolution* **53**:1157–1164 [*262*]

Shyue SK, Hewett-Emmett D, Sperling HG, Hunt DM, Bowmaker JK, Mollon JD & Li WH (1995). Adaptive evolution of color vision genes in higher primates. *Science* **269**:1265–1267 [*298*]

Sigurjónsdóttir H & Gunnarsson K (1989). Alternative mating tactics of arctic charr, *Salvelinus alpinus*, in Thingvallavatn, Iceland. *Environmental Biology of Fishes* **26**:159–176 [*214*]

Silverstein JT & Hershberger WK (1992). Precocious maturation in coho salmon *Oncorhynchus kisutch*: Estimation of the additive genetic variance. *Journal of Heredity* **83**:283–286 [*217*]

Simpson G (1953). *The Major Features of Evolution*. New York, NY, USA: Columbia University Press [*135*]

Singer MC, Ng D, Vasco D & Thomas CD (1992). Rapidly evolving associations among oviposition preferences fail to constrain evolution of insect diet. *The American Naturalist* **139**:9–20 [*233*]

Skúlason S & Smith TB (1995). Resource polymorphisms in vertebrates. *Trends in Ecology and Evolution* **10**:366–370 [*210–212, 214–215, 218*]

Skúlason S, Snorrason SS, Noakes DLG, Ferguson MM & Malmquist HJ (1989a). Segregation in spawning and early life history among polymorphic Arctic charr, *Salvelinus alpinus*, in Thingvallavatn, Iceland. *Journal of Fishery and Biology* **35**:225–232 [*212–213, 221*]

Skúlason S, Noakes DLG & Snorrason SS (1989b). Ontogeny of trophic morphology in four sympatric morphs of arctic charr *Salvelinus alpinus* in Thingvallavatn, Iceland. *Biological Journal of the Linnean Society* **38**:281–301 [*217–218, 221*]

Skúlason S, Antonsson T, Gudbergsson G,

Malmquist HJ & Snorrason SS (1992). Variability in Icelandic Arctic charr. *Iceland Agricultural Sciences* **6**:142–153   [*221*]

Skúlason S, Snorrason SS, Ota D & Noakes DLG (1993). Genetically based differences in foraging behaviour among sympatric morphs of arctic charr (Pisces; Salmonidae). *Animal Behavior* **45**:1179–1192   [*216–218, 221*]

Skúlason S, Snorrason SS, Noakes DLG & Ferguson MM (1996). Genetic basis of life history variations among sympatric morphs of Arctic charr *Salvelinus alpinus*. *Canadian Journal of Fisheries and Aquatic Sciences* **53**:1807–1813   [*217–218, 220–221*]

Skúlason S, Snorrason SS & Jonsson B (1999). Sympatric morphs, populations and speciation in freshwater fish with emphasis on arctic charr. In *Evolution of Biological Diversity*, eds. Magurran AE & May RM, pp. 70–92. Oxford, UK: Oxford University Press   [*210–212, 219–220, 222, 225*]

Slatkin M (1978). Spatial patterns in the distribution of polygenic characters. *Journal of Theoretical Biology* **70**:213–228   [*141*]

Slatkin M (1980). Ecological character displacement. *Ecology* **61**:163–177   [*57*]

Slatkin M (1981). Fixation probabilities and fixation times in a subdivided population. *Evolution* **35**:477–488   [*122, 126*]

Slatkin M (1982). Pleiotropy and parapatric speciation. *Evolution* **36**:263–270   [*36*]

Slobodchikoff CN, ed. (1976). *Concepts of Species*. Stroudburg, PA, USA: Dowden, Hutchinson, and Ross   [*23*]

Smith TB (1993). Disruptive selection and the genetic basis of bill size polymorphism in the African finch, *Pyrenestes*. *Nature* **363**:618–620   [*212*]

Smith TB & Skúlason S (1996). Evolutionary significance of resource polymorphisms in fishes, amphibians, and birds. *Annual Review of Ecology and Systematics* **27**:111–133   [*210–212, 215, 219, 221*]

Smith GR & Todd TN (1984). Evolution of fish species flocks in north temperate lakes. In *Evolution of Fish Species Flocks*, eds. Echelle AA & Kornfield I, pp. 47–68. Orono, ME, USA: University of Maine Press   [*210–211, 213–214*]

Smouse PE, Long J & Sokal RR (1986). Multiple regression and correlation extensions of the Mantel test of matrix correspondence.

Systematic Zoology **35**:627–632   [*327*]

Snorrason SS, Jónasson PM, Jonsson B, Lindem T, Malmquist HJ, Sandlund OT & Skúlason S (1992). Population dynamics of the planktivorous arctic charr *Salvelinus alpinus* ("murta") in Thingvallavatn. *Oikos* **64**:352–364   [*218*]

Snorrason SS, Skúlason S, Jonsson B, Malmquist HJ, Jónasson PM, Sandlund OT & Lindem T (1994a). Trophic specialization in Arctic charr *Salvelinus alpinus* (Pisces: Salmonidae): Morphological divergence and ontogenetic niche shifts. *Biological Journal of the Linnean Society* **52**:1–18   [*210, 212, 215–216, 218–221, 226*]

Snorrason SS, Malmquist HJ, Sandlund OT, Skúlason S, Jonsson B & Jónasson PM (1994b). Modifications in life-history characteristics of planktivorous arctic charr *Salvelinus alpinus* (L.) in Thingvallavatn, Iceland. *Proceedings of the International Association of Theoretical and Applied Limnology* **25**:2108–2112   [*218*]

Sokal RR & FJ Rohlf (1981). *Biometry*, Second Edition. San Francisco, CA, USA: Freeman   [*283*]

Solignac M & Monnerot M (1986). Race formation, speciation, and introgression within *Drosophila simulans*, *D. mauritiana*, and *D. sechellia* inferred from mitochondrial DNA analysis. *Evolution* **20**:531–539   [*235*]

Soltis DE & Soltis PS (1993). Molecular data and the dynamic nature of polyploidy. *Critical Reviews in Plant Sciences* **12**:243–275   [*265*]

Soreghan MJ, Scholtz CA & Wells JT (1999). Coarse-grained deep water sedimentation along a border fault margin of Lake Malawi, Africa: Seismic stratigraphic analysis. *Journal of Sedimentary Research* **69**:832–846   [*377*]

Stanford EH, Laude MM & Booysen P de V (1962). Effects of advance in generation under different harvesting regimes on the genetic composition of Pilgrim Ladino clover. *Crop Science* **2**:497–500   [*272*]

Stanhope MJ, Leighton BJ & Hartwick B (1992). Polygenic control of habitat preference and possible role in sympatric population subdivision in an estuarine crustacean. *Heredity* **69**:279–288   [*236*]

Stanhope MJ, Hartwick B & Baillie D (1993). Molecular phylogenetic evidence for multiple shifts in habitat preference in the diversification of an amphipod species. *Molecular Ecology* **2**:99–112   [*236*]

Stanton ML, Rejmanek M & Galen C (1994). Changes in vegetation and soil fertility along a predictable snowmelt gradient in the Mosquito Range, Colorado, USA. *Arctic and Alpine Research* **26**:364–374   [*273*]

Stearns SC (1992). *The Evolution of Life History Strategies*. Oxford, UK: Oxford University Press   [*315*]

Stearns SC & Crandall RE (1984). Plasticity for age and size at sexual maturity: A life-history response to unavoidable stress. In *Fish Reproduction: Strategies and Tactics*, eds. Potts GW & Wootton RJ, pp. 14–33. London, UK: Academic Press   [*217*]

Steinberg MS (1996). Adhesion in development: An historical overview. *Developmental Biology* **180**:377–388   [*311*]

Steinfartz S, Veith M & Tautz D (2000). Mitochondrial sequence analysis of *Salamandra* taxa suggests old splits of major lineages and postglacial recolonizations of Central Europe from distinct source populations of *Salamandra salamandra*. *Molecular Ecology* **9**:397–410   [*317–318*]

Stenson AG (2000). *Use of Molecular Markers at Different Taxonomic Levels: Evolution of the Northern Lesser Antillean Anole Radiation*. PhD thesis. Bangor, UK: University of Wales   [*335*]

Stenson AG, Malhotra A & Thorpe RS (2000). Highly polymorphic microsatellite loci from the Dominican anole (*Anolis oculatus*) and their amplification in other bimaculatus series anoles. *Molecular Ecology* **9**:1680–1681   [*331*]

Stenson AG, Malhotra A & Thorpe RS (2002). Population differentiation and gene flow in a species displaying pronounced geographic variation in morphology, the Dominican anole (*Anolis oculatus*). *Molecular Ecology* **11**:1679–1688   [*331, 333, 335*]

Stephens DW & Krebs JR (1986). *Foraging Theory*. Princeton, NJ, USA: Princeton University Press   [*249*]

Stewart FM & Levin BR (1973). Partitioning of resources and the outcome of interspecific competition: A model and some general considerations. *The American Naturalist* **107**:171–198   [*282, 289*]

Stiassny MLJ & Jensen JS (1987). Labroid

interrelationships revisited: Morphological complexity, key innovations, and the study of comparative diversity. *Bulletin of the Museum of Comparative Zoology* **151**:269–319   [*187–188*]

Stouthamer R, Breeuwer JAJ & Hurst GDD (1999). *Wolbachia pipientis*: Microbial manipulator of arthropod reproduction. *Annual Review of Microbiology* **53**:71–102   [*252, 259–260*]

Strauss RE (1990). Heterochronic variation in the developmental timing of cranial ossifications in poecilid fishes (Cyprinodontiforums). *Evolution* **44**:1558–1567   [*218*]

Strecker U, Meyer CG, Sturmbauer C & Wilkens H (1996). Genetic divergence and speciation in an extremely young species flock in Mexico formed by the genus *Cyprinodon* (Cyprinodontidae, Teleostei). *Molecular Phylogenetics and Evolution* **6**:143–149   [*362*]

Strong DR, Lawton JH & Southwood TRE (1984). *Insects on Plants: Community Patterns and Mechanisms*. Oxford, UK: Blackwell Scientific Publications   [*233, 236, 252*]

Sturmbauer C, Verheyen HE & Meyer A (1994). Mitochondrial phylogeny of the Lamprologini, the major substrate spawning lineage of cichlid fishes from Lake Tanganyika in Eastern Africa. *Molecular Biology and Evolution* **11**:691–703   [*371*]

Sutherland SD & Vickery RK Jr (1993). On the relative importance of floral color, shape, and nectar rewards in attracting pollinators to *Mimulus*. *Great Basin Naturalist* **53**:107–117   [*268*]

Sutterlin AM & Maclean D (1984). Age at first maturity and the early expression of oocyte recruitment processes in two forms of Atlantic salmon (*Salmo salar*) and their hybrids. *Canadian Journal of Fisheries and Aquatic Sciences* **41**:1139–1149   [*217*]

Svärdson G (1979). *Speciation of Scandinavian Coregonus*. Institute of Freshwater Research, Report No. 57. Lund, Sweden: Bloms Boktrykkeri AB   [*214*]

Svedäng H (1990). Genetic basis of life-history variation of dwarf and normal Arctic charr, *Salvelinus alpinus* (L.) in Stora Rösjön, central Sweden. *Journal of Fish Biology* **36**:917–932   [*217, 221*]

Szentesi A & Jermy T (1990). The role of experience in host plant choice by phytophagous insects. In *Insect–Plant Interactions*, Vol. II, ed. Bernays EA, pp. 39–74. Boca Raton, FL, USA: CRC Press [*255, 257*]

Taberlet P, Fumagalli L, Wust-Saucy AG & Cosson JF (1998). Comparative phylogeography and postglacial colonization routes in Europe. *Molecular Ecology* **7**:453–464 [*319*]

Takimoto G, Higashi M & Yamamura N (2000). A deterministic model for sympatric speciation by sexual selection. *Evolution* **54**:1870–1881 [*45, 100*]

Taper ML & Case TJ (1985). Quantitative genetic models for the coevolution of character displacement. *Ecology* **66**:355–371 [*57*]

Taper ML & Case TJ (1992). Models of character displacement and the theoretical robustness of taxon cycles. *Evolution* **46**:317–333 [*60, 72*]

Tauber CA & Tauber MJ (1989). Sympatric speciation in insects: Perception and perspective. In *Speciation and Its Consequences*, eds. Otte D & Endler JA, pp. 307–344. Sunderland, MA, USA: Sinauer Associates Inc. [*49*]

Tautz D (2003). Evolutionary biology: Splitting in space. *Nature* **421**:225–226 [*166, 335*]

Tautz D & Schmid KJ (1998). From genes to individuals: Developmental genes and the generation of the phenotype. *Philosophical Transactions of the Royal Society of London B* **353**:231–240 [*314*]

Taylor P (1989). Evolutionary stability in one-parameter models under weak selection. *Theoretical Population Biology* **36**:125–143 [*55, 107, 382*]

Taylor EB & Bentzen P (1993). Evidence for multiple origins and sympatric divergence of trophic ecotypes of smelt (*Osmerus*) in northeastern North America. *Evolution* **47**:813–832 [*215–216, 219*]

Taylor EB & McPhail JD (1999). Evolutionary history of an adaptive radiation in species pairs of threespine sticklebacks (*Gasterosteus*): Insights from mitochondrial DNA. *Biological Journal of the Linnean Society* **66**:271–291 [*197, 199–200, 208*]

Taylor EB & McPhail JD (2000). Historical contingency and ecological determinism interact to prime speciation in sticklebacks, *Gasterosteus. Proceedings of the Royal Society of London B* **267**:2375–2384 [*197, 199–200, 208*]

Taylor EB, McPhail JD & Schluter D (1997). History of ecological selection in sticklebacks: Uniting experimental and phylogenetic approaches. In *Molecular Evolution and Adaptive Radiation*, eds. Givinish TJ & Sytsma KJ, pp. 511–534. Cambridge, UK: Cambridge University Press [*197–200*]

Templeton AR (1980). The theory of speciation via the founder principle. *Genetics* **94**:1011–1038 [*119*]

Templeton AR (1989). The meaning of species and speciation: A genetic perspective. In *Speciation and Its Consequences*, eds. Otte D & Endler JA, pp. 3–27. Sunderland, MA, USA: Sinauer Associates Inc. [*10*]

Templeton AR (1996). Experimental evidence for the genetic transilience model of speciation. *Evolution* **50**:909–915 [*114*]

Ten Cate C (2000). How learning mechanisms might affect evolutionary processes. *Trends in Ecology and Evolution* **15**:179–181 [*256*]

Ten Cate C & Vos DR (1999). Sexual imprinting and evolutionary processes in birds: A reassessment. *Advances in the Study of Behavior* **28**:1–31 [*256*]

Théron A & Combes C (1995). Asynchrony of infection timing, habitat preference, and sympatric speciation of schistosome parasites. *Evolution* **49**:372–375 [*49*]

Thomas F, Oget E, Gente P, Desmots D & Renaud F (1999). Assortative pairing with respect to parasite load in the beetle *Timarcha maritima* (Chrysomelidae). *Journal of Evolutionary Biology* **12**:385–390 [*46*]

Thompson V (1986). Synthetic lethals: A critical review. *Evolutionary Theory* **8**:1–13 [*125*]

Thorpe WH (1945). The evolutionary significance of habitat selection. *Journal of Animal Ecology* **14**:67–70 [*257*]

Thorpe JE (1989). Developmental variation in salmonid populations. *Journal of Fish Biology* **35**(Suppl. A):295–303 [*217*]

Thorpe RS (1991). Clines and cause: Microgeographic variation in the Tenerife gecko *Tarentola delalandii. Systematic Zoology* **40**:172–187 [*327*]

Thorpe RS (1996). The use of DNA divergence to help determine the correlates of evolution of morphological characters. *Evolution* **50**:524–531 [*327*]

Thorpe RS (2002). Analysis of color spectra in comparative evolutionary studies: Molecular phylogeny and habitat adaptation in the St Vincent Anole, *Anolis trinitatis*. *Systematic Biology* **51**:554–569 [*344*]

Thorpe RS & Baez M (1993). Geographic variation in scalation of the lizard *Gallotia stehlini* within the island of Gran Canaria. *Biological Journal of the Linnean Society* **48**:75–87 [*327*]

Thorpe RS & Malhotra A (1992). Are *Anolis* lizards evolving? *Nature* **355**:506 [*329*]

Thorpe RS & Malhotra A (1996). Molecular and morphological evolution within small islands. *Philosophical Transactions of the Royal Society of London B* **351**:815–822 [*324, 327, 329–330*]

Thorpe RS & Richard M (2001). Evidence that ultraviolet markings are associated with patterns of molecular gene flow. *Proceedings of the National Academy of Sciences of the USA* **98**:3929–3934 [*166, 335*]

Thorpe RS & Stenson AG (2003). Phylogeny, paraphyly and ecological adaptation of the colour and pattern in the *Anolis roquet* complex on Martinique. *Molecular Ecology* **12**:117–132 [*325, 333–335, 344*]

Thorpe JE, Morgan RIG, Talbot C & Miles MS (1983). Inheritance of developmental rates in Atlantic salmon, *Salmo salar* L. *Aquaculture* **33**:119–128 [*217*]

Thorpe JE, Metcalfe NB & Huntingford FA (1992). Behavioral influences on life-history variation in juvenile Atlantic salmon, *Salmo salar*. *Environmental Biology and Fishery* **33**:331–340 [*220*]

Thorpe RS, McGregor DP, Cumming AM & Jordan WC (1994). DNA evolution and colonization sequence of island lizards in relation to geological history: mtDNA RFLP, cytochrome b, cytochrome oxidase, 12s rRNA sequence, and nuclear RAPD analysis. *Evolution* **48**:230–240 [*331–332*]

Thorpe RS, Malhotra A, Black H, Daltry JC & Wüster W (1995). Relating geographic patterns to phylogenetic processes. *Philosophical Transactions of the Royal Society of London B* **349**:61–68 [*327*]

Thorpe RS, Black H & Malhotra A (1996). Matrix correspondence tests on the DNA phylogeny of the Tenerife lacertid elucidates both historical causes and morphological adaptation. *Systematic Biology* **45**:335–

343 [*327*]

Ting CT, Tsaur SC, Wu ML & Wu CI (1998). A rapidly evolving homeobox at the site of a hybrid sterility gene. *Science* **282**:1501–1504 [*392*]

Travis J (1995). Evaluating the adaptive role of morphological plasticity. In *Ecological Morphology: Integrative Organismal Biology*, eds. Wainwright PC & Reilly SM, pp. 99–122. Chicago, IL, USA: University of Chicago Press [*217*]

Travisano M (1993). *Adaptation and Divergence in Experimental Populations of the Bacterium* Escherichia coli: *The Roles of Environment, Phylogeny and Chance*. PhD thesis. East Lansing, MI, USA: Michigan State University [*285*]

Travisano M (1997). Long-term experimental evolution in *Escherichia coli*. VI. Environmental constraints on adaptation and divergence. *Genetics* **146**:471–479 [*288*]

Travisano M & Lenski RE (1996). Long-term experimental evolution in *Escherichia coli*. IV. Targets of selection and the specificity of adaptation. *Genetics* **143**:15–26 [*280–282, 284–285*]

Travisano M & Rainey PB (2000). Studies of adaptive radiation using model microbial systems. *The American Naturalist* **156**:S35–S44 [*280, 296–297*]

Travisano M, Vasi F & Lenski RE (1995a). Long-term experimental evolution in *Escherichia coli*. III. Variation among replicate populations in correlated responses to novel environments. *Evolution* **49**:189–200 [*280, 282–283*]

Travisano M, Mongold JA, Bennett AF & Lenski RE (1995b). Experimental tests of the roles of adaptation, chance and history in evolution. *Science* **267**:87–90 [*286*]

Tregenza T & Butlin RK (1999). Speciation without isolation. *Nature* **400**:311–312 [*107, 362*]

Tregenza T & Wedell N (2000). Genetic compatibility, mate choice and patterns of parentage. *Molecular Ecology* **9**:1013–1027 [*125*]

Tremblay E & Pennacchio F (1988). Speciation in Aphidiine Hymenoptera. In *Advances in Parasitic Hymenoptera Research*, ed. Gupta V, pp. 139–146. Leiden, Netherlands: EJ Brill [*241*]

Treves DS, Manning S & Adams J (1998). Repeated evolution of an acetate-crossfeeding polymorphism in long-term populations of *Escherichia coli*. *Molecular Biology and Evolution* **15**:789–797 [*290*]

Tsagkarakou A, Navajas M, Papaioannou-Souliotis P & Pasteur N (1998). Gene flow among *Tetranychus urticae* (Acari: Tetranychidae) populations in Greece. *Molecular Ecology* **7**:71–79 [*253*]

Turelli M & Orr HA (2000). Dominance, epistasis and the genetics of postzygotic isolation. *Genetics* **154**:1663–1679 [*124–125*]

Turelli M, Barton NH & Coyne JA (2001). Theory and speciation. *Trends in Ecology and Evolution* **16**:330–343 [*32, 110, 140, 142, 165, 214*]

Turkington R & Harper JL (1979). The growth, distribution, and neighbor relationships of *Trifolium repens* in a permanent pasture. *Journal of Ecology* **67**:245–254 [*272*]

Turner JRGT (1981). Adaptation and evolution in *Heliconius*: A defense of neo-Darwinism. *Annual Review of Ecology and Systematics* **12**:99–122 [*243*]

Turner GF (1994). Speciation mechanisms in Lake Malawi cichlids: A critical review. *Archiv für Hydrobiologie – Advances in Limnology* **44**:139–160 [*362*]

Turner GF (1996). *Offshore Cichlids of Lake Malawi*. Lauenau, Germany: Cichlid Press [*174*]

Turner GF & Burrows MT (1995). A model of sympatric speciation by sexual selection. *Proceedings of the Royal Society of London B* **260**:287–292 [*45, 72, 100, 305, 311*]

Udovic D (1980). Frequency-dependent selection, disruptive selection, and the evolution of reproductive isolation. *The American Naturalist* **116**:621–641 [*37, 42–43, 47, 49, 52, 67, 71, 73, 84–85, 244*]

Underwood G (1959). The anoles of the eastern Caribbean. Part III. Revisionary notes. *Bulletin of the Museum of Comparative Zoology, Harvard* **121**:191–226 [*324*]

Vacquier VD (1998). Evolution of gamete recognition proteins. *Science* **281**:1995–1998 [*125*]

Vala F, Breeuwer JAJ & Sabelis MW (2000). *Wolbachia*-induced "hybrid breakdown" in the two-spotted spider mite *Tetranychus urticae* Koch. *Proceedings of the Royal Society of London B* **267**:1931–1937 [*253*]

Vala F, Egas M, Breeuwer JAJ & Sabelis MW.

Wolbachia affects mating and oviposition behavior of its spider mite host. *Journal of Evolutionary Biology*. In press [*257, 262–263*]

Vamosi SM & Schluter D (1999). Sexual selection against hybrids between sympatric sticklebacks: Evidence from a field experiment. *Evolution* **53**:874–880 [*196*]

Van Berkum FH (1986). Evolutionary patterns of the thermal sensitivity of spring speed in *Anolis* lizards. *Evolution* **40**:495–504 [*338*]

Van der Laan JD & Hogeweg P (1995). Predator–prey coevolution: Interactions across different time scales. *Proceedings of the Royal Society of London B* **259**:35–42 [*73–74*]

Vandermeer JH & Boucher DH (1978). Varieties of mutualistic interactions in population models. *Journal of Theoretical Biology* **74**:549–558 [*94*]

Vandewalle PA, Havard M, Claes G & De Vree F (1992). Mouvements des mâchoires pharyngiennes pendant la prise de nourriture chez le *Serranus scriba* (Linné, 1758) Pisces, Serranidae. *Canadian Journal of Zoology* **70**:145–160 [*187*]

Van Dooren TJM (1999). The evolutionary ecology of dominance–recessivity. *Journal of Theoretical Biology* **198**:519–532 [*67, 391*]

Van Doorn GS & Weissing FJ (2001). Ecological versus sexual selection models of sympatric speciation. *Selection* **2**:17–40 [*72, 101, 390–391*]

Van Doorn GS, Luttikhuizen PC & Weissing FJ (2001). Sexual selection at the protein level drives the extraordinary divergence of sex related genes during sympatric speciation. *Proceedings of the Royal Society of London B* **268**:2155–2161 [*101, 390*]

Van Doorn GS, Dieckmann U & Weissing FJ. Sympatric speciation by sexual selection: A critical re-evaluation. Unpublished [*101–102, 109*]

Van Oppen MJH, Turner GF & Rico C (1997). Unusually fine-scale genetic structuring found in rapidly speciating Malawi cichlid fishes. *Proceedings of the Royal Society of London B* **264**:1803–1812 [*174, 185*]

Van Tienderen PH & de Jong G (1986). Sex ratio under the haystack model: Polymorphism may occur. *Journal of Theoretical Biology* **122**:69–81 [*58*]

Van Valen L (1976). Ecological species, multi-species, and oaks. *Taxon* **25**:233–239 [9]

Vanzolini PE & Williams EE (1981). The vanishing refuge: A mechanisms for ecogeographic speciation. *Papeis Avulsos Zoologici* **34**:251–255 [342]

Vasi F, Travisano M & Lenski RE (1994). Long-term experimental evolution in *Escherichia coli*. II. Changes in life-history traits during adaptation to a seasonal environment. *The American Naturalist* **144**:432–456 [280, 282]

Verdyck P (1998). Genetic differentiation and speciation among four *Phyllotreta* species (Coleoptera: Chrysomelidae). *Biological Journal of the Linnean Society* **64**:463–476 [240]

Vermeij G (1974). Adaptation, versatility and evolution. *Systematic Zoology* **22**:466–477 [187]

Via S (1990). Ecological genetics and host adaptation in herbivorous insects: The experimental study of evolution in natural and agricultural systems. *Annual Review of Entomology* **35**:421–426 [245–246]

Via S (1991). Specialized host plant performance of pea aphid clones is not altered by experience. *Ecology* **72**:1420–1427 [254]

Via S (1999). Reproductive isolation between sympatric races of pea aphids. I. Gene flow restriction and habitat choice. *Evolution* **53**:1446–1457 [49, 242, 249, 254, 257]

Via S (2001). Sympatric speciation in animals: The ugly duckling grows up. *Trends in Ecology and Evolution* **16**:381–390 [221, 243]

Via S (2002). The ecological genetics of speciation. *The American Naturalist* **159**:S1–S7 [243, 245]

Via S & Hawthorne DJ (2002). Correlated gene effects on host use and habitat choice in pea aphids. *The American Naturalist* **159**:S76–S88 [228, 246]

Via S, Bouck AC & Skillman S (2000). Reproductive isolation between divergent races of pea aphids on two hosts. II. Selection against migrants and hybrids in the parental environment. *Evolution* **54**:1626–1637 [254, 256–257]

Vickery RK Jr (1978). Case studies in the evolution of species complexes in *Mimulus*. *Evolutionary Biology* **11**:405–507 [273]

Vickery RK Jr (1992). Pollinator preferences for yellow, orange, and red flowers of *Mimulus verbenaceus* and *M. cardinalis*. *Great Basin Naturalist* **52**:145–148 [268]

Vickery RK Jr (1995). Speciation in *Mimulus*, or, can a simple flower color mutant lead to species divergence? *Great Basin Naturalist* **55**:177–180 [268]

Vincent TL, Cohen Y & Brown JS (1993). Evolution via strategy dynamics. *Theoretical Population Biology* **44**:149–176 [57, 60, 75]

Volpe JP & Ferguson MM (1996). Molecular genetic examination of the polymorphic Arctic charr, *Salvelinus alpinus*, of Thingvallavatn, Iceland. *Molecular Ecology* **5**:763–772 [221]

Vouidibio J, Capy P, Defaye D, Pla E, Sandrin J, Csink A & David JR (1989). Short-range genetic structure of *Drosophila melanogaster* populations in an Afrotropical urban area and its significance. *Proceedings of the National Academy of Sciences of the USA* **86**:8442–8446 [236]

Vrijenhoek RC, Marteinsdóttir G & Schenck R (1987). Genotypic and phenotypic aspects of niche diversification in fishes. In *Community and Evolutionary Ecology of North American Stream Fishes*, eds. Matthews W & Heins D, pp. 245–250. Norman, OK, USA: University of Oklahoma Press [217, 223]

Wada H, Shimada A, Fukamachi S, Naruse K & Shima A (1998). Sex-linked inheritance of the *lf* locus in the medaka fish (*Oryzias latipes*). *Zoological Sciences* **15**:123–126 [181]

Waddington CH (1953a). Genetic assimilation of an acquired character. *Evolution* **7**:118–126 [223, 343]

Waddington CH (1953b). Epigenetics and evolution. *Symposium of the Society of Experimental Biology* **7**:186–199 [343]

Wade MJ & Goodnight CJ (1998). The theories of Fisher and Wright in the context of metapopulations: When nature does many small experiments. *Evolution* **52**:1537–1553 [119]

Wainwright PC (1988). Morphology and ecology: Functional basis of feeding constraints in Caribbean labrid fishes. *Ecology* **69**:635–645 [337]

Wainwright PC (1989). Functional morphology of the pharyngeal jaw apparatus in perciform fishes: An experimental analysis of the Haemulidae. *Journal of Morphology*

200:231–245   [*187–188*]
Wainwright PC, Osenberg CW & Mittelbach GG (1991). Trophic polymorphism in the pumpkinseed sunfish (*Lepomis gibbosus* Linnaeus) effects of environment on ontogeny. *Functional Ecology* **5**:40–55   [*217*]

Waits L, Taberlet P, Swenson JE, Sandegren F & Franzen R (2000). Nuclear DNA microsatellite analysis of genetic diversity and gene flow in the Scandinavian brown bear (*Ursus arctos*). *Molecular Ecology* **9**:421–431   [*320*]

Wallace AR (1889). *Darwinism*. London, UK: Macmillan   [*17*]

Wallace AR (1910). *The World of Life: A Manifestation of Creative Power, Directive Mind and Ultimate Purpose*. London, UK: Chapman & Hall   [*17*]

Walsh BD (1864). On phytophagic varieties and phytophagic species. *Proceedings of the Entomological Society of Philadelphia* **3**:403–430   [*237*]

Wang H, McArthur ED, Sanderson SC, Graham JH & Freeman DC (1997). Narrow hybrid zone between two subspecies of big sagebrush (*Artemisia tridentata*: Asteraceae). IV. Reciprocal transplant experiments. *Evolution* **51**:95–102   [*275*]

Waring GL, Abrahamson WG & Howard DJ (1990). Genetic differentiation among host-associated populations of the gallmaker *Eurosta solidaginis* (Diptera: Tephritidae). *Evolution* **44**(6):1648–1655   [*242*]

Waser NM (1978). Competition for hummingbird pollination and sequential flowering in two Colorado wildflowers. *Ecology* **59**:934–944   [*274*]

Waser NM (1993). Population structure, optimal outbreeding, and assortative mating in angiosperms. In *The Natural History of Inbreeding and Outbreeding: Theoretical and Empirical Perspectives*, ed. Thornhill NW, pp. 173–199. Chicago, IL, USA: University of Chicago Press   [*265*]

Waser NM (1998). Pollination, angiosperm speciation, and the nature of species boundaries. *Oikos* **82**:198–201   [*267, 273*]

Waser NM (2001). Pollinator behavior and plant speciation: Looking beyond the "ethological isolation" paradigm. In *Cognitive Ecology of Pollination*, eds. Chittka L & Thomson JD, pp. 318–335. Cambridge, UK: Cambridge University Press   [*267, 273*]

Waser NM & Price MV (1985a). Reciprocal transplant experiments with *Delphinium nelsonii* (Ranunculaceae): Evidence for local adaptation. *American Journal of Botany* **72**:1726–1732   [*272*]

Waser NM & Price MV (1985b). The effect of nectar guides on pollinator preference: Experimental studies with a montane herb. *Oecologia* **67**:121–126   [*269, 274*]

Waser NM & Price MV (1993). Crossing distance effects on prezygotic performance in plants: An argument for female choice. *Oikos* **68**:303–308   [*265*]

Waser NM & Price MV (1994). Crossing-distance effects in *Delphinium nelsonii*: Outbreeding and inbreeding depression in progeny fitness. *Evolution* **48**:842–852   [*273*]

Waser NM, Chittka L, Price MV, Williams N & Ollerton J (1996). Generalization in pollination systems, and why it matters. *Ecology* **77**:279–296   [*267, 271*]

Waser NM, Price MV & Shaw RG (2000). Outbreeding depression varies among cohorts of *Ipomopsis aggregata* planted in nature. *Evolution* **54**:485–491   [*273*]

Weatherhead PL & Robertson RJ (1979). Offspring quality and the polygyny threshold: "The sexy son hypothesis". *The American Naturalist* **113**:201–208   [*99*]

Webster TP (1977). Geographic variation in *Anolis brevirostris*. In *The Third Anolis Newsletter*, ed. Williams EA, pp. 153–164. Cambridge, MA, USA: Museum of Comparative Zoology   [*339*]

Weiblen GD (2001). Phylogenetic relationships of fig wasps pollinating functionally dioecious figs based on mitochondrial DNA sequences and morphology. *Systematic Biology* **50**:243–267   [*231*]

Weiblen GD (2002). How to be a fig wasp. *Annual Review of Entomology* **47**:299–330   [*232*]

Weller SG & Sakai AK (1999). Using phylogenetic approaches for the analysis of plant breeding system evolution. *Annual Review of Ecology and Systematics* **30**:167–199   [*276*]

Weller SG, Sakai AK, Ashby K, Golonka A, Kutcher B & Rankin AE (1998). Dioecy and the evolution of wind pollination in *Schiedea* and *Alsinidendron* (Caryophyllaceae: Alsinoideae) in the Hawaiian Islands. *American*

Journal of Botany **85**:1377–1388 [*272*]

Werner EE & Gilliam JF (1984). The ontogenetic niche and species interactions in size-structured populations. *Annual Review of Ecology and Systematics* **15**:393–425 [*226*]

Werren JH (1998). *Wolbachia* and speciation. In *Endless Forms: Species and Speciation*, eds. Howard DJ & Berlocher SH, pp. 245–260. Oxford, UK: Oxford University Press [*252, 259, 261*]

Werth CR & Windham MD (1991). A model for divergent, allopatric speciation of polyploid pteridophytes resulting from silencing of duplicate-gene expression. *The American Naturalist* **137**:515–526 [*125*]

Wesslingh RA & Arnold ML (2000). Pollinator behaviour and the evolution of Louisiana iris hybrid zones. *Journal of Evolutionary Biology* **13**:171–180 [*268, 275*]

West-Eberhard MJ (1986). Alternative adaptations, speciation, and phylogeny (a review). *Proceedings of the National Academy of Sciences of the USA* **83**:1388–1392 [*211*]

West-Eberhard MJ (1989). Phenotypic plasticity and the origins of diversity. *Annual Review of Ecology and Systematics* **20**:249–278 [*211, 219, 343*]

Wheeler QD & Meier R, eds. (2000). *Species Concepts and Phylogenetic Theory*. New York, NY, USA: Columbia University Press [*23–24*]

White MJD (1968). Models of speciation. *Science* **159**:1065–1070 [*232*]

White MJD (1978). *Modes of Speciation*. San Francisco, CA, USA: Freeman [*112, 114*]

Whitlock MC (1995). Variance induced peak shift. *Evolution* **49**:252–259 [*119*]

Whitlock MC, Phillips PC, Moore FBG & Tonsor SJ (1995). Multiple fitness peaks and epistasis. *Annual Review of Ecology and Systematics* **26**:601–629 [*125*]

Wilding CS, Butlin RK & Grahame J (2001). Differential gene exchange between parapatric morphs of *Littorina saxatilis* detected using AFLP markers. *Journal of Evolutionary Biology* **14**:611–619 [*166*]

Williams EE (1972). The origin of faunas. Evolution of lizard congeners in a complex island fauna: A trial analysis. *Evolutionary Biology* **6**:47–89 [*337*]

Williams EE (1983). Ecomorphs, faunas, island size, and diverse end points in island radiations of *Anolis*. In *Lizard Ecology: Studies of a Model Organism*, eds. Huey RB, Pianka ER & Schoener TW, pp. 326–370. Cambridge, MA, USA: Harvard University Press [*337*]

Williams EE & Peterson JA (1982). Convergent and alternative designs in the digital adhesive pads of scincid lizards. *Science* **215**:1509–1511 [*323*]

Williams EE & Rand AS (1977). Species recognition, dewlap function, and faunal size. *American Zoologist* **17**:261–270 [*339*]

Wills CJ (1977). A mechanism for rapid allopatric speciation. *The American Naturalist* **111**:603–605 [*125*]

Wilson DS (1989). The diversification of single gene pools by density- and frequency-dependent selection. In *Speciation and Its Consequences*, eds. Otte D & Endler JA, pp. 366–385. Sunderland, MA, USA: Sinauer Associates Inc. [*211*]

Wilson EO (1992). *The Diversity of Life*. New York, NY, USA: WW Norton & Company [*211*]

Wilson EO & Brown WL (1953). The subspecies concept and its taxonomic application. *Systematic Zoology* **2**:97–111 [*21*]

Wilson P & Thomson JD (1996). How do flowers diverge? In *Floral Biology: Studies on Floral Evolution in Animal-Pollinated Plants*, eds. Lloyd DG & Barrett SCH, pp. 88–111. London, UK: Chapman & Hall [*267*]

Wilson DS & Turelli M (1986). Stable underdominance and the evolutionary invasion of empty niches. *The American Naturalist* **127**:835–850 [*39, 42, 67, 222, 244*]

Wimberger PH (1991). Plasticity of jaw and skull morphology in the neotropical cichlids *Geophagus brasiliensis* and *G. steindachneri*. *Evolution* **45**:1545–1563 [*217, 220*]

Wimberger PH (1992). Plasticity of fish body shape, the effects of diet, development family and age in two species of *Geophagus* (Pisces: Cichlidae). *Biological Journal of the Linnean Society* **45**:197–218 [*217, 220*]

Wimberger PH (1994). Trophic polymorphisms, plasticity, and speciation in vertebrates. In *Theory and Application in Fish Feeding Ecology*, eds. Stouder DJ, Fresh KL & Feller RJ, pp. 19–43. Columbia, SC, USA: University of South Carolina Press [*217, 220*]

Winemiller KO (1992). Ecomorphology of freshwater fishes. *National Geographic Research and Exploration* **8**:308–327   [*367*]

Witte F, Barel CDN & van Oijen MJP (1997). Intraspecific variation of haplochromine cichlids from Lake Victoria and its taxonomic implications. *South African Journal of Science* **93**:585–594   [*174*]

Wood CC (1995). Life history variation and population structure in sockeye salmon. In *Evolution and the Aquatic Ecosystem*, ed. Nielsen JL, pp. 195–216. Bethesda, MD, USA: American Fisheries Society   [*222*]

Wood CC & Foote CJ (1990). Genetic differences in early development and growth of sympatric sockeye salmon and Kokanee (*Oncorhynchus nerka*) and their hybrids. *Canadian Journal of Fisheries and Aquatic Sciences* **47**:2250–2260   [*212, 214*]

Wood CC & Foote CJ (1996). Evidence for sympatric genetic divergence of anadromous and nonanadromous morphs of sockeye salmon (*Oncorhynchus nerka*). *Evolution* **50**:1265–1279   [*71, 214, 216*]

Wood TK, Tilmon KJ, Shantz AB, Harris CK & Pesek J (1999). The role of host-plant fidelity in initiating insect race formation. *Evolutionary Ecology Research* **1**:317–332   [*238, 246*]

Woodcock G & Higgs PG (1996). Population evolution on a multiplicative single-peak fitness landscape. *Journal of Theoretical Biology* **179**:61–73   [*129*]

Wooding S & Ward R (1997). Phylogeography and pleistocene evolution in the North American black bear. *Molecular Biology and Evolution* **14**:1096–1105   [*319*]

Worthington EB (1954). Speciation of fishes in African lakes. *Nature* **173**:1064–1067   [*362*]

Wright S (1931). Evolution in Mendelian populations. *Genetics* **16**:97–159   [*70, 118*]

Wright S (1932). The roles of mutation, inbreeding, crossbreeding and selection in evolution. In *Proceedings of the Sixth International Congress on Genetics*, ed. Jones DF, Vol. 1, pp. 356–366. Austin, TX, USA: Genetics Society of America   [*115, 117–118*]

Wu CI (1985). A stochastic simulation study of speciation by sexual selection. *Evolution* **39**:66–82   [*132*]

Wu CI (2001). The genic view of the process of speciation. *Journal of Evolutionary Biology* **14**:851–865   [*221–222*]

Wu CI & Palopoli MF (1994). Genetics of postmating reproductive isolation in animals. *Annual Review of Genetics* **27**:283–308   [*124–125*]

Yamaoka K (1978). Pharyngeal jaw structure in labrid fish. *Publications of the Seto Marine Laboratory* **24**:409–426   [*187*]

Yodzis P (1989). *Introduction to Theoretical Ecology.* New York, NY, USA: Harper & Row   [*306*]

Zahavi A (1975). Mate selection – a selection for a handicap. *Journal of Theoretical Biology* **53**:205–214   [*99*]

Zahavi A (1977). The cost of honesty. Further remarks on the handicap principle. *Journal of Theoretical Biology* **67**:603–605   [*99*]

Zahavi A & Zahavi A (1997). *The Handicap Principle.* Oxford, UK: Oxford University Press   [*99*]

Index

specialization, *229–230*
speciation mechanisms, *230–231*
sympatric speciation with host shifts,
235–243
conditions needed, *243–247*
examples, *237–243*
see also pests, agricultural
intergenomic conflicts, *265, 277*
intragenomic conflicts, *265, 277*
intralacustrine radiations, *173–174, 186–191,
210, 362–379*
invasion fitness, *56–58, 116, 391*
bacteria, *294–295*
sexual populations, *83, 87*
Ipomoea purpurea, 274
Ipomopsis, 268, 275
*Ipomopsis aggregata, 268, 269, 274, 275,
Plate 6*
Ipomopsis tenuituba, 268, 269, 275, Plate 6
Iris, 268, 275
island biogeography theory, *163, 291*
islands
African mountains as, *345, 346, 350*
Anolis lizards *see Anolis* lizards, Caribbean
speciation rates on larger, *163, 164–165*
times for speciation on, *376, 377*
isolating mechanisms, *21–22, 113*
see also reproductive isolation
isolation, geographic *see* geographic isolation

Jadera haematoloma, 242
Jamaica, *336, 338*
juglone, *232*

labroid fishes, pharyngeal jaw apparatus,
187–190
lagoons, marginal, *174*
lakes
timing of speciation and formation,
370–371
see also African Great Lakes;
intralacustrine radiations; northern
freshwater fishes
Larinus weevils, *233*
Laspeyresia, 239
leafhoppers, *238*
leaf rosettes, giant, *346, 347, 352–353, 358*
learning
adaptive, *249, 250–251, 253–256, 262*
effects on evolution, *250–251*
future research, *390*
leks, *176*
Lepidoptera (butterflies), *239, 267*
Lepisosteus, 363
Leptothorax, 241
Lesser Antilles islands, *324, 325*

lice, *231*
Lieiomyza helianthi, 245
light conditions, ambient, *177–178, 190*
limnetic forms
northern freshwater fishes, *212, 215, 218*
threespine sticklebacks, *196–197, 199–203,
218, Plate 2*
lineage sorting, *312–313*
linkage disequilibrium, *52, 71, 81*
in adaptive speciation models, *92*
in ecological speciation, *266*
in evolution of assortative mating, *44, 47,
48–49*
in evolution of divergent habitat preference,
50, 51
in insects, *247*
linkage hitchhiking, *284*
linkage maps, genome-wide, *228*
Liriomyza leaf miners, *246*
lobelias, giant (*Lobelia*), *303, 345–361,
Plate 11*
adaptive speciation, *357–359*
morphological adaptations, *345–346,
351–353*
phylogenetic patterns and biogeography,
353, 354–356
similarities to giant senecios, *351–353*
local adaptation
in agricultural pests, *253*
along environmental gradients, *141, 152,
161*
in BDM models, *125–127*
divergence driven by selection for, *121–123*
pollinator-facilitated, *273*
Lochmaea, 240
Lopidea, 239
Lotka–Volterra models, *57, 94, 96, 143*
Lycaenidae, *239*

macromutations, *276*
Malawi, Lake, *173, 174, 176, 191, 376–377,
378*
see also haplochromine cichlids, African
Mallophaga, *231*
Malus pumila, 237
marcescent foliage, *346, 348, 352, 353, 358*
Martinique, *324, 325, 333–334, 335*
mate choice (or preference)
adaptive, agricultural pests, *256–259,
262–263*
in allopatric speciation, *136*
as by-product of food choice, *257*
condition-dependent, *256, 257*
context-dependent, *251, 257, 258–259*
divergent, evolution, *42–47*
evolution, in absence of niche divergence,
43–45

The International Institute for Applied Systems Analysis

is an interdisciplinary, nongovernmental research institution founded in 1972 by leading scientific organizations in 12 countries. Situated near Vienna, in the center of Europe, IIASA has been producing valuable scientific research on economic, technological, and environmental issues for nearly three decades.

IIASA was one of the first international institutes to systematically study global issues of environment, technology, and development. IIASA's Governing Council states that the Institute's goal is: *to conduct international and interdisciplinary scientific studies to provide timely and relevant information and options, addressing critical issues of global environmental, economic, and social change, for the benefit of the public, the scientific community, and national and international institutions.* Research is organized around three central themes:

– Energy and Technology;
– Environment and Natural Resources;
– Population and Society.

The Institute now has National Member Organizations in the following countries:

Austria
The Austrian Academy of Sciences

China
National Natural Science
Foundation of China

Czech Republic
The Academy of Sciences of the
Czech Republic

Egypt
Academy of Scientific Research and
Technology (ASRT)

Estonia
Estonian Association for
Systems Analysis

Finland
The Finnish Committee for IIASA

Germany
The Association for the Advancement
of IIASA

Hungary
The Hungarian Committee for Applied
Systems Analysis

Japan
The Japan Committee for IIASA

Netherlands
The Netherlands Organization for
Scientific Research (NWO)

Norway
The Research Council of Norway

Poland
The Polish Academy of Sciences

Russian Federation
The Russian Academy of Sciences

Sweden
The Swedish Research Council for
Environment, Agricultural Sciences
and Spatial Planning (FORMAS)

Ukraine
The Ukrainian Academy of Sciences

United States of America
The National Academy of
Sciences